3D Reconstruction and Mobile Mapping in Urban Environments Using Remote Sensing

Editors

San Jiang
Duojie Weng
Jianchen Liu
Wanshou Jiang

MDPI

Basel • Beijing • Wuhan • Barcelona • Belgrade • Novi Sad • Cluj • Manchester

Editors

San Jiang
School of Computer Sciences
China University of Geosciences
Wuhan
China

Duojie Weng
Department of Land
Surveying and
Geo-Informatics
The Hong Kong
Polytechnic University
Hong Kong
China

Jianchen Liu
The College of Geodesy and
Geomatics
Shandong University of
Science and Technology
Qingdao
China

Wanshou Jiang
State Key Laboratory of
Information Engineering in
Surveying, Mapping and
Remote Sensing
Wuhan University
Wuhan
China

Editorial Office
MDPI AG
Grosspeteranlage 5
4052 Basel, Switzerland

This is a reprint of articles from the Special Issue published online in the open access journal *Remote Sensing* (ISSN 2072-4292) (available at: https://www.mdpi.com/journal/remotesensing/special_issues/3D_Reconstruction_and_Mobile_Mapping).

For citation purposes, cite each article independently as indicated on the article page online and as indicated below:

Lastname, A.A.; Lastname, B.B. Article Title. *Journal Name* **Year**, *Volume Number*, Page Range.

ISBN 978-3-7258-2213-3 (Hbk)
ISBN 978-3-7258-2214-0 (PDF)
doi.org/10.3390/books978-3-7258-2214-0

Cover image courtesy of San Jiang

Contents

About the Editors

San Jiang

San Jiang received his B.S. degree in remote sensing science and technology from Wuhan University in 2010, and his M.Sc. and Ph.D. degrees in photogrammetry and remote sensing from Wuhan Univeristy in 2012 and 2018, respectively. From 2012 to 2014, he worked as an assistant engineer in Tianjin Institute of Surveying and Mapping. From 2014 to 2015, he joined the LIESMARS (State Key Laboratory of Information Engineering in Surveying, Mapping and Remote Sensing of Wuhan Univeristy) as a research assistant. Currently, he is an associate professor in the School of Computer Science at China University of GeoSciences (Wuhan). His research interests include image matching, SfM-based aerial triangulation, and 3D reconstruction.

Duoejie Weng

Duoejie Weng received his B.Sc. and M.Sc. degrees in electrical engineering from Hohai University, Nanjing, China, in 2007 and 2010, respectively, and his Ph.D. degree from the Hong Kong Polytechnic University in 2016. He is currently a post-doctoral researcher at the Hong Kong Polytechnic University. His research interests include GNSS integrity monitoring, kinematic GPS, and sensor integration.

Jianchen Liu

Jianchen Liu was born in Jiamusi, Heilongjiang, China, in 1987. He received the M.S. degree in geomatics engineering from the Shandong University of Science and Technology, Qingdao, China, in 2013, and his Ph.D. degree in photogrammetry and remote sensing from the School of Remote Sensing and Information Engineering, Wuhan University, Wuhan, China, in 2017. He is currently an associate professor with the College of Geomatics, Shandong University of Science and Technology. His research interests include UAV photogrammetry, computer stereovision, and 3D modeling by multiview images.

Wanshou Jiang

Wanshou Jiang received his bachelor's and master's degrees in photogrammetry and remote sensing from Wuhan Technical University of Surveying and Mapping, respectively, in 1989 and 1996. In 2004, he received his PhD degree in photogrammetry and remote sensing from Wuhan University. He started his research career in 1989 as a software developer in analytical photogrammetry. In 2000, he joined the LIESMARS (the State Key Laboratory of Information Engineering in Surveying, Mapping and Remote Sensing) as an associate researcher and then he was offered the tenure position of researcher in 2005. His research interests include image registering, image classification, change detection, 3D reconstruction, etc. He made great contributions to the famous digital photogrammetric workstation VirtuoZo and designed a software platform, named OpenRS, for remote sensing image processing.

Preface

In the last decade, remote sensing-based techniques have become a meaningful solution to maintain the orderly evaluation of urban environments. Three-dimensional reconstruction and mobile mapping are two critical roles that are essential in supporting varying applications in urban environments. This Special Issue focuses on the techniques for 3D reconstruction and mobile mapping in urban environments by using remote sensing, including new instruments for rapid data acquisitions, perspective invariant algorithms for reliable feature matching, efficient SfM- and SLAM-based solutions for robust image orientation, and deep learning-based neural networks to reshape the whole pipeline of 3D reconstruction and mobile mapping. For the construction of this Special Issue, we really appreciate the authors who contribute their valuable work and the editors for their passionate assistance, which form the base of the successful organization of this Special Issue.

San Jiang, Duojie Weng, Jianchen Liu, and Wanshou Jiang
Editors

remote sensing

MDPI

Editorial

Editorial on Special Issue "3D Reconstruction and Mobile Mapping in Urban Environments Using Remote Sensing"

San Jiang [1,*], Duojie Weng [2], Jianchen Liu [3] and Wanshou Jiang [4]

1 School of Computer Science, China University of Geosciences, Wuhan 430074, China
2 Department of Land Surveying and Geo-Informatics, The Hong Kong Polytechnic University, Hong Kong 999077, China; wengduojie.lsgi@polyu.edu.hk
3 The College of Geodesy and Geomatics, Shandong University of Science and Technology, Qingdao 266590, China; liujianchen@sdust.edu.cn
4 State Key Laboratory of Information Engineering in Surveying, Mapping and Remote Sensing, Wuhan University, Wuhan 430074, China; jws@whu.edu.cn
* Correspondence: jiangsan@cug.edu.cn

Citation: Jiang, S.; Weng, D.; Liu, J.; Jiang, W. Editorial on Special Issue "3D Reconstruction and Mobile Mapping in Urban Environments Using Remote Sensing". *Remote Sens.* **2024**, *16*, 3236. https://doi.org/10.3390/rs16173236

Received: 16 August 2024
Accepted: 28 August 2024
Published: 31 August 2024

1. Introduction

Both 3D reconstruction and mobile mapping are critical in supporting various applications in urban environments, including but not limited to autonomous driving, smart logistics, pedestrian navigation, and virtual reality. In the last decade, remote sensing-based techniques have emerged as a meaningful solution for ensuring urban environments are evaluated in an orderly fashion, due to the rapid evolution of cutting-edge techniques, e.g., SfM (Structure from Motion), SLAM (Simultaneous Localization and Mapping), and the revolution in deep learning techniques that enhance the entire pipeline, e.g., NeRF (Neural Radiance Field). In conclusion, the explosive development of 3D reconstruction and mobile mapping has been particularly notable in recent years.

This Special Issue comprises high-quality papers focusing on the techniques and applications of 3D reconstruction and mobile mapping in urban environments. A total of 15 papers are published in this Special Issue, covering topics such as image feature matching and dense matching, LiDAR/image-fused SLAM for image orientation and tunnel mapping, NeRF-based scene rendering and orthophoto generation, and other interesting applications, such as InSAR point cloud registration and 3D Ground-Penetrating Radar (3D GPR) for underground imaging and positioning. The details of each paper will be described in the following section.

2. Overview

Reliable feature matching is the first step of 3D reconstruction, determining the success of subsequent processing. Focusing on the feature matching of spherical images, Jiang et al. [1] present an algorithm by combining local geometric rectification with convolutional neural network (CNN) learned descriptors. It addresses the challenge of the geometric distortions inherent in spherical images and improves the performance of 3D reconstruction systems. The method utilized includes a local geometric rectification, a CNN-based descriptor learning network for rectified patches, and a robust essential matrix estimation for outlier removal. The effectiveness of the proposed solution is demonstrated through experiments using real spherical images.

Yao et al. [2] introduce a quasi-dense matching algorithm for oblique stereo images with large viewpoint changes. The core idea of the proposed method relies on the combination of VGG16-UNet-based semantic segmentation with LoFTR-based local feature enhancement. The method involves segmenting multiplanar scenes, performing affine-invariant feature matching, and enhancing weak texture regions to improve the matching accuracy. By using low-altitude stereo images, the experiments demonstrate significant advantages in match quantity, accuracy, and spatial distribution over classical and deep-learning methods.

The 3D reconstruction of ancient buildings plays a critical role in digital city construction. By using recent techniques, Ge et al. [3] present a NeRF-based 3D reconstruction workflow for UAV images with depth supervision. It introduces a multi-resolution hash coding approach to reduce hash conflicts and a truncated signed distance function (TSDF) to improve geometric accuracy. Through the use of collected UAV (Unmanned Aerial Vehicle) images, the test results demonstrate that the proposed solution can render images with clearer structural details and achieves a notable improvement in performance, with a 15% gain on average in the Peak Signal-to-Noise Ratio (PSNR) and a 10% gain in the Structural Similarity Index Measure (SSIM), producing detailed and accurate 3D models that are suitable for the digital preservation of cultural heritage sites.

A digital orthophoto is one of the most important developments, and it has been produced via the standard photogrammetric workflow for many years. Recently, by using the cutting-edge technique, Lv et al. [4] have presented a comparative study of explicit and implicit methods for generating digital orthophotos. The explicit method, termed as TDM (top-view-constrained dense matching), relies on the traditional geometric approach, while the implicit method, namely Instant NGP, employs neural rendering with Neural Radiance Fields. The comparative test concludes that both methods can produce accurate and high-quality orthophotos; due to the usage of the Compute Unified Device Architecture (CUDA) acceleration technique, TDM has significantly higher efficiency. To summarize, the study offers insights for selecting appropriate digital orthophoto generation methods based on efficiency and quality requirements.

Arza-García et al. [5] propose a cost-effective method for assessing the structural stability of a typical 3D model application, rubble mound breakwaters (RMBs), through the combination of UAV photogrammetry and Random Sample Consensus (RANSAC). In the proposed workflow, the photogrammetric point clouds of the RMB are generated via Structure from Motion and Multi-View Stereo (SfM-MVS) from pre- and post-storm flights, and they are fed to RANSAC for plane extraction and segmentation. Finally, by using a spatial proximity criterion, the cuboids of the two time periods are registered. The tests conducted on a breakwater in Porto, Portugal, show that the proposed method successfully identified post-storm structural changes and showcased its potential for monitoring RMB.

For urban 3D modeling, Cui et al. [6] introduce a method to extract urban building heights from Gaofen-7 stereo satellite images. The key technique involves using a contour matching algorithm to accurately determine rooftop elevations and using ground filtering to generate a DEM (Digital Elevation Model) from the DSM (Digital Surface Model). The proposed solution addresses challenges like occlusions, inaccurate ground elevation, and high-rise buildings, and it has been well-verified by using stereo images from three different provinces. The results verify the improved accuracy in building height extraction, especially beneficial for high-rise buildings and sites with complex terrain or vegetation.

For multi-source data fusion, Liu et al. [7] present a robust multi-sensor SLAM system, termed LVI-fusion, that integrates camera, lidar, and IMU data. The proposed mainly consists of a time alignment module to handle varying data frequencies, an image segmentation module for dynamic target removal, and a depth recovery model for feature points. The system uses a sliding window optimization module to achieve real-time pose calculation. The tests, carried out in various environments, demonstrate that the proposed method has high accuracy and robustness, and outperforms the other existing SLAM solutions, particularly in dynamic settings.

Xu et al. [8] present an enhanced Strapdown Inertial Navigation System (SINS) and a LiDAR tightly integrated SLAM system for urban environments with sparse structural features. The method refines an edge point extraction process from the LOAM algorithm and introduces a Kalman filter using line distance error as the primary observation metric to improve the robustness and accuracy of the system. The experimental tests conducted in various environments demonstrate its superior performance, with a 17% enhancement in positioning accuracy, especially in scenarios with limited structural features.

Point cloud registration, which aims to align two 3D point clouds using keypoint correspondences, is essential in photogrammetry and remote sensing. Traditional methods face challenges due to uncertainties in keypoint detection and matching, leading to outliers that reduce efficiency and accuracy. Wang et al. [9] present a new registration method using a compatibility graph and accelerated guided sampling, introducing a minimum subset sampling approach to minimize outlier impact and a preference-based sampling strategy to enhance computational efficiency and accuracy. Using synthetic and real datasets, the test results show that the proposed solution achieves a minimum rotation error of 0.737° and a minimum translation error of 0.0201 m, respectively, compared with existing methods.

In complex scenes with closely adjacent trees and buildings, the accurate extraction of building point clouds is challenging. Su et al. [10] introduce a two-stage method for building-point-cloud extraction based on geometric information. The first stage coarsely extracts building points, which are refined using mask polygons and a region-growing algorithm in the second stage. The method integrates the Alpha Shape algorithm and neighborhood expansion to address missing boundary points and applies mask extraction to the original points to avoid errors in facade identification. The approach shows significant improvements in extraction accuracy, outperforming PointNet by 20.73% in terms of precision and achieving results comparable to the HDL-JME-GGO network on the Urban-LiDAR and Vaihingen datasets.

Mutlti-source data fusion is a key step in the application of vehicle-borne mobile mapping systems (MMSs). Ji et al. [11] propose a method for vehicle-borne laser point cloud and panoramic images based on occlusion removal. The approach involves removing irrelevant points, extracting relevant scenes based on trajectory points, and applying a collinear model with spherical projection for matching. In addition, a vectorial angle selection algorithm is designed, in order to filter out occluded projections. The experimental results show the proposed solution can achieve an average pixel error of 2.82 pixels and a positional error of 4 cm, verifying that it is effective for data fusion applications in navigation, surveying, and mapping.

Cheng et al. [12] introduce an image-aided LiDAR framework for the extraction, classification, and characterization of lane markings from mobile mapping data. The framework addresses road safety by improving lane-marking inventory through a combination of imagery and LiDAR data, enhancing the detection of markings under various conditions. The framework includes road surface identification and color/intensity enhancement, and utilizes a geographic information system for visualization. The study demonstrates the system's effectiveness over an extended road network, showing the potential to improve road safety analyses.

To address the need for the high-precision point cloud mapping of subway trains in long tunnel scenarios, Li et al. [13] introduce a LiDAR and inertial measurement sensor-based map construction method. The approach integrates a tightly coupled front-end odometry system by using Kalman filters with back-end optimization via factor graphs. In the front end, inertial measurements predict filter updates based on LiDAR points and local map planes. A global pose graph, built from inter-frame odometry and constraints, undergoes smoothing optimization for accurate mapping. The experiments show that it achieves a trajectory consistency of 0.1 m and an accumulated error of less than 0.2% compared to ground truth.

Array interferometric synthetic aperture radar (Array InSAR) systems can address shadow issues by performing scans in opposite directions. However, point clouds from two scans must be registered accurately. Cui et al. [14] present a robust registration method for urban Array InSAR point clouds, which uses images to represent 3D data, where pixel positions reflect azimuth and ground range, and pixel intensity represents height. The KAZE algorithm and an enhanced matching approach identify corresponding points to estimate transformation relationships. The experimental results show that it achieves the facade registration with a relative angular difference of less than 0.5°, and ground element registration achieves a Root Mean Square Error (RMSE) of less than 1.5 m.

Three-Dimensional Ground-Penetrating Radar (3D GPR) offers non-destructive and continuous subsurface detection but faces challenges regarding positioning accuracy in complex urban environments. Zhang et al. [15] propose a multi-level robust positioning method to enhance the accuracy of 3D GPR. In areas with strong GNSS signals, differential GNSS technology ensures rapid, precise positioning. For weak GNSS signals, a GNSS/INS tightly coupled solution improves accuracy, while in GNSS-denied environments, SLAM technology integrates INS data and 3D point clouds. This approach achieves a positioning accuracy of better than 10 cm, delivers high-quality 3D images of underground urban structures, and supports urban road surveys and underground disease detection.

3. Conclusions

This Special Issue focuses on the techniques for 3D reconstruction and mobile mapping in urban environments by using remote sensing methods, detailing the rapid development of new instruments for data acquisitions, the perspective invariant algorithms for feature matching, efficient SfM and SLAM-based solutions for image orientation, and deep-learning-based neural networks, all of which enhance the whole pipeline of 3D reconstruction and mobile mapping. Numerous high-quality manuscripts that covered a wide range of hot topics were submitted to this Special Issue. Ultimately, 15 of these papers were published after undergoing strict peer review, ensuring that this Special Issue will provide useful clues to guide further research.

Funding: This research received no external funding.

Acknowledgments: The authors would like to thank authors who have contributed their work to this Special Issue and reviewers who have paid their valuable time and effort to ensure the quality of all accepted papers. Meanwhile, heartfelt thanks to the Editors of this Special Issue, especially to Nicole Ju at MDPI for her careful and patient help in organizing this Special Issue.

Conflicts of Interest: The authors declare no conflicts of interest.

References

1. Jiang, S.; Liu, J.; Li, Y.; Weng, D.; Chen, W. Reliable Feature Matching for Spherical Images via Local Geometric Rectification and Learned Descriptor. *Remote Sens.* **2023**, *15*, 4954. [CrossRef]
2. Yao, G.; Zhang, J.; Zhu, F.; Gong, J.; Jin, F.; Fu, Q.; Ren, X. Quasi-Dense Matching for Oblique Stereo Images through Semantic Segmentation and Local Feature Enhancement. *Remote Sens.* **2024**, *16*, 632. [CrossRef]
3. Ge, Y.; Guo, B.; Zha, P.; Jiang, S.; Jiang, Z.; Li, D. 3D Reconstruction of Ancient Buildings Using UAV Images and Neural Radiation Field with Depth Supervision. *Remote Sens.* **2024**, *16*, 473. [CrossRef]
4. Lv, J.; Jiang, G.; Ding, W.; Zhao, Z. Fast Digital Orthophoto Generation: A Comparative Study of Explicit and Implicit Methods. *Remote Sens.* **2024**, *16*, 786. [CrossRef]
5. Arza-García, M.; Gonçalves, J.A.; Ferreira Pinto, V.; Bastos, G. On-Site Stability Assessment of Rubble Mound Breakwaters Using Unmanned Aerial Vehicle-Based Photogrammetry and Random Sample Consensus. *Remote Sens.* **2024**, *16*, 331. [CrossRef]
6. Cui, Y.; Zhao, S.; Jiang, W.; Yu, G. Urban Building Height Extraction from Gaofen-7 Stereo Satellite Images Enhanced by Contour Matching. *Remote Sens.* **2024**, *16*, 1556. [CrossRef]
7. Liu, Z.; Li, Z.; Liu, A.; Shao, K.; Guo, Q.; Wang, C. LVI-Fusion: A Robust Lidar-Visual-Inertial SLAM Scheme. *Remote Sens.* **2024**, *16*, 1524. [CrossRef]
8. Xu, X.; Guan, L.; Gao, Y.; Chen, Y.; Liu, Z. Enhanced Strapdown Inertial Navigation System (SINS)/LiDAR Tightly Integrated Simultaneous Localization and Mapping (SLAM) for Urban Structural Feature Weaken Occasions in Vehicular Platform. *Remote Sens.* **2024**, *16*, 2527. [CrossRef]
9. Wang, C.; Zheng, Z.; Zha, B.; Li, H. Fast Robust Point Cloud Registration Based on Compatibility Graph and Accelerated Guided Sampling. *Remote Sens.* **2024**, *16*, 2789. [CrossRef]
10. Su, Z.; Peng, J.; Feng, D.; Li, S.; Yuan, Y.; Zhou, G. A Building Point Cloud Extraction Algorithm in Complex Scenes. *Remote Sens.* **2024**, *16*, 1934. [CrossRef]
11. Ji, J.; Wang, W.; Ning, Y.; Bo, H.; Ren, Y. Research on a Matching Method for Vehicle-Borne Laser Point Cloud and Panoramic Images Based on Occlusion Removal. *Remote Sens.* **2024**, *16*, 2531. [CrossRef]
12. Cheng, Y.-T.; Shin, Y.-H.; Shin, S.-Y.; Koshan, Y.; Hodaei, M.; Bullock, D.; Habib, A. Image-Aided LiDAR Extraction, Classification, and Characterization of Lane Markings from Mobile Mapping Data. *Remote Sens.* **2024**, *16*, 1668. [CrossRef]

13. Li, C.; Pan, W.; Yuan, X.; Huang, W.; Yuan, C.; Wang, Q.; Wang, F. High-Precision Map Construction in Degraded Long Tunnel Environments of Urban Subways. *Remote Sens.* **2024**, *16*, 809. [CrossRef]
14. Cui, C.; Liu, Y.; Zhang, F.; Shi, M.; Chen, L.; Li, W.; Li, Z. A Novel Automatic Registration Method for Array InSAR Point Clouds in Urban Scenes. *Remote Sens.* **2024**, *16*, 601. [CrossRef]
15. Zhang, J.; Hu, Q.; Zhou, Y.; Zhao, P.; Duan, X. A Multi-Level Robust Positioning Method for Three-Dimensional Ground Penetrating Radar (3D GPR) Road Underground Imaging in Dense Urban Areas. *Remote Sens.* **2024**, *16*, 1559. [CrossRef]

remote sensing

MDPI

Article

Reliable Feature Matching for Spherical Images via Local Geometric Rectification and Learned Descriptor

San Jiang [1,2], Junhuan Liu [1], Yaxin Li [2], Duojie Weng [2] and Wu Chen [2,*]

[1] School of Computer Science, China University of Geosciences, Wuhan 430074, China; jiangsan@cug.edu.cn (S.J.); liujh@cug.edu.cn (J.L.)

[2] Department of Land Surveying and Geo-Informatics, The Hong Kong Polytechnic University, Hong Kong 999077, China; yaxin.pu.li@connect.polyu.hk (Y.L.); ceweng@polyu.edu.hk (D.W.)

* Correspondence: wu.chen@polyu.edu.hk

Abstract: Spherical images have the advantage of recording full scenes using only one camera exposure and have been becoming an important data source for 3D reconstruction. However, geometric distortions inevitably exist due to the spherical camera imaging model. Thus, this study proposes a reliable feature matching algorithm for spherical images via the combination of local geometric rectification and CNN (convolutional neural network) learned descriptor. First, image patches around keypoints are reprojected to their corresponding tangent planes based on a spherical camera imaging model, which uses scale and orientation data from the keypoints to achieve both rotation and scale invariance. Second, feature descriptors are then calculated from the rectified image patches by using a pre-trained separate detector and descriptor learning network, which improves the discriminability by exploiting the high representation learning ability of the CNN. Finally, after classical feature matching with the ratio test and cross check, refined matches are obtained based on an essential matrix-based epipolar geometry constraint for outlier removal. By using three real spherical images and an incremental structure from motion (SfM) engine, the proposed algorithm is verified and compared in terms of feature matching and image orientation. The experiment results demonstrate that the geometric distortions can be efficiently reduced from rectified image patches, and the increased ratio of the match numbers ranges from 26.8% to 73.9%. For SfM-based spherical image orientation, the proposed algorithm provides reliable feature matches to achieve complete reconstruction with comparative accuracy.

Keywords: spherical image; feature matching; geometric rectification; structure from motion; 3D reconstruction; learned descriptor

Citation: Jiang, S.; Liu, J.; Li, Y.; Weng, D.; Chen, W. Reliable Feature Matching for Spherical Images via Local Geometric Rectification and Learned Descriptor. *Remote Sens.* **2023**, *15*, 4954. https://doi.org/10.3390/rs15204954

Academic Editor: Massimiliano Pepe

Received: 13 September 2023
Revised: 7 October 2023
Accepted: 11 October 2023
Published: 13 October 2023

1. Introduction

Image-based 3D reconstruction has become a critical module in recent photogrammetric systems [1], which has been adopted in varying applications ranging from conventional digital urban construction [2] to the recent archaeological excavation [3] and transmission corridor inspection [4]. Because of the low cost of imaging sensors and the maturity of processing techniques, perspective cameras are the most widely used instruments for data acquisition in image-based 3D reconstruction, especially for aerial photogrammetry. With the increasing demands for 3D reconstruction for street or indoor environments, perspective cameras become inefficient and non-applicable for data acquisition. The main reason is that their limited FOV (field of view) causes significantly more image recording burden to cover the omnidirectional scene.

In contrast to the limited FOV of perspective cameras, spherical cameras, also known as omnidirectional cameras, have the advantage of recording full scenes using only one camera exposure, as they have respectively 360° and 180° FOV in the horizontal and vertical directions [5]. Except for professional spherical cameras, e.g., the LadyBug series that is widely used in mobile mapping systems (MMSs) [6], recent years have also witnessed the explosive development of consumer-grade spherical cameras that feature low costs and light weights, e.g., the Insta360 and Ricoh Theta [7]. For image-based 3D reconstruction, the capability and popularity of spherical cameras have promoted their usage in varying fields, including, but not limited to, damaged building evaluation [8], urban 3D modeling [9] and tunnel rapid mapping [10]. Thus, spherical images are becoming an important data source for 3D reconstruction.

Feature matching is the prerequisite to implementing image-based 3D reconstruction. In the literature, feature matching has been achieved through local feature-based image matching methods that compute descriptors for image patches around detected keypoints and cast image matching as searching nearest neighbors among two sets of descriptors. The pipeline of local feature-based image matching consists of two major steps, i.e., feature detection and matching based on the well-designed descriptors [11,12], and outlier removal based on photometric and geometric constraints [13]. Existing research has promoted the development of feature matching techniques toward the direction of automation and precision. However, the vast majority of existing algorithms are used for perspective images, which differ from spherical images in the camera imaging model [14]. Perspective images use a 2D plane imaging model that projects 3D scene points to 2D image points on the image plane. On the contrary, spherical images are recorded by projecting scene points onto the 3D sphere, which are further flattened to the 2D image plane. Because of the transformation from the 3D sphere to the 2D plane, geometric distortions are inevitably introduced into the recorded spherical images, which become more and more serious in the regions near the equator to the poles [15] as shown in Figure 1. Thus, more attention should be paid to reducing distortions in spherical images.

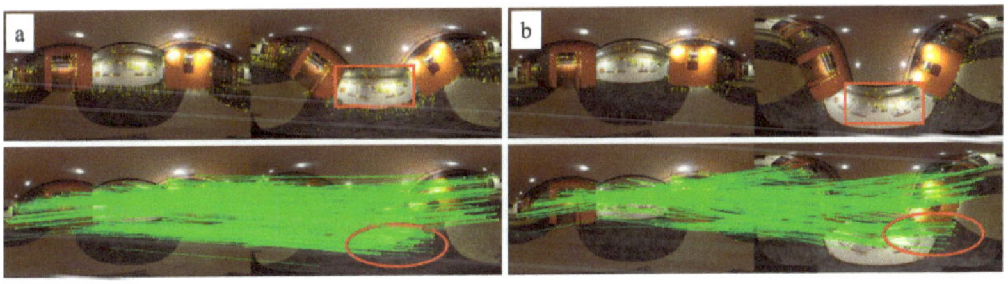

Figure 1. The illustration of geometric distortion in the spherical images. (**a**,**b**) indicate image pairs that are rotated around the X axis with the angles of 45° and 75°, respectively. The red rectangles and ellipses show increasing distortions and decreasing matches.

In the literature, both 2D plane and 3D sphere-based algorithms have been documented to alleviate the geometric distortions in spherical images [16–18]. For 2D plane-based methods, existing solutions can be divided into three groups, i.e., global methods, semi-global methods, and local methods. In global methods, Wang et al. [18] have implemented a SLAM (simultaneous localization and mapping) system, namely CubemapSLAM, in which the cubic-map reprojection solution is used to convert each spherical image into six perspective images that are then processed by using classical feature matching methods. Considering the distribution pattern of image geometric distortions, Taira et al. [17] aimed to execute feature matching on the region near the sphere equator, which is achieved by rotating spherical images around the Y axis and detecting local features from the regions near the equator. Compared with the global cubic-map reprojection solution, it can be seen as a semi-global rectification method. In contrast to the global and semi-global rectification solutions, Chuang and Perng [16] proposed reprojecting the local image patches of keypoints onto the corresponding tangent planes and calculating feature descriptors from rectified image patches. Except for rectification-based methods, other research achieves feature detection and descriptor computation by considering the principle of the spherical imaging model. The proposed solutions are usually designed on the spherical grid for neighbor searching, such as SPHORB [19] and BRISKS [20], instead of the plane grid used in the classical methods. In the above-mentioned solutions, classical heuristic algorithms are widely used for feature detection and description.

In recent years, CNN (convolutional neural network)-based deep learning networks have also been widely used for feature matching due to their powerful representation learning ability [21,22]. According to network tasks, existing CNNs can be divided into three groups, i.e., joint feature and metric learning networks that learn the similarity of image patches [23,24], separate detector and descriptor learning networks that learn to compute descriptors [25,26], and joint detector and descriptor learning networks that learn to detect keypoints and compute descriptors [27,28]. These CNN models have achieved comparative or superior performance for feature matching of perspective images. To avoid the degenerated performance for spherical images, recent research has also attempted to design CNNs that can adapt to geometric distortions in spherical images. The reported solutions can be divided into three groups, i.e., tangent projection methods, CNN kernel shape resizing methods, and CNN sampling point adjustment methods. For the first one, equirectangular images are first projected to undistorted tangent images [29] or divided into quasi-uniform discrete images [30], and existing CNNs are applied to the resulting images. For the second one, CNNs are designed to work on equirectangular images by adjusting the CNN kernel shape [31–33]. In Su and Grauman [32], a CNN termed SPHCONV was proposed to produce results as the output of applying perspective CNNs to the corresponding tangent images. SPHCONV was achieved by defining convolution kernels with varying shapes for pixels in different image rows. Su and Grauman [34] proposed a kernel transformer network (KTN) to learn spherical kernels by taking as input the latitude angle and source kernels for perspective images. For the third one, sampling points of CNN kernels are adjusted based on geometric distortions instead of adjusting the convolution kernel shape. Zhao et al. [33] and Coors et al. [31] designed distortion-aware networks that sample non-regular grid locations according to the distortions of different pixels. The core idea of these networks is to determine the sampling locations based on the spherical projection of a regular grid on the corresponding tangent plane. Due to regular convolution kernels, these frameworks enable the transfer between CNN models for perspective and equirectangular images.

To achieve feature matching for spherical images, both hand-crafted and learning-based methods can provide useful solutions. On the one hand, the redesigned methods can solve the geometric distortions from the camera imaging principle of spherical images. These algorithms, however, cannot leverage existing mature techniques. On the other hand, the methods that use a reprojection strategy can be easily adapted to the algorithms designed for perspective images and cooperated with the representation learning ability of CNNs. Based on the above-mentioned observation, this study proposes a reliable feature matching method for spherical images through the combination of local geometric rectification and CNN learned descriptors. The main contributions are summarized as follows: (1) we design a local geometric rectification algorithm based on the camera imaging model of spherical images and the scale and orientation data from the feature detector; (2) we implement a reliable feature matching workflow for spherical images by using a CNN descriptor learning network for the rectified image patches and a robust essential matrix estimation algorithm for outlier removal in feature matching; and (3) we verify the validation and demonstrate the performance of the proposed solution by using real spherical images in the terms of feature matching and SfM (structure from motion)-based image orientation.

This paper is organized as follows. Section 2 presents the details of the proposed feature matching algorithm, including local geometric rectification, deep learning-based descriptor generation, and outlier removal via essential matrix estimation. Section 3 gives the details of the used datasets and experimental analysis and comparison for feature matching and SfM-based image orientation. Finally, Section 5 presents the conclusions and future studies.

2. Methodology

Figure 2 presents the overall workflow of the proposed algorithm and verification solution. It mainly consists of three steps. First, SIFT (scale invariant feature transform) [12] keypoints are detected mainly because of their wide usage in industrial fields, and the image patches around them are reprojected for local geometric rectification; second, feature descriptors are then calculated from rectified patches based on a pre-trained separate detector and descriptor learning network, which are subsequently fed into the standard SIFT matching module with cross-check and ratio-test constraints; third, refined matches are obtained after outlier removal by using the geometric constraint via the essential matrix estimation. In this study, the proposed algorithm is finally verified in feature matching and SfM-based image orientation by using three real spherical images, which are captured from varying environments and different platforms. The details of the implementation are presented in the following subsections.

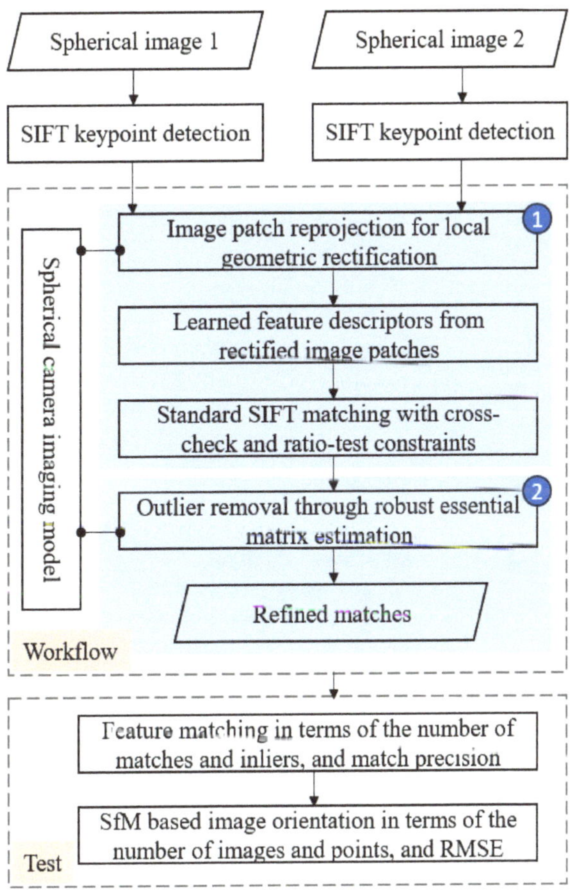

Figure 2. The overall workflow of the proposed algorithm and verification solution.

2.1. Spherical Camera Imaging Model

The camera imaging model defines the geometric relationship between 3D scene points in the object space and their corresponding 2D image points in the image plane. In the literature, the widely used spherical camera imaging model can be categorized into three major groups, i.e., unified camera model [35], general camera model [36], and multi-camera model [37]. Due to the wide usage of multi-camera imaging instruments and the simple formula of the imaging model, the unit sphere camera model that belongs to the multi-camera model is adopted in this study for feature matching and SfM-based image orientation. For the unit sphere camera model, the intrinsic parameters K of a sphere camera include three parameters without other distortion parameters, including one for the focal length f and two for the principal point (c_x, c_y). Generally, the radius r of the unit sphere camera model is set as one. In other words, the focal length of the spherical camera is set as $f = 1$; the principal point coordinates are fixed at the center of images, i.e., $c_x = W/2$ and $c_y = H/2$, in which W and H indicate the image width and height, respectively.

Based on the definition of the spherical camera imaging model, the imaging procedure from the 3D scene points to 2D image points can be illustrated in Figure 3, in which the spherical image is represented in the equirectangular projection (ERP) format. For the imaging procedure, Figure 3a presents the spherical camera imaging model that maps one 3D point P in the object space to the 3D point p on the sphere. Figure 3b shows the transformation between the 3D point p and its corresponding 2D point in the image plane. In this projection, the point p on the unit sphere can be formulated in two coordinate systems, i.e., the geographic coordinate system $O - r\theta\varphi$ and Cartesian coordinate system $O - XYZ$. In the former, the coordinate of point p is represented using the longitude θ and latitude φ; in the latter, the coordinate of point p is represented using three coordinate terms $(x, y, z)^T$.

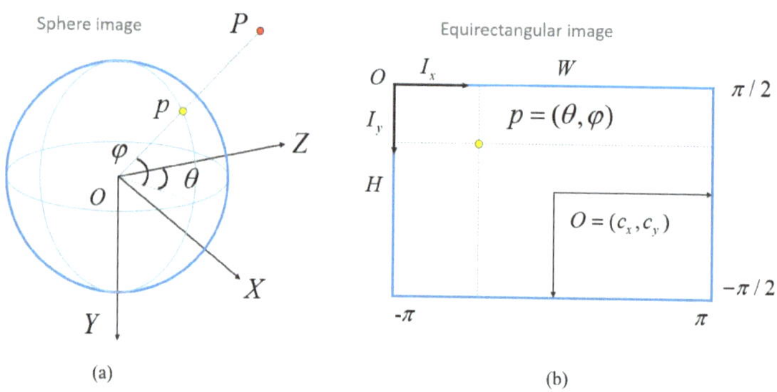

Figure 3. The principle of spherical camera imaging model and coordinate transformation: (**a**) the spherical camera imaging model; (**b**) the coordinate transformation between the spherical image and equirectangular image [5].

According to the coordinate system definition, the transformation from the geographic coordinate system $O - r\theta\varphi$ to the Cartesian coordinate system $O - XYZ$ can be expressed by using Equation (1), in which the sphere radius $r = 1$. In addition, the transformation between 3D geographic coordinates and 2D image coordinates can be formulated as Equation (2), where I_x and I_y are the image coordinates in the ERP image plane. These two equations establish the coordinate transformation between 3D sphere points and 2D image points and form the basic formulas for the subsequent local geometric rectification and outlier removal:

$$\begin{pmatrix} x \\ y \\ z \end{pmatrix} = \begin{pmatrix} \cos(\varphi)\sin(\theta) \\ -\sin(\varphi) \\ \cos(\varphi)\cos(\theta) \end{pmatrix} \quad (1)$$

$$\begin{pmatrix} \theta \\ \varphi \end{pmatrix} = \begin{pmatrix} \frac{I_x - c_x}{W} * 2\pi \\ \frac{c_y - I_y}{H} * \pi \end{pmatrix} \quad (2)$$

2.2. Image Patch Reprojection for Local Geometric Rectification

The geometric distortion in the spherical image seriously degenerates the repeatability of local features due to the appearance difference of image patches around detected keypoints. In this study, image patch reprojection is used to achieve local geometric rectification and alleviate the geometric distortions. The core of image patch reprojection is to project the original patch on the sphere to the corresponding patch on the tangent plane that goes through the keypoint in the geographic coordinate system $O - r\theta\varphi$. The principle of image patch reprojection is illustrated in Figure 4. For the keypoint $I = (I_x, I_y)$ detected from the ERP spherical image, as shown in Figure 4a, its corresponding geographic coordinate $p = (\theta, \varphi)$, as presented in Figure 4b, is first calculated according to Equation (2). By using the normal vector that starts from the origin O to the sphere point p, a tangent plane is then defined as shown by the red line in Figure 4b. Based on the imaging geometry, the local patch around p can be projected onto the tangent plane and generate the rectified patch, as shown in Figure 4c.

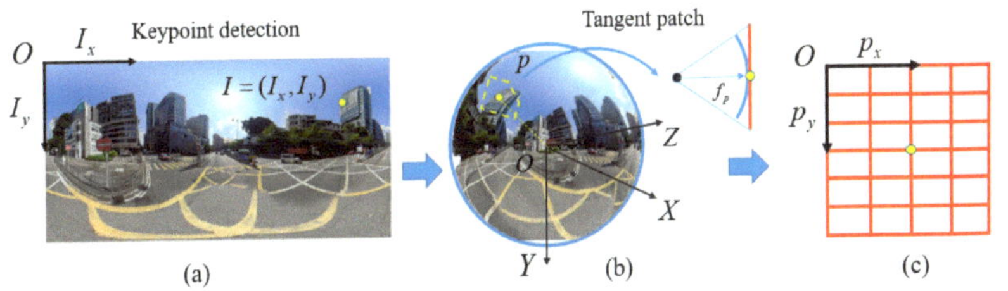

Figure 4. The illustration of image patch reprojection for local geometric rectification: (**a**) the keypoint detected from the ERP spherical image; (**b**) the position of the keypoint is transformed to the spherical coordinate system, in which the tangent plane is defined; (**c**) the image grid defined on the tangent plane.

In the above-mentioned reprojection procedure, the scale *scale* and orientation *ori* parameters should be carefully determined to define image patches since it ensures the scale and rotation invariant for descriptors. Fortunately, the required data can be obtained from widely used feature detection algorithms. In the context of feature detection using SIFT, a feature point f can be represented as $f = (I_x, I_y, scale, ori)$, in which (I_x, I_y) indicates the pixel coordinates; *scale* and *ori* indicate the scale and orientation parameters, respectively. Suppose that the desired width and height of the rectified patch are labeled as W_p and H_p, respectively, for the original image scale. Thus, the patch size for the feature point p can be calculated using Equation (3), in which S_R is the scale ratio between the pyramid layers of the feature point p and the original image. Generally, S_R can be calculated as $scale/scale_0$. For the SIFT used in this study, the original image scale is set as $scale_0 = 1.6$:

$$\begin{cases} W_{sp} = W_p * S_R \\ H_{sp} = H_p * S_R \end{cases} \tag{3}$$

Based on the defined patch size, a pinhole camera model for the rectified patch is defined with the focal length $f_p = W/4$ and principal point $c_{xp} = W_{sp}/2$ and $c_{yp} = H_{sp}/2$ to ensure the same spatial resolution as the original spherical image. In this study, an inverse procedure is utilized to generate the rectified image patch to ensure the desired dimension of output patches. The rectified image patch is computed based on the following steps:

(1) For each image point $p = (p_x, p_y)^T$ in the rectified image patch, as shown in Figure 4c, its homogeneous coordinate $p_h = (p_{xh}, p_{yh}, 1)^T$ is calculated based on Equation (4):

$$\begin{pmatrix} p_{xh} \\ p_{yh} \end{pmatrix} = \begin{pmatrix} \frac{p_x - c_{xp}}{f_p} \\ \frac{p_y - c_{yp}}{f_p} \end{pmatrix} \quad (4)$$

(2) Considering that a unit sphere camera model is used to define the Cartesian coordinate system $O - XYZ$, the homogeneous coordinate p_h is then projected onto the sphere point p_{ls} through the normalization operation presented in Equation (5):

$$p_{ls} = \frac{p_h}{||p_h||} \quad (5)$$

(3) The sphere point p_{ls} is further transformed from the local Cartesian coordinate system of the rectified image patch to the global Cartesian coordinate system $O - XYZ$ by using a transformation matrix $R = R_y(\theta) * R_x(\varphi) * R_z(ori)$, as presented by Equation (6). The transformation matrices $R_z(ori)$, $R_x(\varphi)$ and $R_y(\theta)$ define the rotation around the Z, X, and Y axes with the orientation ori, latitude φ and longitude θ, respectively:

$$p_s = R * p_{ls} \quad (6)$$

(4) According to the transformation between 3D sphere points and 2D image points as presented in Equations (1) and (2), the image point $I = (I_x, I_y)$ in the ERP image is calculated from p_s and used to interpolate the gray values for generating the rectified image patch.

Based on the above-mentioned procedure, the rectified image patches with the size of W_{sp} and H_{sp} can be generated based on the tangent plane reprojection, which is finally resized to the dimension of W_p and H_p. Noticeably, in step (3), the rotation $R_z(ori)$ around the Z axis indicates the transformation from the major orientation of feature point f to the nominal orientation of the Cartesian coordinate system. It is used to achieve the orientation invariant for the subsequently generated descriptors.

2.3. Learned Feature Descriptors from Rectified Image Patches

The rectified image patches are then used to compute descriptors for feature matching. In this study, a separate detector and descriptor learning network is adopted due to two main reasons. On the one hand, image patches are the input of the network, which differs from that for the joint detector and descriptor learning network; on the other hand, this strategy can be easily integrated into the existing workflow for the subsequent feature matching and SfM-based image orientation, instead of the joint feature and metric learning network.

Considering the performance of the existing separate detector and descriptor learning networks [22], a pre-trained HardNet [38] network is selected for the descriptor calculation. Figure 5 shows the network structure and sampling strategy in network training. The network is the same as L2-Net [26]. It consists of seven CNN layers with batch normalization and ReLU activation, except for the last layer without activation. To obtain multi-scale information, the dilated convolution is used in the third and fifth layers. For an input image patch with a size of 32 by 32 pixels, HardNet outputs a 128D descriptor with the same dimension as the widely used SIFT descriptor. In contrast to L2-Net, HardNet adopts a hard negative sampling strategy and triplet margin loss function for network training, which further enhances the discriminative ability of the network. Thus, by using the HardNet network, 128D descriptors are calculated from the rectified image patches and used for the subsequent feature matching.

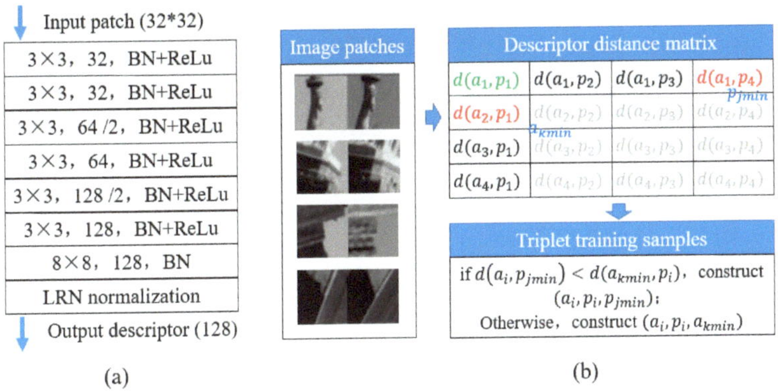

Figure 5. The network structure and sampling strategy of HardNet: (**a**) the network structure of HardNet; (**b**) the sampling strategy used in network training.

2.4. Outlier Removal through Robust Essential Matrix Estimation

To establish correspondence between two images, the initial matches are first obtained based on the standard feature matching strategy. The nearest and second-nearest neighbor searching is executed between two sets of feature descriptors, and the feature points that pass through the ratio test are set as candidate matches. Meanwhile, the cross-checking strategy is also used to further refine the initial matches.

Due to repetitive patterns in images and the limited discriminative ability of local descriptors, false matches are inevitably retained in the initial matches. In this study, the coplanar geometric constraint is utilized to refine the initial matches, which requires that three vectors, i.e., the baseline vector that connects projection centers and two observing vectors that start from projection centers to the scene point, are coplanar. Suppose that the relative orientation of two spherical images is expressed by the relative rotation R and translation T; the intrinsic parameter K of the spherical camera are known. Therefore, an essential matrix $E = [T]_{\times} R$ can be calculated to encode the relative orientation. For two corresponding rays p_1 and p_2, the coplanar constraint is then formulated by Equation (7):

$$p_2^T E p_1 = 0 \tag{7}$$

where p_1 and p_2 are the spherical coordinates of two corresponding image points I_1 and I_2 in the image plane, which are calculated according to Equations (1) and (2). The geometrical meaning of the coplanar constraint is shown in Figure 6. If p_1 and p_2 are a true match, the three vectors Rp_1, p_2 and T are coplanar. In other words, p_2 lies on the circular plane composed of the vector Rp_1 and T with the normal vector $\vec{n}$. Thus, using the estimated essential matrix E, false matches can be identified from the initial matches.

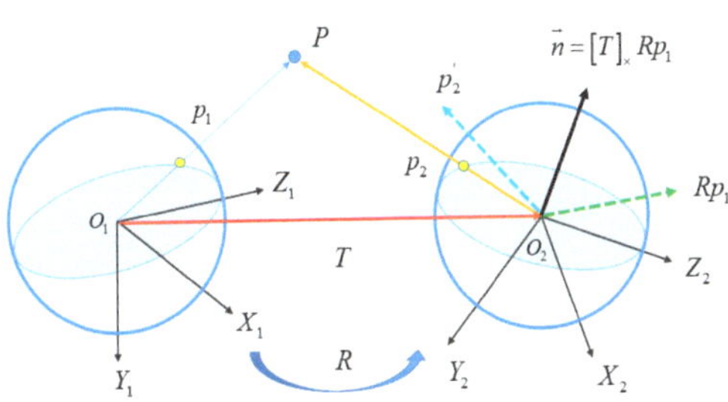

Figure 6. The principle of relative orientation for spherical images.

To achieve a robust estimation of the essential matrix E, the RANSAC-based hypothesis verify framework [39] is used in this study. During the iteration in RANSAC, the error metric e and error threshold e_p are required to label true and false matches. According to the coplanar constraint as shown in Figure 6, the corresponding ray p_2 of p_1 in the left image lies on the circular plane that is defined by the normal vector $\vec{n}$ and the projection center O_2 of the right image. Thus, this study adopts the vector-to-plane geodesic angular error metric [14] as presented in Equation (8):

$$e = abs(\sin^{-1}(p_2^T E p_1)) \tag{8}$$

where $abs(\cdot)$ indicates the absolute value. At the same time, the error threshold e_p in the unit of pixels is converted to spherical angles in the unit of degrees. In this study, the conversion is implemented according to Equation (9):

$$e_a = \frac{2\pi}{\max(W,H)} e_p \tag{9}$$

where $2\pi/\max(W,H)$ indicates the scale factor of these two metrics; e_a is the error threshold in the spherical angles. In conclusion, based on the estimated essential matrix E, the corresponding points p_1 and p_2 are labeled as one inlier if the angular error $e < e_a$. Based on the coplanar constraint, refined matches are obtained from the initial matches.

2.5. Implementation of the Proposed Algorithm

The proposed algorithm is implemented by using the C++ programming language. For SIFT feature detection, the open-source library SIFTGPU [40] with default parameter settings is used due to its hardware-accelerated high efficiency. For descriptor learning, the pre-trained HardNet network released on the official website is directly used due to two main reasons. On the one hand, it is trained using the Brown and HPatches datasets, which have large diversity in terms of viewpoint and illumination; on the other hand, this study aims to achieve feature matching using geometric rectified patches, instead of using spherical images directly. Thus, no retraining is necessary for the utilized network. For nearest neighbor searching-based feature matching, the maximum distance and the ratio test threshold are set as 0.7 and 0.8, respectively. For essential matrix estimation, the 8-point algorithm [41] is used, in which eight corresponding points form eight linear equations, and the linear system is then solved through SVD (singular value decomposition) [42]. In addition, the error threshold e_p is set as 4 pixels.

3. Experiments and Results

In the experiments, three datasets are utilized to evaluate the performance of the proposed algorithm for the feature matching of spherical images. First, the adopted datasets and evaluation metrics are described. Second, the comparison with other algorithms is conducted for feature matching in terms of the number of matches and inliers and the matching precision. Third, the proposed algorithm is integrated with an incremental SfM workflow for image orientation. In this study, all tests are conducted on a Windows desktop computer that is configured with 32 GB memory, an Intel Core i7-8700K 3.7 GHz CPU (central processing unit), and an NVIDIA GeForce GTX 1050Ti GPU (graph processing unit).

3.1. Test Sites and Datasets

Detailed information on the three spherical datasets is presented in Table 1. The datasets are captured by using both consumer-grade and professional sphere cameras, which are fixed on the ground or in a hand-held tripod and mounted on the moving car. The characteristic of each test site and the details for data acquisition are listed as follows.

Table 1. Detailed information of the three spherical datasets.

Item Name	Dataset 1	Dataset 2	Dataset 3
Scene type	Outdoor	Hybrid	Street
Sensor type	Sphere	Sphere	Sphere
Camera model	Garmin VIRB 360	Garmin VIRB 360	Ladybug3
Storage format	Equirectangular	Equirectangular	Equirectangular
Sensor platform	Ground tripod	Hand-held rod	Moving car
Number of images	37	279	1937
Image size (pixel)	5640 × 2820	5640 × 2820	5400 × 2700

- The first dataset is recorded from a campus, which includes a parterre surrounded by high buildings as shown in Figure 7a. For image acquisition, a Garmin VIRB 360 camera is used, which stores images in the equirectangular representation format. The data acquisition is conducted around the central parterre, and there are a total number of 37 images collected with a resolution of 5640 by 2820 pixels.
- The second dataset includes a complex building structure that covers from its rooftop to the inner aisles as shown in Figure 7b. Parterres exist on the rooftop, and the inner aisles connect different layers. For image acquisition, the same Garmin VIRB 360 camera as in dataset 1 is adopted by using a hand-held tripod. A total number of 279 spherical images are collected, which cover the whole inner aisles.
- The third dataset is collected using an MMS system. The test site goes along an urban street, whose length is approximately 7.0 km. Along the street, low residual buildings are located near the two roadsides as shown in Figure 7c. In this test site, a PointGrey Ladybug3 camera that is made of six fisheye cameras is used. By setting the interval distance of 3 m for camera exposure, there are a total number of 1937 spherical images collected from this site.

(a) (b) (c)

Figure 7. The illustration samples of the used spherical datasets: (**a**) dataset 1; (**b**) dataset 2; (**c**) dataset 3.

3.2. Evaluation Metrics

The proposed algorithm would be evaluated in feature matching and SfM-based image orientation. For feature matching, three metrics are utilized, i.e., the number of matches and inliers, and matching precision. The first indicates the number of obtained initial matches; the second indicates the total number of obtained true matches; the third represents the number ratio of true matches and initial matches. In SfM-based image orientation, the obtained matches are then fed into an incremental SfM engine to reconstruct camera poses and scene points. For performance evaluation, three metrics are used, i.e., the number of images and points, and RMSE (root mean square error). The first and second metrics indicate the completeness of the image orientation, which is calculated as the number of registered images and reconstructed 3D points. The third metric is calculated as the reprojection error in BA (bundle adjustment) optimization. The description of used evaluation metrics is listed in Table 2.

Table 2. The description of the used metrics for performance evaluation. Categories 1 and 2 indicate the terms of feature matching and SfM-based image orientation, respectively. RMSE represents the root mean square error in BA optimization.

Category	Metric	Description
1	No. matches	The number of initial matches before outlier removal (large value indicates good results).
	No. inliers	The total number of true matches after outlier removal (large value indicates good results).
	Match precision	The ratio between the numbers of true matches and initial matches (large value indicates good results).
2	No. images	The number of resumed images in SfM-based image orientation (small value indicates good results).
	No. points	The number of reconstructed 3D points in SfM-based image orientation (large value indicates good results).
	RMSE	The RMSE of the bundle adjustment optimization (small value indicates good results).

3.3. The Analysis of the Performance for Local Geometric Rectification

Local geometric rectification via image patch reprojection is the first step in the proposed algorithm. It aims to alleviate appearance differences caused by the spherical camera model. For visual analysis, Figure 8 presents the image patches that are directly cropped from images and geometrically rectified based on tangent plane projection, which are rendered by yellow and green colors, respectively. It is clearly shown that geometric distortions exist in original image patches, such as the curve boundaries of buildings. After geometric rectification, the distortions can be decreased, especially for the regions near the poles.

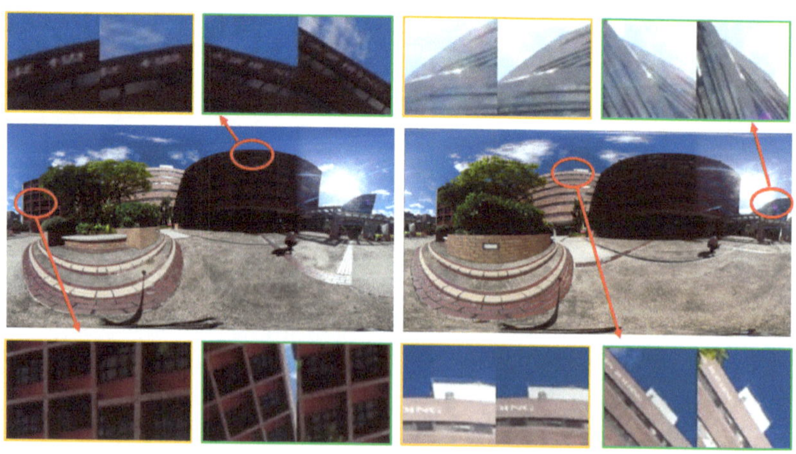

Figure 8. The comparison of extracted local image patches from one image pair in dataset 1. For each item, the left and right items are directly cropped around keypoints and geometrically rectified based on tangent plane reprojection, respectively.

In local geometric rectification, the orientation *ori* and scale *scale* of the output image patches have a great influence on the performance of the subsequent descriptor calculation.

In this study, the scale *scale* and orientation *ori* are obtained from the used SIFT keypoint detectors. Figure 9 shows the comparison of local geometric rectification under different configurations. The geometric rectification can dramatically decrease the appearance differences as the results are presented from Figure 9a to Figure 9b. Although they have high appearance similarity, the generated image patches are not invariant to the changes in orientation and scale. By using the orientation and scale from detected SIFT features, the image patches are then rotated and scaled accordingly as illustrated in Figure 9c,d, respectively. For the visual analysis of the proposed algorithm, Figure 10 illustrates the generated image patches from dataset 3. We can see that the structure and texture of generated patches from the proposed algorithm are more regular as verified by the patches labeled by the red rectangle.

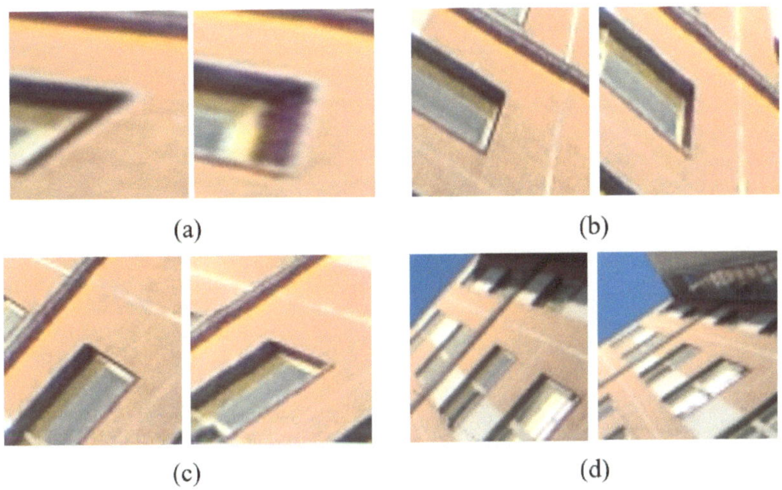

Figure 9. The comparison of local geometric rectification: the image patch (**a**) directly cropped from the spherical image without geometric rectification; (**b**) without orientation and scale; (**c**) with only orientation; and (**d**) with both orientation and scale. Noticeably, the image size is 32 by 32 pixels for all patches.

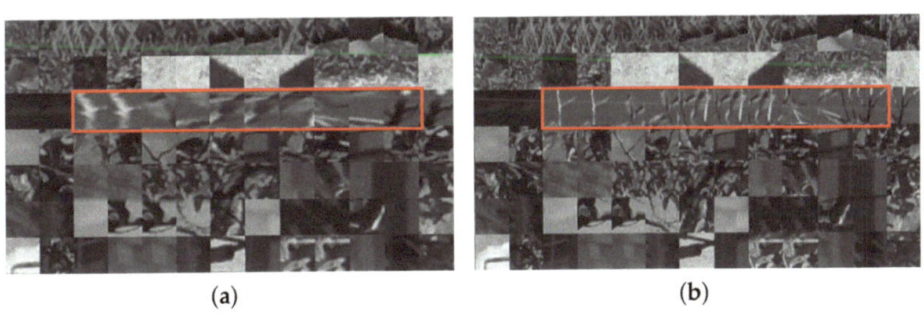

Figure 10. The illustration of generated image patches: image patch (**a**) directly cropped from the spherical image without geometric rectification and (**b**) rectified by the proposed algorithm. The red rectangle indicates the effect of geometric rectification.

To verify the validation of the proposed local geometric rectification solution, three image pairs with varying viewpoints are selected from dataset 2 for tests, and the four configurations presented in Figure 9 are used for image patch extraction and feature matching. The statistical results of the number of inliers are shown in Figure 11, in which the methods with labels 1, 2, 3, and 4 correspond to the four configurations in Figure 9a–d.

It is shown that for all three image pairs, the number of inliers increases obviously for the methods with the label from 1 to 4. For a visual illustration, Figure 12 presents the matching results of image pair 2. We can see that the geometric rectification increase matches near poles as shown in Figure 12b; the introduction of orientation and scale further increases matches over the whole image plane as presented in Figure 12c,d.

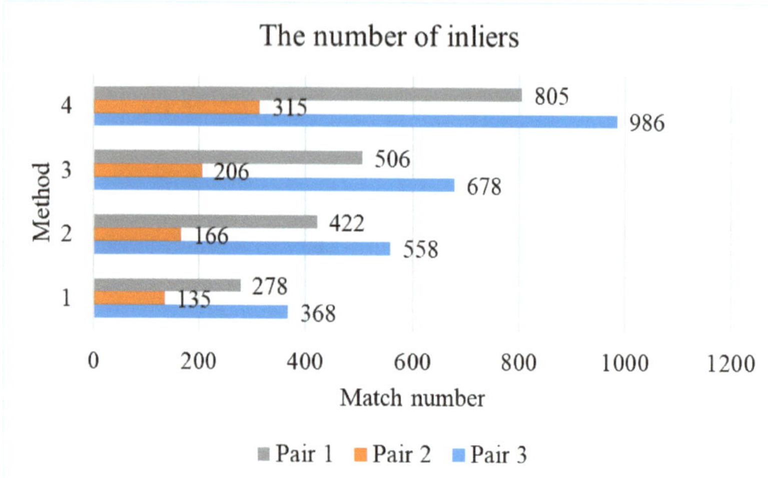

Figure 11. The comparison of the number of inliers of different methods.

(**a**) 135/180 (**b**) 166/258

(**c**) 206/320 (**d**) 315/480

Figure 12. The comparison of different image patch extraction methods for feature matching: image patch (**a**) directly cropped from the spherical image without geometric rectification; geometrically rectified (**b**) without orientation and scale; (**c**) with only orientation; and (**d**) with both orientation and scale.

3.4. The Comparison of Local Feature-Based Matching

Local feature-based matching is then conducted by using the geometrically rectified image patches. In this test, three metrics are used for performance evaluation, including the number of matches, the number of inliers, and match precision. For comparison analysis, four methods are adopted in this study, i.e., SIFT, ASLFeat, NGR-H (HardNet for non-geometric rectified patches), and the proposed algorithm (HardNet for geometric rectified patches). SIFT is used as the baseline algorithm, which has been widely used in the photogrammetry field. ASLFeat is an end-to-end network for feature detection and description [28]. NGR-H is utilized to verify the advantage of deep learning-based descriptors when compared with hand-crafted descriptors. Before feature matching, image pairs are first selected based on the sequential and spatial constraints in the data acquisition. For the three datasets, there are a total number of 157, 4941, and 14,836 image match pairs.

Table 3 presents the statistical results of feature matching for the three datasets. It is shown that compared with separated detection and description methods, i.e., SIFT and NGR-H, the proposed algorithm achieves the best performance under all used metrics, except for the matching precision in dataset 1. In particular, compared with SIFT, the increasing ratio of the number of inliers is 73.9% for dataset 1, which is higher than the values of 34.2% and 26.8% for datasets 2 and 3, respectively. The main reason is that the top region of the images is covered by sky and cloud, as illustrated in Figure 7, and few keypoints are extracted from the region with large distortions. When comparing SIFT and NGR-H, we can see that NGR-H achieves better performance in dataset 1 and comparative performance in datasets 2 and 3. It verifies that the learned descriptor has a high tolerance to image distortions. For the end-to-end network ASLFeat, the number of inliers is obviously lower than the proposed method, which are 83, 198, and 177 for the three datasets, respectively. The main reason is the low position accuracy of detected keypoints from down-sampled feature maps as mentioned in [22].

Table 3. The statistical results of feature matching for the tested algorithms. The mean of each metric is calculated from all selected image pairs for feature matching. The best values are in bold.

Metric	Method	Dataset 1	Dataset 2	Dataset 3
No. matches	SIFT	165	232	296
	ASLFeat	**337**	**385**	253
	NGR-H	248	234	286
	Ours	290	297	**371**
No. inliers	SIFT	111	158	250
	ASLFeat	83	198	177
	NGR-H	168	160	244
	Ours	**193**	**212**	**317**
Match Precision	SIFT	0.57	0.64	0.79
	ASLFeat	0.33	0.51	0.68
	NGR-H	**0.62**	0.59	0.81
	Ours	0.60	**0.67**	**0.82**

For the further visual analysis, Figures 13–15 show the matching results of one selected image pair from the three datasets. We can see that the proposed algorithm achieves the best performance in the number of matches and inliers. In the term of match precision, comparative performance can be observed from image pairs 1 and 3 for the three methods. For image pair 2, the proposed algorithm has better performance to cope with the large distorted regions. Due to the low position accuracy, the number of inliers and match precision of ASLFeat is obviously lower than the other methods. Considering the performance of the evaluated methods, only SIFT, NGR-H, and the proposed algorithm would be further analyzed in the following experiments.

For the overall statistical analysis, Figure 16 presents the statistical results of the number of inliers by using the frequency histogram and accumulative frequency. For each sub-figure, the range of the inlier number is divided into bins with the same width, and the inlier number of all selected image pairs votes for the bins and the accumulative frequency. For interpretation, the point near the value of 90% in the accumulative frequency is highlighted in each sub-figure, and the range of bins and inliers are labeled. It is shown that for the three datasets, the proposed algorithm has a larger span for both bins and inliers when compared with SIFT and NGR-H. It means that more image pairs have a larger number of inliers.

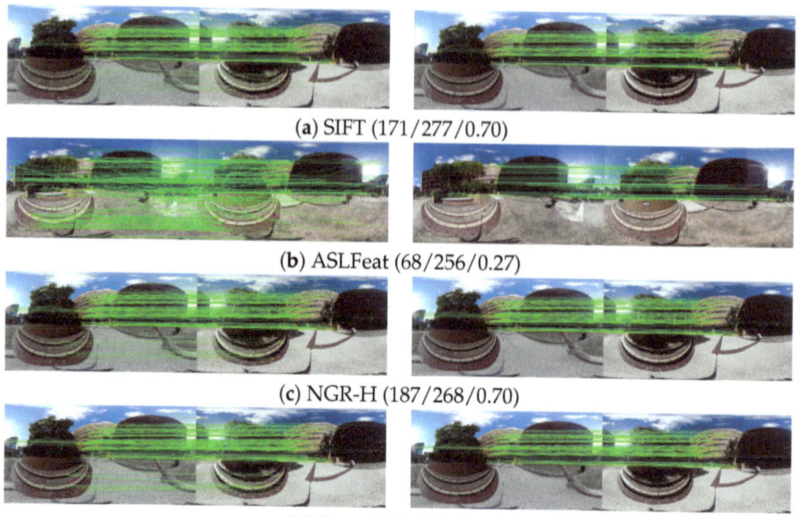

(**a**) SIFT (171/277/0.70)

(**b**) ASLFeat (68/256/0.27)

(**c**) NGR-H (187/268/0.70)

(**d**) Ours (206/298/0.69)

Figure 13. The comparison of feature matching for dataset 1. For each method, the **left** and **right** images represent the results of initial and refined matches. The values in the bracket are the number of inliers and initial matches, and the match precision, respectively.

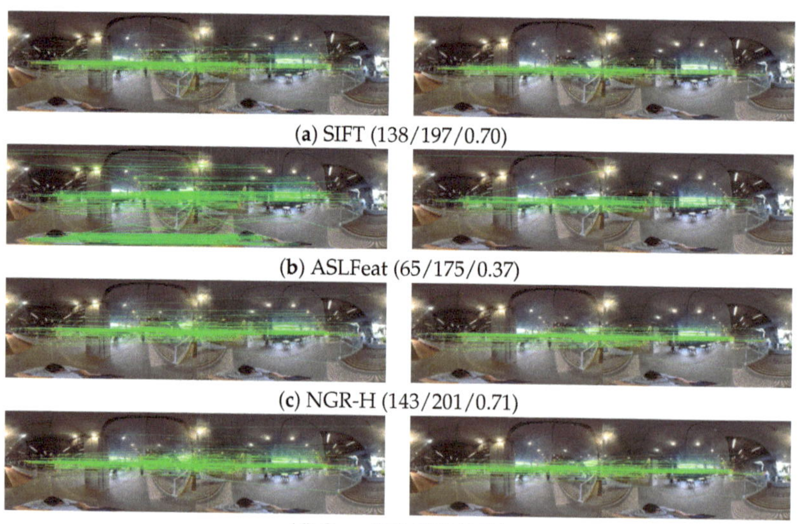

(**a**) SIFT (138/197/0.70)

(**b**) ASLFeat (65/175/0.37)

(**c**) NGR-H (143/201/0.71)

(**d**) Ours (193/263/0.73)

Figure 14. The comparison of feature matching for dataset 2. For each method, the **left** and **right** images represent the results of initial and refined matches. The values in the bracket are the number of inliers and initial matches, and the match precision, respectively.

(a) SIFT (323/416/0.78)

(b) ASLFeat (90/145/0.62)

(c) NGR-H (321/403/0.80)

(d) Ours (413/525/0.79)

Figure 15. The comparison of feature matching for dataset 3. For each method, the **left** and **right** images represent the results of initial and refined matches. The values in the bracket are the number of inliers and initial matches, and the match precision, respectively.

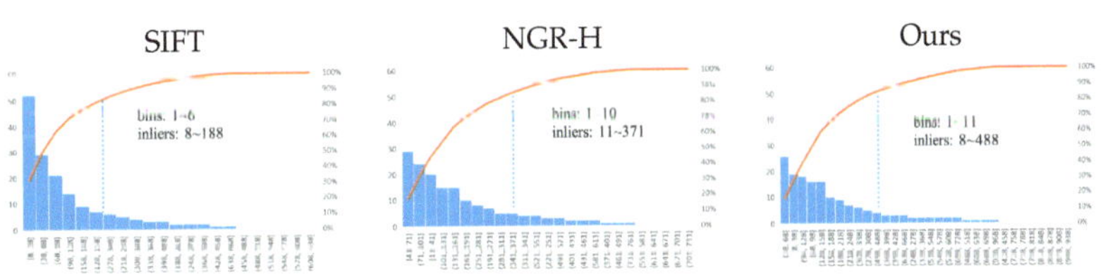

(a) dataset 1

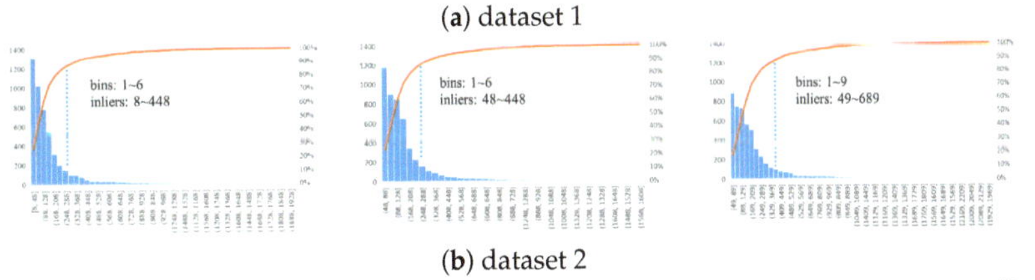

(b) dataset 2

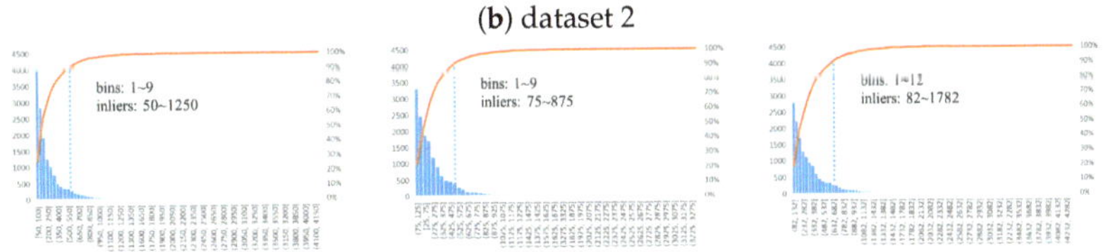

(c) dataset 3

Figure 16. The statistical analysis of the number of inliers for the three datasets. Each figure presents two terms. The **bottom one** is the bin frequency that inlier numbers fall into, which is arranged in descending order; the **top one** indicates the accumulation of the bin frequencies.

3.5. Application in SfM-Based Image Orientation

SfM-based image orientation can be achieved by using the refined feature matches In our previous work, an incremental SfM engine was designed and implemented [5] The inputs of the SfM engine are spherical images in the ERP format. After the sequential execution of SIFT feature matching, essential matrix-based outlier removal, and the iterative bundle adjustment, sparse reconstruction can be obtained, including the oriented images and reconstructed 3D points. Based on the established workflow, the proposed feature matching algorithm is integrated with the SfM engine for image orientation.

Table 4 presents the statistical results of image orientation for the three datasets. We can see that all images can be successfully reconstructed for the three test algorithms. The number of reconstructed 3D points from the proposed algorithm are 4645, 49,252, and 363,371 for the three datasets, respectively, whose increase ratios are approximately 80.8%, 22.8%, and 25.2% when compared with SIFT. It is almost consistent with the increased ratio of feature matching as presented in Section 3.4. Considering the metric RMSE in the BA optimization, SIFT achieves better performance than the proposed algorithm, whose values are 0.74, 0.80, and 0.56 for the three datasets, respectively. It can explain from two aspects. On the one hand, fewer matched points would be involved in the BA optimization, which would decrease the ratio of false matches in SIFT; on the other hand, the distortions near the pole are larger than the other regions, which would further decrease the position accuracy of matched points in the proposed algorithm. In addition, Figure 17 presents the image orientation results of the three datasets based on the SfM engine. It is shown that all images in the three datasets are well reconstructed, which can be used for subsequent 3D reconstruction procedures, e.g., dense matching and texture mapping. Based on the comparison, we can conclude that the proposed algorithm can reconstruct more 3D points and achieves comparative accuracy when compared with other methods.

Table 4. The statistical results of image orientation for the three datasets in terms of the number of oriented images and reconstructed 3D points and precision. The RMSE is in pixels.

Dataset	Method	Images	Points	RMSE
Dataset 1	SIFT	37	2569	0.74
	NGR-H	37	3832	0.80
	Ours	37	4645	0.80
Dataset 2	SIFT	279	40,118	0.80
	NGR-H	279	38,927	0.83
	Ours	279	49,252	0.82
Dataset 3	SIFT	1937	290,240	0.56
	NGR-H	1937	289,681	0.61
	Ours	1937	363,371	0.60

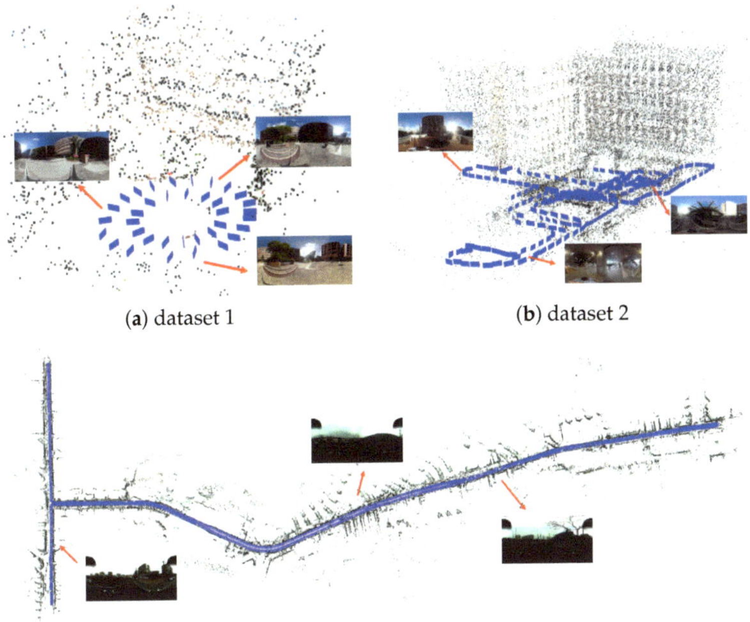

(a) dataset 1 (b) dataset 2

(c) dataset 3

Figure 17. Image orientation results based on the SfM engine. The blue rectangles indicate the oriented images, and reconstructed 3D points are rendered by the color of the images.

4. Discussion

Spherical images are becoming a promising data source for the 3D reconstruction of complex scenes due to their omnidirectional FOV. However, geometric distortions are inevitably added to the recorded images of their spherical camera imaging model. Considering the wide usage of spherical cameras and their promising applications in 3D reconstruction, this study designs and implements a reliable feature matching method for spherical images. The main purpose is to reduce the geometric distortions that are caused by the projection from the 3D sphere to the 2D plane and improve the discriminative power of descriptors by exploiting deep learning-based techniques. The performance of the proposed algorithm is verified by spherical images captured from both consumer-grade and professional cameras.

Compared with existing methods, two major advantages are designed for the proposed algorithm. On the one hand, local geometric rectification is adopted to remove the distortions. For scale and rotation invariance, it is implemented by considering both orientation ori and scale oriof the output image patches since they have a great influence on the subsequent descriptor calculation. Specifically, the scale ori and orientation ori information in the SIFT keypoint detector is used to improve the performance as demonstrated in Section 3.3. On the other hand, the learned descriptor is then utilized to describe rectified patches because they have shown high discriminative power in recent studies, and the results are verified in Section 3.4. In addition, a robust outlier removal method is designed as the final step to refine the initial matches, which is based on the essential matrix estimation in the sphere coordinate system. Based on the designed feature matching method, reliable feature matches can be used to achieve SfM- and SLAM-based image orientation as shown in Section 3.5.

According to the experimental results, some limitations could also be observed in this study. First, the unit sphere camera model is used for image orientation, which consists of three intrinsic parameters, i.e., one for the focal length f and two for the principal point (c_x, c_y). The ideal camera model may not be enough to establish the imaging model for consumer-grade cameras. It can be observed from the RMSE presented in Table 4, in which the RMSE of datasets 1 and 2 is larger than that of dataset 3. Second, the hand-crafted SIFT detector is used to detect keypoints for patch generation. However, compared with aerial images, spherical images are often captured from near-ground streets or indoor rooms that include a majority of low- or non-textured regions. Thus, a few keypoints can be detected from these scenes, which can be verified by the results presented in Figure 14. In future studies, more spherical camera imaging models would be compared in the SfM-based image orientation. Furthermore, deep learning-based detector-free networks can be used to address the second issue.

5. Conclusions

This study implements a reliable feature matching algorithm for spherical images via the combination of local geometric rectification and the CNN learned descriptor. After SIFT-based feature detection, image patches are first reprojected to their corresponding tangent planes for the local geometric rectification, which can achieve scale- and orientation-invariant geometric rectification. Using a pre-trained separate detector and descriptor network, feature descriptors are then generated and used to obtain the initial matches. Finally, refined matches are obtained after outlier removal that is implemented using the essential matrix-based epipolar geometry. The performance is verified by using real spherical images, and experimental results demonstrate that the proposed algorithm can provide reliable feature matches and improve the completeness of SfM-based image orientation.

Author Contributions: Conceptualization, S.J. and W.C.; methodology, S.J. and J.L.; software, S.J. and J.L.; validation, J.L., Y.L. and D.W.; formal analysis, J.L.; resources, Y.L.; data curation, S.J.; writing—original draft preparation, S.J. and J.L.; writing—review and editing, S.J. and J.L.; visualization, J.L.; supervision, W.C.; project administration, S.J. and W.C.; funding acquisition, S.J. All authors have read and agreed to the published version of the manuscript.

Funding: This research was funded by the National Natural Science Foundation of China (Grant No. 42371442), the Hubei Provincial Natural Science Foundation of China (Grant No. 2023AFB568), and the Hong Kong Scholars Program (Grant No. 2021-114).

Data Availability Statement: Research data would be shared from e-mail query.

Acknowledgments: The authors would like to thank authors who have made their algorithms of SiftGPU and ColMap free and open-source software packages, which is helpful to the research in this paper. Meanwhile, heartfelt thanks to the anonymous reviewers and the editors, whose comments and advice improve the quality of the work.

Conflicts of Interest: The authors declare no conflict of interest.

References

1. Jiang, S.; Jiang, W.; Wang, L. Unmanned Aerial Vehicle-Based Photogrammetric 3D Mapping: A survey of techniques, applications, and challenges. *IEEE Geosci. Remote Sens. Mag.* **2022**, *10*, 135–171. [CrossRef]
2. Wu, B.; Xie, L.; Hu, H.; Zhu, Q.; Yau, E. Integration of aerial oblique imagery and terrestrial imagery for optimized 3D modeling in urban areas. *ISPRS J. Photogramm. Remote Sens.* **2018**, *139*, 119–132. [CrossRef]
3. Chiabrando, F.; D'Andria, F.; Sammartano, G.; Spanò, A. UAV photogrammetry for archaeological site survey. 3D models at the Hierapolis in Phrygia (Turkey). *Virtual Archaeol. Rev.* **2018**, *9*, 28–43. [CrossRef]
4. Jiang, S.; Jiang, W.; Huang, W.; Yang, L. UAV-based oblique photogrammetry for outdoor data acquisition and offsite visual inspection of transmission line. *Remote Sens.* **2017**, *9*, 278. [CrossRef]
5. Jiang, S.; You, K.; Li, Y.; Weng, D.; Chen, W. 3D Reconstruction of Spherical Images based on Incremental Structure from Motion. *arXiv* **2023**, arXiv:2306.12770.
6. Torii, A.; Havlena, M.; Pajdla, T. From google street view to 3d city models. In Proceedings of the 2009 IEEE 12th International Conference on Computer Vision Workshops, ICCV Workshops, Kyoto, Japan, 29 September–2 October 2009; pp. 2188–2195.

7. Gao, S.; Yang, K.; Shi, H.; Wang, K.; Bai, J. Review on panoramic imaging and its applications in scene understanding. *IEEE Trans. Instrum. Meas.* **2022**, *71*, 1–34. [CrossRef]
8. Jhan, J.P.; Kerle, N.; Rau, J.Y. Integrating UAV and ground panoramic images for point cloud analysis of damaged building. *IEEE Geosci. Remote Sens. Lett.* **2021**, *19*, 1–5. [CrossRef]
9. Fangi, G.; Pierdicca, R.; Sturari, M.; Malinverni, E. Improving spherical photogrammetry using 360 omni-cameras: Use cases and new applications. *Int. Arch. Photogramm. Remote Sens. Spat. Inf. Sci.* **2018**, *42*, 331–337. [CrossRef]
10. Janiszewski, M.; Torkan, M.; Uotinen, L.; Rinne, M. Rapid photogrammetry with a 360-degree camera for tunnel mapping. *Remote Sens.* **2022**, *14*, 5494. [CrossRef]
11. Bay, H.; Ess, A.; Tuytelaars, T.; Van Gool, L. Speeded-up robust features (SURF). *Comput. Vis. Image Underst.* **2008**, *110*, 346–359. [CrossRef]
12. Lowe, D.G. Distinctive image features from scale-invariant keypoints. *Int. J. Comput. Vis.* **2004**, *60*, 91–110. [CrossRef]
13. Jiang, S.; Jiang, W. Reliable image matching via photometric and geometric constraints structured by Delaunay triangulation. *ISPRS J. Photogramm. Remote Sens.* **2019**, *153*, 1–20. [CrossRef]
14. Pagani, A.; Stricker, D. Structure from motion using full spherical panoramic cameras. In Proceedings of the 2011 IEEE International Conference on Computer Vision Workshops (ICCV Workshops), Barcelona, Spain, 6–13 November 2011; pp. 375–382.
15. Lichti, D.D.; Jarron, D.; Tredoux, W.; Shahbazi, M.; Radovanovic, R. Geometric modelling and calibration of a spherical camera imaging system. *Photogramm. Rec.* **2020**, *35*, 123–142. [CrossRef]
16. Chuang, T.Y.; Perng, N. Rectified feature matching for spherical panoramic images. *Photogramm. Eng. Remote Sens.* **2018**, *84*, 25–32. [CrossRef]
17. Taira, H.; Inoue, Y.; Torii, A.; Okutomi, M. Robust feature matching for distorted projection by spherical cameras. *IPSJ Trans. Comput. Vis. Appl.* **2015**, *7*, 84–88. [CrossRef]
18. Wang, Y.; Cai, S.; Li, S.J.; Liu, Y.; Guo, Y.; Li, T.; Cheng, M.M. CubemapSLAM: A piecewise-pinhole monocular fisheye SLAM system. In Proceedings of the Asian Conference on Computer Vision, Perth, WA, Australia, 2–6 December 2018; pp. 34–49.
19. Zhao, Q.; Feng, W.; Wan, L.; Zhang, J. SPHORB: A fast and robust binary feature on the sphere. *Int. J. Comput. Vis.* **2015**, *113*, 143–159. [CrossRef]
20. Guan, H.; Smith, W.A. BRISKS: Binary features for spherical images on a geodesic grid. In Proceedings of the IEEE Conference on Computer Vision and Pattern Recognition, Honolulu, HI, USA, 21–26 July 2017; pp. 4516–4524.
21. Chen, L.; Rottensteiner, F.; Heipke, C. Feature detection and description for image matching: From hand-crafted design to deep learning. *Geo-Spat. Inf. Sci.* **2021**, *24*, 58–74. [CrossRef]
22. Jiang, S.; Jiang, W.; Guo, B.; Li, L.; Wang, L. Learned local features for structure from motion of uav images: A comparative evaluation. *IEEE J. Sel. Top. Appl. Earth Obs. Remote Sens.* **2021**, *14*, 10583–10597. [CrossRef]
23. Han, X.; Leung, T.; Jia, Y.; Sukthankar, R.; Berg, A.C. Matchnet: Unifying feature and metric learning for patch-based matching. In Proceedings of the IEEE Conference on Computer Vision and Pattern Recognition, Boston, MA, USA, 7–12 June 2015; pp. 3279–3286.
24. Kumar BG, V.; Carneiro, G.; Reid, I. Learning local image descriptors with deep siamese and triplet convolutional networks by minimising global loss functions. In Proceedings of the IEEE Conference on Computer Vision and Pattern Recognition, Las Vegas, NV, USA, 27–30 June 2016; pp. 5385–5394.
25. Luo, Z.; Shen, T.; Zhou, L.; Zhu, S.; Zhang, R.; Yao, Y.; Fang, T.; Quan, L. Geodesc: Learning local descriptors by integrating geometry constraints. In Proceedings of the European Conference on Computer Vision (ECCV), Munich, Germany, 8–14 September 2018; pp. 168–183.
26. Tian, Y.; Fan, B.; Wu, F. L2-net: Deep learning of discriminative patch descriptor in euclidean space. In Proceedings of the IEEE Conference on Computer Vision and Pattern Recognition, Honolulu, HI, USA, 21–26 July 2017; pp. 661–669.
27. Dusmanu, M.; Rocco, I.; Pajdla, T.; Pollefeys, M.; Sivic, J.; Torii, A.; Sattler, T. D2-net: A trainable cnn for joint description and detection of local features. In Proceedings of the IEEE/CVF Conference on Computer Vision and Pattern Recognition, Long Beach, CA, USA, 15–20 June 2019; pp. 8092–8101.
28. Luo, Z.; Zhou, L.; Bai, X.; Chen, H.; Zhang, J.; Yao, Y.; Li, S.; Fang, T.; Quan, L. Aslfeat: Learning local features of accurate shape and localization. In Proceedings of the IEEE/CVF Conference on Computer Vision and Pattern Recognition, Seattle, WA, USA, 13–19 June 2020; pp. 6589–6598.
29. Eder, M.; Shvets, M.; Lim, J.; Frahm, J.M. Tangent images for mitigating spherical distortion. In Proceedings of the IEEE/CVF Conference on Computer Vision and Pattern Recognition, Seattle, WA, USA, 13–19 June 2020; pp. 12426–12434.
30. Shan, Y.; Li, S. Descriptor matching for a discrete spherical image with a convolutional neural network. *IEEE Access* **2018**, *6*, 20748–20755. [CrossRef]
31. Coors, B.; Condurache, A.P.; Geiger, A. Spherenet: Learning spherical representations for detection and classification in omnidirectional images. In Proceedings of the European Conference on Computer Vision (ECCV), Munich, Germany, 8–14 September 2018; pp. 518–533.
32. Su, Y.C.; Grauman, K. Learning spherical convolution for fast features from 360 imagery. *Adv. Neural Inf. Process. Syst.* **2017**, *30*.
33. Zhao, Q.; Zhu, C.; Dai, F.; Ma, Y.; Jin, G.; Zhang, Y. Distortion-aware CNNs for Spherical Images. In Proceedings of the International Joint Conference on Artificial Intelligence , Stockholm, Sweden, 13–19 July 2018; pp. 1198–1204.

34. Su, Y.C.; Grauman, K. Kernel transformer networks for compact spherical convolution. In Proceedings of the IEEE/CVF Conference on Computer Vision and Pattern Recognition, Long Beach, CA, USA, 15–20 June 2019; pp. 9442–9451.
35. Mei, C.; Rives, P. Single view point omnidirectional camera calibration from planar grids. In Proceedings of the Proceedings 2007 IEEE International Conference on Robotics and Automation, Rome, Italy, 10–14 April 2007; pp. 3945–3950.
36. Scaramuzza, D.; Martinelli, A.; Siegwart, R. A flexible technique for accurate omnidirectional camera calibration and structure from motion. In Proceedings of the Fourth IEEE International Conference on Computer Vision Systems (ICVS'06), 4–7 January 2006; pp. 45–45.
37. Ji, S.; Shi, Y.; Shi, Z.; Bao, A.; Li, J.; Yuan, X.; Duan, Y.; Shibasaki, R. Comparison of two panoramic sensor models for precise 3d measurements. *Photogramm. Eng. Remote Sens.* **2014**, *80*, 229–238. [CrossRef]
38. Mishchuk, A.; Mishkin, D.; Radenovic, F.; Matas, J. Working hard to know your neighbor's margins: Local descriptor learning loss. In Proceedings of the 31st International Conference on Neural Information Processing Systems, Long Beach, CA, USA, 4–9 December 2017; Volume 30.
39. Fischler, M.A.; Bolles, R.C. Random sample consensus: A paradigm for model fitting with applications to image analysis and automated cartography. *Commun. ACM* **1981**, *24*, 381–395. [CrossRef]
40. Wu, C. SiftGPU: A GPU Implementation of Sift. 2007. Available online: http://cs.unc.edu/~{}ccwu/siftgpu (accessed on 10 October 2023).
41. Hartley, R.; Zisserman, A. *Multiple View Geometry in Computer Vision*; Cambridge University Press: Cambridge, UK, 2003.
42. Umeyama, S. Least-squares estimation of transformation parameters between two point patterns. *IEEE Trans. Pattern Anal. Mach. Intell.* **1991**, *13*, 376–380. [CrossRef]

remote sensing

MDPI

Article

Quasi-Dense Matching for Oblique Stereo Images through Semantic Segmentation and Local Feature Enhancement

Guobiao Yao [1,2], Jin Zhang [1], Fengqi Zhu [3], Jianya Gong [2,*], Fengxiang Jin [4], Qingqing Fu [1] and Xiaofang Ren [1]

1 School of Surveying and Geo-Informatics, Shandong Jianzhu University, Jinan 250101, China; 13565@sdjzu.edu.cn (G.Y.); 2021160106@stu.sdjzu.edu.cn (J.Z.); 13622@sdjzu.edu.cn (Q.F.); 11812@sdjzu.edu.cn (X.R.)
2 School of Remote Sensing and Information Engineering, Wuhan University, Wuhan 430070, China
3 Shandong Provincial Institute of Land Surveying and Mapping, Jinan 250101, China; zhufengqi@shandong.cn
4 College of Geomatics, Shandong University of Science and Technology, Qingdao 266590, China; fxjin@sdjzu.edu.cn
* Correspondence: gongjy@whu.edu.cn; Tel.: +86-027-6877-8003

Abstract: This paper proposes a quasi-dense feature matching algorithm that combines image semantic segmentation and local feature enhancement networks to address the problem of the poor matching of image features because of complex distortions, considerable occlusions, and a lack of texture on large oblique stereo images. First, a small amount of typical complex scene data are used to train the VGG16-UNet, followed by completing the semantic segmentation of multiplanar scenes across large oblique images. Subsequently, the prediction results of the segmentation are subjected to local adaptive optimization to obtain high-precision semantic segmentation results for each planar scene. Afterward, the LoFTR (Local Feature Matching with Transformers) strategy is used for scene matching, enabling enhanced matching for regions with poor local texture in the corresponding planes. The proposed method was tested on low-altitude large baseline stereo images of complex scenes and compared with five classical matching methods. Results reveal that the proposed method exhibits considerable advantages in terms of the number of correct matches, correct rate of matches, matching accuracy, and spatial distribution of corresponding points. Moreover, it is well-suitable for quasi-dense matching tasks of large baseline stereo images in complex scenes with considerable viewpoint variations.

Keywords: oblique stereo images; deep learning; semantic segmentation; weak texture feature matching; quasi-dense matching

Citation: Yao, G.; Zhang, J.; Zhu, F.; Gong, J.; Jin, F.; Fu, Q.; Ren, X. Quasi-Dense Matching for Oblique Stereo Images through Semantic Segmentation and Local Feature Enhancement. *Remote Sens.* **2024**, *16*, 632. https://doi.org/10.3390/rs16040632

Academic Editor: Shuying Li

Received: 27 October 2023
Revised: 18 January 2024
Accepted: 25 January 2024
Published: 8 February 2024

1. Introduction

In recent years, obtaining high-resolution multiview images of ground scenes has become increasingly easier with the development of ground mobile wide-baseline photography, UAV oblique photography, and other technologies [1]. However, under large viewpoint conditions, substantial changes in the main optical axis can lead to substantial distortions or masking in the scale, orientation, surface brightness, and neighborhood information of the same spatial target in stereo images. In addition, the existence of a large number of weak texture areas poses a great challenge for dense image matching and automated image processing [2–4].

Classical image matching methods can be divided into two categories: grayscale and feature matching. Grayscale-based matching algorithms use the grayscale information of the image to determine the similarity of the image for matching. Common grayscale algorithms include normalized cross-correlation (NCC) [5], mean absolute difference (MAD) [6], and least square matching (LSM) [7]. These grayscale matching algorithms have high accuracy but require high computation and are sensitive to noise. Feature-based matching methods first detect features in the image, subsequently extract feature descriptors, and finally determine matching features based on the Euclidean distance of the descriptors. Scale-invariant feature matching methods, represented by SIFT, exhibit good scale invariance but are difficult to adapt to considerable changes in viewpoint [8]. Reference [9] optimized SIFT feature points using the NCC method, improving the matching accuracy. Reference [10] constructed a feature extraction method combining filter decomposition and phase consistency rules, and employed a Gaussian mixture model to determine matching points. Reference [11] proposed an affine-invariant oblique image matching method that estimates the initial affine transformation based on image orientation parameters, corrects the image based on the affine transformation, and finally performs SIFT matching on the corrected image. Reference [12] simulated the full-range viewpoint change of the image and performed SIFT feature matching. This method exhibits good affine invariance; however, obtaining matching features in weak texture areas is challenging.

With the rapid development of computer software and hardware, deep learning methods based on convolutional neural networks (CNNs) have opened up a new way for realizing image matching. Deep learning matching is a data-driven image matching method that can autonomously learn the deep-level representation of object features from a large amount of image data. Currently, deep learning matching is classified into dense and sparse matching. The former achieves pixel-by-pixel dense correspondence in overlapping areas by predicting the disparity map of stereo images and the latter is oriented toward feature extraction, description, and matching for staged training and optimization with high matching reliability, such as the classical L2-Net [13]. HardNet [14,15] enhances the differentiation between descriptors by constraining the distance between nonsynonymous descriptors through a loss function based on L2-Net. AffNet proposed in reference [16] uses multiscale Hessian to detect feature point locations, followed by HardNet and its loss function to estimate the affine neighborhood. R2D2 achieved improvements in network architecture, training strategy, and visualization methods as well as improved the computational efficiency and robustness through separable convolutional layers [17]. Inspired by SuperGlue [18], reference [19] introduced the position encoding and attention mechanism using the Transformer network to construct a model called LoFTR, which has texture enhancement capabilities. This method considerably improved the matching performance in weak texture areas; however, adapting to changes in the viewpoint of the images is challenging. Reference [20] proposed a performance baseline for deep feature matching called DFM. It adopts a two-stage approach, where the initial transformation is performed using feature information containing rich deep semantic information. Then, through hierarchical matching from deep to shallow and coarse to fine levels, the final matching pairs are obtained. Similarly inspired by SuperGlue [18], the GlueStick uses a depth map neural network to unify the descriptors of points and lines into one framework, and employs the information between points to glue the lines from the matching images, improving the joint matching efficiency of the model. This indicates that the complementary performance of using two features in a single framework greatly improves performance [21]. Furthermore, reference [22] proposed an end-to-end deep learning network and its weighted average loss function for wide-baseline image matching with high inclination angles. This approach allows nonmatching similarity descriptors to participate in training through weighting, improving the discriminability of nonmatching descriptors and matching performance of matching descriptors. However, adapting to images with multiple planar scenes and oblique perspectives is difficult. VGG16 is a classic deep CNN model comprising 16 convolutional and three fully connected layers with powerful feature extraction capabilities [23]. UNet is a deep learning model for semantic segmentation tasks that comprises symmetric encoder and decoder parts and can achieve pixel-level image segmentation [24]. Reference [25] proposed an integrated VGG16-UNet, which has demonstrated some reliability in image classification and segmentation tasks and provided a feasible method for image segmentation and matching in multiplanar complex scenes.

In summary, for oblique stereo images with complex scenes and geometric distortions, it is difficult to achieve more reliable dense matching results using both classical feature matching algorithms and deep learning matching strategies. Deep learning segmentation models and texture-enhanced convolutional networks are expected to be the breakthrough in solving such image matching problems. Therefore, this paper proposes a reliable quasi-dense feature matching algorithm that combines image semantic segmentation and local feature enhancement network, which integrates the VGG16-UNet multiplanar semantic segmentation and LoFTR local feature enhancement network. The proposed algorithm first segments and extracts the corresponding planar scene and then applies the weak texture enhancement strategy in the planar scene to obtain quasi-dense feature matching. The effectiveness of the proposed method is verified using actual stereo images of complex scenes with large viewpoints.

2. Materials and Methods

For stereo images of multiplanar scenes with large viewpoints, we first train the VGG16-UNet model using typical segmented data of oblique multiplanar scenes, achieving preliminary segmentation of complex scenes into individual planes. Subsequently, we employ a neighborhood search-based adaptive thresholding strategy to optimize the segmented local regions. Afterward, we use affine-invariant feature matching to recognize corresponding planes and apply the LoFTR method with local feature transformation to extract weak texture features for each identified plane. Finally, we fuse the results of local plane matching and obtain a semi-dense matching result. Figure 1 shows the technical approach of the proposed algorithm.

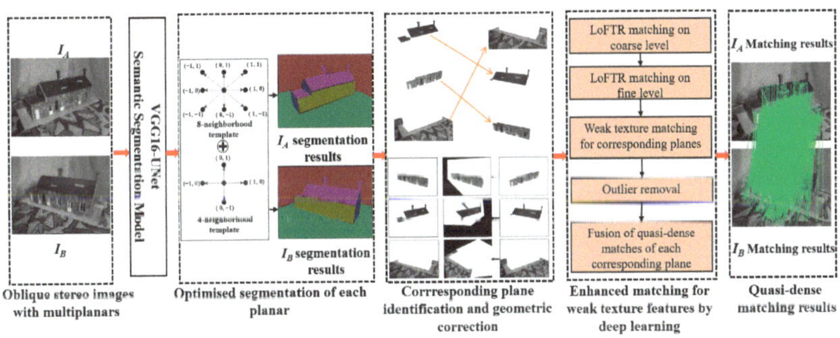

Figure 1. Technical approach of the proposed algorithm.

2.1. Automatic Semantic Segmentation Strategy

2.1.1. Multiplanar Semantic Segmentation Model

VGG16-UNet is a deep CNN model based on the fusion of VGG16 and Unet models. It combines the powerful feature extraction capability of VGG16 and pixel-level semantic segmentation capability of Unet. To cope with the quasi-dense matching task of complex scenes, we propose to apply VGG16-UNet to the semantic segmentation of multiplanar scenes. Figure 2 shows the model structure and the design of each parameter. In the encoding stage, the first 13 convolutional layers of VGG are used as the feature extraction network, and a 3 × 3 size convolutional kernel is used to compress the input image from 512 × 512 pixels to 32 × 32 × 512 pixels after four down-sampling steps to achieve the feature extraction from multiplanar scenes. In the corresponding decoding part, up-sampling and feature fusion are used to complete the segmentation of each plane, and the decoder restores the final output layer size to 512 × 512 pixels through continuous up-sampling and convolutional stacking, and subsequently outputs the segmentation map.

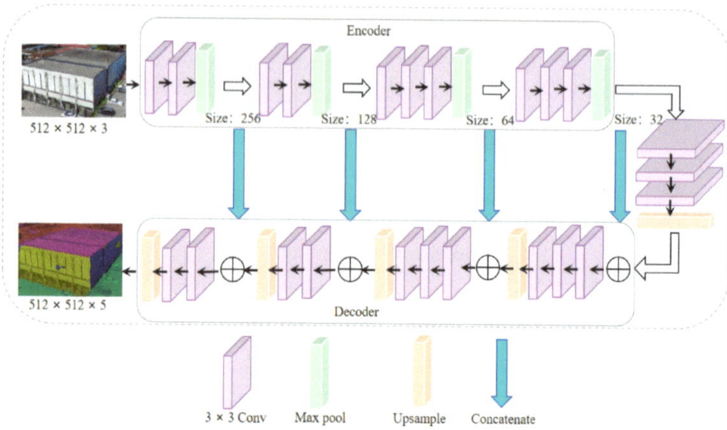

Figure 2. VGG16-UNet network architecture diagram.

2.1.2. Training Data

Extensive testing has revealed that VGG16-UNet has strong feature extraction capabilities and good transfer learning performance. Therefore, to fully train the VGG16-UNet model, we carefully selected 80 typical building image data of various types. These data are taken from low-altitude oblique views, and due to the presence of occlusion factors, the buildings in the images show one top and two side views, with a paucity of texture on the scene surface (Figure 3). These data are manually labeled into five sections: building top (pink), building facade (yellow or purple), ground (green), and background (dark red), corresponding to the 80 labeled images. Figure 3 shows an example of the training data.

Figure 3. Example of training dataset for image segmentation used in this paper.

2.1.3. Image Segmentation and Adaptive Optimization

The trained VGG16-UNet automatically predicts a set of pixels located in the same plane, extracts the mask map of each plane, and completes the initial segmentation of the local plane. However, some mask maps may contain holes, and the use of segmentation results at this point will inevitably affect subsequent matching results. Therefore, we propose an adaptive optimization method. Figure 4 shows a schematic of adaptive optimization, which mainly includes discrete region removal outside the main plane region and the filling of the hole region in the main plane region. Removal of the discrete region eliminates segmentation noise outside the main plane, whereas filling the hole eliminates the noise inside the main plane.

$$r = \frac{1}{2max(R)}, \; R = \cup_{i=1}^{n} S_i, \tag{1}$$

where R represents the area of the connected region, n denotes the number of iterations, and S_i represents the area of the region obtained by expanding in the i-th iteration. The maximum value of R corresponds to the area of the main plane region.

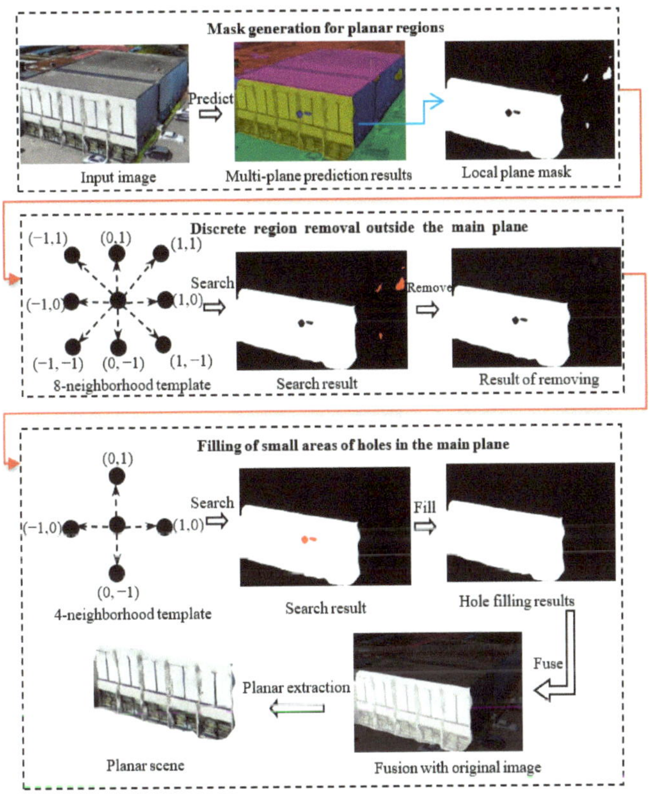

Figure 4. Adaptive optimization for local plane segmentation.

Next, to obtain the optimized results inside and outside the main region, the locally connected regions are color-inverted based on the adaptive threshold r. Considering the optimization accuracy and efficiency, an eight-neighborhood template is used to retrieve small discrete areas, whereas a four-neighborhood template is used to fill small hole areas. When segmenting, the local plane segmentation can be achieved by performing a Bitwise-AND operation between the mask map and the original image. This operation results in an image content containing only the mask region. Figure 5 shows the effect of each plane segmentation optimization. It shows that the proposed strategy achieves adaptive optimization of the planar scene by correcting the noise in the internal and external body regions, ensuring the reliability of segmentation and laying the foundation for subsequent quasi-dense matching of the planar scenes.

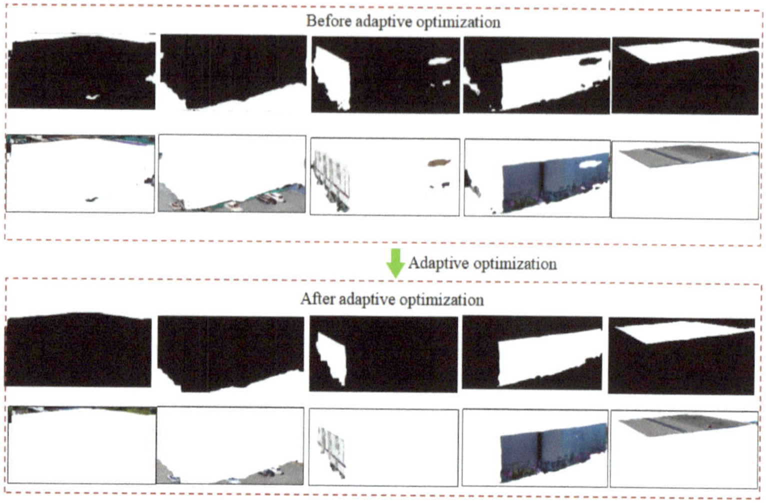

Figure 5. Comparison before and after plane segmentation optimization.

2.2. *Quasi-Dense Matching Method*

2.2.1. Automatic Identification of Corresponding Planes

Before performing the enhancement matching of weak texture features, it is necessary to first pair and recognize the corresponding plane scenes in the left and right images for obtaining corresponding planes. The affine-invariant feature matching algorithm described in reference [22] can robustly extract corresponding features from plane scenes with large viewpoint variations. Therefore, in this section, we employ this algorithm to automatically recognize corresponding planes. The process can be briefly described as follows: extract any plane from the left image, match it with each plane in the right image, identify the corresponding plane with the most corresponding features, and $>m$ (matching points, set to 8) is identified as the corresponding plane. Similarly, we iterate through all the planes in the left image, complete the feature matching with each plane of the right image and discriminate, and finally obtain each corresponding plane pair. Figure 6 shows the automatic identification process.

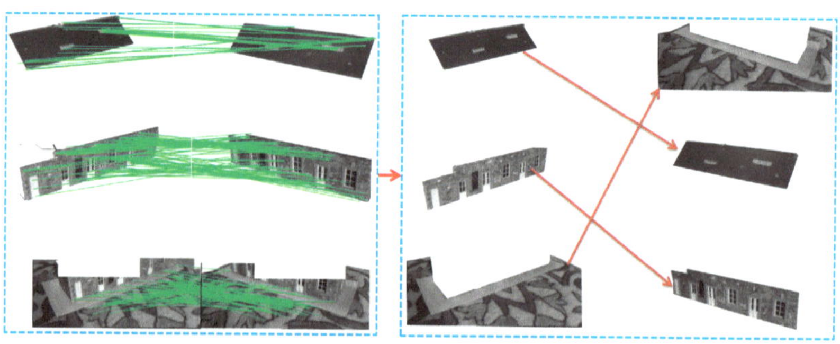

Figure 6. Process of automatic identification of corresponding planes.

2.2.2. LoFTR-Based Weak Texture Feature Enhancement Matching

The LoFTR strategy proposed in reference [19] can effectively enhance feature distinctiveness in weak texture regions; however, it struggles to adapt to affine deformations between images. Therefore, in this section, we first estimate the perspective transformation matrix based on the obtained corresponding planes and their corresponding features to minimize the geometric deformations between corresponding planes. Subsequently, we apply the LoFTR algorithm to extract weak texture features from the corresponding planes. Figure 7 shows the specific matching process, which primarily comprises the five key steps outlined below.

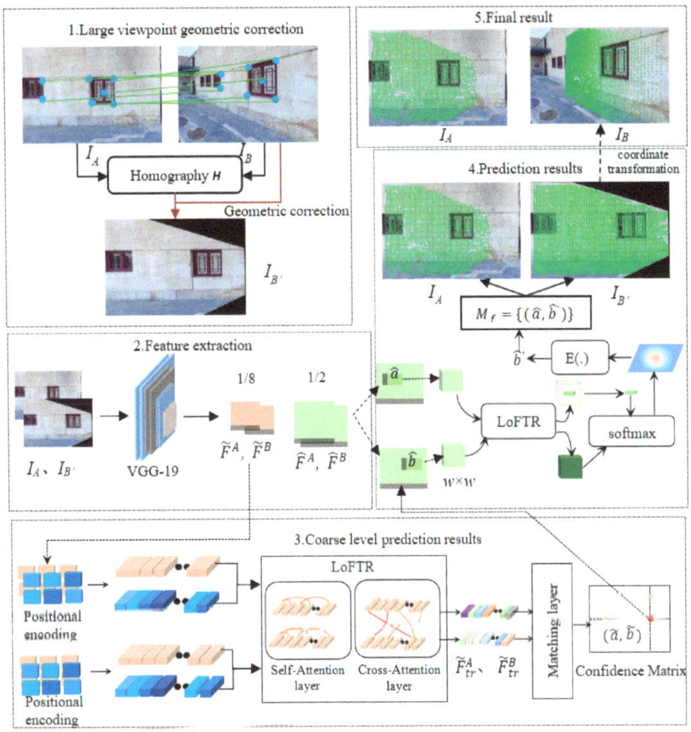

Figure 7. Weak texture feature enhancement matching for large viewpoint transformation.

(1) Large viewpoint correction: First, using the I_A and I_B matching points obtained in the previous plane recognition process, we estimate the projection transformation matrix H based on Equation (2) and the random sample consensus (RANSAC) algorithm as follows:

$$\begin{cases} x' = \frac{h_{11}x+h_{12}y+h_{13}}{h_{31}x+h_{32}y+h_{33}} \\ y' = \frac{h_{21}x+h_{22}y+h_{23}}{h_{31}x+h_{32}y+h_{33}} \end{cases}, \quad (2)$$

where (x, y) and (x', y') represent the feature matching points in I_A and I_B, respectively, and $h_{11}, h_{12} \ldots, h_{33}$ represent the nine projection transformation parameters in H. Subsequently, according to Equation (3), the right image is corrected through projective transformation as follows:

$$\begin{bmatrix} x'' \\ y'' \\ 1 \end{bmatrix} = \begin{bmatrix} h_{11}h_{12}h_{13} \\ h_{21}h_{22}h_{23} \\ h_{31}h_{32}h_{33} \end{bmatrix} \begin{bmatrix} x' \\ y' \\ 1 \end{bmatrix}, \quad (3)$$

where (x', y') and (x'', y'') represent the pixel coordinates of I_B and $I_B{}'$ before and after the correction of the projective deformation in the right image, respectively. After correcting the right image through projective deformation, the affine distortion of the corresponding region is considerably improved, and the geometric consistency of the plane tends to be better than before. Thus, the LoFTR strategy is introduced for matching.

(2) Feature extraction: For the image pair I_A and $I_B{}'$, feature extraction is first performed using VGG CNN, resulting in 1/8 coarse feature and 1/2 fine feature maps for both images.

(3) Generating coarse-level feature prediction results: The coarse extracted feature maps $\tilde{F}^A$ and $\tilde{F}^B$ are flattened into one-dimensional vectors, and position encoding is added to each vector. These vectors with position encoding are then inputted into the LoFTR module, which comprises N (N = 4) self-attention and cross-attention layers. The LoFTR module utilizes a self-attention mechanism to capture the correlations between different positions within the image, learning the importance of local features and enhancing the discriminative ability of the convolutional model for different texture features. After processing through this module, two enhanced texture feature maps with higher discriminability, labeled as $\tilde{F}^A_{tr}$ and $\tilde{F}^B_{tr}$, are outputted. Subsequently, the similarity between these two feature maps is calculated to perform the matching of corresponding features.

(4) Outputting the prediction results: For any coarse-level matching prediction $\left(\tilde{a}, \tilde{b}\right)$ $\in Mc$, local corresponding windows of size $w \times w$ (w = 5) are cropped from the fine feature maps $\hat{a}$,$\hat{b}$. Second, a smaller LoFTR module then transforms the cropped features within each window, yielding two transformed local feature maps, $\hat{F}^A_{tr}(\hat{a})$ and $\hat{F}^B_{tr}\left(\hat{b}\right)$, centered at $\hat{a}$ and $\hat{b}$, respectively. Third, we correlate the center vector of $\hat{F}^A_{tr}(\hat{a})$ with all vectors in $\hat{F}^B_{tr}\left(\hat{b}\right)$ and thus produce a heatmap that represents the matching probability of each pixel in the neighborhood of $\hat{a}$ with $\hat{b}$, and the location $\hat{b}'$ is obtained by calculating the expectation of the probability distribution. Finally, all coarse-level matches are refined within the local windows of the fine level, resulting in the fine-level matching predictions M_f for I_A and $I_B{}'$.

(5) Outputting the final result: Finally, the coordinates of the fine-level matching points on $I_B{}'$ are normalized to the original coordinate system of the right image I_B using Equation (3), representing the final result of weak texture feature-enhanced matching.

3. Results

3.1. Experimental Environment

In the experiment, we used RTX2080ti GPU, 9-9900K processor, 64 GB RAM, and Ubuntu18.04 operating system. The software platform is PyCharm (v 2023.3.2). The training dataset of weak texture feature based on LoFTR is adopted from the open-source MegaDepth dataset. During the training process, the Transformer loop count N is set to four, the LoFTR module feature transformation count N_f is set to one, and the window size w for extracting patches from the fine-level feature map is set to five. The threshold θ_C for coarse-level matching prediction is set to 0.2. The training is completed after 30 iterations using the gradient descent algorithm.

3.2. Evaluation Metrics

(1) Number of correct matching points, k_{ε_0}: Fifteen pairs of uniformly distributed corresponding points are manually selected from the stereo images. The fundamental matrix F_0 is estimated using the least-squares method and considered as the ground truth. Using the well-known fundamental matrix F_0, the error of any matching point is calculated using Equation (4). A threshold ε_0 (set to 3.0) is set and imposed for

the error. If the error was less than ε_0, the pair of points is a correct pair of matching points and is included in the count of correct matching points, k_{ε_0}:

$$\varepsilon_i = \sqrt{\left(x_i'^T F_0 x_i\right)^2 / ((F_0 x_i)_1^2 + (F_0 x_i)_2^2}. \quad (4)$$

(2) Match correct rate, α: This is defined by $\alpha = k_{\varepsilon_0}/k$, where k denotes the total number of matching points.

(3) Matching root-mean-squared error (RMSE) ε_{RMSE} (pixel). This is calculated using Equation (5):

$$\varepsilon_{\text{RMSE}} = \sqrt{\frac{1}{k}\sum_{i=1}^{k} \varepsilon_i^2}, \quad (5)$$

where k represents the total number of matches and ε_i is calculated using Equation (4).

(4) Matching spatial distribution quality, $\hat{D}$: References [26,27] generated Delaunay triangulation based on the matching points. They evaluated the spatial distribution quality of the matching points by considering the areas and shapes of each triangle, as well as the global and local distribution of the matching points. This is calculated using Equation (6):

$$\left.\begin{array}{c} \mathrm{D} = \mathrm{D_A} \times \mathrm{D_S} = \sqrt{\frac{\sum_{i=1}^{n}((\mathrm{A_i}/\overline{\mathrm{A}})-1)}{n-1}} \times \sqrt{\frac{\sum_{i=1}^{n}(\mathrm{S_i}-1)}{n-1}},\ \overline{\mathrm{A}} = \frac{\sum_{i=1}^{n}\mathrm{A_i}}{n},\ \mathrm{S_i} = \frac{3\max(\mathrm{J_i})}{\pi} \\ \hat{D} = \frac{\mathrm{D}}{\mathrm{D_G}},\ D_G = (\sum_{i=1}^{n} A_i)/A_I \end{array}\right\}, \quad (6)$$

where n represents the total number of generated triangles; $\mathrm{A_i}$ and $\max(\mathrm{J_i})$ represent the area and maximum arc of the i-th triangle, respectively; $\overline{\mathrm{A}}$ represents the average area of the triangles; $\mathrm{D_A}$ represents the uniformity of the areas of each triangle; and $\mathrm{D_S}$ represents the uniformity of the internal angles of the triangles. The lower the D value, the higher the geometric uniformity of the local triangles. $\mathrm{A_I}$ represents the area of the image and $\mathrm{D_G}$ represents the coverage of matching points in the global image. A higher $\mathrm{D_G}$ value indicates a wider spatial distribution of matching points in the image. Therefore, this model can fully reflect the quality of the matching point spatial distribution, and the quality of the matching point spatial distribution increases with decreasing $\hat{D}$.

3.3. Experimental Methods and Data

To fully validate the advantages of our proposed method, we used six methods for comparative testing. (1) DFM: This method achieves high accuracy by performing coarse-to-fine matching of images at different hierarchical levels of features. (2) AffNet: This method uses an affine-invariant estimation network to learn affine parameters. It enhances the distinctiveness between descriptors using the HardNet loss function, making it suitable for scenes with viewpoint changes. (3) SuperGlue: This method constructs an image information aggregation model based on attention mechanisms. The loss function of the model is established using graph neural networks. (4) LoFTR: This method combines position encoding and attention mechanisms in the Transformer, generating a model suitable for weak texture matching. (5) GlueStick: A GNN architecture is designed to be able to combine the contextual information of all features to improve the accuracy of the matching. (6) Our proposed method. To objectively evaluate these six methods, the RANSAC algorithm is used to remove outliers, and the inlier coordinates for each method are outputted. As shown in Figure 8, six groups of low-altitude large viewpoint building scene images (a–f) are selected as the test data.

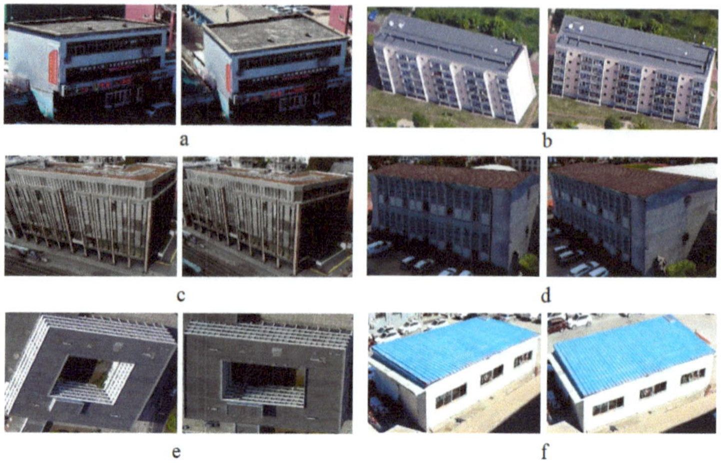

Figure 8. Test images. (**a–f**) six groups of large oblique stereo images with building scene.

3.4. Experimental Results and Analysis

Figures 9–14 show the matching results of six groups of data based on DFM, AffNet, SuperGlue, LoFTR, GlueStick, and our proposed method, respectively. Table 1 presents the quantitative experimental results of the six methods. Here, k_{ε_0} and α represent the number of correctly matched points and the correct rate of matching, respectively. ε_{RMSE} represents the RMSE of matching, and $\hat{D}$ represents the quality of spatial distribution of matching points. The optimal test results of each group of data in the table are represented in bold.

Table 1. The contrast of test results using six methods. The best values are highlighted in bold.

Test Data	Evaluation Metrics	Ours	DFM	AffNet	SuperGlue	LoFTR	GlueStick
(a)	k_{ε_0}/(Pair)	**2751**	1199	832	618	1695	336
	a/(%)	**0.80**	0.61	0.41	0.61	0.58	0.58
	ε_{RMSE}/(Pixel)	1.20	0.36	**0.35**	1.83	0.65	0.65
	$\hat{D}$	56.9	59.2	87.2	64.9	59.2	**33.7**
(b)	k_{ε_0}/(Pair)	**898**	31	82	537	520	259
	a/(%)	0.49	**0.53**	0.14	0.52	0.22	0.40
	ε_{RMSE}/(Pixel)	0.35	**0.18**	0.37	0.39	0.38	0.81
	$\hat{D}$	32.6	37.5	**17.13**	54.33	40.9	39.8
(c)	k_{ε_0}/(Pair)	**2751**	602	1100	618	1695	393
	a/(%)	**0.68**	0.34	0.33	0.61	0.58	0.47
	ε_{RMSE}/(Pixel)	0.99	**0.35**	0.37	1.24	2.07	0.71
	$\hat{D}$	45.5	27.7	45.9	**23.3**	46.5	32.6
(d)	k_{ε_0}/(Pair)	**2254**	237	296	330	1291	241
	a/(%)	**0.82**	0.40	0.20	0.50	0.60	0.52
	ε_{RMSE}/(Pixel)	0.86	**0.35**	0.36	2.9	0.57	0.67
	$\hat{D}$	41.1	32.0	28.2	26.8	43.3	**24.4**
(e)	k_{ε_0}/(Pair)	**1530**	56	125	196	1015	179
	a/(%)	**0.64**	0.43	0.22	0.24	0.46	0.48
	ε_{RMSE}/(Pixel)	0.99	**0.32**	0.36	1.88	1.06	0.69
	$\hat{D}$	40.1	24.1	**18.8**	23.5	43.3	23.0
(f)	k_{ε_0}/(Pair)	**2059**	915	974	273	1034	226
	a/(%)	**0.69**	0.46	0.42	0.39	0.47	0.44
	ε_{RMSE}/(Pixel)	0.49	0.37	**0.35**	1.88	1.46	0.75
	$\hat{D}$	**27.5**	83.3	28.2	32.5	43.3	34.6

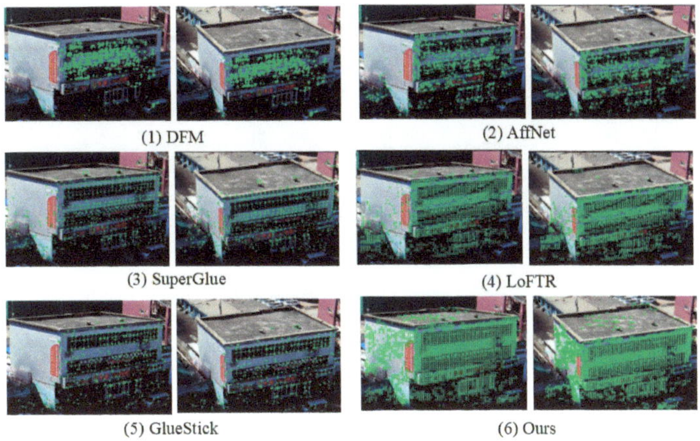

Figure 9. Test results of group images (a).

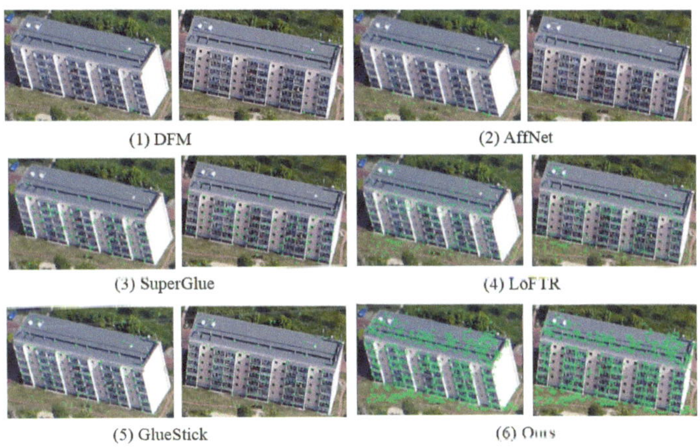

Figure 10. Test results of group images (b).

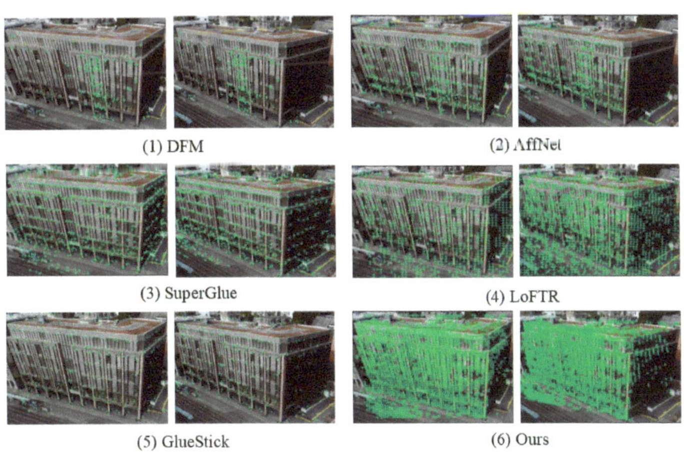

Figure 11. Test results of group images (c).

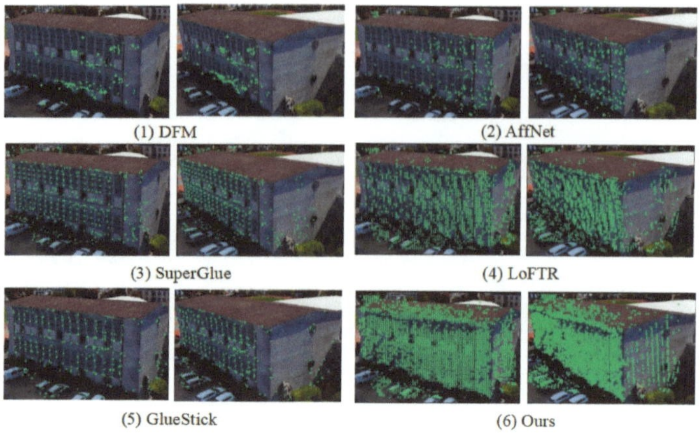

Figure 12. Test results of group images (d).

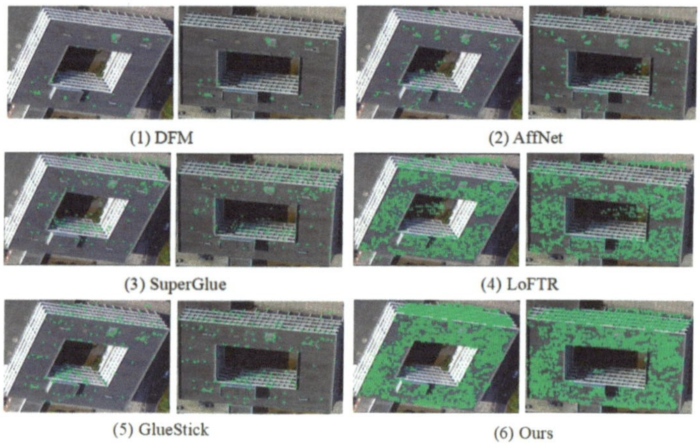

Figure 13. Test results of group images (e).

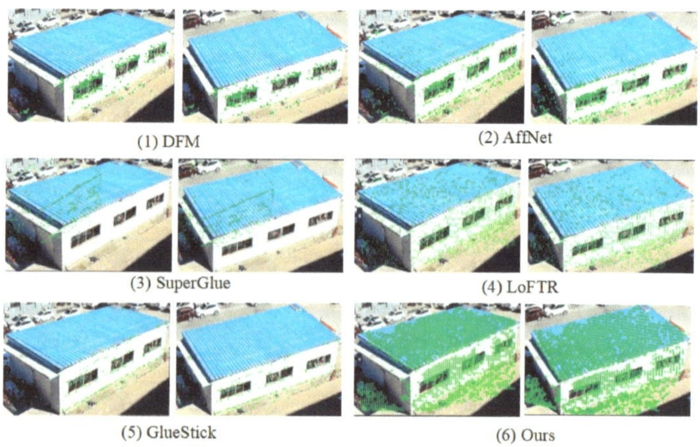

Figure 14. Test results of group images (f).

4. Discussion

(1) The proposed method has significant advantages in terms of the number of correctly matched points. Table 1 presents the quantitative experimental results for six groups of large viewpoint stereo images in architectural scenes that show the highest number of correctly matched points obtained using the proposed method. As shown in Figures 9–14, our proposed method can achieve accurate and dense matching results in each group of images, especially for matching a large number of corresponding points on the top and facades of buildings, which provides sufficient tie points for image orientation and three-dimensional (3D) reconstruction. The reasons are twofold. First, the multiplane segmentation and corresponding plane matching method proposed in this paper can transform the matching of complex 3D scenes into simple plane scene matching. Second, the LoFTR texture enhancement strategy introduced in this paper effectively improves the problem of weak texture on the top and facades of buildings, leading to accurate and dense matching results.

(2) According to the above experimental results, DFM has advantages in accuracy, but its effect on affine changes is poor. Compared with DFM, SuperGlue is more capable of handling large viewpoint affine transformations and single-texture regions; however, the number of matching points is much less than that obtained using our method. The LoFTR algorithm, which is based on the SuperGlue method, uses Transformer positional encoding and attention mechanisms to significantly enhance the texture features of building facades. GlueStick has not improved or even decreased in quantity compared to SuperGlue, but has improved in spatial distribution quality and matching accuracy. However, obtaining a sufficient number of matching points due to the influence of image distortion is challenging.

(3) Our method also demonstrates some advantages in terms of matching accuracy and precision. Table 1 shows that our method achieves high matching correctness rates for most of the test data (a, b, d–f), and sub-pixel matching precision for test data (b–f). The reasons behind this are as follows. First, our method performs individual matching for each planar scene and utilizes strict homography geometric transformations for distortion correction and constrained matching, effectively ensuring matching correctness and precision. Second, during the quasi-dense matching process, the proposed method first conducts coarse level matching prediction and then refines the matches at a finer level, ensuring the accurate positioning of matching points.

(4) The proposed method exhibits good spatial distribution quality for the matching points. Figures 9–14 show that the distribution area of the matching points of our method in image space has significantly improved. Table 1 demonstrates that our method outperforms DFM and LoFTR algorithms in terms of the spatial distribution quality of matching points. Our method has good spatial distribution quality for matching points.

5. Conclusions

In this study, we propose a matching algorithm that combines image semantic segmentation and local feature enhancement networks for stereo images in complex scenes with significant viewpoint changes. The proposed algorithm first employs an automatic semantic segmentation method to extract the planes of different scenes. The LoFTR strategy is then used to enhance the weak texture features of each local plane, enabling accurate and dense feature matching. The experimental results demonstrate that the proposed method has advantages in terms of the number of correctly matched points, matching accuracy, matching precision, and spatial distribution quality of matched points. It is suitable for the dense matching of wide-baseline oblique stereo images. In future work, we plan to integrate a line feature matching algorithm to achieve more complementary feature matching along building structure edges. This can be applied to the fine-scale 3D reconstruction of urban building scenes.

Author Contributions: Conceptualization, J.G. and G.Y.; methodology, G.Y. and F.Z.; software, G.Y and J.Z.; data curation, J.Z. and G.Y.; validation, G.Y., J.Z. and F.J.; formal analysis, G.Y. and J.G., writing—original draft preparation, G.Y. and Q.F.; writing—review and editing, J.G., F.J. and X.R.; supervision, J.G. and F.J. All authors have read and agreed to the published version of the manuscript.

Funding: This work was supported by the National Natural Science Foundation of China with Project No. 42171435, the Shandong Provincial Natural Science Foundation with Project No. ZR2021MD006, the China Postdoctoral Science Foundation with Project No. 2023M732686, and the Undergraduate Education and Teaching Reform Foundation of Shandong Province with Project No. Z2021014. This work was also funded by the high quality graduate course of Shandong Province with Project No. SDYKC2022151.

Data Availability Statement: Data are available upon request due to restrictions.

Acknowledgments: The authors would like to thank Jiaming Sun and Zhuxuan Wu for providing their key algorithms.

Conflicts of Interest: The authors declare no conflicts of interest.

References

1. Ge, Y.; Guo, B.; Zha, P.; Jiang, S.; Jiang, Z.; Li, D. 3D Reconstruction of Ancient Buildings Using UAV Images and Neural Radiation Field with Depth Supervision. *Remote Sens.* **2024**, *16*, 473. [CrossRef]
2. Yao, G.B.; Yilmaz, A.; Meng, F.; Zhang, L. Review of wide-baseline stereo image matching based on deep learning. *Remote Sens.* **2021**, *13*, 3247. [CrossRef]
3. Ji, S.; Luo, C.; Liu, J. A Review of Dense Stereo Image Matching Methods Based on Deep Learning. *Geomat. Inf. Sci. Wuhan Univ.* **2021**, *46*, 193–202. [CrossRef]
4. Liu, J.; Ji, S.P. Deep learning based dense matching for aerial remote sensing images. *Acta Geod. Cartogr. Sin.* **2019**, *48*, 1141–1150. [CrossRef]
5. Luo, S.D.; Chen, H.B. Stereo matching algorithm of adaptive window based on region growing. *J. Cent. South Univ. Technol.* **2005**, *36*, 1042–1047.
6. Fritz, C.O.; Morris, P.E.; Richler, J.J. Effect size estimates: Current use, calculations, and interpretation. *Exp. Psychol. Gen.* **2012**, *141*, 2–18. [CrossRef]
7. Yang, H.; Zhang, S.; Zhang, Q. Least Squares Matching Methods for Wide Base-line Stereo Images Based on SIFT Features. *Acta Geod. Cartogr. Sin.* **2010**, *39*, 187–194. [CrossRef]
8. David, G.L. Distinctive image features from scale-invariant keypoints. *Int. J. Comput. Vis.* **2004**, *60*, 91–110. [CrossRef]
9. Yang, H.; Zhang, S.; Wang, L. Robust and precise registration of oblique images based on scale-invariant feature transformation algorithm. *IEEE Geosci. Remote Sens. Lett.* **2012**, *9*, 783–787. [CrossRef]
10. Zhang, Q.; Wang, Y.; Wang, L. Registration of images with affine geometric distortion based on maximally stable extremal regions and phase congruency. *Image Vis. Comput.* **2015**, *36*, 23–39. [CrossRef]
11. Xiao, X.W.; Guo, B.X.; Li, D.R.; Zhao, X.A. Quick and affine invariance matching method for oblique images. *Acta Geod. Cartogr. Sin.* **2015**, *44*, 414–442. [CrossRef]
12. Morel, J.-M.; Yu, G. Asift: A new framework for fully affine invariant image comparison. *SIAM J. Imaging Sci.* **2009**, *2*, 438–469. [CrossRef]
13. Tian, Y.R.; Fan, B.; Wu, F.C. L2-net: Deep learning of discriminative patch descriptor in euclidean space. In Proceedings of the IEEE Conference on Computer Vision and Pattern Recognition, Honolulu, HI, USA, 21–26 July 2017; pp. 661–669. [CrossRef]
14. Mishchuk, A.; Mishkin, D.; Radenovic, F. Working hard to know your neighbor's margins: Local descriptor learning loss. *Adv. Neural Inf. Process. Syst.* **2017**, *1*, 4826–4837. [CrossRef]
15. Zhang, C.; Yao, G.; Man, X.; Huang, P.; Zhang, L.; Ai, H. Affine invariant feature matching of oblique images based on multi-branch network. *Acta Geod. Cartogr. Sin.* **2021**, *50*, 641–651. [CrossRef]
16. Mishkin, D.; Radenovic, F.; Matas, J. Repeatability is not enough: Learning affine regions via discriminability. In Proceedings of the 2018 Computer Vision, Munich, Germany, 8–14 September 2018; pp. 287–304. [CrossRef]
17. Revaud, J.; Weinzaepfel, P.; De Souza, C.; Pion, N.; Csurka, G.; Cabon, Y.; Humenberger, M. R2D2: Repeatable and reliable detector and descriptor. *arXiv* **2019**, arXiv:1906.06195.
18. Sarlin, P.-E.; DeTone, D.; Malisiewicz, T.; Rabinovich, A. SuperGlue: Learning feature matching with graph neural networks. In Proceedings of the IEEE 2020 Conference on Computer Vision and Pattern Recognition (CVPR), Seattle, WA, USA, 14–19 June 2020. [CrossRef]
19. Sun, J.; Shen, Z.; Wang, Y.; Bao, H.; Zhou, X. LoFTR: Detector-free local feature matching with transformers. *arXiv* **2021**, arXiv:2104.00680.
20. Efe, U.; Ince, K.; Alatan, A. DFM: A Performance Baseline for Deep Feature Matching. *arXiv* **2021**, arXiv:2106.07791.
21. Pautrat, R.; Suárez, I.; Yu, Y.; Pollefeys, M.; Larsson, V. Gluestick: Robust image matching by sticking points and lines together. *arXiv* **2023**, arXiv:2304.02008.

22. Yao, G.B.; Yilmaz, A.; Zhang, L.; Meng, F.; Ai, H.B.; Jin, F.X. Matching large baseline oblique stereo images using an end-to-end convolutional neural network. *Remote Sens.* **2021**, *13*, 274. [CrossRef]
23. Simonyan, K.; Zisserman, A. Very Deep Convolutional Networks for Large-Scale Image Recognition. *arXiv* **2014**, arXiv:1409.1556. [CrossRef]
24. Ronneberger, O.; Fischer, P.; Brox, T. U-net: Convolutional networks for biomedical image segmentation. In Proceedings of the Medical Image Computing and Computer-Assisted Intervention–MICCAI 2015: 18th International Conference, Munich, Germany, 5–9 October 2015. [CrossRef]
25. Wu, Z.; Han, X.; Lin, Y.L.; Uzunbas, M.G.; Goldstein, T.; Lim, S.N.; Davis, L.S. Dcan: Dual channel-wise alignment networks for unsupervised scene adaptation. In Proceedings of the European Conference on Computer Vision (ECCV), Munich, Germany, 8–14 September 2018; pp. 518–534. [CrossRef]
26. Zhu, Q.; Wu, B.; Xu, Z.X. Seed point selection method for triangle constrained image matching propagation. *IEEE Geosci. Remote Sens. Lett.* **2006**, *3*, 207–211. [CrossRef]
27. Yao, G.; Zhang, J.; Gong, J.; Jin, F. Automatic Production of Deep Learning Benchmark Dataset for Affine-Invariant Feature Matching. *ISPRS Int. J. Geo-Inf.* **2023**, *12*, 33. [CrossRef]

remote sensing

Article

3D Reconstruction of Ancient Buildings Using UAV Images and Neural Radiation Field with Depth Supervision

Yingwei Ge [1], Bingxuan Guo [1,*], Peishuai Zha [1], San Jiang [2], Ziyu Jiang [3] and Demin Li [1,4]

1 The State Key Laboratory of Information Engineering in Surveying, Mapping and Remote Sensing, Wuhan University, Wuhan 430079, China
2 School of Computer Science, China University of Geosciences, Wuhan 430074, China; jiangsan@cug.edu.cn
3 School of Geodesy and Geomatics, Wuhan University, Wuhan 430079, China
4 School of Artificial Intelligence, Zhejiang College of Security Technology, Wenzhou 325016, China
* Correspondence: b.guo@whu.edu.cn

Abstract: The 3D reconstruction of ancient buildings through inclined photogrammetry finds a wide range of applications in surveying, visualization and heritage conservation. Unlike indoor objects, reconstructing ancient buildings presents unique challenges, including the slow speed of 3D reconstruction using traditional methods, the complex textures of ancient structures and geometric issues caused by repeated textures. Additionally, there is a hash conflict problem when rendering outdoor scenes using neural radiation fields. To address these challenges, this paper proposes a 3D reconstruction method based on depth-supervised neural radiation fields. To enhance the representation of the geometric neural network, the addition of a truncated signed distance function (TSDF) supplements the existing signed distance function (SDF). Furthermore, the neural network's training is supervised using depth information, leading to improved geometric accuracy in the reconstruction model through depth data obtained from sparse point clouds. This study also introduces a progressive training strategy to mitigate hash conflicts, allowing the hash table to express important details more effectively while reducing feature overlap. The experimental results demonstrate that our method, under the same number of iterations, produces images with clearer structural details, resulting in an average 15% increase in the Peak Signal-to-Noise Ratio (PSNR) value and a 10% increase in the Structural Similarity Index Measure (SSIM) value. Moreover, our reconstruction model produces higher-quality surface models, enabling the fast and highly geometrically accurate 3D reconstruction of ancient buildings.

Keywords: 3D reconstruction; UAV images; neural radiation field; deep supervision; hash coding

Citation: Ge, Y.; Guo, B.; Zha, P.; Jiang, S.; Jiang, Z.; Li, D. 3D Reconstruction of Ancient Buildings Using UAV Images and Neural Radiation Field with Depth Supervision. *Remote Sens.* **2024**, *16*, 473. https://doi.org/10.3390/rs16030473

Academic Editor: Riccardo Roncella

Received: 14 November 2023
Revised: 20 January 2024
Accepted: 23 January 2024
Published: 25 January 2024

1. Introduction

The utilization of 3D reconstruction techniques not only facilitates the restoration of the original structure and color of ancient buildings but also enables the digital preservation of these historical treasures [1,2]. Through 3D reconstruction, meticulous digital replicas can be generated to safeguard and document these invaluable cultural legacies [3,4]. This paper employs the neural radiance fields (NeRF) technique [5] in the 3D reconstruction of ancient buildings, aiming to explore a swift and highly precise method for reconstructing buildings through neural rendering.

Unmanned Aerial Vehicles (UAVs) are known for their mobility, flexibility, speed and cost-effectiveness. Utilizing UAVs as aerial photography platforms enables the rapid acquisition of high-quality, high-resolution images, holding significant promise for the production of geographic mapping data [6,7]. With the advancement of tilt photogrammetry, techniques for dense point cloud generation and the construction of 3D triangular grid models from 2D images have matured, incorporating sparse reconstruction (Structure from Motion, SFM) [8] and dense reconstruction (Multiple View Stereo, MVS) [8,9]. This has made 3D solid building reconstruction a reality. However, existing tilt photogrammetry-based 3D reconstruction methods are slow and entail substantial time overheads [10]. Dense reconstruction, which involves matching all or most of the pixels in multiple images, demands extensive data processing and often redundant computations, resulting in an overall low reconstruction efficiency. These limitations hinder its real-time applications [11]. Additionally, this method necessitates a complex process involving feature extraction, feature matching, depth fusion and Poisson reconstruction [12,13], which can introduce errors at various stages and lead to incomplete or flawed final results. This paper addresses the following issues that need to be resolved: (1) The conventional approach to reconstructing the surface model of ancient buildings is hampered by the slow processing speed. (2) The intricate surface textures found on ancient buildings, coupled with the presence of repetitive textures, can have a detrimental impact on the geometric accuracy of the model reconstruction.

In recent years, the NeRF technique, based on neural rendering, has gained extensive use in the field of 3D reconstruction. NeRF leverages neural implicit representation, employing neural networks to implicitly learn 3D scene features. It reconstructs triangular mesh models by combining these learned features with the Marching Cubes algorithm [14]. However, NeRF faces efficiency challenges due to the use of computationally intensive large Multilayer Perceptrons (MLPs), requiring hours or even days for training. Additionally, NeRF represents geometry by predicting the object density through neural networks, which lacks a strong physical foundation. This leads to the generation of triangular mesh models with rough surfaces, low geometric accuracy and suboptimal quality, limiting its applications [15]. Recent research has introduced new ideas based on NeRF, such as PlenOctrees [16] and Instant Neural Graphics Primitives (Instant-ngp) [17], aimed at accelerating NeRF network model training to minutes. However, these methods often compromise geometric accuracy, resulting in rough surface meshes that do not faithfully represent real-world physical structures. Subsequently, the Instant-NSR method [18] emerged, combining the approaches of Instant-ngp and NeuS [19], enhancing the model's geometrical structure. While this approach has improved the results, it may still exhibit depressions and uneven surface pits. Mip-NeRF [20] effectively resolves NeRF's challenges with high-frequency detail aliasing and distortion by refining the encoding of the sampling points, yet it still requires a considerable amount of time for network training. Neuralangelo [21] enhances the network architecture, but this advancement comes at the cost of increased computational demands and prolonged training periods due to additional sampling requirements. Meanwhile, 3D Gaussian splatting (3D GS) [22] introduces Gaussian functions for scene representation, offering increased adaptability in scene portrayal. However, its utility is somewhat constrained, as it struggles to accommodate images captured at varying scales.

In modern times, the 3D reconstruction of ancient buildings, achieved through the utilization of UAVs and various data collection methods, seeks to create more comprehensive models by integrating vast amounts of information. However, these data-rich approaches often lead to a significant computational burden in traditional 3D reconstruction, which places added strain on computers and prolongs the reconstruction process. Consequently, this paper proposes to improve the accuracy and training speed of reconstructions by combining the truncated signed distance function (TSDF) with sparse point cloud depth supervision, as well as implementing a progressive training strategy. This technique is introduced into the field of the three-dimensional reconstruction of ancient buildings to address the challenges of extensive computational demands and slow reconstruction speeds in traditional methods. This paper aims to enhance the geometric accuracy of NeuS-reconstructed models through two methods of geometric optimization. The primary contributions of this paper are as follows:

- Combined network training with the TSDF and depth supervision: Our approach combines the TSDF and depth supervision in network training. Integrating the TSDF into the signed distance function (SDF) neural network to improve geometric representation within the neural network. Simultaneously, this study utilizes sparse point

cloud depth information to supervise the training of the SDF neural network, further enhancing the geometric accuracy of three-dimensional mesh models.
- A progressive training method that gradually enhances the resolution of hash coding during the training process has been designed. This approach focuses on improving the characteristics of the scene and hash coding, effectively utilizing the feature hash table's capacity. By doing so, it mitigates hash conflicts within the mesh feature hash table under multi-resolution conditions. The ultimate goal is to produce rendered images with clear, detailed textures, enriching the visual quality.

This paper aims to enhance the accuracy of the NeuS-reconstructed geometric model through two geometric optimization methods. The first method involves the incorporation of the TSDF into the SDF neural network, which results in an improved geometric representation within the neural network. The second method utilizes depth information to supervise the neural network training, further enhancing the geometric accuracy of the reconstruction model using data from a sparse point cloud. In outdoor scenes, where large hash conflicts are common, this paper proposes a progressive training method based on multi-resolution hash coding technology to alleviate these conflicts and improve the expressive capabilities of the neural network.

2. Related Work

In a range of fields including mapping, remote sensing and computer vision, the NeRF technique has enabled the rendering and reconstruction of 3D scenes [23]. Despite its groundbreaking capabilities, NeRF still grapples with issues related to model generation efficiency, quality and scalability. One of the primary concerns is its computational intensity, both in terms of the number of sampling points and the time required for training, particularly due to the utilization of two large MLPs containing eight hidden layers [24]. Moreover, NeRF's reliance on straightforward volume rendering and direct density prediction through density MLP neural networks, lacking a robust physical foundation, often results in a rough surface and low geometric accuracy in the generated triangular mesh model [25]. In light of these challenges, researchers worldwide are dedicating efforts to improve and innovate the NeRF model.

In traditional geometric reconstruction, the literature [26–28] all focuses on the optimization of dense point clouds to enhance their quality. The literature [28] leverages images from multiple viewpoints, combines scene geometry constraints and estimates depths for sparse points to achieve high-quality dense reconstruction. The literature [26] proposes the sparse voxel DAGs method, efficiently reconstructing point clouds by establishing a sparse voxel data structure and employing dynamic adaptive mesh refinement and local region. The literature [27] presents a progressive 3D point set upsampling method based on localized blocks, gradually increasing the point density by utilizing the geometric and normal information among these blocks, thereby enhancing the point cloud details and resolution. However, due to the substantial memory requirements of these methods, they are more suited for small-scale reconstruction projects, where they tend to yield better results.

To address the issues of clarity and realism in NeRF technology, numerous researchers have conducted in-depth explorations into various aspects of the technology process, achieving significant improvements. To enable NeRF to handle a wider range of image situations and reduce its requirements for image sources, the literature [29] addresses the issue of NeRF producing poorer results with low-quality images by simulating the blurring process to synthesize blurred views, thereby improving NeRF's robustness to blurred input images. The literature [20] introduces Mip-NeRF, which transforms the original NeRF point sampling method into cone sampling, enriching the details of the sampling and considering the changes in the scale of the observation distance in ray sampling. The NeRF++ [30] model divides the scene into foreground and background parts. The foreground sampling method is consistent with NeRF, but background sampling involves projecting light onto a unit sphere, thus controlling the depth of light within a defined range. Similarly, we have adopted this method in ancient architectural scenes, specializing in the encoding of foreground targets. The literature [31] integrates NeRF++ and Mip-NeRF concepts, ensuring positional relevance is maintained as sampling points extend to infinity. The literature [30,31] extends NeRF to large scene domains, but the increase in sampling information adds to the network training burden. To tackle the challenges of rough 3D models and noise low-fidelity geometric approximations, researchers both domestically and internationally have integrated deeper physical foundations into the geometric expression of neural networks to improve the accuracy. The literature [32] introduces UNISURF, using an occupancy network to represent implicit surfaces, assigning each sampling point as 0 or 1 to indicate the presence of a surface. The literature [33] presents Plenoxels, emphasizing the critical role of micro-voxel renderers in the evolution of NeRF technology. Plenoxels depart from using neural networks, focusing instead on optimizing the density and color parameters of voxel grid vertices through derivative-based solutions. This method achieves a training speed 3000 times faster than traditional NeRF. The literature [19] discusses NeuS, providing a mathematical explanation for NeRF's low geometric accuracy and employing SDF values to create an unbiased density function, thereby rectifying inherent biases in volumetric rendering formulas. To accelerate network training and reduce memory usage, the literature [34] presents NSVF, a strategy that manages scene data through a sparse voxel octree, selectively excluding irrelevant voxels during light sampling to speed up the process and minimize data overheads. The literature [17] proposes Instant-ngp, using a multi-resolution hash encoding (MHE) model [35] to encode the spatial information of 3D points, allowing for smaller MLP networks in training and rendering, marking a considerable advance in the NeRF training speed, reducing it from hours to just a few seconds. However, the need for pre-allocating fixed memory for data storage could lead to conflicts and impact the quality of results when training data volumes increase. To enhance the training efficacy, some researchers have integrated supervisory mechanisms during training. Point-NeRF [36] merges traditional MVS methods with NeRF, introducing a point cloud-based NeRF. The literature [37] uses MVS-generated depth maps to supervise SDF network training. Nerfing MVS [38] uses depth information from the NeRF network to train depth networks, then creates predicted depth maps to inversely guide NeRF network training. These methods, however, are time-consuming in generating depth information, leading to longer overall process times. Our approach, in contrast, does not use depth maps but instead employs sparse point clouds to gather depth information, considerably shortening the total process duration.

Despite the ongoing advancements in neural radiation field research, there remain certain unresolved issues: (1) The accuracy of neural radiation field reconstruction surfaces is not yet at a desirable level. (2) The training speed of the NeRF model remains relatively slow. To address these challenges, this study introduces a novel approach for surface representation based on multi-resolution hash coding using symbolic distance functions. Additionally, it also replaces the SDF with the TSDF to enhance model stability and employs sparse point cloud supervision to improve the depth expression within the model. Furthermore, this study advocates for the adoption of incremental training, aiming to significantly improve both the accuracy of the model reconstruction and training speed overall.

3. Methods

This paper integrates the multi-resolution hash position coding method and NeuS with the concept of a signed distance function into the NeRF framework for volume rendering. The optimization of the TSDF neural network, combined with sparse point cloud depth supervision, is utilized to reconstruct models of ancient buildings in outdoor environments from UAV images. The technology roadmap is depicted in Figure 1.

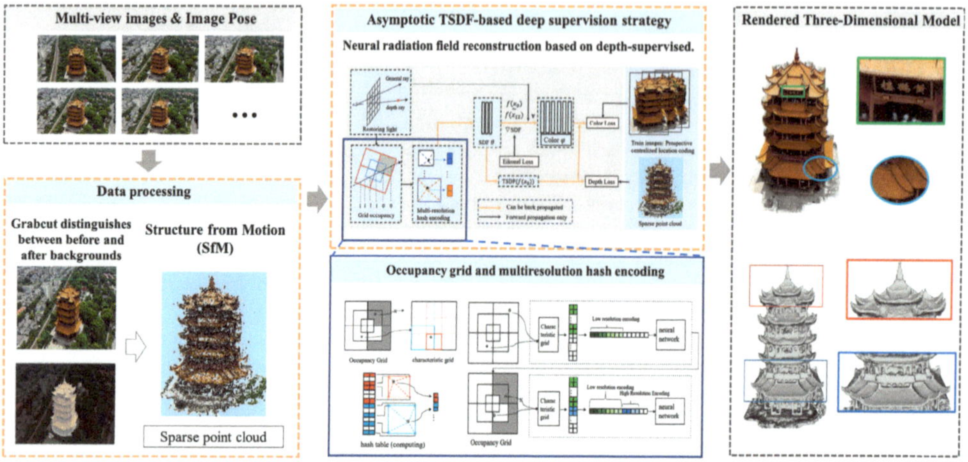

Figure 1. Flowchart of the algorithm of neural radiation field reconstruction based on depth supervision.

In this paper, the method is outlined as follows: starting from a pixel in an image and the light is recovered. The light passes through a multi-resolution hash grid and the internal hash features of the grid can be obtained using interpolation methods. These hash features are then combined with their positions in an SDF network. The SDF network provides SDF values and geometric features. These values, along with the viewing direction, are input into a color network to generate RGB values. The network is optimized by minimizing the difference between the output RGB values and the actual image pixel values. For pixels corresponding to sparse point clouds, the point cloud depth information is computed to supervise the optimization of the 3D model structure by weighting the pixel depths obtained from the TSDF values.

3.1. Data Processing

The fast retrieval feature of hash feature coding, as demonstrated in reference [17], has significantly reduced the training time of NeRF networks from hours to seconds. While multi-resolution hash coding provides computational efficiency by trading a larger memory footprint, the constraint is the finite memory and hash table size. This study introduces two methods to minimize conflicts when dealing with limited hash tables: (1) foreground centralized positional coding and (2) progressive multi-resolution hash coding, which will be detailed in Section 3.2.

Foreground centralized positional coding tackles the issue of growing scene content that exceeds the limited and fixed storage capacity of the 3D feature mesh. This overage results in severe hash conflicts in position encoding, which surpass the neural network's capacity to resolve. The surrounding environmental data can cause training neglect and result in image blurring.

In the wrap-around tilt photography approach, the scene is divided into foreground and background, as depicted in Figure 2; the foreground is our target object, while the background is the surrounding scene environment. The application of Grabcut [39] enables the distinction between the foreground (comprising the target building and the central region of interest) and the background (encompassing non-target scene elements along the image periphery). To enhance the neural network's grasp of vital target information, this paper primarily feeds the network with foreground information, while diminishing the influence of background data at the image edges. This approach curtails the feature overlap between critical information and edge information in the hash table, thereby reinforcing the network's attentional mechanism.

Figure 2. Grabcut distinguishes between before and after backgrounds.

Our depth supervision information is derived from a sparse point cloud, obtained through sparse reconstruction. Sparse reconstruction, also known as SfM, involves feature extraction from the input multi-view images, followed by feature matching to obtain homonymous image points between the images. Based on these homonymous image points, SfM can estimate the internal and external orientation elements of each image more accurately via methods such as forward rendezvous and backward rendezvous and obtain the sparse point cloud in the object-side space and use the depth information of the corresponding pixels of the point cloud as the a priori information for depth supervision.

3.2. Progressive Multi-Resolution Hash Coding

This paper employs progressive multi-resolution hash coding, as depicted in Figure 3, where blue represents low-resolution encoding grids, used for extracting low-resolution features, while pink represents high-resolution encoding grids, used for extracting high-resolution features. Hash coding can lead to data volume and hash conflict challenges. Progressive multi-resolution hash coding is adopted in this study, allowing low-resolution mesh features to capture scene or object outlines and similarities, while high-resolution mesh features prioritize detailed scene or object information.

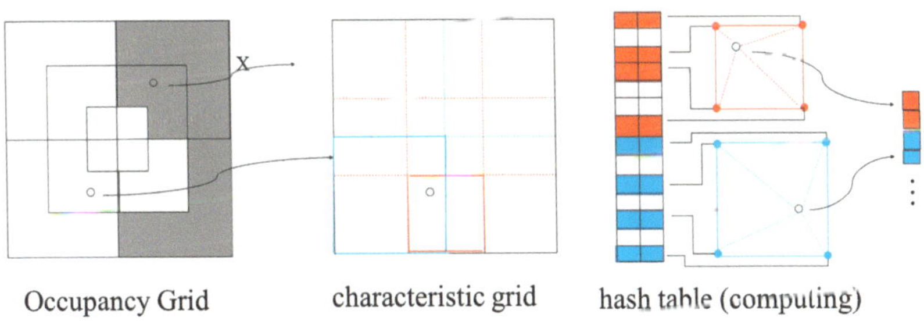

Figure 3. Occupancy grid and multi-resolution hash encoding.

Instant-ngp combines low-resolution and high-resolution feature encoding for all scene points, which results in hash conflicts and partial blurring of image details. Progressive multi-resolution hash coding, depicted in Figure 4, aims to prevent non-critical points from affecting high-resolution mesh features. This approach enhances the speed and accuracy of 3D building reconstruction for neural rendering.

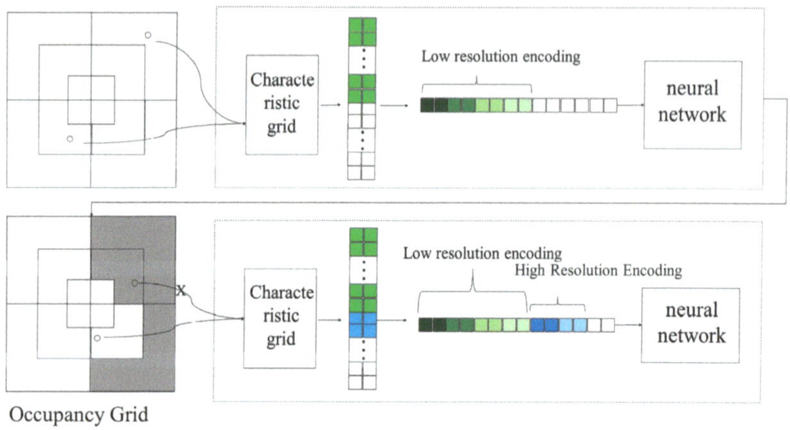

Figure 4. Asymptotic multi-resolution hash coding technology roadmap.

The proposed coding method follows a "from coarse to fine" principle. Initially, during network pre-training, high-resolution feature coding information is masked, while low-resolution hash feature coding is preserved to represent the model's general outline and location. Additionally, the low-resolution feature information is utilized to eliminate empty grid cells, speeding up light sampling and reducing interference from blank areas. As training progresses, the masking of high-resolution feature-encoding information is gradually reduced to enhance the model's surface representation. This encoding approach maximizes the utilization of the high-resolution hash feature table, mitigating hash conflicts to some extent. As a result, it leads to enhanced clarity in image rendering and a significant improvement in the detail of the geometric model.

3.3. Asymptotic TSDF-Based Deep Supervision Strategy

NeuS has exposed inherent errors in NeRF's volume rendering formulation, specifically related to the polar inconsistency of the density and weight values, which results in low geometric accuracy in the neural radiation field. This paper incorporates the concept of the SDF constraint network from NeuS and introduces the TSDF, a form of three-dimensional implicit expression. The TSDF represents an enhancement of the SDF concept, introducing truncation to create values within the range of $[-1, 1]$. The formula for the TSDF is depicted in Figure 5.

$$tsdf_i(x) = \max(-1, \min(1, \frac{tsdf_i(x)}{t})) \tag{1}$$

where t denotes the truncation distance and the TSDF will truncate to 1 or -1 when the absolute value of the SDF is greater than t. The TSDF reduces the variance between the data, increases the stability and makes it easier for the loss to converge in network training, while removing voxels that are farther away from the surface, reducing spurious airborne floats and decreasing the memory size of the reconstructed mesh.

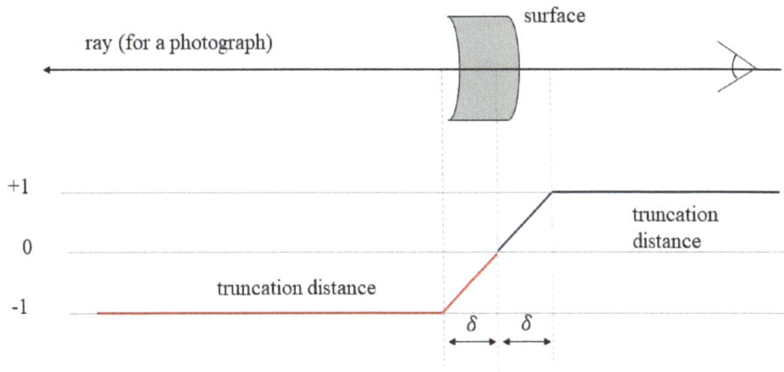

Figure 5. Truncated symbol distance function.

The TSDF is not differentiable at its truncation points, which makes it less suitable for neural network learning. In this paper, the Tanh function is introduced as an approximation of the TSDF. The computational formula is given in Equation (2), where "S" is a trainable hyperparameter and "Z" represents the value of the symbolic distance function. This function bears a resemblance to the TSDF, as both are monotonically increasing odd functions with a value range of $[-1, 1]$. During network training, the value of "S" is initially set to a smaller value, retaining the volume density of points farther from the surface. As the training progresses and the network's scene perception improves, "S" gradually increases, reducing the TSDF truncation distance, thereby focusing on preserving the volume density of points in closer proximity to the surface, which is critical for effective volume rendering.

$$TSDF = \frac{e^{SZ} - e^{-sz}}{e^{SZ} + e^{-sz}} \tag{2}$$

The TSDF neural network is established based on the SDF neural network, as depicted in the optimization flow chart in Figure 6, where the TSDF is introduced for truncation after the network outputs the SDF values, converting them into density values. Light-sampled spatial points are first filtered through the occupancy grid to retain points with high occupancy probabilities. These selected points undergo multi-resolution hash coding. The result of this coding is then fed into the SDF neural network, which produces a multidimensional feature vector where the first dimension represents the SDF value. The color neural network takes this feature vector along with additional information, such as the direction and normal vectors of the points output by the SDF neural network, and it outputs the RGB values. Each valid sampled point is assigned a density value, synthesized by the TSDF value and an RGB value. Points along the same ray are grouped together and their colors are combined according to an unbiased volume rendering formula to obtain the pixel's color value. During training, this paper employs network supervision for the RGB truth values, while the TSDF values are used to update the occupancy of the occupancy mesh. This explicit adjustment brings the voxels of the occupancy mesh close to the object's surface, effectively sieving out points that are far from the reconstructed surface or have no impact on the surface, thus enhancing the light sampling efficiency.

NeRF inputs are only image data and corresponding bitmap information. The rendering and reconstruction of the 3D scene are achieved solely based on the pixel values as supervision, which leads to a significantly constrained geometric representation within the neural network. On the one hand, there is an inherent error in the volume density values obtained by NeRF due to biased volume rendering formulas. On the other hand, there is a lack of supervision regarding the 3D information. In response to this situation, this paper introduces sparse depth information to supervise network training, aiming to enhance the neural network's capability to represent geometric structures.

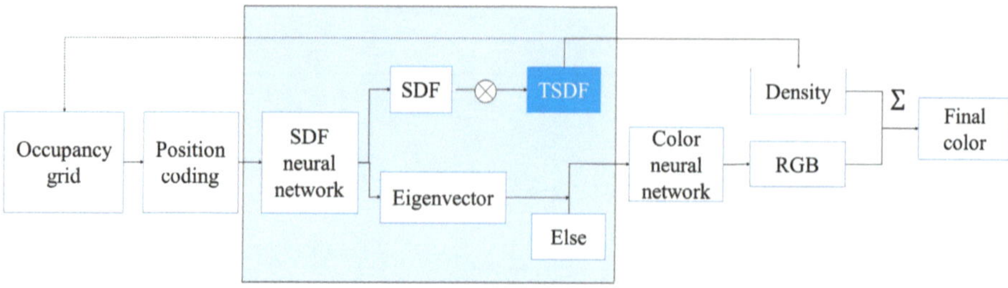

Figure 6. Optimization flow of TSDF neural network framework.

The sparse point cloud used in this paper is not for all pixels of all images, so the training of the deep supervised network is not for all rays. During the training process of the deep supervised network, this paper divides the training rays into two categories, which are ordinary rays and depth rays. As shown in Figure 7, ordinary rays are randomly extracted from all training images, while depth rays are extracted from the pixels corresponding to the sparse point cloud.

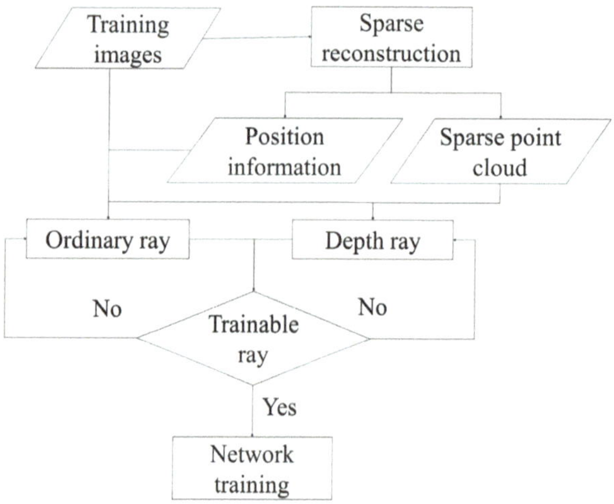

Figure 7. Recovery of normal and deep light training flowchart.

In this paper, the TSDF values obtained from network training are converted into weight values. This weight value can not only synthesize the color, but also the depth. Knowing the position and step spacing of all sampling points on the ray, it is easy to obtain the distance of each point from the origin, which is the depth value. By performing a weighted sum using the depth value and its corresponding weight value, the depth value for this specific ray can be accurately determined. As depicted in Figure 8, the neural network consists of two fully connected MLP networks: the SDF neural network and the color neural network. The SDF neural network comprises one hidden layer, while the color neural network comprises three hidden layers.

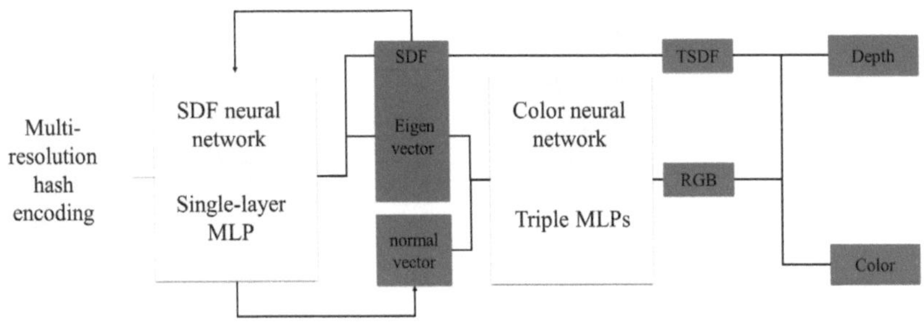

Figure 8. Flowchart of forward propagation of deeply supervised neural radiation field.

The inputs and outputs of the two networks are different. The input of the SDF neural network comprises three-dimensional point coordinates (x, y, z), which are encoded utilizing a multi-resolution hash position encoding methodology. The output from the SDF neural network is a feature vector of 13 dimensions. The foremost dimension of this vector signifies the SDF value, which can be further convertible into the TSDF value. The inputs of the color neural network are the 13-dimensional feature vectors, including the direction vector and the normal vector information of the point, where the normal vector can be obtained by finding the gradient of the SDF function or approximated by Equation (3). The output produced by the color neural network is a tri-dimensional vector, representing the RGB components.

$$\vec{n} = \begin{bmatrix} f(x+\varepsilon, y, z) - f(x-\varepsilon, y, z) \\ f(x, y+\varepsilon, z) - f(x, y-\varepsilon, z) \\ f(x, y, z+\varepsilon) - f(x, y, z-\varepsilon) \end{bmatrix} \tag{3}$$

To train the neural network, three loss functions are constructed in this paper, which are the color loss, SDF loss and depth loss. The color loss is calculated as follows:

$$\mathcal{L}_{color} = \frac{1}{m}\sum_k \mathcal{R}(\hat{C}_k, C_k) + \frac{1}{m}\sum_k MSE(\hat{C}_k, C_k) \tag{4}$$

where m denotes the number of rays per batch, $\mathcal{R}$ denotes the L1 loss, MSE denotes the mean square error loss and $\hat{C}_k$ and C_k denote the predicted and true color values.

The SDF loss is the Eikonal loss, which is used to constrain the symbolic distance function and is calculated as follows:

$$\mathcal{L}_{Eikonal} = \frac{1}{nm}\sum_{k,i}(\|\vee f(\hat{P}_{k,i})\|_2 - 1)^2 \tag{5}$$

where n denotes the number of all sampling points, m denotes the number of rays per batch and $\vee f(\hat{P}_{k,i})$ denotes the derivative of the SDF function, which can also be interpreted as the normal vector of the sampling points.

The depth loss is used to supervise the depth value of a depth ray and the depth loss of a general ray is calculated as follows:

$$\mathcal{L}_{depth} = \frac{1}{m}\sum_k MSE(\hat{D}_k, D_k) \tag{6}$$

where MSE denotes the mean square error loss, and $\hat{D}_k$ and D_k denote the predicted depth value and the true depth value.

4. Experiments

4.1. Experimental Data

In order to verify the effectiveness of the algorithm, three sets of DTU building datasets are used for the experiments in this paper; each set of data contain image data, mask data, empty three-file data, etc., and the description of the datasets is shown in Table 1. When collecting the DTU data, the position of the camera is placed on a sphere with a radius of 50 cm and the camera is roughly 35 cm from the surface of the object.

Table 1. Description of the DTU dataset.

Dataset	Numbers of Image	Data Content
DTU15	49	Resolution (of a photo) 1600 × 1200 Camera parameters Mask data Point cloud data
DTU24	49	Resolution (of a photo) 1600 × 1200 Camera parameters Mask data Point cloud data
DTU40	49	Resolution (of a photo) 1600 × 1200 Camera parameters Mask data Point cloud data

The other set of experimental data are the UAV-acquired building image data, one set of Pix4d sample data and one set of self-collected data from the Yellow Crane Building, as shown in Table 2; the two sets of data are acquired by flying in a circular manner around the building. The third set of data are from Huayan Temple, consisting of five camera shots, with the shooting angle being from above the Huayan Temple tower.

Table 2. Drone image data.

Dataset	Number of Images	Image Size
Pix4d sample Data	36	4592 × 3056
Yellow Crane Data	60	3965 × 2230
Huayan Temple Data	40	6000 × 4000

4.2. Evaluation Indicators

The Peak Signal-to-Noise Ratio (PSNR), which can be used to measure the difference between two images, is calculated as shown in Equation (7).

$$\text{PSNR} = 10 \cdot \log_{10}\left(\frac{MAX_G^2}{MSE}\right) \tag{7}$$

where MAX_G^2 is the maximum pixel value appearing in the truth image. Usually, if the pixel value is represented by B-bit binary, then $MAX_G = 2^B - 1$. MSE is the mean square error between the true value image G and the rendered image R of the same size. This paper uses color images, so it is necessary to calculate the PSNR of the three channels of RGB separately and take the average, as the final PSNR value. The higher the PSNR value, it means that the image is closer to the original image.

The Structural Similarity Index Measure [40] (SSIM) is a full-reference image quality evaluation index, which can better reflect the subjective perception of the human eye. The calculation is relatively complex, respectively, from the brightness L, contrast C and structure S, which are three aspects of the measure of image similarity. The formulas for the three functions are as follows:

$$\text{L}(\text{x}, \text{y}) = \frac{2\mu_x\mu_y + C_1}{u_x^2 + \mu_y^2 + C_1} \tag{8}$$

$$C(x,y) = \frac{2\sigma_x\sigma_y + C_2}{\sigma_x^2 + \sigma_y^2 + C_2} \tag{9}$$

$$S(x,y) = \frac{\sigma_{xy} + C_3}{\sigma_x\sigma_y + C_1} \tag{10}$$

where μ denotes the mean, σ denotes the variance and C_1, C_2 and C_3 denote the constants used to keep the formula stable; the $\sigma_x\sigma_y$ in the above formula is calculated as follows:

$$\sigma_x\sigma_y = \frac{1}{N-1}\sum_{i=1}^{N}(x_i - \mu_x)(y_i - \mu_y) \tag{11}$$

SSIM combines the three functions, and the final formula is as follows:

$$SSIM(x,y) = [L(x,y)]^{\alpha} \cdot [C(x,y)]^{\beta} \cdot [S(x,y)]^{\gamma} \tag{12}$$

where $\alpha > 0$, $\beta > 0$ and $\gamma > 0$ denote the weight values of each metric, which are generally equal weights.

$SSIM \in [0,1]$, the larger the SSIM value, the smaller the image distortion and closer to the original image it is. In practical applications, the image can be chunked using sliding windows so that the total number of chunks is N. Considering the influence of the window shape on the chunks, Gaussian weighting is used to compute the mean, variance and covariance of each window and then the structural similarity of the corresponding chunks is computed as the SSIM and, finally, the mean value is used as the structural similarity measure of the two images, i.e., the average SSIM.

4.3. Hash Coding Experiment

The experimental platform was an ubuntu system with 32 G of RAM, a GeForce RTX 3080Ti graphics card with 12 G of video memory and a 12th Gen Intel@CoreTM i7-12700KF × 20 processor. The number of network training iterations for Instant-ngp, NeuS and the method in this paper were 100,000, 50,000 and 50,000, respectively.

4.3.1. Qualitative Experimental Analysis

This paper employs progressive multi-resolution hash coding and primarily focuses on comparing and analyzing the results of two methods, Instant-ngp and NeuS. Instant-ngp utilizes multi-resolution hash coding, while NeuS employs frequency coding in NeRF. Figure 9 illustrates the comparison of the rendering results for the three algorithms on DTU15, DTU24 and DTU40, respectively.

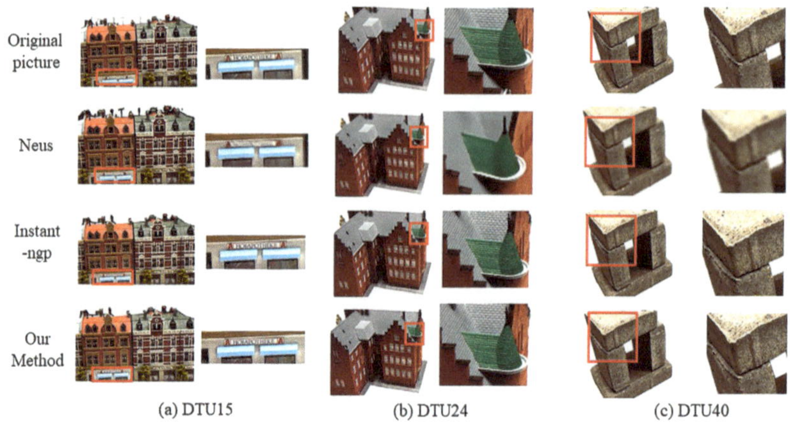

Figure 9. Rendering results of different methods. (**a**) Shows the DTU dataset scene 15 rendering results; (**b**) shows scene 24; (**c**) shows scene 40.

As a whole, NeuS has the most iterations, but has the worst rendering quality and cannot render the image clearly; both Instant-ngp and this paper's method can synthesize the viewpoints better and the image obtained via this paper's method is clearer in comparison between the two. In the DTU15 dataset, the method proposed in this paper is clearer and more realistic than the Instant-ngp method, particularly evident in the billboard letters shown in Figure 9a, which is closer to the original image. In the roof surface part of the DTU24 dataset, the results of this paper's method are clearer than the Instant-ngp texture structure, more granular and three-dimensional. In the DTU40 dataset, there is no significant difference between the results of Instant-ngp and this paper's method, but it is clearer than NeuS.

4.3.2. Quantitative Experimental Analysis

This subsection evaluates Instant-ngp, NeuS and the method of this paper using two metrics, the PSNR and SSIM. After the network is trained to a certain extent, this paper randomly selects a number of images from the image dataset to be used for testing and obtains the corresponding rendered images. Then, the PSNR value and SSIM value between the rendered image and the original image are calculated and the average is taken as the final evaluation value. Table 3 shows the comparison of the PSNR value of the rendered images of the three methods, and six rendered images and the original image are randomly selected from each method for comparison. It can be observed that for the rendered images of the three datasets, the NeuS method exhibits the lowest PSNR values, which are 20.9014, 22.0228 and 27.8526, indicating a lower proximity to the original image and a large amount of blurring. In contrast, the average PSNR values of the method proposed in this paper are 22.2156, 24.3423 and 28.7186, respectively. These values are notably higher than those achieved via the Instant-ngp method, exceeding Instant-ngp's PSNR values by more than 25%. This suggests that the application of low-conflict progressive multi-resolution hash coding can enhance the detail expression capability of the neural network, leading to rendered images that, consequently, are clearer and more closely resemble the original image.

Table 3. PSNR evaluation table of the rendered image results of the three methods.

		Instant-ngp	NeuS	Ours
DTU15	1	21.5906	17.8316	24.5007
	2	22.8636	16.7975	21.3661
	3	20.4145	16.8967	23.5825
	4	20.2797	18.2014	21.3009
	5	19.2271	16.1140	20.8408
	6	21.0331	18.9674	21.7025
	Average	20.9014	17.4681	22.2156
DTU24	1	23.8592	19.6505	24.0335
	2	19.9375	19.8496	21.9429
	3	23.5673	21.7333	24.5284
	4	25.3783	17.9581	29.2147
	5	21.3128	18.5247	22.9470
	6	18.0817	18.3397	23.3875
	Average	22.0228	19.3427	24.3423
DTU40	1	26.8330	21.1750	29.2166
	2	26.9306	20.4683	29.2910
	3	27.3707	21.6330	28.7746
	4	27.7076	19.8074	28.3549
	5	28.7993	19.5547	28.1579
	6	29.4745	21.3349	28.5163
	Average	27.8526	20.6622	28.7186

Table 4 shows the comparison of the SSIM values of the rendered images of the three different methods. From the table, it can be seen that the NeuS method shows a relatively low image structure similarity, with values around 0.7, which suggests that the images produced using NeuS are not adequately trained, leading to an incomplete expression of detailed structures. However, the method discussed in this paper exhibits the highest structural similarity value for the rendered images. Following closely is Instant-ngp and both these methods achieve SSIM values generally in the range of 0.9, which is significantly higher compared to NeuS. This comparison further demonstrates the effectiveness of multi-resolution hash coding in the fine-grained representation of structures.

Table 4. Evaluation table of SSIM values of rendered image results for the three methods.

		Instant-ngp	NeuS	Ours
DTU15	1	0.8540	0.7951	0.8883
	2	0.8975	0.5711	0.9267
	3	0.9107	0.6188	0.9142
	4	0.8301	0.7983	0.8395
	5	0.9002	0.9002	0.9076
	6	0.8497	0.8497	0.8666
	Average	0.8450	0.6953	0.8809
DTU24	1	0.9313	0.7350	0.8795
	2	0.6199	0.7510	0.9290
	3	0.9090	0.7978	0.9299
	4	0.9164	0.6847	0.9471
	5	0.8806	0.7028	0.9176
	6	0.8055	0.8079	0.7687
	Average	0.8438	0.7465	0.8953
DTU40	1	0.9193	0.7186	0.9246
	2	0.9210	0.6985	0.9228
	3	0.9179	0.6346	0.9020
	4	0.9119	0.7381	0.9193
	5	0.9041	0.7309	0.9324
	6	0.9025	0.7215	0.9437
	Average	0.9128	0.7070	0.9275

Table 5 shows the training efficiency comparison between the NeuS method represented by frequency position coding and Instant-ngp represented by multi-resolution hash coding. It is obvious from the table that multi-resolution hash coding has an absolute advantage in time and Instant-ngp is almost 50 times faster than NeuS. For the rendered images obtained via different methods, NeuS needs at least 8 h to obtain the corresponding rendering results, but the rendered image has a large gap with the original image and the clarity is not high, while Instant-ngp only needs about 10 min to obtain the rendered image with relatively good quality.

Table 5. Evaluation table of training time for the three methods.

	Ours Method/min	Instant-ngp/min	NeuS/min
DTU15	10.1	10	497
DTU24	10.3	10	501
DTU40	10.2	10	494

The method in this paper is based on multi-resolution hash coding and the training time is similar to Instant-ngp for the same number of iterations. The training efficiency is also significantly improved compared to the NeuS method.

4.4. Depth-Supervised Ablation Experiments on Ancient Buildings

The Instant-ngp, NeuS and Colmap methods are compared in this section of experiments. Among them, the number of NeuS iterations is 100,000 times and the number of Instant-ngp and the method in this paper is 50,000 times. The experimental platform is the ubuntu system with 32 G of RAM, GeForce RTX 3080Ti with 12 G of video memory and 12th Gen Intel@CoreTM i7-12700KF $\times$ 20 processor.

4.4.1. Qualitative Experimental Analysis

The qualitative experiment is divided into two parts, a comparison of the rendering quality of the methods and a comparison of the reconstruction models between the methods. (1) Rendering quality comparison. The three columns in Figure 10, respectively, show the rendered images and local magnification effects of NeuS, Instant-ngp and the method presented in this paper. As a whole, NeuS can only render the general structure and outline of the model and cannot capture the detail information, which is due to the insufficient network expression of NeuS and the need for a longer training time; Instant-ngp and the method in this paper have better rendering results and both of them have the ability to express detail.

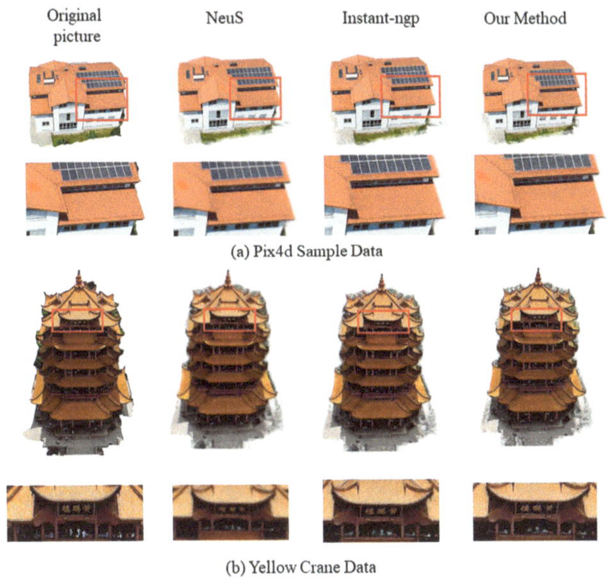

Figure 10. Comparison of rendering results of different methods. (**a**) Shows Pix4d sample data rendering results; (**b**) shows the Yellow Crane Tower data rendering results.

For the Pix4d sample data, the rendering result of NeuS can only vaguely express the shape and appearance of the building and fails to adequately render the detailed structure, such as the tile structure on the roof, three rows of solar panels, etc. Instant-ngp and the method described in this paper are both capable of quickly rendering the detailed structure of the building in a short time. However, the method presented in this paper outperforms Instant-ngp by producing a clearer rendering and more pronounced texture, resulting in a rendered image with enhanced clarity and a more distinct structural representation.

For the Yellow Crane Tower data, the difference in the rendering quality between the three different methods is even more obvious. From the perspective of the plaque of the Yellow Crane Tower, NeuS does not render the shape and content of the plaque because of insufficient training and the complexity of the structure of the Yellow Crane Tower itself; Instant-ngp and this paper's method can directly render the shape of the plaque and the three words "Yellow Crane Tower" and the two methods have a significant improvement in rendering quality compared with NeuS. Both of them have a significantly improved rendering quality compared with NeuS. Compared with Instant-ngp, this paper shows that under the same resolution and the same number of training times, the method in this paper renders the "Yellow Crane Tower" with a higher clarity. Similarly, the image obtained via this method is more detailed and can significantly represent the arrangement of the tiles.

(2) Reconstructing geometric contrasts. This paper proposes two geometric optimization methods: one is TSDF optimization and the other is the introduction of a depth supervision method based on TSDF optimization. This paper compares the Instant-ngp, NeuS and Colmap methods and analyzes the differences between the reconstruction models of each method.

Figure 11 shows the comparison of the reconstructed models of the Instant-ngp, NeuS, TSDF and Colmap methods. The geometric reconstruction quality of Instant-ngp is lower and cannot reconstruct the surface well; NeuS and the TSDF method in this paper can reconstruct the closed watertight model, but the surface of the TSDF optimization method in this paper is flatter and the reconstruction effect is slightly better.

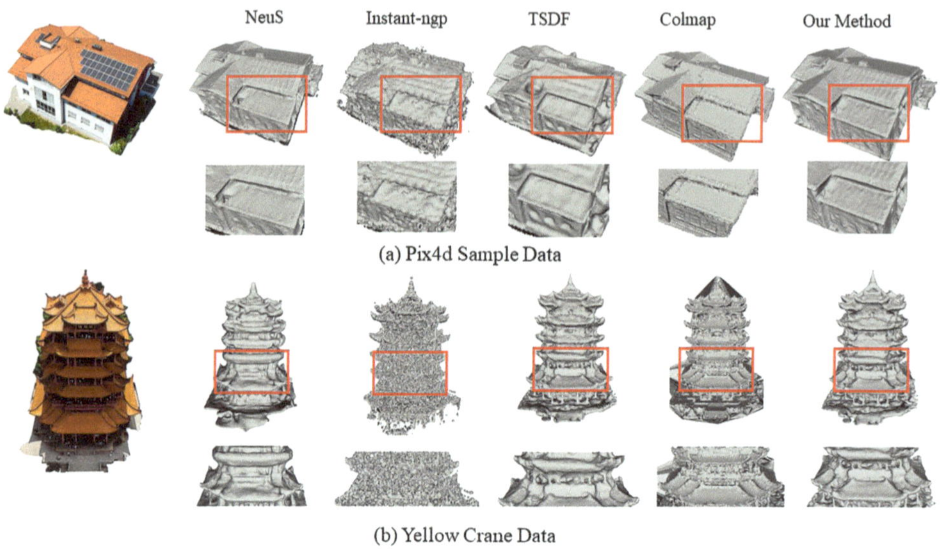

Figure 11. Comparison of TSDF optimization and reconstruction effect of each method. (**a**) Shows the reconstruction results for Pix4d sample data; (**b**) shows the reconstruction results for the Yellow Crane Tower data.

As shown in Figure 11, the Instant-ngp method results in a relatively sparse and fragmented reconstructed model for both the Pix4d sample data and the Yellow Crane Building data, failing to form a satisfactory surface model. While the NeuS method is capable of reconstructing the surface, it falls short in adequately expressing the geometric structure of the building over a certain period, leading to structural errors or imperfections in some areas, such as sunken roofs and uneven solar panels, etc. The TSDF method presented in this paper offers a more comprehensive reconstruction than both Instant-ngp and NeuS, particularly for buildings with simpler structures like those in Pix4d. For complex structures, such as the Yellow Crane Tower, the results are superior to other methods, but the visualization still does not meet the criteria for high precision.

Figure 11 shows the reconstruction model and local method effects of the TSDF method, Colmap method and the addition of the depth supervision method in this paper. For the complex structure of the Yellow Crane Tower data, the surface refinement achieved via the TSDF method is inadequate. However, the reconstruction quality significantly improves after adding the depth supervision on the basis of the TSDF optimization method. The eave edges of the Yellow Crane Tower exhibit a fine and even structure, with sharp protruding edges and a flat, smooth eave surface. Compared with the Colmap reconstruction model, the surface of the model of this paper's method is smooth, avoiding the problem of surface noise and the detailed parts are also more prominent, such as the corridors, columns and other structures of the Yellow Crane Tower in the local zoomed-in image.

For the Pix4d building, the model after adding depth supervision can show the staggered feeling of the roof tile structure. This effect is attributed to a portion of the sparse point cloud on the roof, which constrains the geometric representation in the neural network. However, the solar panels appear uneven due to the intense light reflection on their surfaces, leading to deviations in the point cloud position and thus the unevenness of the reconstructed surface. The surface of the model of the Colmap method is too smooth and many structures are not fully expressed, such as the eaves of the tiles and their appendage structures, etc.

As shown in Figure 12, for the complex Huayan Temple data, using the SDF method did not achieve sufficient surface refinement. Adding depth supervision to the TSDF method significantly improved the reconstruction, resulting in finely detailed roof edges, sharp and prominent edge parts and a smooth eave surface. Compared to Colmap and NeuS, our method produced a model with a smoother surface, avoiding noise issues and more pronounced details.

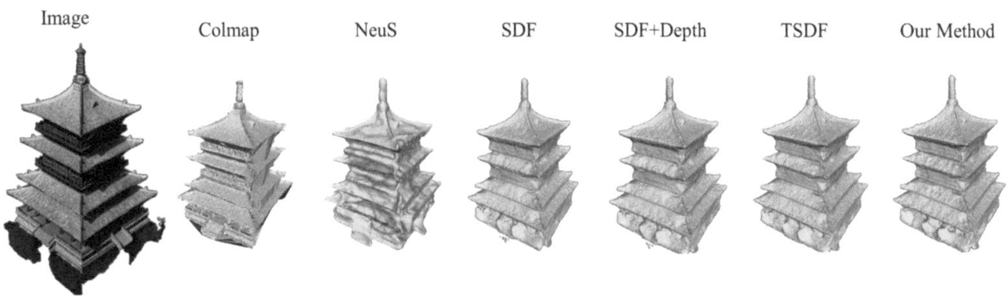

Figure 12. Comparison of the reconstruction effect. The SDF, SDF + Depth supervision, TSDF and the method in this paper are the results of ablation experiments; Colmap and NeuS methods are the results of comparison experiments.

4.4.2. Quantitative Experimental Analysis

This part of the quantitative analysis focuses on the quality analysis of the rendered images and the overall modeling efficiency analysis. The quality of the rendered image represents the expressive ability of the neural network and, to a certain extent, it can also indicate the geometric effect of the reconstruction. Table 6 shows the comparison of the PSNR indexes of the rendered images of Instant-ngp, NeuS and the method in this paper.

It can be seen from Table 6 that the NeuS method renders the worst image quality, with the average PSNR values for the two datasets being 21.2128 and 22.0479, respectively. Although NeuS demonstrates superior geometric expression capabilities, its training efficiency is suboptimal, resulting in inadequately rendered images over a short period. Compared to the rendering quality of the Instant-ngp method, the PSNR values of this paper's method are higher at 24.0229 and 25.5023.

Table 7 shows the comparison of the structural similarity index of the results of each method. From the data in the table, it can be seen that the rendered image of this paper's method has a higher degree of restoration and a clearer texture structure.

Table 6. PSNR evaluation of rendered images via different methods.

		Instant-ngp	NeuS	Our Method
Pix4d	1	25.4210	21.5347	25.8437
	2	24.4684	22.8885	25.4387
	3	24.8765	21.9155	25.8641
	4	24.7463	22.7518	26.5812
	5	24.7451	21.5997	24.8237
	6	24.1549	21.7302	24.4624
	Average	24.7353	22.0701	25.5023
Yellow Crane	1	22.0467	20.5486	23.9559
	2	22.3473	19.5063	24.2902
	3	21.7972	22.2307	23.7182
	4	21.6883	22.6365	23.7127
	5	22.3321	20.8539	24.3762
	6	22.0755	21.5008	24.0845
	Average	22.0479	21.2128	24.0229

Table 7. SSIM evaluation of different methods for rendering images.

		Instant-ngp	NeuS	Our Method
Pix4d	1	0.9469	0.9024	0.9470
	2	0.9517	0.9100	0.9518
	3	0.9437	0.9052	0.9535
	4	0.9465	0.9127	0.9559
	5	0.9404	0.9082	0.9493
	6	0.9395	0.9070	0.9563
	Average	0.9448	0.9076	0.9523
Yellow Crane	1	0.9276	0.8943	0.9406
	2	0.9278	0.8982	0.9397
	3	0.9255	0.8867	0.9394
	4	0.9264	0.8999	0.9389
	5	0.9288	0.8922	0.9427
	6	0.9274	0.8972	0.9416
	Average	0.9273	0.8948	0.9405

The average SSIM values of the two datasets of this paper's method are 0.9405 and 0.9523, respectively. In contrast, the rendered images of the NeuS method are more blurred and lack detail in parts, resulting in the lowest quality scores of 0.8948 and 0.9076. The SSIM values of the rendered images using the Instant-ngp method are 0.9273 and 0.9448, in which the structural similarity of the Pix4d data is quite close to that of the method proposed in this paper, because the structure and texture of the building are relatively simple, thus minimizing the differences. However, from the data of the Yellow Crane Building, we can see that this paper's method demonstrates superior rendering capabilities in more complex scenes.

Table 8 shows the comparison of the training time for NeuS, Instant-ngp, Colmap and the method in this paper.

Table 8. Training schedule for different methods.

Dataset	Instant-ngp/min	NeuS/min	Colmap/min	Our Method/min
Pix4d	9	504	41	16
Yellow Crane	10	517	44	16

The data presented in the table indicate that the NeuS method exhibits the longest reconstruction time, with training durations exceeding 8 h. Despite 100,000 iterations of learning, the neural network's expressive capability remains suboptimal. Followed by Colmap, the reconstruction time is 40 min to 50 min. The method in this paper, while marginally longer in training duration compared to Instant-ngp, significantly enhances both the rendering quality and the geometric precision of the reconstruction. Consequently, the training time for the method delineated in this paper is considered within an acceptable threshold. The PSNR and SSIM in the ablation experiments are shown in Tables 9 and 10, respectively:

Table 9. PSNR evaluation of rendered images via different methods.

		NeuS	SDF	SDF + Depth	TSDF	Our Method
Huayan temple	1	20.0790	19.7807	18.6804	22.0038	21.4941
	2	21.4474	21.8201	19.8980	20.3897	23.0139
	3	20.1258	20.5362	21.3039	20.1366	21.4459
	4	19.4199	19.5288	20.8696	21.6823	22.2220
	5	19.3783	19.7992	18.2688	19.7121	20.1800
	6	18.2056	21.7851	20.9672	22.3610	21.0538
	Average	19.7760	20.5417	19.9980	21.0476	21.5683

Table 10. SSIM evaluation of different methods for rendering images.

		NeuS	SDF	SDF + Depth	TSDF	Our Method
Huayan Temple	1	0.8131	0.8327	0.8915	0.9105	0.9012
	2	0.8512	0.7858	0.8854	0.8654	0.8733
	3	0.8859	0.8069	0.7965	0.8421	0.9102
	4	0.7964	0.7934	0.8701	0.8369	0.9171
	5	0.7842	0.8610	0.8531	0.8554	0.8760
	6	0.8701	0.8714	0.8068	0.9024	0.8821
	Average	0.8335	0.8252	0.8506	0.8514	0.8933

The comparison of the training as well as reconstruction durations is shown in Table 11.

Table 11. Training schedule for different methods.

Dataset	Colmap/min	NeuS/min	SDF/min	SDF + Depth/min	TSDF/min	Our Method/min
Huayan Temple	35	311	23	24	22	23

Based on Tables 9 and 10, it can be observed that the average PSNR and SSIM metrics in this paper are superior to those of other experiments. However, the difference is not very significant, mainly due to issues with the aerial perspective and the presence of certain occlusions. The effect is not as good as surround shooting. Nevertheless, through ablation experiments using the method employed in this paper, it can be seen that the accuracy is still better than other algorithms.

From Table 11, it can be deduced that the NeuS method has the longest reconstruction time, exceeding 5 h of training time. After 100,000 iterations, the neural network's expressive capability is insufficient. Next is Colmap, with a reconstruction time of 35 min. When compared to the ablation experiments, the rendering quality of the method in this paper has significantly improved. This paper's method is on par with the SDF, SDF depth supervision, TSDF and it outperforms Colmap in terms of rendering speed.

5. Discussion

This study proposes a deep-learning-based method for the 3D reconstruction of ancient buildings from UAV-captured images. The method comprises three main steps: processing sampling points using multi-resolution hash coding, introducing the TSDF for threshold truncation during training and integrating depth information for supervised training. The innovations and characteristics of this research can be summarized as follows: (1) Progressive multi-resolution hash coding: This study focuses on target objects in large scenes, implementing centralized foreground position coding and adopting a "coarse-to-fine" progressive multi-resolution hash coding strategy. In the initial phase of network training, high-resolution feature-encoding information is masked, retaining only the low-resolution hash feature encoding. As the training progresses, the masking of high-resolution feature-encoding information is gradually reduced, thereby optimizing feature expression. (2) Progressive TSDF-based depth supervision strategy: The Tanh function is used instead of the traditional piecewise distance function in the TSDF and the truncation distance of the TSDF is set to decrease progressively with the training time Additionally, depth information from sparse point clouds generated by SfM is introduced as prior knowledge, enhancing the network's capability to express 3D geometric structures.

This paper utilizes a dataset of building images collected by UAVs conducting a comparative analysis with several classical neural radiance field technology-based methods to validate the practicality of the proposed algorithm. From Figure 9, it is evident that, compared to classical neural radiance field methods, the rendered images from this paper's method exhibit enhanced detail richness and superior texture clarity. In comparison with NeuS, the improved method in this paper not only ensures the quality of the rendered images but also significantly enhances the network training time. When contrasted with Instant-ngp, the rendered image details in this paper's method are more distinct. Furthermore, as seen in Figures 10 and 11, the 3D implicit reconstruction method in this paper demonstrates a higher accuracy compared to other methods. Finally, as shown in Table 8, compared to Instant-ngp and Colmap, this method is capable of reconstructing high-quality 3D models more swiftly compared to Instant-ngp and Colmap. Despite taking slightly longer than Instant-ngp for reconstruction, it is within an acceptable range.

The main reasons for the improvements in the rendered image quality, model geometric structure and network training efficiency of the proposed method are analyzed as follows:

(1) Reasons for improvement in rendered image quality: In this study, the images were preprocessed during the model training phase, employing a strategy of masking the background area to reduce the interference from background noise. Additionally, the adoption of progressive multi-resolution hash coding combined with occupying a three-dimensional grid fully exploits the high-resolution feature space in the hash table. Such a strategy allows the high-resolution grid to more accurately and intensively represent the detailed structure of the scene. This not only effectively resolves hash conflicts but also substantially improves the quality of the rendered images, leading to a more precise and detailed visual output.
(2) Reasons for improvement in model geometric structure: The integration of the TSDF values in this method ensures that the voxels in the occupied grid more closely adhere to the object's surface. This mechanism effectively filters out key points that significantly impact the reconstructed surface while eliminating points with little or no effect. Furthermore, the incorporation of depth supervision information enhances the model's depth representation capability, significantly improving the geometric structure of the generated model.
(3) Reasons for improvement in network training efficiency: At the initial stage of training, this study employed progressive multi-resolution hash coding, accelerating the ray sampling process by eliminating ineffective grids in the occupied grid. As the training progresses, the strategic application of the TSDF values for the threshold truncation continuously updates the occupancy of the grid, further speeding up the ray sampling efficiency. Moreover, integrating depth supervision information into the training regimen significantly hastens the model's convergence towards high-quality outcomes, ensuring the rapid attainment of superior results.

Therefore, the method proposed in this study is suitable for processing 3D ancient buildings data reconstruction, especially in scenarios requiring rapid and high-precision reconstruction. Not only can this method quickly reconstruct high-quality 3D models, but it also excels in maintaining the clarity of details and textures in rendered images.

6. Conclusions

This paper introduces a low-conflict multi-resolution hash feature location coding method that alleviates hash conflicts through background masking and progressive training. The initial step involves masking the background region in the scene, followed by a "from coarse to fine" approach where low-dimensional position encoding is applied prior to high-dimensional position encoding. This reduction in hash conflicts within high-dimensional features and the mitigation of aliasing in high-dimensional features not only enhances the quality of neural radiance field rendering but also ensures efficient network training, thereby facilitating subsequent geometric optimization. This paper tackles two main issues: (1) The development of a TSDF representation for surface reconstruction and model training supervision through the use of sparse point clouds. This approach serves to stabilize model training and enhance the model's depth representation, thereby significantly enhancing the overall model accuracy. (2) The introduction of an asymptotic training strategy based on multi-resolution hash grids. This strategy gradually refines the details of the reconstructed model, boosting model convergence and expediting the model training process.

Furthermore, this paper introduces an advanced geometric optimization technique for TSDF networks. The native NeRF relies on a biased volume rendering formulation that synthesizes colors solely through density and color, resulting in noisy reconstructed surfaces and low geometric accuracy. To address this, the SDF value is introduced as a weight for color synthesis instead of the original density value. The SDF is asymptotically truncated to obtain the TSDF using the SDF-MLP network, thereby enhancing the geometric constraints of the network and improving the geometric accuracy and detail expression in the reconstructed model. Additionally, a geometric optimization method is employed for deep-information supervised neural networks. Sparse reconstruction estimates the bitmap information from the input image and acquires a sparse point cloud for the depth information. In this approach, training rays are divided into depth rays and ordinary rays, both of which are input into the neural network simultaneously. The depth rays are supervised by depth information during training, enhancing the network's geometric expression capabilities. This method fully utilizes the depth information from sparse reconstruction, facilitating the accurate reconstruction of intricate architectural structures. Through experimental comparisons, this method outperforms the Colmap 3D reconstruction method in terms of reconstruction efficiency and quality.

This paper introduces an improved neural radiance field technique into the field of the 3D reconstruction of ancient architecture, capable of performing centralized multi-resolution hash coding for large-scale ancient architectural scenes captured by UAVs. This method effectively eliminates irrelevant background information, minimizing redundant data encoding, thus significantly enhancing the rendering quality of ancient architectural images. Additionally, this paper proposes a progressive TSDF depth supervision network, providing robust support for the geometric optimization of ancient buildings. Compared to traditional NeRF methods, which may suffer from surface noise and insufficient geometric accuracy in processing ancient buildings, our proposed approach can reconstruct the geometric structure and surface details of ancient architecture more precisely, greatly improving the accuracy in the preservation and restoration of cultural relics. Through this advanced 3D reconstruction technology, a new perspective and methodology are offered for the digital preservation and study of ancient buildings, aiding in the better conservation and heritage of these precious cultural assets.

The 3D reconstruction of ancient architecture using NeRF with depth map supervision is a method that utilizes neural networks and deep-learning techniques. Despite achieving certain effects, there are still limitations in data quality: the reconstruction quality heavily relies on the quality of the input data. If the resolution of the depth map data is low, contains a significant amount of noise or lacks diversity, it may result in the model being unable to accurately capture the details of the building. Subsequent measures, such as using UAVs and ground-level supplementary captures, can be employed to achieve a more refined 3D reconstruction.

Author Contributions: B.G. and Y.G. conceived and designed the whole procedure of this paper S.J. and Y.G. contributed to the introduction, system model sections and manuscript writing. P.Z performed and analyzed the computer simulation results and drew partial figures. Z.J. and D.L reviewed and amended writing. All authors have read and agreed to the published version of the manuscript.

Funding: This work is supported by the Research Program of Wuhan University-Huawei Geoinformatics Innovation Laboratory [grant No. K22-4201-011], the Key Laboratory of Urban Land Resources Monitoring and Simulation, Ministry of Land and Resources [grant No. KF-2022-07-003] and the CRSRI Open Research Program [grant No. CKWV20231167/KF].

Data Availability Statement: The data that support the findings of this study are available from the corresponding author, upon reasonable request.

Acknowledgments: The Research Program of Wuhan University-Huawei Geoinformatics Innovation Laboratory and The Project Supported by the Open Fund of Key Laboratory of Urban Land Resources Monitoring and Simulation, Ministry of Natural Resources.

Conflicts of Interest: The authors declare no conflicts of interest.

References

1. Klimkowska, A.; Cavazzi, S.; Leach, R.; Grebby, S. Detailed three-dimensional building façade reconstruction: A review on applications, data and technologies. *Remote Sens.* **2022**, *14*, 2579. [CrossRef]
2. Geiger, A.; Ziegler, J.; Stiller, C. Stereoscan: Dense 3d reconstruction in real-time. In Proceedings of the 2011 IEEE Intelligent Vehicles Symposium (IV), Baden, Germany, 5–9 June 2011; pp. 963–968.
3. Wang, T.; Zhao, L. Virtual reality-based digital restoration methods and applications for ancient buildings. *J. Math.* **2022**, *2022*, 2305463. [CrossRef]
4. Qu, Y.; Huang, J.; Zhang, X. Rapid 3D reconstruction for image sequence acquired from UAV camera. *Sensors* **2018**, *18*, 225. [CrossRef]
5. Mildenhall, B.; Srinivasan, P.P.; Tancik, M.; Barron, J.T.; Ramamoorthi, R.; Ng, R. Nerf: Representing scenes as neural radiance fields for view synthesis. *Commun. ACM* **2021**, *65*, 99–106. [CrossRef]
6. Shang, L.; Wang, C. Three-Dimensional Reconstruction and Protection of Mining Heritage Based on Lidar Remote Sensing and Deep Learning. *Mob. Inf. Syst.* **2022**, *2022*, 2412394. [CrossRef]
7. Pepe, M.; Alfio, V.S.; Costantino, D.; Scaringi, D. Data for 3D reconstruction and point cloud classification using machine learning in cultural heritage environment. *Data Brief* **2022**, *42*, 108250. [CrossRef] [PubMed]
8. Schonberger, J.L.; Frahm, J.-M. Structure-from-motion revisited. In Proceedings of the IEEE Conference on Computer Vision and Pattern Recognition, Las Vegas, NV, USA, 27–30 June 2016; pp. 4104–4113.
9. Pepe, M.; Alfio, V.S.; Costantino, D. UAV platforms and the SfM-MVS approach in the 3D surveys and modelling: A review in the cultural heritage field. *Appl. Sci.* **2022**, *12*, 12886. [CrossRef]
10. Pei, S.; Yang, R.; Liu, Y.; Xu, W.; Zhang, G. Research on 3D reconstruction technology of large-scale substation equipment based on NeRF. *IET Sci. Meas. Technol.* **2023**, *17*, 71–83. [CrossRef]
11. Lee, J.Y.; DeGol, J.; Zou, C.; Hoiem, D. Patchmatch-rl: Deep mvs with pixelwise depth, normal, and visibility. In Proceedings of the IEEE/CVF International Conference on Computer Vision, Montreal, BC, Canada, 11 October 2021; pp. 6158–6167.
12. Schönberger, J.L.; Price, T.; Sattler, T.; Frahm, J.-M.; Pollefeys, M. A vote-and-verify strategy for fast spatial verification in image retrieval. In Proceedings of the Computer Vision–ACCV 2016: 13th Asian Conference on Computer Vision, Taipei, Taiwan, 20–24 November 2016; Revised Selected Papers, Part I 13, 2017; pp. 321–337.
13. Dang, W.; Xiang, L.; Liu, S.; Yang, B.; Liu, M.; Yin, Z.; Yin, L.; Zheng, W. A Feature Matching Method based on the Convolutional Neural Network. *J. Imaging Sci. Technol.* **2023**, *67*, 030402. [CrossRef]
14. Cubes, M. A high resolution 3d surface construction algorithm/william e. *Lorensen Harvey E. Cline–SIG* **1987**, *87*, 76.
15. Tolstikhin, I.O.; Houlsby, N.; Kolesnikov, A.; Beyer, L.; Zhai, X.; Unterthiner, T.; Yung, J.; Steiner, A.; Keysers, D.; Uszkoreit, J. Mlp-mixer: An all-mlp architecture for vision. *Adv. Neural Inf. Process. Syst.* **2021**, *34*, 24261–24272.
16. Yu, A.; Li, R.; Tancik, M.; Li, H.; Ng, R.; Kanazawa, A. Plenoctrees for real-time rendering of neural radiance fields. In Proceedings of the IEEE/CVF International Conference on Computer Vision, Montreal, BC, Canada, 11 October 2021; pp. 5752–5761.
17. Müller, T.; Evans, A.; Schied, C.; Keller, A. Instant neural graphics primitives with a multiresolution hash encoding. *ACM Trans. Graph. (ToG)* **2022**, *41*, 102. [CrossRef]
18. Zhao, F.; Jiang, Y.; Yao, K.; Zhang, J.; Wang, L.; Dai, H.; Zhong, Y.; Zhang, Y.; Wu, M.; Xu, L. Human performance modeling and rendering via neural animated mesh. *ACM Trans. Graph. (TOG)* **2022**, *41*, 1–17. [CrossRef]
19. Wang, P.; Liu, L.; Liu, Y.; Theobalt, C.; Komura, T.; Wang, W. Neus: Learning neural implicit surfaces by volume rendering for multi-view reconstruction. *arXiv* **2021**, arXiv:2106.10689.

20. Barron, J.T.; Mildenhall, B.; Tancik, M.; Hedman, P.; Martin-Brualla, R.; Srinivasan, P.P. Mip-nerf: A multiscale representation for anti-aliasing neural radiance fields. In Proceedings of the IEEE/CVF International Conference on Computer Vision, Montreal, BC, Canada, 11 October 2021; pp. 5855–5864.
21. Li, Z.; Müller, T.; Evans, A.; Taylor, R.H.; Unberath, M.; Liu, M.-Y.; Lin, C.-H. Neuralangelo: High-Fidelity Neural Surface Reconstruction. In Proceedings of the IEEE/CVF Conference on Computer Vision and Pattern Recognition, Vancouver, BC, Canada, 18–22 June 2023; pp. 8456–8465.
22. Kerbl, B.; Kopanas, G.; Leimkühler, T.; Drettakis, G. 3D Gaussian Splatting for Real-Time Radiance Field Rendering. *ACM Trans. Graph.* **2023**, *42*, 1–14. [CrossRef]
23. Condorelli, F.; Rinaudo, F.; Salvadore, F.; Tagliaventi, S. A comparison between 3D reconstruction using nerf neural networks and mvs algorithms on cultural heritage images. *Int. Arch. Photogramm. Remote Sens. Spat. Inf. Sci.* **2021**, *43*, 565–570. [CrossRef]
24. Lehtola, V.V.; Koeva, M.; Elberink, S.O.; Raposo, P.; Virtanen, J.-P.; Vahdatikhaki, F.; Borsci, S. Digital twin of a city: Review of technology serving city needs. *Int. J. Appl. Earth Obs. Geoinf.* **2022**, *114*, 102915. [CrossRef]
25. Gao, K.; Gao, Y.; He, H.; Lu, D.; Xu, L.; Li, J. Nerf: Neural radiance field in 3d vision, a comprehensive review. *arXiv* **2022**, arXiv:2210.00379.
26. Villanueva, A.J.; Marton, F.; Gobbetti, E. SSVDAGs: Symmetry-aware sparse voxel DAGs. In Proceedings of the 20th ACM SIGGRAPH Symposium on Interactive 3D Graphics and Games, Redmond, WA, USA, 27–28 February 2016; pp. 7–14.
27. Verbin, D.; Hedman, P.; Mildenhall, B.; Zickler, T.; Barron, J.T.; Srinivasan, P.P. Ref-nerf: Structured view-dependent appearance for neural radiance fields. In Proceedings of the 2022 IEEE/CVF Conference on Computer Vision and Pattern Recognition (CVPR), New Orleans, LA, USA, 18–24 June 2022; pp. 5481–5490.
28. Newcombe, R.A.; Izadi, S.; Hilliges, O.; Molyneaux, D.; Kim, D.; Davison, A.J.; Kohi, P.; Shotton, J.; Hodges, S.; Fitzgibbon, A. Kinectfusion: Real-time dense surface mapping and tracking. In Proceedings of the 2011 10th IEEE International Symposium on Mixed and Augmented Reality, Basel, Switzerland, 26–29 October 2011; pp. 127–136.
29. Ma, L.; Li, X.; Liao, J.; Zhang, Q.; Wang, X.; Wang, J.; Sander, P.V. Deblur-nerf: Neural radiance fields from blurry images. In Proceedings of the IEEE/CVF Conference on Computer Vision and Pattern Recognition, New Orleans, LA, USA, 18–24 June 2022; pp. 12861–12870.
30. Zhang, K.; Riegler, G.; Snavely, N.; Koltun, V. Nerf++: Analyzing and improving neural radiance fields. *arXiv* **2020**, arXiv:2010.07492.
31. Barron, J.T.; Mildenhall, B.; Verbin, D.; Srinivasan, P.P.; Hedman, P. Mip-nerf 360: Unbounded anti-aliased neural radiance fields. In Proceedings of the IEEE/CVF Conference on Computer Vision and Pattern Recognition, New Orleans, LA, USA, 18–24 June 2022; pp. 5470–5479.
32. Oechsle, M.; Peng, S.; Geiger, A. Unisurf: Unifying neural implicit surfaces and radiance fields for multi-view reconstruction. In Proceedings of the IEEE/CVF International Conference on Computer Vision, Montreal, BC, Canada, 11 October 2021; pp. 5589–5599.
33. Fridovich-Keil, S.; Yu, A.; Tancik, M.; Chen, Q.; Recht, B.; Kanazawa, A. Plenoxels: Radiance fields without neural networks. In Proceedings of the IEEE/CVF Conference on Computer Vision and Pattern Recognition, New Orleans, LA, USA, 18–24 June 2022; pp. 5501–5510.
34. Liu, L.; Gu, J.; Zaw Lin, K.; Chua, T.-S.; Theobalt, C. Neural sparse voxel fields. *Adv. Neural Inf. Process. Syst.* **2020**, *33*, 15651–15663.
35. Huang, X.; Alkhalifah, T. Efficient physics-informed neural networks using hash encoding. *arXiv* **2023**, arXiv:2302.13397. [CrossRef]
36. Xu, Q.; Xu, Z.; Philip, J.; Bi, S.; Shu, Z.; Sunkavalli, K.; Neumann, U. Point-nerf: Point-based neural radiance fields. In Proceedings of the IEEE/CVF Conference on Computer Vision and Pattern Recognition, New Orleans, LA, USA, 18–24 June 2022; pp. 5438–5448.
37. Zhang, J.; Yao, Y.; Quan, L. Learning signed distance field for multi-view surface reconstruction. In Proceedings of the IEEE/CVF International Conference on Computer Vision, Montreal, BC, Canada, 11 October 2021; pp. 6525–6534.
38. Wei, Y.; Liu, S.; Rao, Y.; Zhao, W.; Lu, J.; Zhou, J. Nerfingmvs: Guided optimization of neural radiance fields for indoor multi-view stereo. In Proceedings of the IEEE/CVF International Conference on Computer Vision, Montreal, BC, Canada, 11 October 2021; pp. 5610–5619.
39. Rother, C.; Kolmogorov, V.; Blake, A. "GrabCut" interactive foreground extraction using iterated graph cuts. *ACM Trans. Graph. (TOG)* **2004**, *23*, 309–314. [CrossRef]
40. Wang, Z.; Bovik, A.C.; Sheikh, H.R.; Simoncelli, E.P. Image quality assessment: From error visibility to structural similarity. *IEEE Trans. Image Process.* **2004**, *13*, 600–612. [CrossRef] [PubMed]

remote sensing

MDPI

Technical Note

Fast Digital Orthophoto Generation: A Comparative Study of Explicit and Implicit Methods

Jianlin Lv †, Guang Jiang *, Wei Ding † and Zhihao Zhao

School of Telecommunication Engineering, Xidian University, Xi'an 710065, China;
jianlinlv@stu.xidian.edu.cn (J.L.); weiding@stu.xidian.edu.cn (W.D.); 20011210455@stu.xidian.edu.cn (Z.Z.)
* Correspondence: gjiang@mail.xidian.edu.cn
† These authors contributed equally to this work.

Abstract: A digital orthophoto is an image with geometric accuracy and no distortion. It is acquired through a top view of the scene and finds widespread applications in map creation, planning, and related fields. This paper classifies the algorithms for digital orthophoto generation into two groups: explicit methods and implicit methods. Explicit methods rely on traditional geometric methods, obtaining geometric structure presented with explicit parameters with Multi-View Stereo (MVS) theories, as seen in our proposed Top view constrained Dense Matching (TDM). Implicit methods rely on neural rendering, obtaining implicit neural representation of scenes through the training of neural networks, as exemplified by Neural Radiance Fields (NeRFs). Both of them obtain digital orthophotos via rendering from a top-view perspective. In addition, this paper conducts an in-depth comparative study between explicit and implicit methods. The experiments demonstrate that both algorithms meet the measurement accuracy requirements and exhibit a similar level of quality in terms of generated results. Importantly, the explicit method shows a significant advantage in terms of efficiency, with a time consumption reduction of two orders of magnitude under our latest Compute Unified Device Architecture (CUDA) version TDM algorithm. Although explicit and implicit methods differ significantly in their representation forms, they share commonalities in the implementation across algorithmic stages. These findings highlight the potential advantages of explicit methods in orthophoto generation while also providing beneficial references and practical guidance for fast digital orthophoto generation using implicit methods.

Keywords: digital orthophoto; neural radiance fields; unmanned aerial vehicles

Citation: Lv, J.; Jiang, G.; Ding, W.; Zhao, Z. Fast Digital Orthophoto Generation: A Comparative Study of Explicit and Implicit Methods. *Remote Sens.* **2024**, *16*, 786. https://doi.org/10.3390/rs16050786

Academic Editors: Wanshou Jiang, San Jiang, Duojie Weng and Jianchen Liu

Received: 17 December 2023
Revised: 18 February 2024
Accepted: 22 February 2024
Published: 24 February 2024

1. Introduction

A digital orthophoto is a remote sensing image that has undergone geometric correction, possessing both map geometric accuracy and image characteristics. It accurately portrays the terrain and landforms of a scene and can be utilized for measuring real distances. It plays a crucial role in various fields, such as land surveying, urban planning, resource management, and emergency response. It aids in monitoring urban development and changes, tracking alterations in land cover and land use. Additionally, in times of natural disasters, time is of the essence. The fast generation of digital orthophotos enables rescue personnel to quickly understand the situation in disaster-stricken areas, enhancing efficiency in responding to emergencies.

The core of digital orthophoto generation lies in obtaining the elevation and texture information of objects within the spatial scene. In order to obtain the elevation and texture of the spatial objects's surface, as shown in Figure 1, the traditional method of generating digital orthophoto mainly draws inspiration from the concept of MVS. It involves reprojecting three-dimensional objects onto different images using the camera's intrinsic and extrinsic parameters. By extracting two image patches centered around the reprojection point, this method then infers the likelihood of the object being at the current elevation based on a quantitative assessment of the similarity between these scenes. Consequently, it reconstructs the necessary spatial structural information of the scene, and the final results are obtained through top-view projection. We define such algorithms that utilize traditional geometry-based approaches to acquire explicit three-dimensional spatial structures and subsequently generate digital orthophotos as explicit methods. The generation process of many types of commercial software, such as Pix4D (version 2.0.104), is carried out using explicit methods. For example, Liu et al. [1] proposed a post-processing method based on Pix4D for digital orthophoto generation. Some works [2–4] are optimized for linear structures in structured scenes.

As a rapidly advancing emerging neural rendering method, NeRF [5] has gained significant attention and shown great potential in recent years. NeRF-related methods inherently offer arbitrary viewpoints, theoretically making them applicable for digital orthophoto generation. They can be used in any scene as long as sparse reconstruction is completed. Therefore, we specifically focused on the feasibility of NeRF in digital orthophoto generation. As shown in Figure 1, NeRF initiates the rendering process by sampling a series of points along targeted rays (represented by the black dots), then estimates the volume density and radiance at specific viewpoints (represented by the circles with an orange outline) for these sample points with neural networks $F(\Theta)$; finally, it applies volume rendering to produce the pixel values. As a specific viewpoint of the scene, the digital orthophoto can be rendered using NeRF by employing a set of parallel rays that are orthogonal to the ground. In this paper, we define the digital orthophoto generation methods based on neural rendering, which do not rely on traditional three-dimensional reconstruction, as implicit methods.

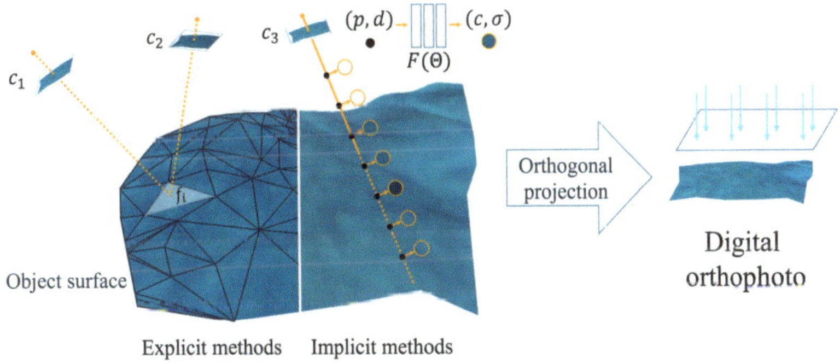

Figure 1. We categorize digital orthophoto generation methods into two types: explicit methods and implicit methods. The typical workflow of explicit methods involves obtaining the geometric structure with explicit parameters like mesh. The implicit methods are based on neural rendering, approximating the geometric structure with implicit neural networks. Both of them generate digital orthophoto through orthogonal projection.

In this paper, we will compare the algorithmic processes and performance of explicit and implicit methods in digital orthophoto generation. Within the explicit methods, we selected the TDM algorithm [6], known for its exceptional speed performance. To unleash its potential, we conducted CUDA version porting and optimization modifications, significantly enhancing the generation efficiency. For implicit methods, we implemented orthographic view rendering based on NeRF and selected the speed-optimized Instant NGP [7] as a representative experiment. The experimental results reveal that the explicit method demonstrates notably high efficiency in generation speed. Both explicit and implicit methods yield acceptable levels of measurement accuracy and exhibit comparable rendering quality.

2. Related Work

2.1. DigitalOrthophoto Generation Methods

In digital photogrammetry, a mature workflow of digital orthophoto generation is presented in [8]. A general digital orthophoto generation approach often relies on 3D reconstruction. Schonberger et al. [9] proposed a complete structure-from-motion (SfM) pipeline. Shen et al. [10] proposed a patch-based dense reconstruction method, allowing for the integration with the SfM pipeline to achieve an entire 3D reconstruction process. Some works [11,12] used smartphone sensors to generate the 3D models. After the 3D reconstruction is completed, it can be orthogonally projected onto a horizontal plane to obtain the digital orthophoto. A digital orthophoto generation method with the assistance of Pix4D is proposed in [1]; they also propose post-processing methods based on Pix4D for digital orthophoto generation. Many efforts are being made to accelerate digital orthophoto generation, but these works are usually focused on specific scenarios. Some works have optimized digital orthophoto generation in structured scenes. For instance, Wang et al. [4] extracted and matched lines from the original images and then transformed these matched lines into the 3D model, reducing the computational cost of pixel-by-pixel matching in dense reconstruction. Li et al. [13] used deep learning methods to obtain a topological graph in the scenes, enhancing the accuracy at the edges of buildings. Lin et al. [2] arranged ground control points at the edges of buildings to ensure the accuracy of these edges. Some studies have made improvements for more specialized scenes. For instance, Lin et al. [14] focused on agricultural surveying scenarios, utilizing the spectral characteristics of vegetation to determine its location, thereby achieving fast digital orthophoto generation in agricultural mapping contexts. Zhao et al. [15] assumed the target scene to be a plane, employing simultaneous localization and mapping (SLAM) for real-time camera pose estimation and projecting the original images onto the imaging plane of the digital orthophoto. These methods speed up the digital orthophoto generation by sacrificing the generality of the algorithms. Some methods [16,17] utilize Digital Elevation Model (DEM) to accelerate the digital orthophoto generation, but this approach is constrained by the acquisition speed of the DEM.

Zhao et al. [6] were the first to propose a process for digital orthophoto generation directly using sparse point clouds. This approach eliminates the redundant computations that occur in the dense reconstruction phase of the standard 3D reconstruction-based digital orthophoto generation methods, significantly increasing the speed of generation.

2.2. NeRF with Sparse Parametric Encodings

In recent years, methods for novel view image synthesis on neural rendering have rapidly evolved. Mildenhall et al. [5] introduced NeRF, which represents a scene as a continuous neural radiance field. NeRF optimizes a fully connected deep network as an implicit function to approximate the volume density and view-dependent emitted radiance from 5D coordinates (x, y, z, θ, ϕ), with σ representing the volume density at a spatial point. To render an image from a specific novel viewpoint, NeRF initially (1) generates camera rays traversing the scene and samples a set of 3D points along these rays, (2) inputs the sampled points and viewing directions into the neural network to obtain a collection of densities RGB values, and (3) employs differentiable volume rendering to synthesize a 2D image.

Many recent works have incorporated sparse parametric encoding into NeRF for enhancement, generally aiming to pre-construct a series of auxiliary data structures with encoded features within the scene. We summarize these NeRFs with sparse parametric encoding into four stages in Figure 2: (1) scene representation, (2) radiance prediction, (3) differentiable renderer, and (4) loss function. For the first stage in Figure 2a, numerous sparse parametric encoding techniques have been proposed, such as dense and multi-resolution grids [7,18,19], plannar factorization [20–22], point clouds [23], and other formats [24,25]. The central concept behind these methods is to decouple local features of the scene from the MLP, thereby enabling the use of more flexible network architectures.

They are typically represented by a grid, as shown in Figure 2a, resulting in the local encoding feature lookup table shown in the orange part. For the second stage in Figure 2b, a coarse–fine strategy is often used to sample along rays, and a cascaded MLP is typically used to predict volume density and view-dependent emitted radiance. Several studies have attempted to enhance rendering quality by improving sampling methods [22,26,27]; some have employed occupancy grids to achieve sampling acceleration [28]; others have focused on adjusting the MLP structure to facilitate easier network training [29]. For the third stage in Figure 2c, the figure exemplifies the most commonly used volume rendering, but other differentiable rendering methods are also employed [30], with Nvdiffrast [31] providing efficient implementations of various differentiable renderers. For the fourth stage in Figure 2d, the figure presents the most commonly used mean squared error loss between rendered and ground truth images, with some works introducing additional supervision, such as methods incorporating depth supervision [32,33]. With different scene representations, various loss functions are incorporated to constrain the network. Neural radiance fields can achieve photorealistic rendering quality and lighting effects, but it often takes hours to optimize the network parameters, and the training process is computationally expensive.

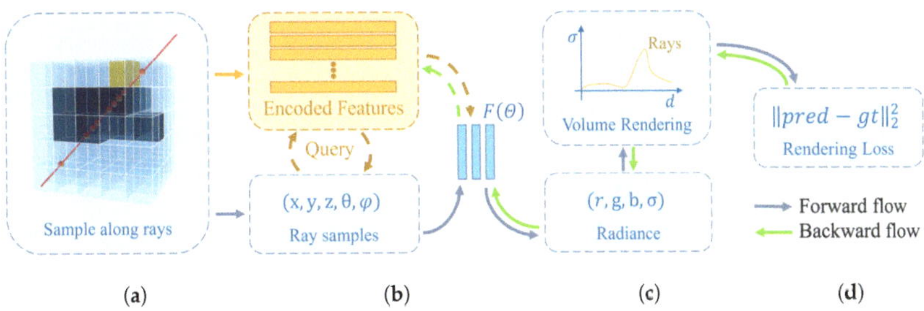

Figure 2. A schematic representation of NeRF with sparse parametric encoding. The process is divided into four stages: (**a**) scene representation, primarily defining auxiliary data structures for a scene's sparse parametric encoding; (**b**) radiance prediction, where queried encoded features (orange arrows) and embedded sampling points are represented as feature embeddings and the radiance at these points is obtained through the function $F(\Theta)$; (**c**) differentiable rendering, rendering meaningful pixel RGB values based on the radiance of sampling points; (**d**) loss computation, calculating the loss based on the rendering results, followed by backpropagation (green arrows) to optimize network parameters.

Both explicit and implicit methods require the initial step of SfM to obtain sparse point clouds and camera poses. The former predicts depth using multi-view geometry theories and describes the geometric structure of the scene using explicit parameters such as mesh, voxel, raster, etc. In contrast, implicit methods gradually fit to the real scene through implicit neural representation during the training process. Finally, both methods render digital orthophoto images from an orthographic viewpoint.

3. Method

An explicit digital orthophoto generation method typically involves the SfM and MVS processes. The TDM method facilitates the fast generation of digital orthophotos directly from sparse point clouds. Unlike MVS, the computation process of TDM is specifically tailored toward the final output of digital orthophotos. Factors unrelated to digital orthophotos are not involved in the computation, facilitating faster generation of digital orthophotos. So we selected the TDM algorithm as the representative explicit method for fast digital orthophoto generation and Instant NGP as the representative implicit method.

An implicit digital orthophoto generation method typically involves optimizing a group of parameters with posed images. This optimization process often takes several hours or even dozens of hours. Instant NGP [7] represents a speed-optimized neural radiance field, achieving the shortest optimization time among current radiance field methodologies. Hence, we select Instant NGP as the representative implicit method for fast digital orthophoto generation.

Both methods rely on the sparse reconstruction results from SfM. To generate digital orthophotos, both methods require prior information of accurate ground normal vectors. By using the Differential Global Positioning System (DGPS) information as a prior for sparse reconstruction, we can obtain accurate ground normal vectors while also recovering the correct scale of the scene.

3.1. Explicit Method—TDM

The TDM algorithm, when generating digital orthophotos, essentially processes information for each pixel, equivalent to raster data processing. To achieve the final rendering, the key lies in accurately estimating the elevation values and corresponding textures for each raster. The following will introduce the algorithm flow of our CUDA-adapted and optimized version of the TDM algorithm in this paper.

Raster Attribute Initialization: by specifying the spatial resolution R_s, the raster image G to be generated is obtained with dimensions $W \times H$, where each raster represents a pixel in the final digital orthophoto image. Each raster unit possesses five attributes: (1) raster color $C_o = ({R^i}_g, {G^i}_g, {B^i}_g)$; (2) raster elevation Z_g; (3) raster normal vector $\vec{n} = (n_x, n_y, n_z)$. (4) Confidence score of raster elevation S_g. (5) The camera group to which the raster belongs C_g. As shown in Figure 3, the algorithm traverses through all three-dimensional point clouds and performs orthographic projection to obtain the raster unit g^i corresponding to a certain three-dimensional point $(X_i, Y_i, Z_i)^T$. Subsequently, the elevation Z_g of that raster is initialized.

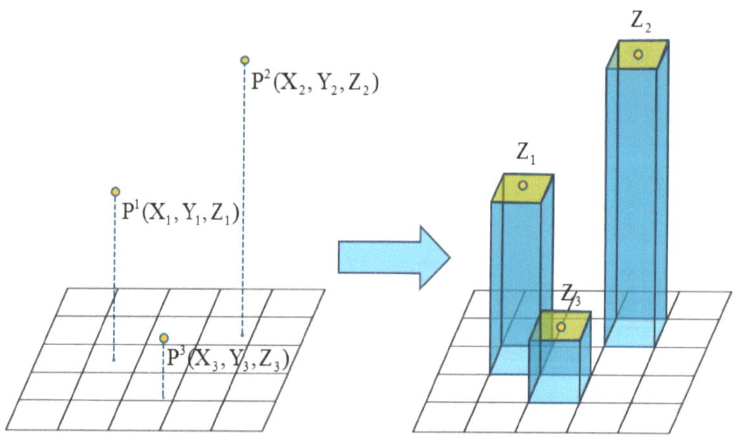

Figure 3. The figure illustrates the process of raster elevation initialization. In the initial stage, there are some points in three-dimensional space. Following the initialization process, rectangular raster units are obtained. The height of the vertical column represents the elevation values of each cell.

Then, we will search for the corresponding camera group C_g for each raster containing an elevation value. This camera group retains the best-angle cameras in all eight directions that can see this raster, facilitating the subsequent process of finding and determining the views. As shown in Figure 4, the space is first evenly divided into eight regions. Then, the camera with the highest view score S_v in each of the eight directions that can see the raster g^i is retained.

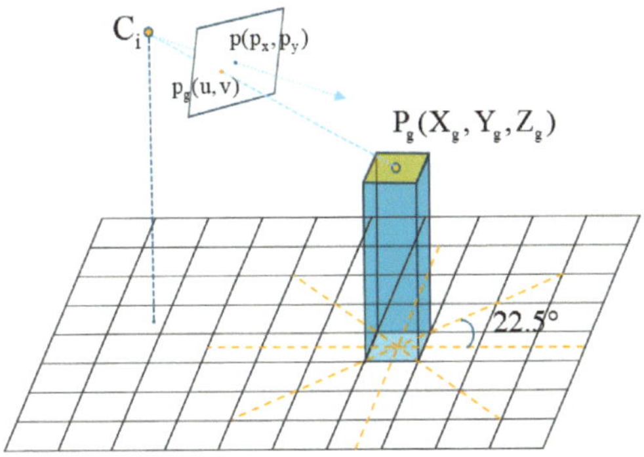

Figure 4. The figure illustrates the process of initializing the camera group. Based on the projection relationship of the pinhole camera, the projection p_g of the raster's world center point P_g on the image plane is obtained. The original space is evenly divided into eight regions, and the camera with the highest S_v in each region is found to be the camera group for this raster.

We denote the principal point coordinates of the image as $(p_x, p_y)^T$, and the world coordinates P_g of the raster g^i, after being projected by the camera, as p_g. The optimal view score S_v is then calculated as

$$S_v = \frac{1}{(u-p_x)^2 + (v-p_y)^2 + \varepsilon} \tag{1}$$

Finally, the eight cameras with the highest scores in each direction are selected as a camera group for the raster. The camera with the highest score is considered as the optimal view camera C_b for the raster unit.

Elevation Propagation: Given rasters with known elevation are considered as seed units g_s, and the propagation starts iteratively from these seed units. Each iteration propagates the elevation information Z from the seed raster unit to all raster units within a patch. Subsequently, the adjustment of Z_g occurs via the random initialization of the normal vector $\vec{n}$, as shown in Figure 5. We project the i-th raster of the raster support plane S^k onto a corresponding image I^j in the camera group ${C_g}^i$, and the corresponding pixel color is denoted as $(R^{jki}, G^{jki}, B^{jki})$. The average color of the nine raster units projected onto the image I^j in the raster support plane S^k is denoted as $(\overline{R^{jk}}, \overline{G^{jk}}, \overline{B^{jk}})$. Define a color vector $\mathbf{V}^{jk}$ to represent the color information of S^k:

$$\mathbf{V}^{jk} = (R^{jk1} - \overline{R^{jk}}, G^{jk1} - \overline{G^{jk}}, B^{jk1} - \overline{B^{jk}}, \ldots, R^{jk9} - \overline{R^{jk}}, G^{jk9} - \overline{G^{jk}}, B^{jk9} - \overline{B^{jk}})^T \tag{2}$$

The number of cameras in the camera group of the i-th raster unit is denoted as ${N^i}_C$, and the number of color vectors that S^k possesses is ${N^k}_C$. Therefore, the equation ${N^k}_C = \sum_{i=1}^{9} {N^i}_C$ can be derived. The color vector corresponding to the optimal view camera C_{bs} of the seed raster is taken as the reference vector. To measure the consistency of ${N^k}_C$ color vectors, the average cosine distance between the reference vector and other vectors is defined as the matching score ${M^k}_s$ for S^k:

$$ {M^k}_s = \frac{\sum_{j=1\ldots {N^k}_C, j\neq s, j\notin {\Phi^k}_o} \frac{{\mathbf{V}^{jk}}^T \mathbf{V}^{sk}}{\|\mathbf{V}^{jk}\|\|\mathbf{V}^{sk}\|}}{{N^k}_C - {N^k}_o} \tag{3}$$

where $\frac{{\mathbf{V}^{jk}}^T \mathbf{V}^{sk}}{\|\mathbf{V}^{jk}\|\|\mathbf{V}^{sk}\|}$ represents the cosine distance, $\mathbf{V}^{sk}$ is the reference vector, ${\Phi^k}_o$ is the set of occluded images, and ${N^k}_o$ is the number of images in ${\Phi^k}_o$.

Furthermore, we can evaluate the reasonableness of depth information through color consistency. Once the computed ${M^k}_s$ exceeds the confidence threshold η, the current depth information is considered reasonable. Then, we update the Z_g, S_g and $\overrightarrow{n}$ of all raster units within the patch. During an iteration, there will be multiple random initializations of $\overrightarrow{n}$. If the matching score remains below η, elevation information will not be propagated.

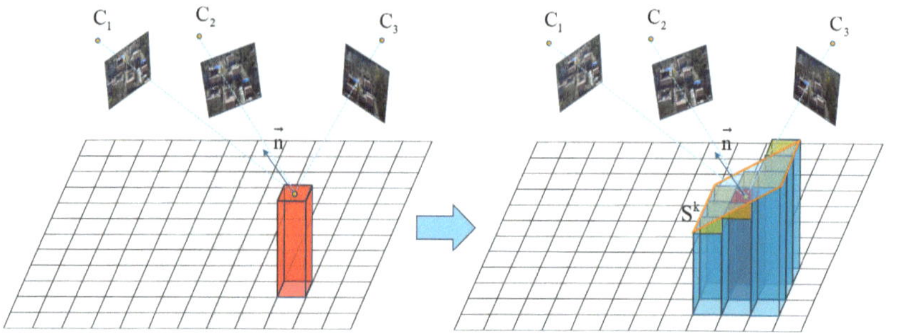

Figure 5. The figure demonstrates the elevation propagation process. The red rectangular raster represents the seed unit. The seed unit with a known elevation and the surrounding eight raster units with unknown elevation form a raster support plane. The raster support plane calculates a matching score based on color consistency. If the score meets the threshold, the elevations of other raster units will be initialized based on the normal vector of the seed raster unit.

Multi-resolution Interpolation-based Elevation Filling: The original algorithm gradually reduces η after each iteration until the elevation propagation is complete. This will result in subsequently obtaining a lower confidence score for Z_g and wasting a considerable amount of time. To efficiently reduce iteration time, we propose a multi-resolution interpolation-based elevation filling algorithm to acquire elevations of raster units with low confidence scores. When the initial value of η is η_0, it gradually decreases with the increase in iteration count until it equals η_e. At this stage, we utilize the proposed algorithm to assign values to raster units g^i without elevations, as shown in Figure 6.

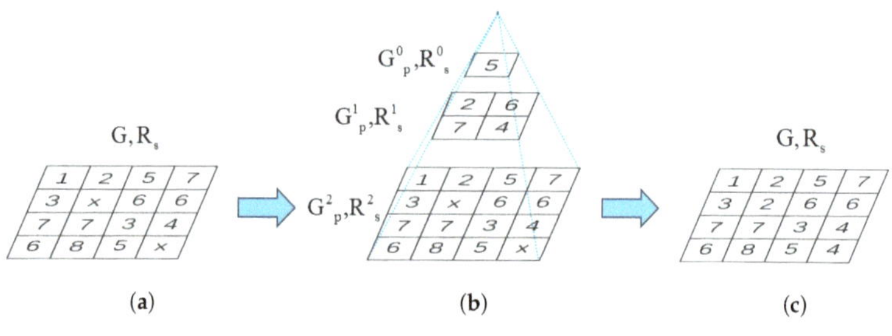

Figure 6. The figures illustrate the multi-resolution interpolation-based elevation filling process. (**a**) The raster image obtained after elevation propagation contains raster units with unknown elevations. (**b**) The process of generating the multi-resolution interpolation raster images. (**c**) The resulting raster image after elevation filling using the multi-resolution interpolation raster images.

After the elevation propagation, the initial seed ${g^i}_s$ for this filling algorithm is derived from raster units g^i within the raster image G, where the confidence measure S_g exceeds η_0. The spatial resolution ${R^i}_{sp}$ of the filling raster image ${G^i}_p$ for the i-th layer of this multi-resolution raster is as follows:

$$ {R^i}_{sp} = \frac{\min[(X_{\max} - X_{\min}), (Y_{\max} - Y_{\min})]}{2^i} \quad (4) $$

where $X_{\max}$ and $Y_{\max}$ are the maximum values of the X- and Y-coordinates in this area. Likewise, $X_{\min}$ and $Y_{\min}$ are the minimum values. When multiple ${g^i}_s$ fall into the same raster unit g^i of ${G^i}_p$, we set the average of these points as the elevation value for that raster unit. If no points fall within a specific raster unit, we will retrieve the elevation value corresponding to the raster position from multi-resolution interpolation raster image ${G^{i-1}}_p$ and set it as the elevation value for g^i. If ${R^i}_{sp} < R_s$, the process is repeated, continuously constructing ${G^i}_p$ as described above. Eventually, there are some raster units that have not been assigned elevation values in G. We will then search for the elevation values corresponding to the raster positions in the highest resolution interpolation raster image ${G^f}_p$ and assign them accordingly.

Texture Mapping: In each image, certain objects might be occluded by other objects, leading to erroneous texture mappings. Occlusion detection is necessary in such cases.

Subsequently, texture mapping is performed based on g^i and the corresponding projection relationship with the optimal view camera C_b, obtaining color information $C_o = ({R^i}_g, {G^i}_g, {B^i}_g)$ for the raster unit. Finally, the generation of the final digital orthophoto is completed.

3.2. Implicit Method—Instant NGP

We will use the most representative Instant NGP [7] as an example to illustrate the process of digital orthophoto generation using implicit methods. As a neural radiance field utilizing sparse parametric encodings, Instant NGP introduces multi-resolution hash encoding to address the $O(N^3)$ parameter complexity associated with dense voxel grids; Figure 7 illustrates this multiresolution hash encoding process in 2D.

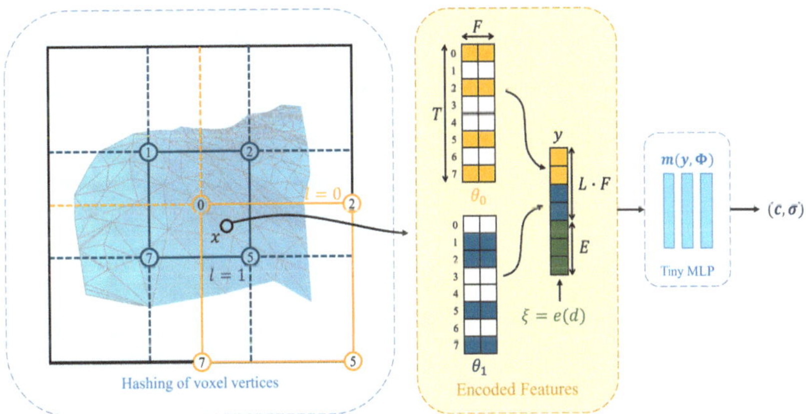

Figure 7. Illustration of the multiresolution hash encoding in 2D. For a given coordinate x, the method queries the encoded features on the surrounding voxels' vertices (blue and orange circles) with the hashing result (numbers in the circles) and performs interpolation on the encoded features in θ_l across L levels. For a given direction d, the embedding function $e(d)$ is applied to generate auxiliary inputs ξ. Subsequently, the encoded features at each level and auxiliary inputs will be concatenated as the final MLP embedding input $y \in \mathbb{R}^{LF+E}$ to obtain the radiance (c, σ). Its optimizable parameters consist of L hash tables θ_l and tiny MLP $m(y; \Phi)$.

In practice, Instant NGP divides the scene into voxel grids with L levels of resolution. For each level of the resolution grids, a compact spatial hash table θ_l of a fixed size T is used to store the F-dimensional feature vectors on that resolution level's grid. When querying the feature vector of a spatial coordinate x in Instant NGP, the process first identifies grid corners spatially close to x on each resolution layer. Then, the feature vectors of adjacent grid corners are looked up in θ_l. Next, linear interpolation is performed to obtain the feature vector of the spatial coordinate x at that resolution level. This process is executed across all L resolution levels. Subsequently, these feature vectors from different resolution layers are concatenated with auxiliary inputs $\xi \in \mathbb{R}^E$, forming the final MLP embedding input $y \in \mathbb{R}^{LF+E}$. Finally, Instant NGP uses a Tiny MLP $m(y;\Phi)$ to obtain the radiance (c,σ) for the spatial coordinate x. This process also aligns with the generalized description of neural radiance fields based on sparse parametric encoding, as shown in Figure 2. Instant NGP can achieve a balance between performance, storage, and efficiency by selecting appropriate hash table sizes T.

As mentioned in Section 1, digital orthophotos can be rendered with neural approaches. In contrast to the typical pinhole camera imaging model, digital orthophotos are rendered using a set of parallel light rays perpendicular to the ground, as shown in Figure 1. To ensure that Instant NGP achieves a rendering quality comparable to explicit methods in scalable scenes, we adopted the largest scale model recommended in the paper.

4. Experiments and Analysis

The data utilized in this study were acquired from the Unmanned Aerial Vehicle (UAV) following a serpentine flight path pattern. A CW-25 Long Endurance Hybrid Gasoline & Battery VTOL drone was used in this data collection. It has a long service life, is fast, has a large payload, and is structurally stable and reliable. It is equipped with the RIY-DG4Pros five-lens camera, providing 42 million pixels and a resolution of 7952 × 5304 pixels. We established the drone ground station GCS1000. The UAV is equipped with the Novatel617D dual-antenna satellite positioning differential board card on board. Subsequently, through DGPS, the UAV can accurately capture changes in the ground station's position, speed, and heading in real time.

We selected the TDM algorithm as the representative explicit method for digital orthophoto generation. Similarly, we used Instant NGP as the representative implicit method for digital orthophoto generation. The commercial software Pix4D is widely used and performs exceptionally well in digital orthophoto generation. Therefore, we have chosen its generated results as the benchmark for measuring accuracy. Pix4D, being an explicit method, requires the full process of traditional 3D reconstruction during digital orthophoto generation. Hence, for the time comparison test, we selected the TDM algorithm, which eliminates redundant computations during the dense reconstruction.

As described in this section, we initially conducted digital orthophoto generation tests on three common scenes: buildings, roads, and rivers. The objective was to demonstrate the image generation quality and algorithm robustness of both explicit and implicit methods across various scenes. Subsequently, to assess the accuracy of the two methods, comparisons were made with the commercial software Pix4D regarding measurement precision. Finally, to evaluate the efficiency of both methods, we measured the time required for generating scenes of different sizes.

4.1. Test on Various Scenes

Figure 8 shows a set of original village photo data used for testing, including numerous scenes of slanted roofs of houses, trees, and other objects. We performed sparse reconstruction in conjunction with the camera's DGPS information, enabling the recovery of accurate scale information and spatial relationships. The resultant 3D sparse point cloud, as shown in Figure 9, and camera poses served as prior information for subsequent explicit and implicit methods in digital orthophoto generation. The resulting digital orthophotos after the final processing through the explicit and implicit methods are shown in Figure 10.

Figure 8. The original images of some village scenes captured by unmanned aerial vehicles.

Figure 9. The point cloud of the village scenery obtained after sparse reconstruction.

As shown in Figure 11, we conducted digital orthophoto generation tests for various scenes using both explicit and implicit methods. Figure 11a,b show that TDM may lead to inaccuracies in areas experiencing sudden height variations, for example, the roof edges of houses, while Instant NGP can accurately depict sudden height variations. Figure 11c,d show that moving objects within the scene may induce ghostly artifacts in the results of Instant NGP but have a minimal impact on TDM. Figure 11e,f show that the clarity of the outputs of Instant NGP does not match that of TDM. The imaging quality of implicit methods is predominantly influenced by the model scale, whereas TDM is directly dictated by the clarity of the original image.

(a) Explicit method (TDM) (b) Implicit method (Instant NGP)

Figure 10. The figure illustrates the digital orthophoto generation results from two methods within the same village scene. (**a**) depicts the output derived from the explicit method. (**b**) depicts the output obtained from the implicit method.

To quantitatively analyze the quality of the digital orthophoto generated using the two methods, we employed two no-reference image quality assessment techniques, Brisque [34] and NIQE [35]. The results in Table 1 show that, in the majority of scenarios, the quality generated by the explicit method (TDM) surpasses that of the implicit method (Instant NGP).

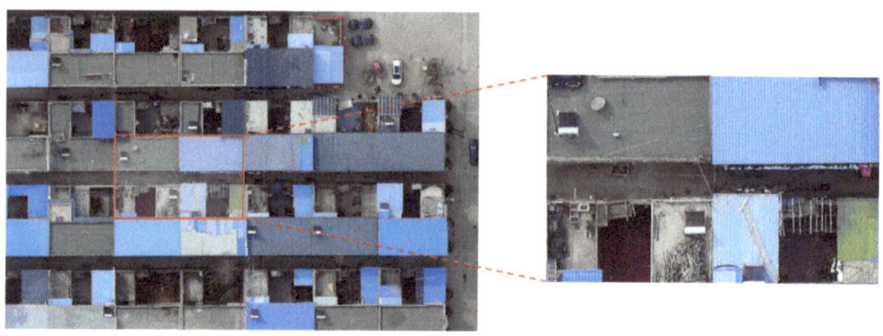

(**a**)

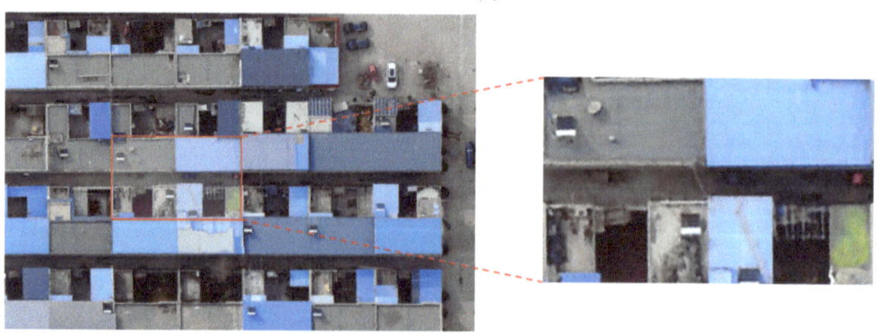

(**b**)

Figure 11. *Cont.*

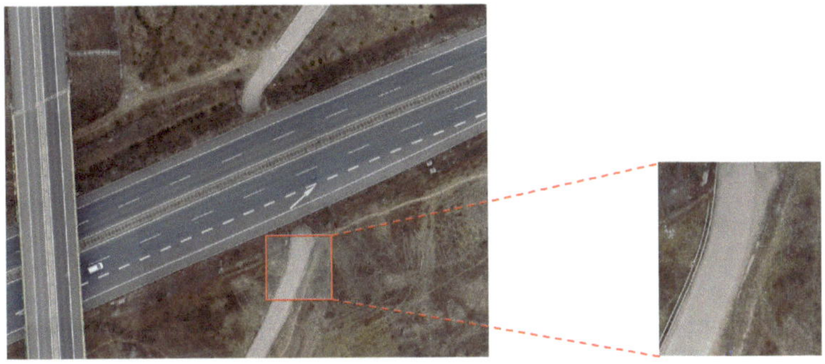

(c)

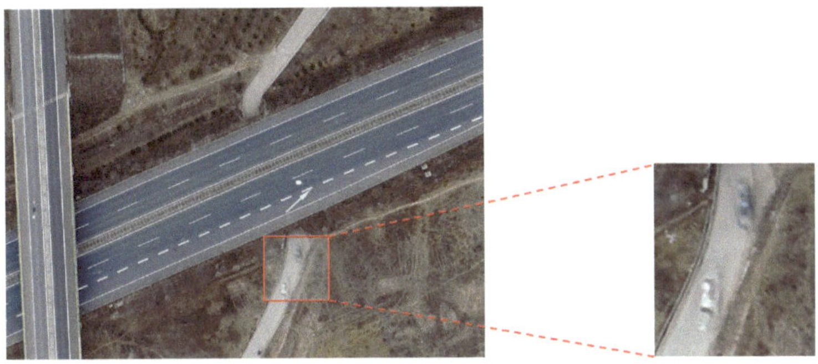

(d)

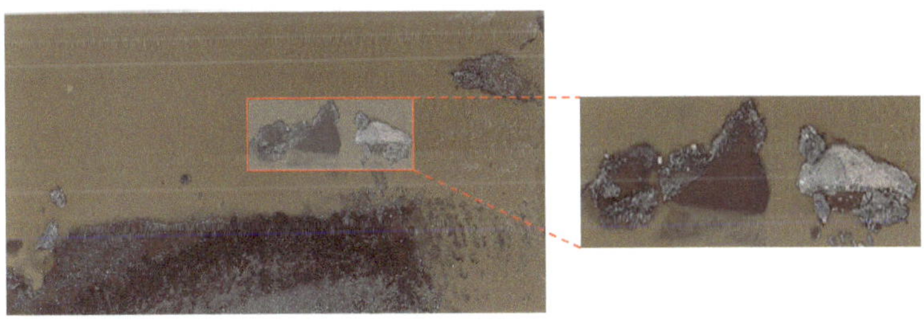

(e)

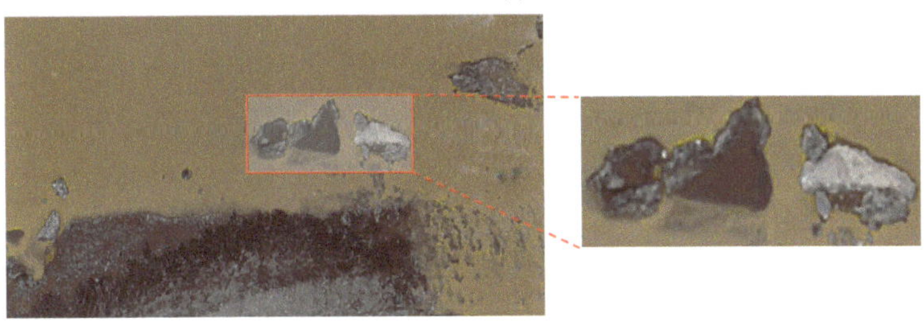

(f)

Figure 11. The figure shows digital orthophoto of scenes "houses", "bridges", and "rivers" generated using two different methods. Images (**a**,**c**,**e**) were generated using the explicit method (TDM), while images (**b**,**d**,**f**) are generated using the implicit method (Instant NGP).

Combining qualitative and quantitative analysis, it can be concluded that the TDM algorithm exhibits superior imaging clarity but demonstrates inaccuracies in areas experiencing sudden height variations. Conversely, Instant NGP is capable of capturing the majority of the scene's structure accurately, yet its imaging clarity is constrained by the scale of the model and may produce ghostly artifacts. Both methods are capable of generating usable digital orthophoto.

Table 1. Quality assessment of images generated using two methods in different scenes and comparisons with the real image. The ↓ means lower is better.

Scenes Method | Metric	Houses		Bridges		River	
	Brisque↓	**NIQE↓**	**Brisque↓**	**NIQE↓**	**Brisque↓**	**NIQE↓**
TDM (cuda)	12.96	2.77	7.88	2.33	12.90	5.01
Instant NGP	50.93	5.47	60.26	7.43	23.66	3.99
Real Images	6.72	1.67	5.91	1.72	7.74	1.74

4.2. Evaluation of Accuracy

An important characteristic of digital orthophotos is map geometric accuracy, so the accuracy of distance measurements is crucial. To validate the measurement accuracy of different digital orthophoto generation methods, we selected a specific area within the city for subsequent testing scenes. We utilized explicit methods (TDM), implicit methods (Instant NGP), and commercial software (Pix4D) to generate digital orthophoto, followed by comparing length measurements, as shown in Figure 12. The box plot displays the differences in distance measurements in digital orthophotos, as can be seen from Figure 13. The median of the box plot generated from Pix4D-to-TDM is 0.0376 m, while the other median from Pix4D-to-Instant NGP is 0.0442 m, both around 0.04 m. In comparison with Pix4D, this study concludes that both the explicit method (TDM) and the implicit method (Instant NGP) for digital orthophoto generation meet the requirements for mapping purposes.

(a)

Figure 12. *Cont.*

(**b**)

(**c**)

Figure 12. Digital orthophotos generated by TDM, Instant NGP and Pix4D.The segments with consistent colors and corresponding values represent identical distances measured across the three results. (**a**) Distance measurement of the explicit method (TDM), (**b**) Distance measurement of the implicit method (Instant NGP), (**c**) distance measurement of Pix4D.

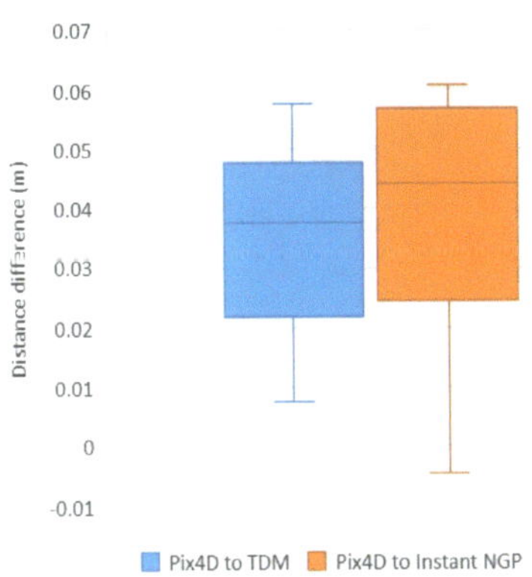

Figure 13. The box plot shows the differences in distance measurements between the explicit method (TDM) and the implicit method (Instant NGP) compared to Pix4D in the same scene, as depicted in Figure 12.

Furthermore, because the brightness of the color difference map can represent the degree of difference between digital orthophoto generated by different algorithms at the same location, in order to further measure the accuracy for digital orthophoto generation, this paper establishes the color difference maps between those methods. As shown in Figure 14, the color difference map shows that explicit methods (TDM) and implicit methods (Instant NGP) produced the same measurability and visibility of the generated digital orthophoto as those generated by the commercial software (Pix4D). In general, the accuracy of the two methods is acceptable according to the comparison with commercial software (Pix4D).

(a)

(b)

Figure 14. Color difference map between each of the two results. (**a**) Color difference map between Pix4D and TDM. (**b**) Color difference map between Pix4D and Instant NGP.

4.3. Evaluation of Efficiency

To verify the generation efficiency between explicit and implicit methods, in this section, we conducted tests on the generation time of digital orthophotos in five different size scenes. These two types of methods were run on a personal computer with an Intel (R) Core (TM) i7-12700 CPU @ 4.90 GHz and an NVIDIA GeForce RTX 3090.

Table 2 illustrates the time consumption for digital orthophoto generation using TDM and Instant NGP at different scene sizes. For TDM, the time measurement ranges from obtaining sparse reconstruction results to the generation of digital orthophotos. For Instant NGP, it starts from acquiring sparse reconstruction results, proceeds through model training, and culminates in rendering digital orthophoto. Across five different scene sizes, the TDM algorithm exhibits superior speed performance compared to Instant NGP, with its runtime reduced by two orders of magnitude. Therefore, the explicit method currently holds a significant advantage over the implicit method in terms of efficiency in digital orthophoto generation.

Table 2. Efficiency comparison of three methods of various scene sizes.

Scene Size (m) @Images	Method	
	TDM	**Instant NGP**
150 × 150 @ 78	36 s	10,243 s
200 × 200 @ 130	60 s	16,931 s
250 × 250 @ 256	88 s	33,210 s
300 × 250 @ 281	103 s	36,454 s
300 × 300 @ 333	129 s	43,576 s

5. Conclusions

In this paper, we categorized the methods for digital orthophoto generation into explicit and implicit methods, exploring the potential of using NeRF for implicit digital orthophoto generation. We selected the most representative fast algorithms from the two categories: the TDM algorithm and Instant NGP. Additionally, we adapted and optimized TDM algorithm to a CUDA version, significantly enhancing the efficiency of digital orthophoto generation.

In both explicit and implicit methods, an initial step involves sparse reconstruction to obtain camera poses, point clouds, and other prior information. The former employs an elevation propagation process that explicitly integrates the local color consistency of images with multi-view geometry theories to acquire scene elevation information and corresponding textures. Conversely, in NeRF, the loss function is designed as the color difference between rendered and real images. Throughout the training process, the neural network gradually fits into the real scene, implicitly capturing the surfaces and textures of scene objects and synthesizing novel view images through differentiable rendering. Finally, both methods complete the entire process to generate digital orthophoto.

We conducted tests on explicit and implicit methods for digital orthophoto generation in various scenes, measuring the generation efficiency and result quality. We employed the commercial software Pix4D as a standard for assessing measurement accuracy and reliability, evaluating both methods. The results indicate that currently, explicit methods exhibit higher efficiency and lower computational resource requirements in generation compared to implicit methods, achieving results with respective advantages and disadvantages. Moreover, both methods meet the requirements for measurement accuracy. In our future work, we aim to further explore the development of implicit methods for digital orthophoto generation, accelerating the generation speed and enhancing the clarity of implicit methods by adding more constraints suitable for digital orthophoto generation.

Author Contributions: Conceptualization of this study, Methodology, Algorithm implementation, Experiment, Writing—Original draft preparation: J.L. and W.D.; Methodology, Supervision of this study, Data curation: G.J.; Algorithm implementation: Z.Z. All authors have read and agreed to the published version of the manuscript.

Funding: This research received no external funding.

Data Availability Statement: Data are contained within the article.

Conflicts of Interest: The authors declare no conflicts of interest.

References

1. Liu, Y.; Zheng, X.; Ai, G.; Zhang, Y.; Zuo, Y. Generating a high-precision true digital orthophoto map based on UAV images. *ISPRS Int. J. Geo-Inf.* **2018**, *7*, 333. [CrossRef]
2. Lin, T.Y.; Lin, H.L.; Hou, C.W. Research on the production of 3D image cadastral map. In Proceedings of the 2018 IEEE International Conference on Applied System Invention (ICASI), Tokyo, Japan, 13–17 April 2018; pp. 259–262.
3. Barazzetti, L.; Brumana, R.; Oreni, D.; Previtali, M.; Roncoroni, F. True-orthophoto generation from UAV images: Implementation of a combined photogrammetric and computer vision approach. *ISPRS Ann. Photogramm. Remote Sens. Spat. Inf. Sci.* **2014**, *2*, 57–63. [CrossRef]
4. Wang, Q.; Yan, L.; Sun, Y.; Cui, X.; Mortimer, H.; Li, Y. True orthophoto generation using line segment matches. *Photogramm. Rec.* **2018**, *33*, 113–130. [CrossRef]
5. Mildenhall, B.; Srinivasan, P.P.; Tancik, M.; Barron, J.T.; Ramamoorthi, R.; Ng, R. Nerf: Representing scenes as neural radiance fields for view synthesis. *Commun. ACM* **2021**, *65*, 99–106. [CrossRef]
6. Zhao, Z.; Jiang, G.; Li, Y. A Novel Method for Digital Orthophoto Generation from Top View Constrained Dense Matching. *Remote Sens.* **2022**, *15*, 177. [CrossRef]
7. Müller, T.; Evans, A.; Schied, C.; Keller, A. Instant neural graphics primitives with a multiresolution hash encoding. *ACM Trans. Graph. (ToG)* **2022**, *41*, 1–15. [CrossRef]
8. DeWitt, B.A.; Wolf, P.R. *Elements of Photogrammetry (with Applications in GIS)*; McGraw-Hill Higher Education: New York, NY, USA, 2000.
9. Schonberger, J.L.; Frahm, J.M. Structure-from-motion revisited. In Proceedings of the IEEE Conference on Computer Vision and Pattern Recognition, Las Vegas, NV, USA, 27–30 June 2016; pp. 4104–4113.
10. Shen, S. Accurate multiple view 3D reconstruction using patch-based stereo for large-scale scenes. *IEEE Trans. Image Process.* **2013**, *22*, 1901–1914. [CrossRef]
11. Fang, K.; Zhang, J.; Tang, H.; Hu, X.; Yuan, H.; Wang, X.; An, P.; Ding, B. A quick and low-cost smartphone photogrammetry method for obtaining 3D particle size and shape. *Eng. Geol.* **2023**, *322*, 107170. [CrossRef]
12. Tavani, S.; Granado, P.; Riccardi, U.; Seers, T.; Corradetti, A. Terrestrial SfM-MVS photogrammetry from smartphone sensors. *Geomorphology* **2020**, *367*, 107318. [CrossRef]
13. Li, Z.; Wegner, J.D.; Lucchi, A. Topological map extraction from overhead images. In Proceedings of the IEEE/CVF International Conference on Computer Vision, Seoul, Republic of Korea, 27 October–2 November 2019; pp. 1715–1724.
14. Lin, Y.C.; Zhou, T.; Wang, T.; Crawford, M.; Habib, A. New orthophoto generation strategies from UAV and ground remote sensing platforms for high-throughput phenotyping. *Remote Sens.* **2021**, *13*, 860. [CrossRef]
15. Zhao, Y.; Cheng, Y.; Zhang, X.; Xu, S.; Bu, S.; Jiang, H.; Han, P.; Li, K.; Wan, G. Real-time orthophoto mosaicing on mobile devices for sequential aerial images with low overlap. *Remote Sens.* **2020**, *12*, 3739. [CrossRef]
16. Hood, J.; Ladner, L.; Champion, R. Image processing techniques for digital orthophotoquad production. *Photogramm. Eng. Remote Sens.* **1989**, *55*, 1323–1329.
17. Fu, J. DOM generation from aerial images based on airborne position and orientation system. In Proceedings of the 2010 6th International Conference on Wireless Communications Networking and Mobile Computing (WiCOM), Chengdu, China, 23–25 September 2010; pp. 1–4.
18. Sun, C.; Sun, M.; Chen, H.T. Direct voxel grid optimization: Super-fast convergence for radiance fields reconstruction. In Proceedings of the IEEE/CVF Conference on Computer Vision and Pattern Recognition, New Orleans, LA, USA, 18–24 June 2022; pp. 5459–5469.
19. Fridovich-Keil, S.; Yu, A.; Tancik, M.; Chen, Q.; Recht, B.; Kanazawa, A. Plenoxels: Radiance fields without neural networks. In Proceedings of the IEEE/CVF Conference on Computer Vision and Pattern Recognition, New Orleans, LA, USA, 18–24 June 2022; pp. 5501–5510.
20. Chen, A.; Xu, Z.; Geiger, A.; Yu, J.; Su, H. Tensorf: Tensorial radiance fields. In Proceedings of the European Conference on Computer Vision, Tel Aviv, Israel, 23–27 October 2022; Springer: Berlin/Heidelberg, Germany, 2022; pp. 333–350.
21. Fridovich-Keil, S.; Meanti, G.; Warburg, F.R.; Recht, B.; Kanazawa, A. K-planes: Explicit radiance fields in space, time, and appearance. In Proceedings of the IEEE/CVF Conference on Computer Vision and Pattern Recognition, Vancouver, BC, Canada, 17–24 June 2023; pp. 12479–12488.
22. Hu, W.; Wang, Y.; Ma, L.; Yang, B.; Gao, L.; Liu, X.; Ma, Y. Tri-miprf: Tri-mip representation for efficient anti-aliasing neural radiance fields. In Proceedings of the IEEE/CVF International Conference on Computer Vision, Paris, France, 2–3 October 2023; pp. 19774–19783.
23. Xu, Q.; Xu, Z.; Philip, J.; Bi, S.; Shu, Z.; Sunkavalli, K.; Neumann, U. Point-nerf: Point-based neural radiance fields. In Proceedings of the IEEE/CVF Conference on Computer Vision and Pattern Recognition, New Orleans, LA, USA, 18–24 June 2022; pp. 5438–5448.
24. Yu, A.; Li, R.; Tancik, M.; Li, H.; Ng, R.; Kanazawa, A. Plenoctrees for real-time rendering of neural radiance fields. In Proceedings of the IEEE/CVF International Conference on Computer Vision, Montreal, BC, Canada, 11–17 October 2021; pp. 5752–5761.
25. Kulhanek, J.; Sattler, T. Tetra-NeRF: Representing Neural Radiance Fields Using Tetrahedra. *arXiv* **2023**, arXiv:2304.09987.

26. Barron, J.T.; Mildenhall, B.; Tancik, M.; Hedman, P.; Martin-Brualla, R.; Srinivasan, P.P. Mip-nerf: A multiscale representation for anti-aliasing neural radiance fields. In Proceedings of the IEEE/CVF International Conference on Computer Vision, Montreal, BC, Canada, 11–17 October 2021; pp. 5855–5864.
27. Barron, J.T.; Mildenhall, B.; Verbin, D.; Srinivasan, P.P.; Hedman, P. Mip-nerf 360: Unbounded anti-aliased neural radiance fields. In Proceedings of the IEEE/CVF Conference on Computer Vision and Pattern Recognition, New Orleans, LA, USA, 18–24 June 2022; pp. 5470–5479.
28. Li, R.; Tancik, M.; Kanazawa, A. Nerfacc: A general nerf acceleration toolbox. *arXiv* **2022**, arXiv:2210.04847.
29. Zhang, K.; Riegler, G.; Snavely, N.; Koltun, V. Nerf++: Analyzing and improving neural radiance fields. *arXiv* **2020**, arXiv:2010.07492.
30. Kerbl, B.; Kopanas, G.; Leimkühler, T.; Drettakis, G. 3D gaussian splatting for real-time radiance field rendering. *ACM Trans. Graph. (ToG)* **2023**, *42*, 1–14. [CrossRef]
31. Laine, S.; Hellsten, J.; Karras, T.; Seol, Y.; Lehtinen, J.; Aila, T. Modular Primitives for High-Performance Differentiable Rendering. *ACM Trans. Graph.* **2020**, *39*, 1–14. [CrossRef]
32. Roessle, B.; Barron, J.T.; Mildenhall, B.; Srinivasan, P.P.; Nießner, M. Dense depth priors for neural radiance fields from sparse input views. In Proceedings of the IEEE/CVF Conference on Computer Vision and Pattern Recognition, New Orleans, LA, USA, 18–24 June 2022; pp. 12892–12901.
33. Deng, K.; Liu, A.; Zhu, J.Y.; Ramanan, D. Depth-supervised nerf: Fewer views and faster training for free. In Proceedings of the IEEE/CVF Conference on Computer Vision and Pattern Recognition, New Orleans, LA, USA, 18–24 June 2022; pp. 12882–12891.
34. Mittal, A.; Moorthy, A.K.; Bovik, A.C. No-Reference Image Quality Assessment in the Spatial Domain. *IEEE Trans. Image Process.* **2012**, *21*, 4695–4708. [CrossRef]
35. Mittal, A.; Soundararajan, R.; Bovik, A.C. Making a "Completely Blind" Image Quality Analyzer. *IEEE Signal Process. Lett.* **2013**, *20*, 209–212. [CrossRef]

remote sensing

MDPI

Article

On-Site Stability Assessment of Rubble Mound Breakwaters Using Unmanned Aerial Vehicle-Based Photogrammetry and Random Sample Consensus

Marcos Arza-García [1,2,*], José Alberto Gonçalves [2], Vladimiro Ferreira Pinto [2] and Guillermo Bastos [1]

[1] CIGEO—Civil & Geomatics Research Group, Higher Polytechnic School of Engineering (EPSE), University of Santiago de Compostela, 27002 Lugo, Spain; inardesign.gbastos@uvigo.es

[2] CICGE—Geospatial Sciences Research Centre, Faculty of Sciences (FCUP), University of Porto, 4169-007 Porto, Portugal; jagoncal@fc.up.pt (J.A.G.); geral@topogonal.com (V.F.P.)

* Correspondence: m.arza@usc.es

Abstract: Traditional methods for assessing the stability of rubble mound breakwaters (RMBs) often rely on 2.5D data, which may fall short in capturing intricate changes in the armor units, such as tilting and lateral shifts. Achieving a detailed analysis of RMB geometry typically requires fully 3D methods, but these often hinge on expensive acquisition technologies like terrestrial laser scanning (TLS) or airborne light detection and ranging (LiDAR). This article introduces an innovative approach to evaluate the structural stability of RMBs by integrating UAV-based photogrammetry and the random sample consensus (RANSAC) algorithm. The RANSAC algorithm proves to be an efficient and scalable tool for extracting primitives from point clouds (PCs), effectively addressing challenges presented by outliers and data noise in photogrammetric PCs. Photogrammetric PCs of the RMB, generated using Structure-from-Motion and MultiView Stereo (SfM-MVS) from both pre- and post-storm flights, were subjected to the RANSAC algorithm for plane extraction and segmentation. Subsequently, a spatial proximity criterion was employed to match cuboids between the two time periods. The methodology was validated on the detached breakwater of Cabedelo do Douro in Porto, Portugal, with a specific focus on potential rotations or tilting of Antifer cubes within the protective layer. The results, assessing the effects of the Leslie storm in 2018, demonstrate the potential of our approach in identifying and quantifying structural changes in RMBs.

Keywords: drone; RMB; groins; in-field inspection; photogrammetry; SfM-MVS; random sample consensus

Citation: Arza-García, M.; Gonçalves, J.A.; Ferreira Pinto, V.; Bastos, G. On-Site Stability Assessment of Rubble Mound Breakwaters Using Unmanned Aerial Vehicle-Based Photogrammetry and Random Sample Consensus. *Remote Sens.* **2024**, *16*, 331. https://doi.org/10.3390/rs16020331

Academic Editors: Wanshou Jiang, San Jiang, Duojie Weng and Jianchen Liu

Received: 3 November 2023
Revised: 10 January 2024
Accepted: 11 January 2024
Published: 14 January 2024

1. Introduction

The protection of coastal zones and harbors from wave damage is crucial to prevent severe economic and ecological consequences. One of the most commonly employed structures for this purpose are the rubble-mound breakwaters (RMBs). These structures are built using various materials and can be adapted to diverse underwater topographies, designed to withstand different wave conditions [1]. However, these protective structures are susceptible to damage and require repairs throughout their lifespan [2]. Therefore, understanding the performance of rubble-mound armor in terms of hydraulic stability is essential for designing new structures and upgrading existing ones, particularly in light of climate change effects such as sea level rise and increased wave storminess [3,4].

Detecting local defects in coastal defense structures, like displacements, breakage, or removals of concrete armor units (CAUs), is crucial to prevent potential threats to the safety of breakwaters. Damages observed in these structures can lead to sliding, settlement, or toppling, causing the displacement, breakage, or removal of CAUs. Additionally, scouring at dike foundations may occur and result in severe damages under extreme wave forces from storms. Identifying local shifts in the elements of these structures would be beneficial for studying their performance and mitigating damages caused by potential defects [5].

Techniques for assessing and measuring lab-scale physical models, with a focus on examining the behavior of RMBs under wave action, have advanced significantly in recent years. Laboratory investigations exploring this subject employ a range of scanning devices, including structured light scanners, infrared scanners, and laser scanners, along with conventional profilers [6]. Some studies have also integrated devices with additional depth measurement capabilities, such as time-of-flight (ToF) and RGB-D cameras (e.g., Kinect) [7–10], providing the advantage of collecting information from the submerged portions of RMB models. Image-based methodologies are also increasingly utilized [11], with a specific subset of studies concentrating on photogrammetric reconstruction [12,13] to achieve a detailed three-dimensional representation of the slopes.

However, unlike laboratory testing of physical models, the field of on-site monitoring of RMBs still has ample room for further advancement in both research and the development of assessment methodologies. Currently, the evaluation of the current maintenance conditions in RMBs and groins typically relies heavily on visual inspections to assess structural damage. Whether conducted during routine monitoring campaigns or post-storm assessments, these inspections provide essential insights for improving breakwater design and maintenance, ensuring their long-term effectiveness. Nevertheless, developing standardized and more efficient methodologies for the in-field evaluation of RMB damage remains a challenging task due to the significant variability in construction from site to site, as well as associated costs and safety considerations.

1.1. SfM-MVS Photogrammetry in RMB Inspection

Terrestrial photogrammetry [14] and terrestrial laser scanning (TLS) [5] have proven to be efficient techniques for examining changes in small dikes or RMBs using point clouds (PCs) derived from different epochs. Nevertheless, the advent of uncrewed aerial vehicles (UAVs) has revolutionized coastal monitoring, offering a cost-effective, flexible, and high-resolution approach to data collection across large areas [15]. While UAVs can be equipped with various sensors, such as UAV-borne LIDAR, affordable and lightweight RGB cameras have become the standard for remote sensing and photogrammetric research. In this context, the photogrammetric applications of these tools in the field are diverse, encompassing tasks such as investigating near-shore hydrodynamics [16], mapping and quantifying volumetric changes on beaches [17], and inspecting offshore civil infrastructures [18]. The integration of UAVs into such tasks not only establishes a robust toolkit for detailed photogrammetric reconstructions and analyses but also introduces real-time monitoring capabilities, particularly crucial after severe events [19]. UAV-based photogrammetry has also proven to be a useful reverse engineering technique, providing data on actual morphologies that can be translated into numerical analyses in different applications (e.g., flooding risk assessment [20], slope stability analysis [21], erosion and accretion studies [22], etc.). However, it is important to acknowledge that the use of UAV-based photogrammetry in water-related studies is not without challenges. Addressing concerns such as limitations in flight time, payload capacity considerations, legal issues, drone security, and varying data acquisition conditions remains critical in some applications [15].

Specific applications of UAV-based surveys in rubble mound groins can be found in previous research, such as in the work of Henriques et al., 2017 [23], which generated photogrammetric orthomosaics and PCs to obtain data about the most exterior protection layer of breakwaters. Gonçalves et al. 2022a and 2022b [24,25] expanded the photogrammetric workflow by incorporating UAV-based real-time kinematic (RTK) data to accurately map the geometry of rubble mound groins. They also conducted an accuracy assessment using independent techniques (i.e., GNSS and TLS). These previous studies demonstrate the potential of UAV-based photogrammetry in monitoring the structural integrity by generating three-dimensional (3D) geometric reconstructions of RMBs, achieving accuracies in some cases better than 3 cm of error in checkpoints. In this regard, UAV-based photogrammetry has proven to be a highly suitable technique to obtain detailed and precise 3D reconstructions, particularly advantageous when dealing with large, complex, and potentially hazardous structures like these. However, to the best of our knowledge there is still insufficient research on multi-temporal monitoring of RMBs using UAV-based photogrammetry. More specifically, there is a clear shortage of studies exploring possible methods of automatic change detection.

1.2. Change Detection Analysis in RMBs

Change detection poses a critical challenge in various remote sensing applications Historically, studies within this domain have relied on 2D information from remote sensing images to address large-scale issues, such as forest monitoring or urban sprawl. Previous research has dedicated significant efforts to developing new methods for detecting changes from images, starting with traditional/classical pixel-based methods that primarily focus on spectral values [26]. More recently, methods in geographic object-based image analysis (GEOBIA) have emerged [27,28], introducing innovative segmentation and classification techniques that consider spatial context along with spectral, topographical, textural, and morphological properties. The emergence of new detectors and feature descriptors has gone beyond the limitations of traditional top-view 2D pixel/object-based analyses [29,30], playing a pivotal role in applications like security and surveillance, infrastructure monitoring, or precision agriculture [31]. However, as image resolution advances to finer levels, several challenges arise when employing 2D image-based methods. Issues like spectral variability and perspective distortion become prominent. In response to these challenges, the incorporation of 3D data in finer-scale studies introduces a different modality for analysis, enabling highly detailed geometric analysis [32].

Among the common techniques used to identify changes from 3D datasets acquired at different time intervals, cross-sectional assessment or the simple comparison of digital surface models (DSM), also known as the DEM of Difference (DoD) method [33], are frequently employed. However, these techniques still predominantly rely on 2.5D information as they primarily operate within a 2D spatial framework, despite considering the elevation or height of objects [34]. In contrast, the damage progression along a sloping coastal structure like a rubble-mound breakwater (RMB) is fundamentally a 3D process. It is crucial to recognize that objects may undergo vertical shifts, rotations, or tilting, emphasizing the need for approaches that can capture and analyze changes in the full 3D spatial context.

On the other hand, methods lacking full 3D spatial information often struggle to differentiate individual armor units, reducing possibilities for subsequent analysis. While the simpler DoD approach can be useful for estimating erosion volume in the breakwater, a more detailed assessment can be carried out at the individual block level. This approach yields more precise statistics and provides a more reliable count of the displaced armor units [6].

Earlier investigations aimed to refine methods for estimating poses of individual blocks, though there is a relative absence of applications on dense PCs obtained directly from on-site photogrammetric surveys. In a study conducted by Puente et al. in 2014 [5], changes in RMBs were examined using TLS PCs from different time periods. To estimate the rigid body transformation parameters, they employed K-means clustering to identify planar segments representing the faces of the cuboids. Bueno Esposito et al., in 2015 [35], presented an approach for reconstructing wave-dissipating blocks from incomplete PCs of RMBs captured by airborne LiDAR. Their method used segmentation based on normal vectors and prior knowledge about the properties of the cuboids to refine the segmentation and define the boundaries of individual armor units. Xu et al., in 2022 [36], presented a deep-learning-based approach for block pose identification that could even identify CAUs with complex shapes, such as tetrapods or clinger blocks. However, this method entails the need for extensive training datasets to feed the convolutional neural network and substantial computational resources, which may present implementation difficulties and necessitate site-specific fine-tuning.

Although not specifically applied to RMBs, Shen et al., in 2018 [37], presented a methodology to extract individual brick poses from a laser scan PC of a cluttered pile of cuboid bricks. Their proposed workflow includes connected component analysis, principal component analysis, and a voting scheme to reconstruct bricks individually. Shen, Wang, and Puente in 2020 [38] proposed a method for detecting changes in masonry walls using TLS PCs with a regular distribution of bricks, a case study analogous to cube-armored breakwaters with a regular placement pattern. They utilized the TLS intensity attribute to differentiate between materials of mortar and bricks, followed by a 3D connected components algorithm to extract and label individual bricks.

1.3. The RANSAC Approach

While there are multiple valid strategies in the statistics field for determining block pose through surface extraction, several of the reviewed solutions may suffer from practical limitations, such as computational intensity, implementation complexity, and sensitivity to data noise [39]. For instance, the M-estimator, L-estimator, R-estimator [40], and Least Median of Squares (LMedS) [41] methods approached regression with outliers as a minimization problem, akin to the least square method that minimizes the sum of squared error values. However, they employed nonlinear and intricate loss functions instead of the square of the error. LMedS aimed to minimize the median of the error values, requiring a numerical optimization algorithm to solve such nonlinear minimization problems. The 3D Hough [42] method transforms spatial data (e.g., 3D points corresponding to a plane) from the 3D data space into a parameter space (e.g., normal vector components and distance from the origin). The most prevalent point in the parameter space is identified as its estimation, demanding a significant amount of memory to represent the parameter space. As stated before, deep learning methods, such as convolutional neural networks (CNNs) for 3D pose estimation [43], have also gained popularity. However, they may face challenges such as high computational resource requirements, the need for large training datasets, and complexity in adapting to different scenarios.

In contrast to the aforementioned methods, the random sample consensus (RANSAC) algorithm [44] simplifies the process into two steps: generating a hypothesis from random samples and verifying it against the data. This approach eliminates the need for complex optimization algorithms and large memory allocations. In that sense, RANSAC can robustly work in a wide range of applications and with several sources of data (e.g., TLS, photogrammetry), even if these data include more than 50% of outliers [45]. Besides its enhanced computational efficiency, RANSAC presents another important advantage in its scalability concerning the size of the input PC and the number and size of the shapes within the data.

This algorithm has already been validated in other applications, such as the automatic extraction of building elements (e.g., roof planes and walls) [46,47], structural planes of rocky slopes [48], water-level planes [49], etc. In these studies, the application of RANSAC has demonstrated efficient performance, even in photogrammetric PCs, which are typically noisier than those obtained through TLS or LiDAR.

This article introduces a novel methodology for monitoring the structural stability of wave-dissipating cuboids of RMBs using UAV-based photogrammetric surveys and RANSAC-based segmentation. To assess the practicality and performance of this approach, we conduct a case study application on a detached breakwater that has experienced damage due to a severe maritime storm. This case study aims to evaluate the effectiveness of the proposed RANSAC-based approach in comparison to traditional methods such as the Difference of Digital Surface Models (or DoD), particularly in the context of detecting and quantifying changes like tilting in individual armor units. The methodology enables the generation of quantitative insights into the extent of damage and the overall structural integrity of the RMB, facilitating the conduction of a zonal stability analysis for this breakwater.

2. Materials and Methods

2.1. Study Site

Located at the mouth of the Douro River in Porto, Portugal, the Cabedelo do Douro area (41°08′N, 8°40′W) has a detached breakwater designed for shore protection due to the significant wave energy in this coastal zone. The RMB plays a crucial role in shielding the Douro River estuary from the Atlantic waves. Its strategic location reinstates the protective function of the sand spit, ensuring the safety of ships and boats navigating through the area [50].

The RMB was constructed with a curved shape, spanning approximately 450 m in the southeast to northwest direction, and its concavity faces the land (Figure 1). The relatively low crest elevation, standing at +6.0 m above mean sea level (AMSL), minimizes its visual impact on the landscape. The structure comprises a rockfill core, overlaid by secondary layers of granite blocks, featuring filter functions and an outer protective layer. This protective layer consists of high-density concrete grooved cuboids (Antifer type) weighing 8 kN, initially arranged in a regular placement pattern.

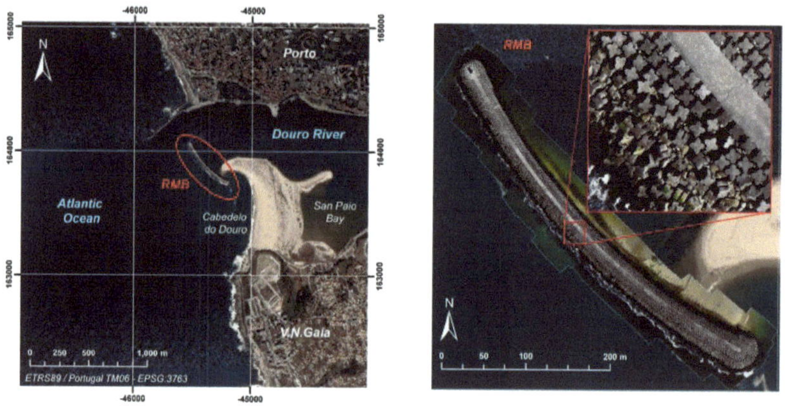

(a) (b)

Figure 1. Location of the structure within the mouth of Douro River (**a**), and (**b**) detailed view of the RMB.

2.2. Field Campaigns

Two flight campaigns were conducted on 12 September and 27 November 2018, with the aim of capturing potential displacements of the armor units. While the interval between the dates may appear short for detecting significant displacements, this period allows for the analysis of the impact of Hurricane Leslie (13–14 October 2018) on the structure. The hurricane, also known as Leslie storm in Spain and Portugal once in the extratropical category, was the most powerful cyclone to reach the Iberian Peninsula since 1842 and one of the longest-lasting Atlantic hurricanes over time. In this sense, the test field provides an excellent environment for validating the methods and detecting potential movements in CAUs.

The aerial images were captured using a UAV Phantom 4 Pro v.2 equipped with a built-in camera (Table 1). All flights were planned using Pix4DCapture (Pix4D, Lausanne, Switzerland) v.4.2.0 following a grid pattern along the breakwater and the adjacent coast, capturing overlapping images (Table 2). The flight speed was set to an intermediate value in the Pix4D app, which, after calculations, resulted in approximately 2.3 m/s.

Table 1. Specifications of UAV Phantom 4 Pro quadcopter.

Weight	1388 g
Max Wind Speed Resistance	10 m/s
Max Flight Time	Approx. 30 min
GNSS Positioning	GPS/GLONASS
Hover Accuracy Range	Vertical: 0.5 m (GPS positioning) Horizontal: 1.5 m (GPS positioning)
Camera resolution	20 megapixels
Sensor size	1-inch CMOS

Table 2. Flight planning parameters.

Altitude for Mapping Mission	30 m
Frontlap	80%
Sidelap	60%
Ground Sample Distance	<1 cm
Speed of Flight	2.3 m/s
Mission Area	2.33 ha

The required level of detail is commonly associated with the concept of ground sampling distance (GSD), which represents the real-world size of an element represented by a single pixel. The GSD can be calculated based on the focal length (f), shooting distance (d), and pixel size (p), as shown in Equation (1) [51]. According to that equation and the camera specifications, the flight altitude was set at 30 m above the ground level to obtain images with GSD values less than 1 cm.

$$\mathrm{GSD} = \frac{d}{f} \cdot p \quad (1)$$

The field operations involving the marking and measurement of ground control points (GCPs) and checkpoints (CPs) were carried out on the same day, immediately preceding each flight. The points were marked on the ground using paint. For the georeferencing of each point, three readings were recorded, and an average was calculated. This process was conducted in real-time kinematic (RTK) mode using double-frequency GNSS equipment with centimetric precision (Leica GNSS Smart Rover 1200). Differential corrections were obtained from the Portuguese DGT's ReNEP reference stations. Topologically, the scene's geometry is a linear acquisition, and such image distribution tends to produce the bending or "dome" effect in photogrammetry. To mitigate this effect, a total of 48 points, forming 7 groups/rows distributed along the central corridor, were selected and measured as illustrated in Figure 2. Subsequently, 8 of these points were chosen as CPs to validate accuracies.

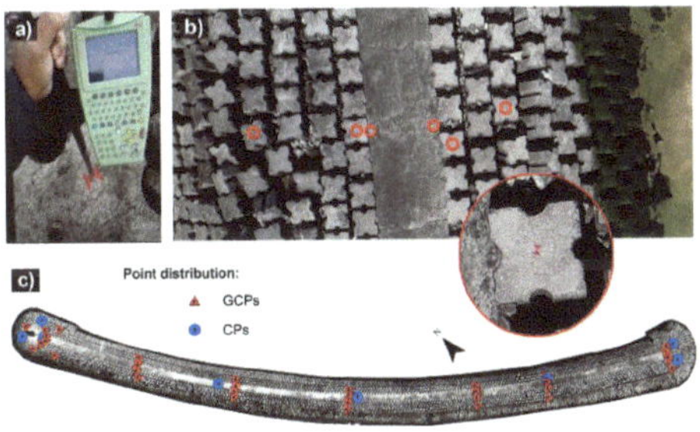

Figure 2. Ground control. (**a**) GNSS receiver and distribution of the GCPs and checkpoints along (**b**) rows and (**c**) the whole RMB.

2.3. Flowchart of the Process

The methodological flow of this study, depicted in Figure 3, is based on the conventional Structure-from-Motion and MultiView Stereo (SfM-MVS) photogrammetric pipeline. Subsequently, RANSAC is employed for plane extraction, as detailed in the following sections.

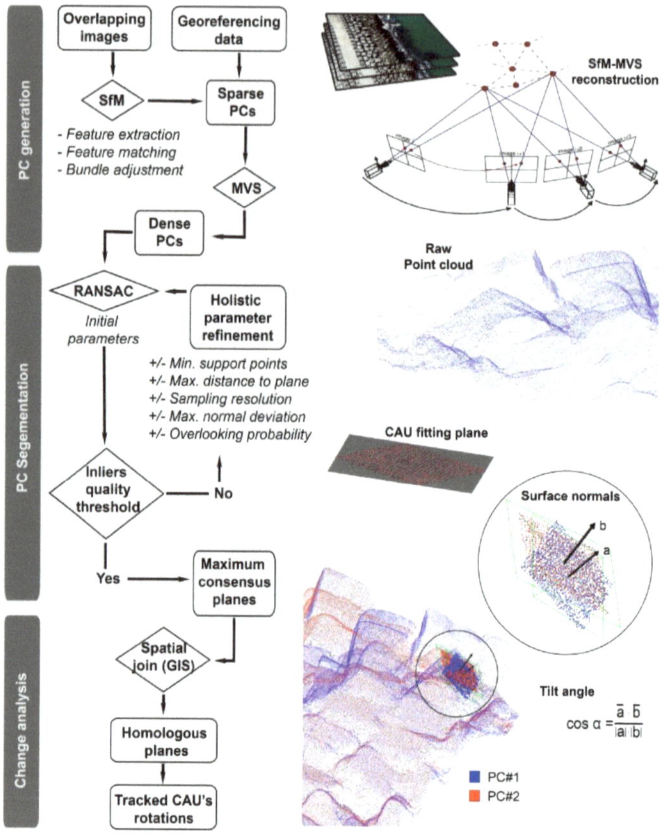

Figure 3. Workflow of the methodology implemented in this study.

2.4. Photogrammetric Reconstruction

SfM-MVS photogrammetry is recognized as a pivotal technique for reconstructing 3D scenes from a set of overlapping 2D images [52,53]. Image processing based on these algorithms involves a series of steps, collectively referred to as the photogrammetric workflow, which facilitates the generation of dense PCs. The core of the photogrammetric workflow lies in the SfM reconstruction (see Figure 3), commencing with (i) key feature extraction, where distinctive (key) points are identified in the input images. These features are extracted by the software, using techniques like Scale-Invariant Feature Transform (SIFT) or Speeded-Up Robust Features (SURF) and serve as the foundation for subsequent stages. Following feature extraction, (ii) a feature matching process is undertaken to establish correspondences between key points across different images. The matched points (namely, tie points) extracted from the images enable the determination of the initial camera positions and matched points in 3D space. These initial estimates are then refined using (iii) bundle adjustment, which iteratively adjusts the camera poses to minimize inconsistencies and enhance the overall accuracy of the reconstruction. By incorporating GCPs into bundle adjustment, external calibration sources are introduced, aligning the reconstruction with real-world coordinates. This alignment compensates for distortions from factors like lens aberrations and sensor imprecisions, thereby enhancing the overall reliability of the reconstruction.

Another component of the workflow is (iv) Multi-View Stereo (MVS), where the initial sparse PCs undergo further refinement and densification to generate a dense PC. Multi-view stereo algorithms, such as Semi-Global Matching (SGM) or PatchMatch Stereo, leverage the geometry and photometric information across multiple views to produce detailed and high-density PCs representing the scene geometry.

To process the aerial datasets of the RMB, we implemented the photogrammetric pipeline using Metashape software (Agisoft, St. Petersburg, Russia) v2.0.2 within a cloud-based infrastructure configured with 64vCPU, @2.3 GHz, 488 GB RAM, 4x Nvidia Tesla M60/32 GB. This setup ensures the computational power necessary for the efficient processing of the aerial datasets. Table 3 outlines the key photogrammetric processing parameters employed in Metashape.

Table 3. Photogrammetric processing parameters.

Image Alignment Method	Adaptive camera model
Alignment Accuracy	High (original image size)
Key Point Limit	50,000
Tie Point Limit	10,000
Depth Maps Quality	High
Filtering Mode	Aggressive

2.5. RANSAC-Based Segmentation

The RANSAC method is a robust algorithm commonly employed for model fitting and segmentation in PCs, enabling their partitioning into simple shapes such as planes, spheres, cylinders, cones, tori, etc. The algorithm operates by iteratively selecting a random subset of points from the input data and fitting a model to these points. The model is then evaluated by counting the number of inliers, which are points that align with the model within a certain threshold [45].

The objective was to use this algorithm to extract planar patches representing the upper faces of the Antifer cuboids of the RMB. Therefore, parameters corresponding to the mathematical model and termination conditions were defined before the iteration process, depending on the characteristics of the PCs. The regularity of the cuboids played an important role, allowing the fine-tuning of parameters based on the results until a certain level of correctness and completeness was achieved. These parameters include the minimum number of points required to form a plane and other thresholds for inlier selection, such as the maximum distance to the plane, the maximum angular deviation of the plane's normal, etc.

A consensus solution was obtained as the best result after k iterations, approximately determined as a function of the desired probability, according to the following equation [54]:

$$k = \frac{\log(1 - z)}{\log(1 - w^n)} \quad (2)$$

where z represents the minimal probability of success in finding at least one proper set of observations, w denotes the percentage probability of observations allowed to be incorrect, and n is the minimal number of points necessary for computing the model.

Once the best model has been identified (i.e., the one with the largest number of inliers), the corresponding consensus planes were extracted by selecting all the inliers consistent with the models. This process was executed on the two photogrammetric PCs using the RANSAC Shape Detection algorithm implemented in CloudCompare software (GNU GPL), v.2,13, with the same parameters, resulting in two segmented PCs.

2.6. RMB Change Analysis

The RANSAC algorithm functions as a surface extraction process, identifying planar segments representing the upper faces of each CAU one by one. The results are then exported as separate entities, with the detected planes having associated attributes, including coordinates defining their centers {Cx, Cy, Cz} and normal vectors {Nx, Ny, Nz}. However, at this stage, there is no direct plane-to-plane correspondence between the cuboids of PC#1 and their counterparts in PC#2. To establish this correspondence, we employed the criterion of proximity, utilizing a GIS tool called "spatial join". This tool assigns each entity with all attributes of the corresponding one in the layer being joined that is closest to it.

By comparing the resulting planes between the two datasets, it becomes possible to quantify the angular deviations or tilting that occurred over time at the individual cuboid level. These deviations were then analyzed in-depth to evaluate the structural changes or shifting within the breakwater.

3. Results and Discussion

3.1. Photogrammetric Reconstruction

The workflow outlined in Section 2.3 was applied to the two datasets obtained in their respective flight campaigns. All processing steps, as described in the preceding sections, were executed in ETRS89 (European Terrestrial Reference System 1989) with rectangular coordinates PTTM06 (Portugal Transverse Mercator 2006), EPSG: 3763. The orthometric height is referenced to the geoid model for mainland Portugal, GeodPT08 [55].

Both image orientation and the subsequent densification of the PC were performed within the automated pipeline of Metashape, selecting the "high" quality setting controls. With this option, the software operates with the original size of the photos, allowing for more detailed and accurate geometry, albeit at the cost of longer processing times. Table 4 summarizes the key characteristics of the photogrammetric processing for both time periods.

Although the flight planning files used in both flights were not exactly identical, the number of images and flight altitude remained reasonably consistent. This consistency is crucial when comparing data across multiple time periods, and whenever possible, equivalent parameters should be maintained, ideally by using the same waypoint file. This approach ensures that image resolutions, and consequently the resolutions of derived PCs, remain relatively uniform. Furthermore, employing the same technique for generating PCs and equivalent GCPs for georeferencing contributes to positional consistency in the resulting photogrammetric products. When PC sources are different, preprocessing steps are often required before applying any change detection algorithm [56]. In contrast, in this case, intermediate co-registration processes can be skipped, making PC data from different time periods directly comparable.

Table 4. Summary of photogrammetric processing results.

			Flight #1	Flight #2
		# of images	249	237
		Mean flight height (m)	30.9	29.6
		GSD (mm)	8.7	8.0
 PC#1		Key points	206,882	212,619
		Dense cloud size (points)	23,119,079	23,645,284
		X	3.8	6.5
	Residuals from GCPs (mm)	Y	6.1	17.8
		Z	7.6	9.7
 PC#2		X	7.2	6.8
	Accuracy from CPs (mm)	Y	8.9	11.4
		Z	6.6	14.3
		DEM resolution (mm/pix)	34.9	32.5

As a reference for further comparison with the proposed RANSAC-based method, we also generated the DoD map (Figure 4) by deriving the differences between the two DEMs, each with resolutions better than 3.5 cm/pix, as shown in Table 4. The DoD provides a straightforward representation of surface elevation changes, making it rather easy to detect and visualize areas experiencing severe alterations. However, even in these cases, obtaining a precise interpretation of the number of shifted blocks is challenging using this approach.

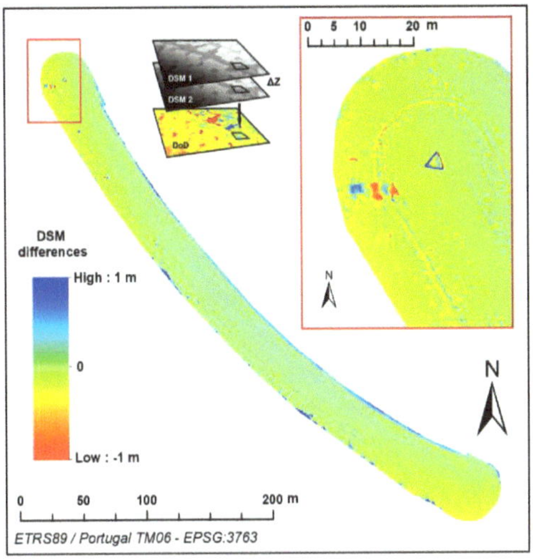

Figure 4. Global DoD. The detail view illustrates the NW head of the RMB showing some accreted and eroded areas that hint at the displacement of some CAUs.

3.2. RANSAC-Based Analysis

3.2.1. PC Segmentation

The PC segmentation process was implemented following the methodology described above to fit planes to the PC data (Figure 5). While applying the RANSAC algorithm, the largest planes in the original PC (Figure 5a), corresponding to the top concrete platform of the RMB, were also detected, as illustrated by the pink, orange, and green patches in Figure 5b. These planes were subsequently removed from the classified data, retaining only the planes representing the CAUs.

Additional challenges associated with the use of RANSAC are depicted in Figure 5c,d. In some instances, planes were fitted across the surfaces of multiple cuboids due to their proximity or similar elevations. This phenomenon is predominantly observed in the upper zone of the RMB, where the CAUs were initially placed level, and due to the stability of these areas, they remain mostly level. Moreover, the narrow gaps between neighboring armor units often go unsampled, consolidating several wave-dissipating block poses into a singular representation, as highlighted by previous studies utilizing alternative methodologies [57,58]. Conversely, there are cases wherein finding a suitable plane representing specific cuboids proves challenging. This occurs predominantly at the lower levels of the RMB, where the PC exhibits lower quality and increased noise due to degraded texture of the cuboids in these areas and the presence of water, algae, etc.

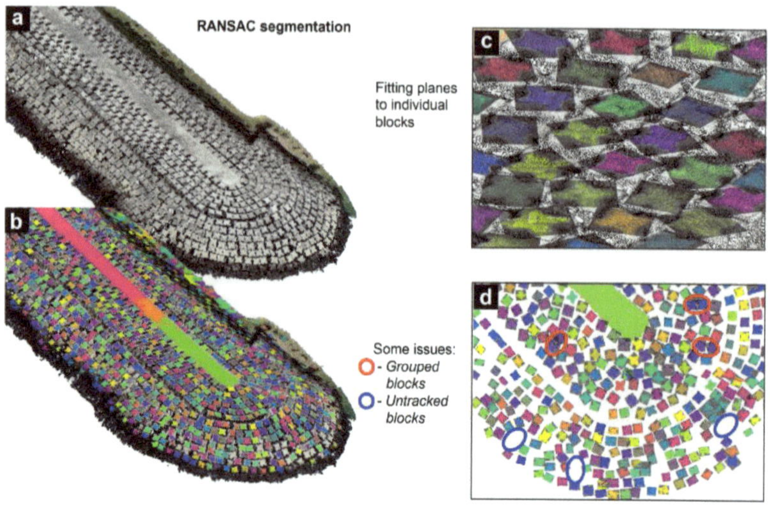

Figure 5. RANSAC segmentation: (**a**) RAW point cloud; (**b**,**c**) point cloud segmented into planes, and (**d**) examples of how some parameter settings produce issues affecting the precision of the segmentation. The colors of planes are assigned arbitrarily for differentiation purposes.

The process of determining appropriate fitting parameters has been carried out iteratively, involving trials with gradual refinement until reasonably satisfactory results were achieved in terms of meaningful interpretation and comprehensiveness. The best outcomes, based on these criteria, were obtained with a minimum support points per plane of 200 and a maximum distance to the fitting plane of 0.005 m. The maximum allowable deviation in the normal direction of the plane from the estimated normal was set to 5°. The overlooking probability value was set to 0.0001, aiming to work with a low probability of missing outliers during the RANSAC plane fitting process.

To evaluate the accuracy of the RANSAC results, we chose a representative sampling area in the southeast quadrant of the RMB, encompassing approximately 1037 CAUs, which accounts for roughly a quarter of the total number of armor units. To prevent the inclusion of flooded areas, cuboids situated at elevations lower than 0 m AMSL were excluded from the sampling. As illustrated in Figure 6, a manual sampling of this zone was performed to verify the correct classification of planes in both time periods.

Items classified as True Positives (TP) correspond to actual cuboids correctly modeled by a plane. False Positives (FP) refer to detected planes that do not precisely correspond to the top face of an individual wave-dissipating block. A significant portion of items falling into this category consists of planes fitted to the lateral faces of some cuboids. False Negatives represent actual CAUs that were not detected as planes by the RANSAC fitting, so they were manually added to account for their number. A single plane fitting two (or more) cuboids has been considered in terms of counting as two (or more) FNs. Lastly, the concept of True Negative (TN) is somewhat more abstract and includes non-cuboids correctly classified as such. As shown in Figure 6, this class includes manually added elements like large rocks within inter-block spaces, which the algorithm correctly identified as non-cuboids.

Table 5 shows the confusion matrix containing TP, TN, FP, and FN values. These values are components of the confusion matrix which defines actual and predicted classes.

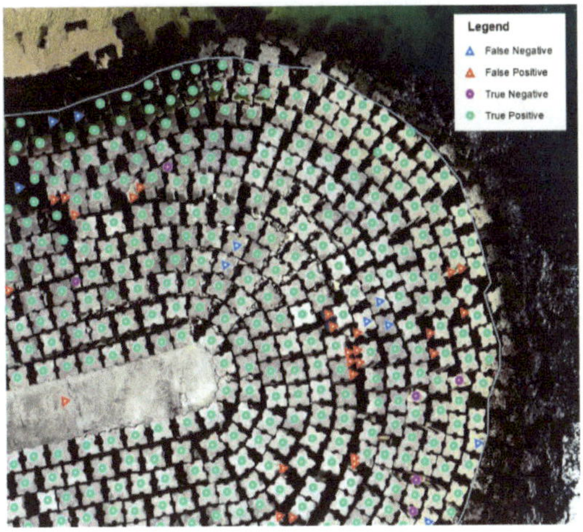

Figure 6. Post-classification accuracy assessment.

Table 5. Confusion matrix.

	Predicted Class					
Actual class	True positives		False negatives		Positive	
	PC#1	932	PC#1	41	PC#1	973
	PC#2	882	PC#2	68	PC#2	950
		1814		109		1923
	False positives		True negatives		Negative	
	PC#1	51	PC#1	12	PC#1	63
	PC#2	66	PC#2	21	PC#2	87
		117		33		150
	Positive		Negative			
	PC#1	983	PC#1	53		
	PC#2	948	PC#2	89		
		1931		142		

Sensitivity, specificity, precision, negative predictive value, and accuracy can be easily derived from the confusion matrix values, with the formulas mentioned in Table 6:

Table 6. Performance indicators based on the TP, FP, TN, and FN parameters [59].

$$Sensitivity = \frac{TP}{TP+FN} \quad (3)$$

$$Specificity = \frac{TN}{TN+FP} \quad (4)$$

$$Precision = \frac{TP}{TP+FP} \quad (5)$$

$$Accuracy = \frac{TP+TN}{N} \quad (6)$$

Sensitivity, representing the percentage of positive cases, is 94%, while specificity, the percentage of negative cases, is 22% in our experiment. Precision achieved 94%, and accuracy, indicating the percentage of correctly identified cases, is 89%.

While these results are promising, there is potential for improvement in the method, especially in reducing FPs associated with detecting lateral faces on the CAUs. Moreover, the count of FNs is relatively high, mainly due to planes fitted to multiple Antifer blocks simultaneously. Conducting lower altitude flights with higher PC resolution could potentially enhance the detection of discontinuities between cuboids and improve sensitivity to detect outliers based on the distance to the planes.

3.2.2. RMB Stability Assessment

The maximum consensus planes obtained by applying the RANSAC algorithm to each of the dense PCs were cross-referenced through a proximity-based criterion. Through this spatial join or alignment process, a total of 3697 pairs of corresponding planes were identified across the entire surface of the breakwater.

While the correlation method used here is advantageous due to its inherent simplicity, it is not without its drawbacks. The effectiveness of this method relies significantly on the precision of the RANSAC algorithm in detecting and segmenting planes. Any inaccuracies in the segmentation of either PC, such as FPs or FNs, directly impact the subsequent plane matching phase. Essentially, an orphan plane, which exists in one dataset without a counterpart in the other, may be matched with the nearest available plane. This could result in semantic inconsistencies and distort subsequent analyses, although it does provide the advantage of generating a comprehensive and continuous dataset. To address these issues, the spatial join tool introduces a distance field within the outcome, representing the spatial closeness of linked geometries. This enables the definition of a specific tolerance threshold to prevent these inconsistencies.

The normal vectors of the fitted planes for each corresponding pair of block faces can be acquired to estimate the tilt angle within a single block. Figure 7 illustrates the overall inclination values obtained for each cuboid in the RMB, categorized based on their magnitudes.

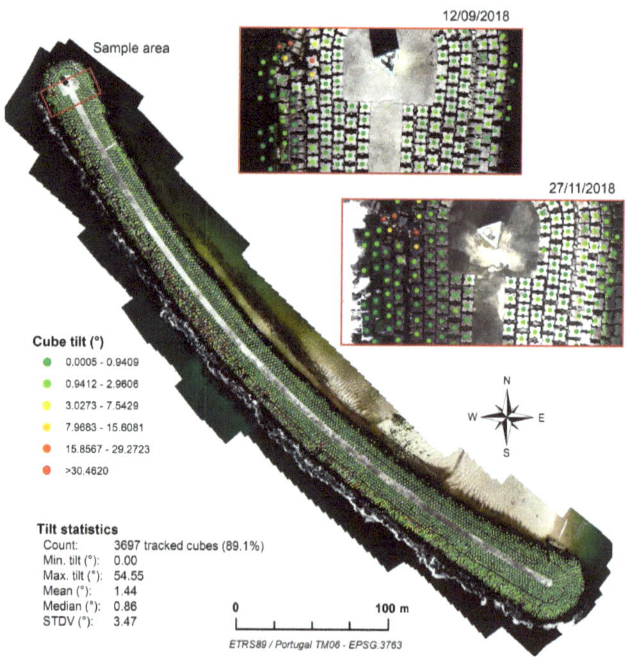

Figure 7. Overall inclination map based on the RANSAC-based method. The colors represent the magnitude of rotations.

In the graphics presented in Figure 8, a more detailed breakdown of these inclinations, considering both magnitude and inclination direction, is provided. Analysis of these figures allows us to deduce that the most significant instabilities of the blocks occur in the predominant southwest (SW) direction, aligning with the most exposed flank of the breakwater. Some tilting of the cuboids is also noticeable in the northeast (NE) body of the RMB, although the movements detected here are generally much smaller. In terms of magnitude, it is observed that 61.1% of the wave-dissipating blocks undergo rotations of zero or less than 1°, and 91.7% experience movements of less than 2° based on data derived from the RANSAC method. However, it is worth noting that rotations of certain CAUs can, in specific cases, exceed 50°.

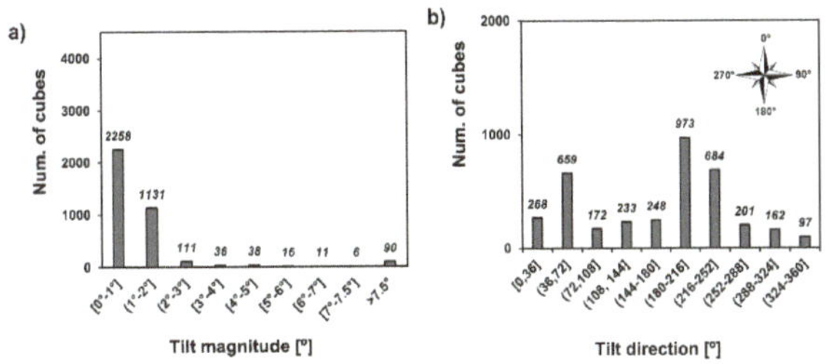

Figure 8. Overall tilt analysis of the identified cuboids: (**a**) magnitude and (**b**) direction of rotations.

In Figure 9, an illustrative region displaying relatively stable blocks within the southwest (SW) body is presented. Evaluating displacements or rotations solely through a visual examination of orthophotos from two different time periods poses a significant challenge. Factors like variations in imaging texture due to cuboid shading, the presence of biofilm, algae, etc., add complexity to the visual comparison of the orthophotos. Nonetheless, careful observation may suggest some rotation in the lower-right cuboid of the image.

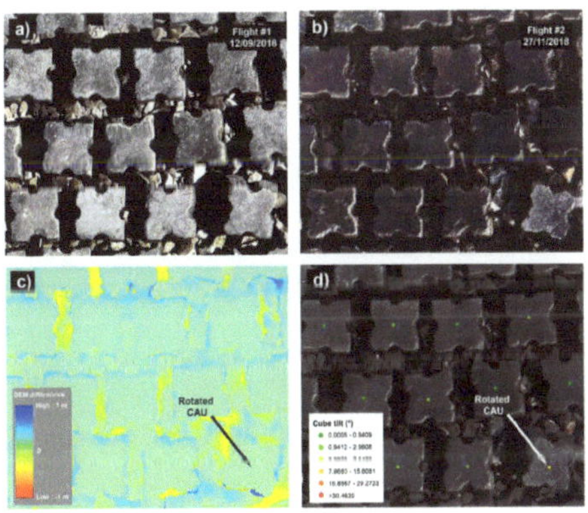

Figure 9. Comparison of DoD vs. RANSAC-based methods. Orthophotos corresponding to (**a**) the first flight (pre-storm) and (**b**) the second flight (post-storm), (**c**) DSM subtraction (DoD), and (**d**) tilting results measured with the RANSAC-based approach.

In the case of DoD, discerning any form of displacement or rotation is challenging, especially for cuboids with significant displacements. Unlike cuboids with noticeable movements, it is difficult to infer any changes in the elevations of rotated CAUs, as these elevations may remain relatively stable despite the rotation. Some elevation fluctuations are observable within inter-block spaces, potentially attributed to the movement of small stones in the underlayer or artifacts in the DSMs caused by occluded regions. The DoD represents elevation changes on a cell-by-cell basis, typically along a predefined direction, often the Z-axis (vertical direction). While the simplicity of the DoD method is advantageous, it does have limitations in intricate contexts, such as overhangs and nearly vertical slopes, where vertical differences may not provide comprehensive insights. Similar findings have been reported in previous studies [60]. Consequently, the precision of interpreting elevation differences along the edges of each CAU is not entirely accurate when using a traditional

2.5D method for change detection like DoD. As illustrated in Figure 9d, the RANSAC-based plane-fitting method demonstrates increased sensitivity, showcasing its effectiveness even in more stable regions of the RMB model.

Operating at the level of individual cuboids, the proposed methodology allows for a more detailed analysis of the structure. The graphs in Figure 10 present both a global analysis (Figure 10a) and a zonal breakdown of cuboid counts against their detected inclinations. The zonal analysis divides the total count of CAUs into five principal zones characterizing the RMB. The crest of the detached breakwater, referred to as RMB top, spans its entire length and includes three rows of wave-dissipating blocks on each side of the central platform. Due to the substantial number of blocks within this area, it exhibits a relatively low occurrence of CAU inclinations, as illustrated in Figure 10b. In the breakwater heads (Figure 10c,d), a limited number of units exhibit relatively high shifts, primarily found in the northwest (NW) head. Furthermore, the southwest (SW) body zone shows significantly higher CAU inclinations than the inner breakwater region (Figure 10e,f), which is consistent with its exposure to wave action. Beyond this simplified examination, the results underscore the potential of these methods to provide quantitative assessments of the extent of damage.

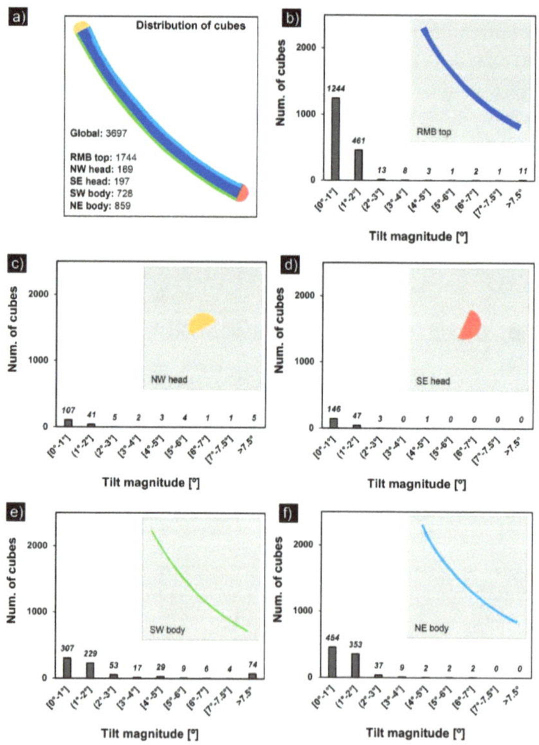

Figure 10. Zonal stability analysis of breakwater. (**a**) Overall distribution of tracked Antifer cubes and zonal tilt measurements in (**b**) RMB top, (**c**) NW head, (**d**) SE head, (**e**) SW body, and (**f**) NE body.

4. Conclusions and Future Remarks

The evaluation of structural changes and tilting in coastal structures, particularly breakwaters, is crucial for ensuring their long-term stability. The integration of aerial imagery, photogrammetric reconstruction, and RANSAC-based segmentation provides an intriguing tool for the continuous monitoring and assessment of breakwater stability.

While DoD remains a prevalent method for analyzing multi-temporal changes due to its simplicity, it has limitations. DEMs inherently lack complete 3D spatial information and may struggle to differentiate individual armor units, leading to reduced accuracy in change detection. This limitation becomes particularly evident in scenarios involving vertical shifts, rotations, or tilting of individual cuboids. The results of this study demonstrate that the proposed approach based on RANSAC is more effective than DEM-based methods in detecting even subtle tilting. This approach provides a detailed and localized understanding of the structural integrity of the breakwater. By enhancing the ability to detect and comprehend structural changes in the RMB over time, it contributes to improved coastal infrastructure management and resilience.

Further improvement and validation of the methodology should focus on obtaining unambiguous matches between CAUs in different epochs. It would also be desirable to refine segmentation accuracy, possibly by integrating the RANSAC method with image-based approaches, such as using detectors and feature extractors for block edges. Additionally, exploring the adaptability of this approach to more intricate shapes of the armor units by fitting other geometric primitives presents an interesting avenue for research.

Author Contributions: Conceptualization, M.A.-G., J.A.G., V.F.P. and G.B.; methodology, M.A.-G., J.A.G., V.F.P. and G.B.; software, M.A.-G. and G.B.; investigation, M.A.-G., J.A.G., V.F.P. and G.B., field data acquisition, V.F.P. and J.A.G.; writing—original draft preparation, M.A.-G.; writing—review and editing, J.A.G. and G.B.; supervision, J.A.G. All authors have read and agreed to the published version of the manuscript.

Funding: This work was supported by the regional government of Galicia (Xunta de Galicia, Spain) with the postdoctoral grant awarded to Marcos Arza García (ED481B-2022/075) for the project UAV-Based Optical RS for Structural Inspection and Monitoring in Coastal Engineering.

Data Availability Statement: Data are contained within the article.

Conflicts of Interest: The authors declare no conflicts of interest.

References

1. Akarsh, P.K.; Chaudhary, B. Review of Literature on Design of Rubble Mound Breakwaters. In Proceedings of the International Conference on Civil Engineering Trends and Challenges for Sustainability, Bapatla, India, 24–25 September 2021; pp. 775–796.
2. Nguyen, D.V.; Van Gelder, P.H.A.J.M.; Verhagen, H.J.; Vrijling, J.K. Optimal inspection strategy for rubble-mound breakwaters with time-dependent reliability analysis. In *Relaibility, Risk and Safety*; Taylor & Francis Group: London, UK, 2010; pp. 1409–1415.
3. Etemad-Shahidi, A.; Bali, M.; van Gent, M.R.A. On the stability of rock armored rubble mound structures. *Coast. Eng.* **2020**, *158*, 103655. [CrossRef]
4. der Meer, J.W. Conceptual design of rubble mound breakwaters. In *Advances in Coastal And Ocean Engineering*; World Scientific: London, UK, 1995, Volume 1, pp. 221–315.
5. Puente, I.; Lindenbergh, R.; González-Jorge, H.; Arias, P. Terrestrial laser scanning for geometry extraction and change monitoring of rubble mound breakwaters. *ISPRS Ann. Photogramm. Remote Sens. Spat. Inf. Sci.* **2014**, *2*, 289–295. [CrossRef]
6. Campos, Á.; Molina-Sanchez, R.; Castillo, C. Damage in rubble mound breakwaters. Part II: Review of the definition, parameterization, and measurement of damage. *J. Mar. Sci. Eng.* **2020**, *8*, 306. [CrossRef]
7. Musumeci, R.E.; Moltisanti, D.; Foti, E.; Battiato, S.; Farinella, G.M. 3-D monitoring of rubble mound breakwater damages. *Meas. J. Int. Meas. Confed.* **2018**, *117*, 347–364. [CrossRef]
8. Lemos, R.; Peña, E.; Santos, J.; Sande, J.; Figuero, A.; Alvarellos, A.; Laiño, E.; Reis, M.T.; Fortes, C.J.; Kerpen, N.B.; et al. 3D Survey Modelling for Damage Assessment in Rubble-Mound Breakwaters Under Oblique Wave Incidence. *Ital. J. Eng. Geol. Environ.* **2020**, *20*, 73–85. [CrossRef]
9. Lemos, R.; Fortes, C.; Silva, G.; Pinheiro, L. An estimation of the damage of scale models of breakwaters using the time of flight method. *Rev. Mecânica Exp.* **2022**, *35*, 49–58.
10. Lemos, R.; Santos, J.A.; Fortes, C.J.E.M. Damage Evolution Prediction during 2D Scale-Model Tests of a Rubble-Mound Breakwater: A Case Study of Ericeira's Breakwater. *Modelling* **2022**, *4*, 1–18. [CrossRef]
11. Vieira, F.; Taveira-Pinto, F.; Rosa-Santos, P. Damage evolution in single-layer cube armoured breakwaters with a regular placement pattern. *Coast. Eng.* **2021**, *169*, 103943. [CrossRef]
12. Lemos, R.; Loja, M.A.R.; Rodrigues, J.; Rodrigues, J.A. Photogrammetric analysis of rubble mound breakwaters scale model tests. *AIMS Environ. Sci.* **2016**, *3*, 541–559. [CrossRef]
13. Fortes, C.J.E.M.; Lemos, R.; Mendonça, A.; Reis, M.T. Damage progression in rubble-mound breakwaters scale model tests, under a climate change storm sequence. *Res. Eng. Struct. Mater.* **2019**, *5*, 415–426. [CrossRef]

14. Marino, S.; Galantucci, R.A.; Saponieri, A. Measuring rock slope damage on rubble mound breakwater through digital photogrammetry. *Meas. J. Int. Meas. Confed.* **2023**, *211*, 112656. [CrossRef]
15. Mishra, V.; Avtar, R.; Prathiba, A.P.; Mishra, P.K.; Tiwari, A.; Sharma, S.K.; Singh, C.H.; Chandra Yadav, B.; Jain, K. Uncrewed Aerial Systems in Water Resource Management and Monitoring: A Review of Sensors, Applications, Software, and Issues. *Adv. Civ. Eng.* **2023**, *2023*, 3544724. [CrossRef]
16. Robin, N.; Levoy, F.; Anthony, E.J.; Monfort, O. Sand spit dynamics in a large tidal-range environment: Insight from multiple LiDAR, UAV and hydrodynamic measurements on multiple spit hook development, breaching, reconstruction, and shoreline changes. *Earth Surf. Process. Landforms* **2020**, *45*, 2706–2726. [CrossRef]
17. Gonçalves, J.A.; Henriques, R. UAV photogrammetry for topographic monitoring of coastal areas. *ISPRS J. Photogramm. Remote Sens.* **2015**, *104*, 101–111. [CrossRef]
18. Stagnitti, M.; Musumeci, R.E.; Foti, E. Surface roughness measurement for the assessment of damage dynamics of existing and upgraded cube-armored breakwaters. *Coast. Eng.* **2023**, *179*, 104226. [CrossRef]
19. King, S.; Leon, J.; Mulcahy, M.; Jackson, L.A.; Corbett, B. Condition survey of coastal structures using UAV and photogrammetry. In Proceedings of the Australasian Coasts & Ports Conference, Twin Waters, Australia, 15–18 August 2017; pp. 704–710.
20. Rezaldi, M.Y.; Yoganingrum, A.; Hanifa, N.R.; Kaneda, Y.; Kushadiani, S.K.; Prasetyadi, A.; Nugroho, B.; Riyanto, A.M. Unmanned aerial vehicle (Uav) and photogrammetric technic for 3d tsunamis safety modeling in cilacap, indonesia. *Appl. Sci.* **2021**, *11*, 11310. [CrossRef]
21. Layek, S.; Villuri, V.G.K.; Koner, R.; Chand, K. Rainfall & Seismological Dump Slope Stability Analysis on Active Mine Waste Dump Slope with UAV. *Adv. Civ. Eng.* **2022**, *2022*, 5858400. [CrossRef]
22. Kim, K.; Francis, O. Integration of In-Situ, Laboratory and Computer Models for Coastal Risk Assessment, Planning and Development. In Proceedings of the Ocean 2018 MTS/IEEE Charleston, Charleston, SC, USA, 22–25 October 2018; pp. 1–5. [CrossRef]
23. Henriques, M.J.; Lemos, R.; Capitão, R.; Fortes, C.J. The monitoring of rubble mound breakwaters. An assessment of UAV technology. In Proceedings of the 7th International Conference on Engineering Surveying, Lisbon, Portugal, 18–20 October 2017; pp. 1–8.
24. Gonçalves, D.; Gonçalves, G.; Pérez-Alvávez, J.A.; Andriolo, U. On the 3D Reconstruction of Coastal Structures by Unmanned Aerial Systems with Onboard Global Navigation Satellite System and Real-Time Kinematics and Terrestrial Laser Scanning. *Remote Sens.* **2022**, *14*, 1485. [CrossRef]
25. Gonçalves, D.; Gonçalves, G.; Pérez-Alvárez, J.; Cunha, M.C.; Andriolo, U. Combining Unmanned Aerial Systems and Structure from Motion Photogrammetry To Reconstruct the Geometry of Groins. *Int. Arch. Photogramm. Remote Sens. Spat. Inf. Sci. ISPRS Arch.* **2022**, *43*, 1003–1008. [CrossRef]
26. Hussain, M.; Chen, D.; Cheng, A.; Wei, H.; Stanley, D. Change detection from remotely sensed images: From pixel-based to object-based approaches. *ISPRS J. Photogramm. Remote Sens.* **2013**, *80*, 91–106. [CrossRef]
27. Blaschke, T.; Hay, G.J.; Kelly, M.; Lang, S.; Hofmann, P.; Addink, E.; Queiroz Feitosa, R.; van der Meer, F.; van der Werff, H.; van Coillie, F.; et al. Geographic Object-Based Image Analysis—Towards a new paradigm. *ISPRS J. Photogramm. Remote Sens.*p **2014**, *87*, 180–191. [CrossRef]
28. Kundal, S.; Chowdhury, A.; Bhardwaj, A.; Garg, P.K.; Mishra, V. GeoBIA-based semi-automated landslide detection using UAS data: A case study of Uttarakhand Himalayas. *SPIE Future Sens. Technol.* **2023**, *12327*, 79.
29. Sharma, S.K.; Jain, K.; Shukla, A.K. A Comparative Analysis of Feature Detectors and Descriptors for Image Stitching. *Appl. Sci.* **2023**, *13*, 6015. [CrossRef]
30. Chen, L.; Rottensteiner, F.; Heipke, C. Feature detection and description for image matching: From hand-crafted design to deep learning. *Geo-Spatial Inf. Sci.* **2021**, *24*, 58–74. [CrossRef]
31. Forero, M.G.; Mambuscay, C.L.; Monroy, M.F.; Miranda, S.L.; Méndez, D.; Valencia, M.O.; Selvaraj, M.G. Comparative analysis of detectors and feature descriptors for multispectral image matching in rice crops. *Plants* **2021**, *10*, 1–24. [CrossRef] [PubMed]
32. Qin, R.; Tian, J.; Reinartz, P. 3D change detection—Approaches and applications. *ISPRS J. Photogramm. Remote Sens.* **2016**, *122*, 41–56. [CrossRef]
33. Williams, R.D. DEMs of difference. *Geomorphol. Tech.* **2012**, *2*, 1–17.
34. Lemos, R.; Capitão, R.; Fortes, C.; Henriques, M.; Silva, L.G.; Martins, T. A methodology for the evaluation of evolution and risk of breakwaters. Application to Portimão harbor and of Faro-Olhão inlet. *J. Integr. Coast. Zone Manag.* **2020**, *20*, 103–119. [CrossRef]
35. Bueno Esposito, M.; Díaz-Vilariño, L.; Martínez-Sánchez, J.; González -Jorge, H.; Arias, P. 3D reconstruction of cubic armoured rubble mound breakwaters from incomplete lidar data. *Int. J. Remote Sens.* **2015**, *36*, 5485–5503. [CrossRef]
36. Xu, Y.; Kanai, S.; Date, H.; Sano, T. Deep-Learning-Based Three-Dimensional Detection of Individual Wave-Dissipating Blocks from As-Built Point Clouds Measured by UAV Photogrammetry and Multibeam Echo-Sounder. *Remote Sens.* **2022**, *14*, 5575. [CrossRef]
37. Shen, Y.; Lindenbergh, R.; Wang, J.; Ferreira, V.G. Extracting individual bricks from a laser scan point cloud of an unorganized pile of bricks. *Remote Sens.* **2018**, *10*, 11709. [CrossRef]
38. Shen, Y.; Wang, J.; Puente, I. A Novel Baseline-Based Method to Detect Local Structural Changes in Masonry Walls Using Dense Terrestrial Laser Scanning Point Clouds. *IEEE Sens. J.* **2020**, *20*, 6504–6515. [CrossRef]
39. Choi, S.; Kim, T.; Yu, W. Performance evaluation of RANSAC family. *J. Comput. Vision* **1997**, *24*, 271–300. [CrossRef]

40. Huber, P.J.; Ronchetti, E.M. *Robust Statistics*; Wiley: Hoboken, NJ, USA, 1981; Volume 1.
41. Rousseeuw, P.J. Least median of squares regression. *J. Am. Stat. Assoc.* **1984**, *79*, 871–880. [CrossRef]
42. Borrmann, D.; Elseberg, J.; Lingemann, K.; Nüchter, A. The 3d hough transform for plane detection in point clouds: A review and a new accumulator design. *3D Res.* **2011**, *2*, 1–13. [CrossRef]
43. Mahendran, S.; Ali, H.; Vidal, R. 3D Pose Regression Using Convolutional Neural Networks. In Proceedings of the IEEE International Conference on Computer Vision Workshops, Honolulu, HI, USA, 21–26 July 2017; pp. 494–495. [CrossRef]
44. Fischler, M.A.; Bolles, R.C. Random sample consensus: A Paradigm for Model Fitting with Applications to Image Analysis and Automated Cartography. *Commun. ACM* **1981**, *24*, 381–395. [CrossRef]
45. Schnabel, R.; Wahl, R.; Klein, R. Efficient RANSAC for point-cloud shape detection. *Comput. Graph. Forum* **2007**, *26*, 214–226. [CrossRef]
46. Gonultas, F.; Atik, M.E.; Duran, Z. Extraction of Roof Planes from Different Point Clouds Using RANSAC Algorithm. *Int. J. Environ. Geoinformatics* **2020**, *7*, 165–171. [CrossRef]
47. Li, Z.; Shan, J. RANSAC-based multi primitive building reconstruction from 3D point clouds. *ISPRS J. Photogramm. Remote Sens.* **2022**, *185*, 247–260. [CrossRef]
48. Wang, S.; Zhang, Z.; Wang, C.; Zhu, C.; Ren, Y. Multistep rocky slope stability analysis based on unmanned aerial vehicle photogrammetry. *Environ. Earth Sci.* **2019**, *78*, 260. [CrossRef]
49. Giulietti, N.; Allevi, G.; Castellini, P.; Garinei, A.; Martarelli, M. Rivers' Water Level Assessment Using UAV Photogrammetry and RANSAC Method and the Analysis of Sensitivity to Uncertainty Sources. *Sensors* **2022**, *22*, 5319. [CrossRef] [PubMed]
50. Iglesias, I.; Venâncio, S.; Pinho, J.L.; Avilez-Valente, P.; Vieira, J.M.P. Two models solutions for the Douro estuary: Flood risk assessment and breakwater effects. *Estuaries Coasts* **2019**, *42*, 348–364. [CrossRef]
51. Jiménez-Jiménez, S.I.; Ojeda-Bustamante, W.; Marcial-Pablo, M.D.J.; Enciso, J. Digital terrain models generated with low-cost UAV photogrammetry: Methodology and accuracy. *ISPRS Int. J. Geo-Information* **2021**, *10*, 285. [CrossRef]
52. Calì, M.; Ambu, R. Advanced 3D photogrammetric surface reconstruction of extensive objects by UAV camera image acquisition. *Sensors* **2018**, *18*, 2815. [CrossRef] [PubMed]
53. Ortiz-Sanz, J.; Gil-Docampo, M.; Rego-Sanmartín, T.; Arza-García, M.; Tucci, G. A PBeL for training non-experts in mobile-based photogrammetry and accurate 3-D recording of small-size/non-complex objects. *Meas. J. Int. Meas. Confed.* **2021**, *178*, 109338. [CrossRef]
54. Carrilho, A.C.; Galo, M. Extraction of building roof planes with stratified random sample consensus. *Photogramm. Rec.* **2018**, *33*, 363–380. [CrossRef]
55. Catalão, J.; Sevilla, M.J. Mapping the geoid for Iberia and the Macaronesian Islands using multi sensor gravity data and the GRACE geopotential model. *J. Geodyn.* **2009**, *48*, 6–15. [CrossRef]
56. Arza-García, M.; Gil-Docampo, M.; Ortiz-Sanz, J. A hybrid photogrammetry approach for archaeological sites: Block alignment issues in a case study (the Roman camp of A Cidadela). *J. Cult. Herit.* **2019**, *38*, 195–203. [CrossRef]
57. Soares, F.; Henriques, M.J.; Rocha, C. Concrete Block Tracking in Breakwater Models Concrete Block Tracking in Breakwater Models. In Proceedings of the FIG Working Week 2017, Helsinki, Finland, 29 May—2 June 2017; pp. 1–14.
58. Henriques, M.J.; Brás, N.; Roque, D.; Lemos, R.; Fortes, C.J.E.M. Controlling the Damages of Physical Models of Rubble-Mound Breakwaters by Photogrammetric Products-Orthomosaics and Point Clouds. In Proceedings of the Proceedings of the 3rd Joint International Symposium on Deformation Monitoring, Viena, Austria, 30 March–1 April 2016; p. 8.
59. Powers, D.M.W. Evaluation: From Precision, Recall and F-Factor to ROC, Informedness, Markedness & Correlation. Technical Report SIE-07-001. *arXiv* **2007**, arXiv:2010.16061v1.
60. Kharroubi, A.; Poux, F.; Ballouch, Z.; Hajji, R.; Billen, R. Three Dimensional Change Detection Using Point Clouds: A Review. *Geomatics* **2022**, *2*, 457–485. [CrossRef]

remote sensing

MDPI

Article

Urban Building Height Extraction from Gaofen-7 Stereo Satellite Images Enhanced by Contour Matching

Yunfan Cui [1], Shuangming Zhao [1,*], Wanshou Jiang [2] and Guorong Yu [3]

1 School of Remote Sensing Information Engineering, Wuhan University, Wuhan 430079, China; yunfancui@whu.edu.cn
2 State Key Laboratory of Information Engineering in Surveying Mapping and Remote Sensing, Wuhan University, Wuhan 430079, China; jws@whu.edu.cn
3 School of Computer Science and Technology, Wuhan University of Science and Technology, Wuhan 430065, China; yuguorong190@wust.edu.cn
* Correspondence: smzhao@whu.edu.cn

Abstract: The traditional method for extracting the heights of urban buildings involves utilizing dense matching algorithms on stereo images to generate a digital surface model (DSM). However, for urban buildings, the disparity discontinuity issue that troubles the dense matching algorithm makes the elevations of high-rise buildings and the surrounding areas inaccurate. The occlusion caused by trees in greenbelts makes it difficult to accurately extract the ground elevation around the building. To tackle these problems, a method for building height extraction from Gaofen-7 (GF-7) stereo images enhanced by contour matching is presented. Firstly, a contour matching algorithm was proposed to extract accurate building roof elevation from GF-7 images. Secondly, a ground filtering algorithm was employed on the DSM to generate a digital elevation model (DEM), and ground elevation can be extracted from this DEM. The difference between the rooftop elevation and the ground elevation represents the building height. The presented method was verified in Yingde, Guangzhou, Guangdong Province, and Xi'an, Shaanxi Province. The experimental results demonstrate that our proposed method outperforms existing methods in building height extraction concerning accuracy.

Keywords: building height extraction; contour matching; Gaofen-7 satellite imagery; urban 3D reconstruction

Citation: Cui, Y.; Zhao, S.; Jiang, W.; Yu, G. Urban Building Height Extraction from Gaofen-7 Stereo Satellite Images Enhanced by Contour Matching. *Remote Sens.* **2024**, *16*, 1556. https://doi.org/10.3390/rs16091556

Academic Editor: John Trinder

Received: 15 March 2024
Revised: 21 April 2024
Accepted: 24 April 2024
Published: 27 April 2024

1. Introduction

A recent study on urban growth typology shows that there has been a large increase in high-rise buildings in China [1]. The building height information holds significant application value in various fields, such as urban local climate [2,3], building energy consumption evaluation [4,5], urban pollution dispersion [6,7], urban carbon emissions evaluation [8,9], earthquake perception [10], and urban 3D reconstruction [11]. Therefore, building height extraction over large regions is essential for a comprehensive understanding of an urban development.

Remote sensing technology is the most commonly used method for building height extraction. Typically, building heights are extracted through three approaches: airborne light detection and ranging (LiDAR), side-looking radar imagery, and high-resolution optical imagery. Airborne LiDAR allows high accuracy measurements [12]. These algorithms extract buildings and their heights through point cloud classification algorithms [13,14] or utilize building footprints from digital maps to reconstruct buildings in three dimensions [15]. However, airborne LiDAR has limitations in coverage and high costs. Algorithms utilizing side-looking radar imagery often require building footprints obtained from digital maps or other sources [16–19]. Nevertheless, with the side-looking geometry, radar images usually record signals from a mixture of different microwave scattering mechanisms, leading to relatively high uncertainties of building height extraction [20].

In contrast, optical satellite imagery has high acquisition efficiency and offers abundant spatial details, hence being widely applied in building height extraction. For single optical satellite images, the shadow-based method is commonly employed to extract building heights. This method utilizes the relationship between the sun, satellite, building rooftops, and shadows in the imagery to extract building height [21–25]. However, the shadow-based method faces difficulties in building height extraction when buildings are short or when shadows are occluded by other objects [26].

For stereo images, a common method involves generating a DSM through dense matching and projecting building footprints or rooftops onto the DSM to extract building heights. Liu et al. [27] utilized semi-global matching (SGM) [28] to generate a DSM, employed morphological filtering [29] on the DSM to generate DEM, and finally derived the normalized DSM (nDSM) using the maximum values within the nDSM as the building heights. Wang et al. [30] improved DEM generation with the more precise cloth simulation filter (CSF) method [31]. To address the issue of missing rooftop elevations in a DSM generated by the SGM algorithm, Zhang et al. [26] proposed a contour-constrained rooftop matching algorithm for building height extraction.

With the rapid development of deep learning, deep learning methods have been widely applied in dense matching [32–34], opening up new possibilities for building height extraction. For instance, Chen et al. [35] utilized a DSM generated by deep learning algorithms in building height extraction. End-to-end deep learning methods have also been proposed for building height extraction in stereo images. Cao et al. [36] designed the M^3net network to extract buildings and their heights from multi-view, multi-spectral images. This method does not rely on dense matching algorithms but requires known building height data for training.

The GF-7 satellite is capable of capturing panchromatic stereo images spanning 20 km in width with a resolution finer than 0.8 m. Its backward camera holds a tilt angle of −5 degrees, while the forward camera tilts at 26 degrees, maintaining a favorable balance between minimized occlusion and a wider stereo intersection angle. It offers valuable data for building height extraction. However, limitations in resolution and the forward camera tilt angle challenge the application of current dense matching algorithms, hindering their accuracy in building height extraction. Relevant research indicates that many 3D breaklines are modeled as more or less smooth transitions from ground level to building level [37]. Figure 1a,b illustrates the impact of this problem on building height extraction. This DSM is generated by the algorithm of He et al. [32] using GF-7 stereo images of Xi'an. In Figure 1a, inaccuracies in the ground elevation around the building are evident. While the actual ground elevation is 355 m, the DSM shows elevations higher than the reality. Figure 1b shows inaccuracies in high-rise buildings. The actual building height is 350 m, with a rooftop elevation of 702 m. There are substantial differences in shape and elevation between the reconstructed buildings and their actual counterparts. Figure 1c illustrates occlusion caused by trees in Guangzhou. Detailed data for both Xi'an and Guangzhou are provided in Section 3.1. These challenges lead to difficulties for algorithms relying on a DSM in accurately extracting the building heights.

To improve the building height estimation accuracy, we proposed a contour matching enhanced building height extraction method. Instead of overlaying the building contours on the DSM directly, we used a contour matching algorithm to obtain more accurate rooftop elevation and ground filtering to generate a DEM from the DSM for more robust ground elevation. Firstly, the given building contours, which can be in ground space or on a GF-7 backward image, are matched to GF-7 forward images with a contour matching, and the rooftop elevation can be extracted using the geometric relationship between the matched building rooftop. Secondly, the ground elevation around the building can be extracted from the DEM, which filters the DSM generated from GF-7 stereo images. GF-7 multispectral images are utilized to improve the accuracy of ground filtering. Finally, the difference between the rooftop elevation and the ground elevation represents the building height.

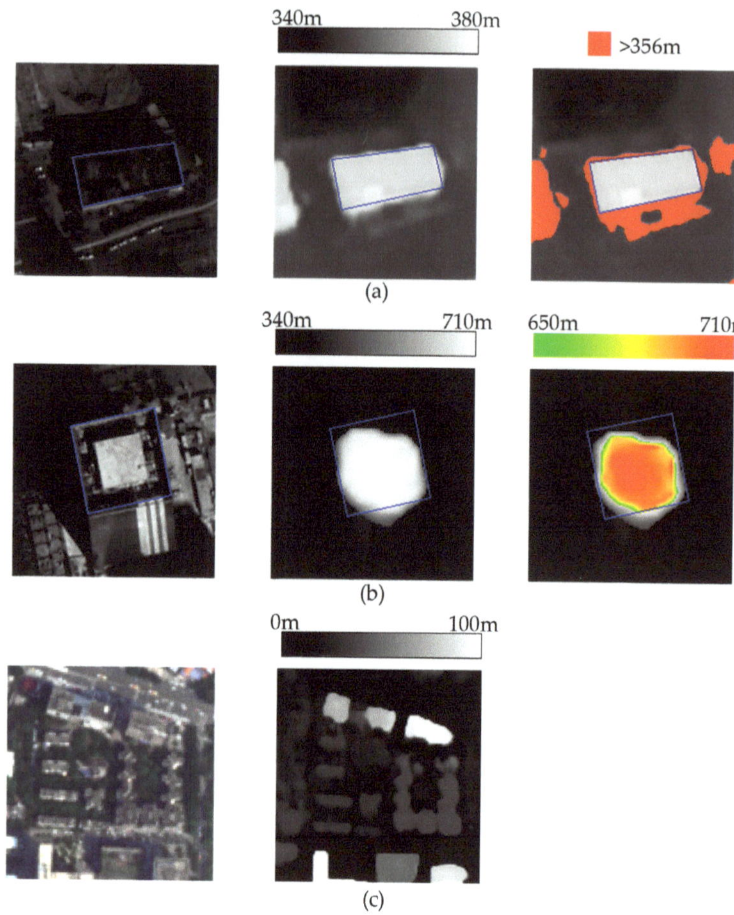

Figure 1. The problems faced in building heights extraction from the DSM. (**a**) Inaccuracies in the ground elevation; (**b**) Inaccuracies in high-rise buildings; (**c**) Occlusion caused by trees. (Left: buildings in the GF-7 backward images, middle: corresponding DSM for the buildings, right: inaccurate elevations in the DSM).

The main contributions of this paper are as follows:

- An object-level contour matching algorithm is proposed to extract the rooftop plane elevation. Contrary to the pixel-level dense matching, which can generate smooth transitions in the DSM, the proposed algorithm, taking the rooftop as an object, can overcome the complex detail interruption of the rooftop.
- A ground filtering considering ground types is proposed for ground elevation extraction. Most existing ground filtering algorithms, which are designed for LiDAR cloud points with multi-echo, will not generate good DEM when applied directly to a satellite-DSM-generated DSM. In our new algorithm, we use multi-spectral imagery to assist in identifying non-ground points and inaccurate ground points in ground filtering algorithms.

Our paper is organized as follows: Section 2 of the paper extensively describes the extraction of the building height and discusses scenarios where multiple elevations exist for building rooftops. Section 3 of the paper demonstrates the effectiveness of this approach through experiments. The proposed algorithm is discussed in Section 4. Finally, Section 5 concludes this paper.

2. Methodology

The algorithm workflow for building height extraction is illustrated in Figure 2. The known data required in this algorithm include the GF-7 images, DSM generated from GF-7 stereo images, building footprints in the geographic coordinate system, or building rooftop contours in GF-7 backward images. The contour matching algorithm for building footprints (CM-F) is described in Algorithm 1. The building rooftop contours in GF-7 backward images may have unclear edges or may encompass podium buildings and building sides. Our algorithm utilizes the backward images to reduce the impact of unclear edges. Furthermore, it is possible to use differences between the forward and backward images to identify building sides and podium buildings. The contour matching algorithm for building rooftop contours (CM-R) is described in Algorithm 2.

Algorithm 1. The contour matching algorithm for building footprint (CM-F)

- Input: GF-7 forward image I_{fwd}, building footprint B_f, DSM.
- Output: Building height H.
- Estimate the elevation search range of rooftop $[Z_{lb}, Z_{ub}]$. (Section 2.5)
- Extract contours in I_{fwd}.(Section 2.1)
- for all Z_i in $[Z_{lb}, Z_{ub}]$
- Obtain candidate building rooftop contour in I_{fwd}, denoted as B^i_f.
- Generate building contour template based on B^i_f. (Section 2.2)
- Calculate the weighted contour matching degree WCM_i. (Section 2.3)
- Obtain building rooftop elevation E_{roof} based on WCM_i. (Section 2.4)
- Extract the ground elevation around the building E_{ground}. (Section 2.5)
- Calculate the building height H.

Algorithm 2. The contour matching algorithm for building rooftop contour (CM-R)

- Input: Stereo pair images I_{bwd} and I_{fwd}, building rooftop contour B_r, DSM.
- Output: Building height H.
- Generate epipolar images EI_{bwd} and EI_{fwd} from I_{bwd} and I_{fwd}.
- Extract contours from EI_{bwd} and EI_{fwd}. (Section 2.1)
- Estimate the disparity search range of rooftop in the epipolar image $[Dis_{lb}, Dis_{ub}]$. (Section 2.5)
- Generate building contour template based on B_r. (Section 2.2)
- Calculate the contour matching degree on EI_{bwd}, denoted as CM_{bwd}. (Section 2.3) And obtain the set of matched building edges S_{bwd}. (Section 2.6)
- Correct the building contour template. (Section 2.3)
- for all Dis_i in $[Dis_{lb}, Dis_{ub}]$
- Calculate the weighted contour matching degree, denoted as WCM_i. (Section 2.3)
- Obtain the building rooftop elevation E_{roof} based on WCM_i. (Section 2.4)
- Calculate the contour matching degree in EI_{fwd}, denoted as CM_{fwd}. And obtain the set of matched building edges S_{fwd}. (Section 2.6)
- Input S_{fwd}, S_{bwd}, CM_{bwd}, CM_{fwd} into Algorithm 3 to identify the building side and podium building.
- Extract the ground elevation around the building E_{ground}. (Section 2.5)
- Calculate the building height H.

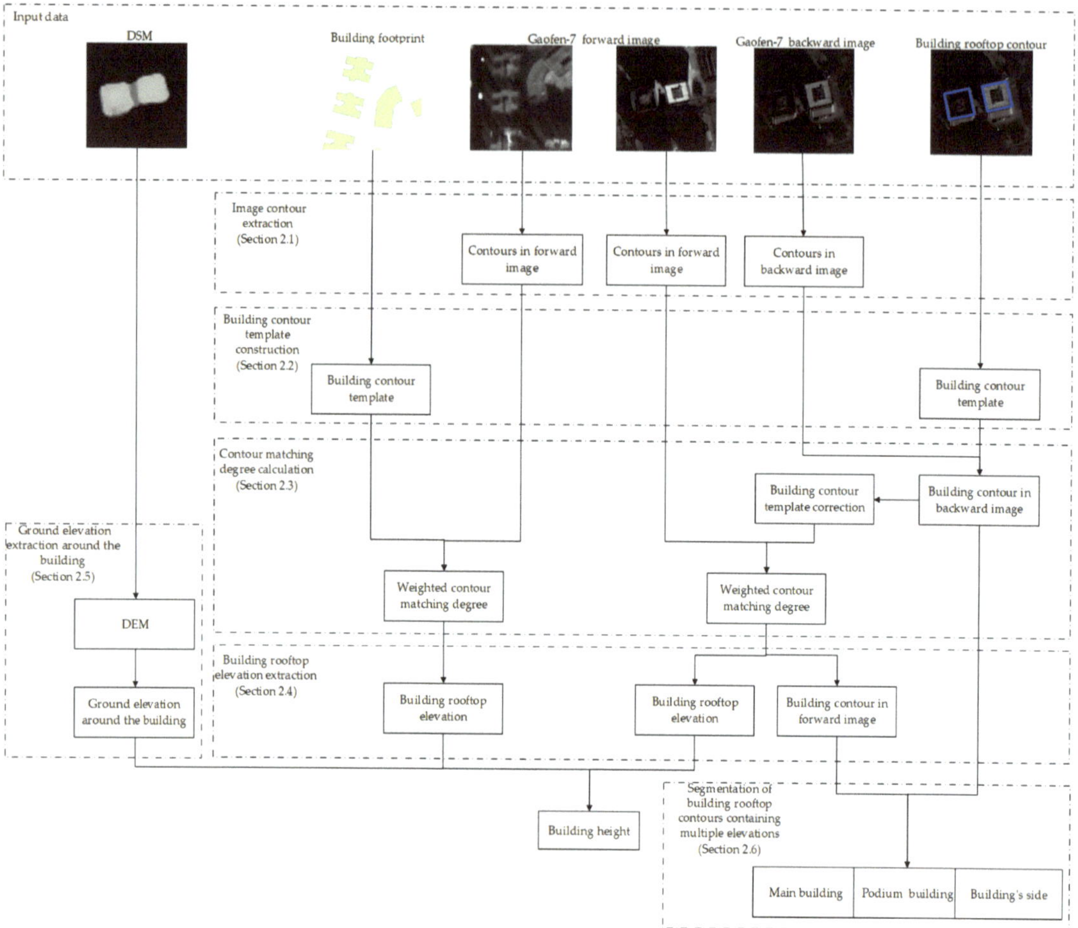

Figure 2. Building height extraction workflow.

2.1. Image Contour Extraction

Building contour consists of a collection of edges formed by continuous curves or lines, which match with the edges extracted from the image in contour match. The Canny edge detection algorithm [38] is utilized to extract edges in the image as contour points. The gradient direction of the image is calculated as the contour point direction, as shown in Equation (1):

$$\alpha = tan^{-1}\left(G_y / G_x\right) \tag{1}$$

where G_x and G_y represent the gradients in the horizontal and vertical directions, respectively. In the arctan function, the signs of G_x and G_y are used to ensure that the gradient direction ranges from $[-\pi, \pi]$.

This study extends the range of contour point direction values from the $[0, \pi]$ as in conventional methods [39] to $[-\pi, \pi]$. Due to the parapet walls at the rooftop, there are two adjacent indistinguishable edges in the image. By expanding the range of gradient direction, these two edges can be distinguished based on their positive or negative gradient directions. An example is provided in Figure 3.

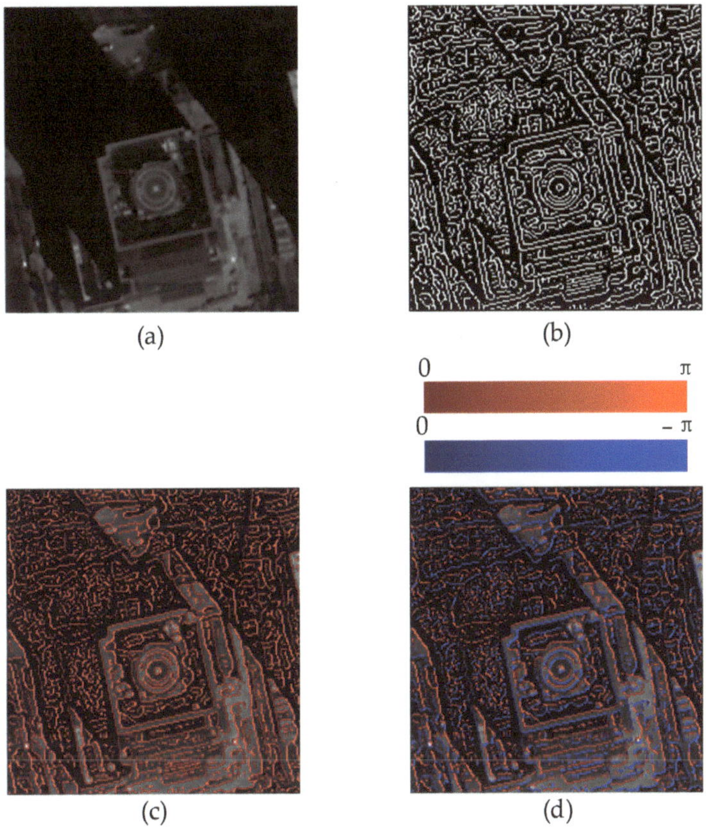

Figure 3. Edge detection results and contour point directions (**a**) original image; (**b**) edge detection results; (**c**) contour point direction in conventional method; (**d**) contour point direction in our method.

2.2. Building Contour Template Construction

Building contour templates are constructed to describe building rooftops. Figure 4 illustrates the process of building contour template construction. The vector polygon of the building is simplified by the Douglas algorithm [40]. Then, we created buffer zones for the edges of the vector polygon. The pixels within the buffer zone are considered potential contour points that constitute building contour templates. Their weights are calculated by the distance to the building edges, as shown in Equation (2).

$$dw = \begin{cases} 1 - |d|/D_{max}, & |d| < D_{max} \\ 0, & |d| \geq D_{max} \end{cases} \tag{2}$$

Here, D_{max} represents the buffer distance; d denotes the distance from the point to the edge in pixels, where d is negative when the point is inside the building contour.

The potential contour point direction is perpendicular to the corresponding edges of the polygon. As buildings in remote sensing images are generally brighter than other features [41], we set the potential contour point direction points inside the polygon. For any point Pt_i on the edge, draw a perpendicular line to the edge. The potential contour points that the perpendicular line passes through are grouped as a set, denoted by G_i. In contour matching, the matched contour of Pt_i is found within the range of G_i.

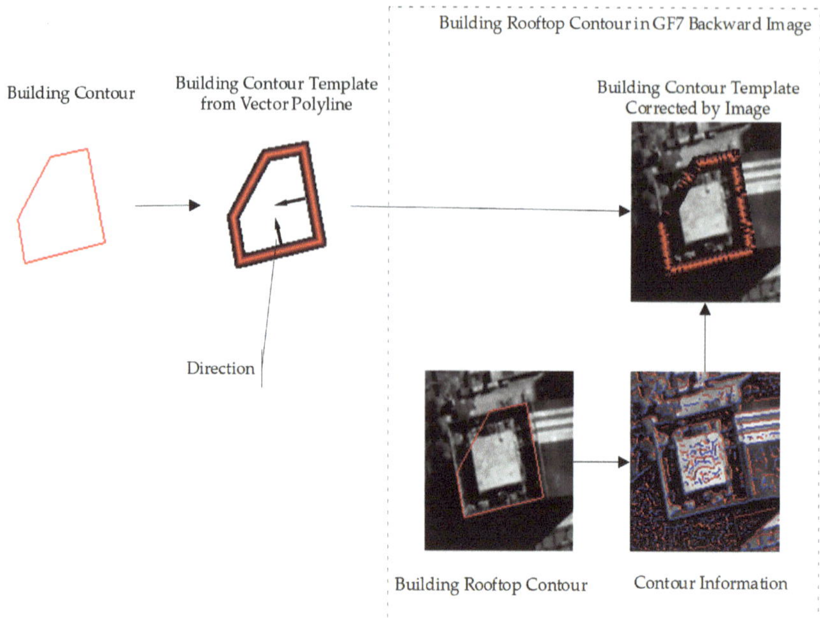

Figure 4. Building contour template construction. (The colors in building contour template represent the weight of the contour point).

2.3. Contour Matching Degree Calculation and Building Contour Template Correction

The contour matching degree represents the similarity between the building rooftop and the contours within the image. The contour matching degree is calculated as follows: The building contour template is moved to the location of the candidate building rooftop in the image, and each potential contour point can correspond to an image pixel. When a corresponding image pixel is a contour point extracted from the image, the angle between the potential contour point direction and contour point direction is calculated, denoted as θ. Then, the weight of the contour point in the image is calculated using Equation (3).

$$dw_{mp} = \begin{cases} dw & \theta \leq 15^\circ \\ dw \times p & \theta \geq 165^\circ \\ 0 & 15^\circ < \theta < 165^\circ \text{ or without corresponding contour point} \end{cases} \tag{3}$$

In this equation, p represents a penalty coefficient. In our study, p is assigned a value of 0.5 experientially.

In set G_i, the contour point with the maximum weight is matched with the Pt_i, denoted as Pt_i^{max}. We denote this maximum weight as $\max_{G_i}(dw_{mp})$, and the contour matching degree can be calculated using Equation (4). When the candidate building rooftop is changed, the building rooftop contour in the image will move along the epipolar line. Therefore, the building edges perpendicular to the epipolar line play an important role in roof elevation extraction. Consequently, by increasing the weights of contour points in these edges, more accurate rooftop elevations can be obtained, and the weighted contour matching degree is computed using Equation (5).

$$CM = \frac{\sum_{i=1}^{num_g} \left(\max_{G_i}(dw_{mp})\right)}{C} \tag{4}$$

$$WCM = \frac{\sum_{i=1}^{num_g} \left(\max_{G_i} \left(dw_{mp} \right) \times f(G_i) \right)}{C} \quad (5)$$

In this context, num_g represents the total number of sets G_i, and C denotes the circumference of the building contour in pixels. The value of the weight function $f(G_i)$ is determined by the edge where Pt_i is located. When the angle between the edge and the epipolar line exceeds 60 degrees, $f(G_i) = 2$; otherwise, $f(G_i) = 1$.

In practical application, the input building rooftop contours extracted by the building extraction algorithm may have unclear edges. Building contour template correction can improve the accuracy of the algorithm in this case. By computing the contour matching degree between the building rooftop contour and the GF-7 backward image, the matched contour points in the backward image are found and used to recalculate the weights of the potential contour point. The corrected weights of the potential contour point are calculated as follows: for any set G_i, if $\max_{G_i}(dw_{mp}) > 0$, then the distance d' between potential contour points within G_i and Pt_i^{max} is calculated. Subsequently, d' is used in Equation (2) to recalculate dw. If $\max_{G_i}(dw_{mp}) = 0$, the dw values of potential contour points in G_i are set to 0. The correction results are illustrated in Figure 4.

2.4. Building Rooftop Elevation Extraction

The principle of building rooftop elevation extraction is illustrated in Figure 5. According to known building contour, multiple candidate building rooftops can be obtained within the elevation search range of rooftop. These candidate rooftops are projected onto the GF-7 forward image using the rational function model and verified by contour matching.

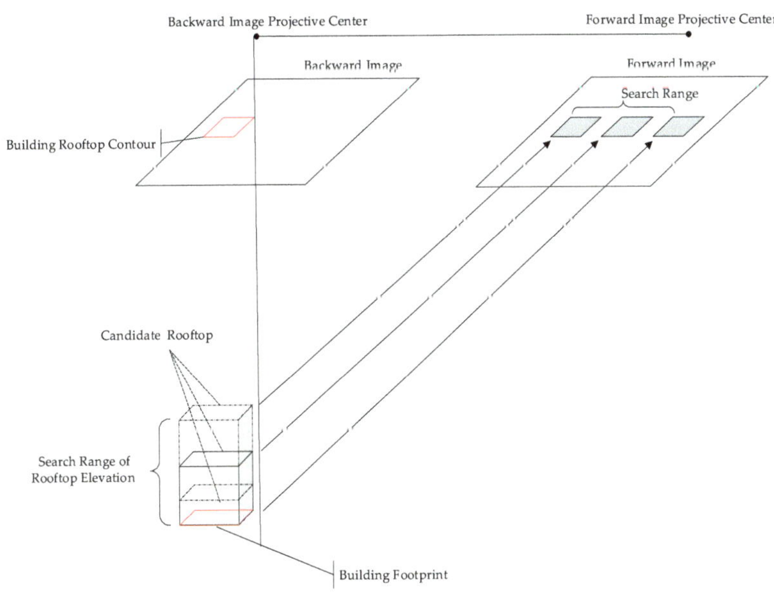

Figure 5. The principle of building rooftop elevation extraction.

The conventional contour matching method [39] sets a threshold for the contour matching degree and obtains the matched building contour based on the maximum value of the contour matching degree. In rooftop elevation extraction, multiple local maximum values of contour matching degree are caused by similar buildings or unclear building edges. The local maximum values lead to mismatches and significant errors. Therefore, our study utilizes elevation information from the DSM to filter out the local maximum values with significant errors.

The curve of contour matching degree versus candidate rooftop elevation is acquired at first. The elevation search range of the rooftop can be estimated using Equation (6).

$$[Z_{lb}, Z_{ub}] = [Z'_{min},\ Z'_{min} + BH_{max}] \tag{6}$$

where BH_{max} is set to be slightly greater than the estimated maximum building height, and Z'_{min} is the minimum elevation within the building buffer zone.

For building footprints in geographic coordinates, the variation between adjacent candidate rooftop elevations is set based on image resolution and stereo intersection angle. For each Z_i within the range $[Z_{lb}, Z_{ub}]$, the candidate building rooftop is projected onto the GF-7 forward image, and WCM_i can be calculated using the method mentioned earlier. For the building rooftop contours in the backward image, the elevation search range of rooftop is transformed to the disparity search range of rooftop, designated as $[Dis_{lb}, Dis_{ub}]$. For each integer Dis_i within the range $[Dis_{lb}, Dis_{ub}]$, the WCM_i and rooftop elevation is calculated, allowing us to acquire the curve of WCM_i versus rooftop elevation.

The minimum elevation Z'_{min} and maximum elevation Z'_{max} within the building buffer zone in the DSM are utilized to filter the local maximum value of contour matching degree. The local maximum values of contour matching degree are sorted in descending order, denoted as $WCM^1_{LM}, WCM^2_{LM}, \ldots WCM^j_{LM} \ldots$, and their corresponding rooftop elevations are denoted as $Z^1_{LM}, Z^2_{LM}, \ldots Z^j_{LM} \ldots$. If condition $CMW^1_{LM} \times 0.7 > CMW^2_{LM}$ is satisfied, it means the contour matching degree has a significant maximum value, and Z^1_{LM} is the rooftop elevation. In the absence of a significant maximum value, two situations need to be distinguished. If any local maximum value satisfies $CMW^j_{LM} > CMW^1_{LM} \times 0.7$, and the rooftop elevation satisfies $\left|Z^j_{LM} - Z'_{max}\right| < 5$ m, then Z^j_{LM} is considered as the rooftop elevation. If condition $Z'_{max} - Z'_{min} < 3$ m is satisfied, it is considered that the corresponding building rooftop does not exist in the GF-7 forward image. This indicates that the building is occluded in the forward image or that the known building differs from reality.

2.5. Ground Elevation Extraction around the Building

Our proposed method utilizes the results of GF-7 multispectral image classification to enhance the accuracy of the DEM generated by the ground filtering algorithm. GF-7 multispectral images are employed to compute the normalized difference vegetation index (NDVI) and the normalized difference water index (NDWI), allowing for the classification of vegetation and water from the image. By projecting input buildings into the DSM, the building can be classified from the DSM. The non-ground points such as vegetation and buildings are removed from the DSM. Additionally, large water bodies lacking texture that tend to cause mismatches are also removed from the DSM.

Subsequently, inaccurate ground points around buildings and trees are removed. In Figure 6a, profile comparisons of DSMs from LiDAR and stereo images are presented for a building in Guangzhou. The red lines represent the DSM from stereo images, and the black represents the DSM from LiDAR. In the ground pointed by the arrow, the DSM from the stereo image is higher than the DSM from LiDAR. These points should be removed from the ground filter. Figure 6b illustrates the method for identifying inaccurate ground points. For each window near the building, we calculated the elevation change along four lines. If $h_1 > 1.5 \times h_2$, the points on this line are considered as inaccurate points. Figure 6c shows a partial multispectral image of Guangzhou, Figure 6d shows the removed points in this image. This process ensures that the elevation of the occluded ground is estimated from nearby ground.

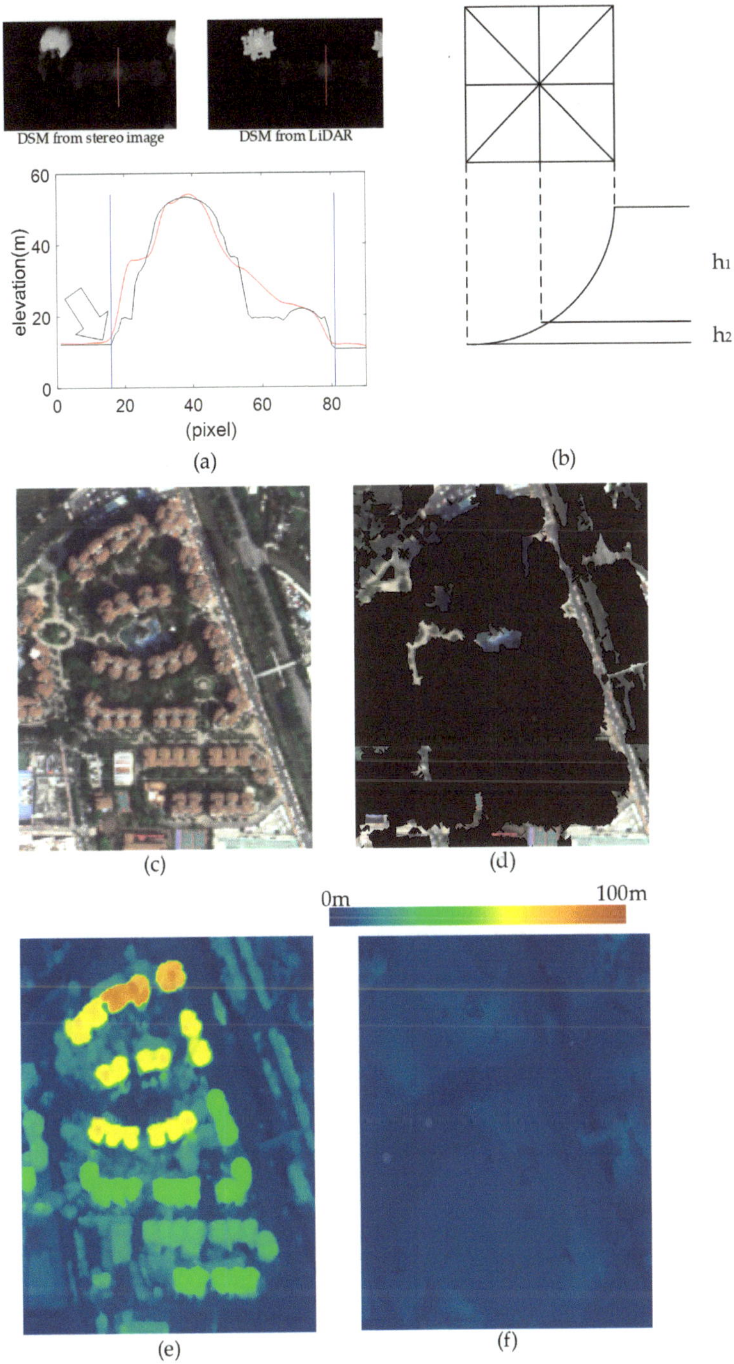

Figure 6. Ground filtering algorithm. (**a**) Profile comparisons of DSMs from LiDAR and the stereo image. (**b**) The way of searching inaccurate ground points. (**c**) Multispectral image. (**d**) Inaccurate points. (**e**) Input DSM. (**f**) Generated DEM.

Finally, the progressive TIN densification algorithm [42] is employed to filter the ground points in the DSM. Figure 6e is the input DSM, and Figure 6f is the generated DEM. The mean elevation around the buildings in the DEM is used as the ground elevation, denoted as E_{ground}.

2.6. Segmentation of Building Rooftop Contours Containing Multiple Elevations

The material of the podium building and building side is similar to that of the main building rooftop, making it difficult to distinguish them in remote sensing images. Consequently, some building rooftop contours in input data encompass the podium building and building side. To address this problem, differences in building contour between forward and backward images are utilized to segment these building rooftop contours. The algorithm process is as follows:

Algorithm 3. Building rooftop contour segmentation process

- Input: Epipolar images EI_{bwd} and EI_{fwd}, building rooftop contour B_r, matched building edge sets S_{bwd} and S_{fwd}, contour matching degrees CM_{fwd} and CM_{bwd}.
- Output: Building rooftop contours B_m, B_p.
- Identify building contours that need to be segmented based on S_{bwd}, S_{fwd}, CM_{fwd}, CM_{bwd}.
- Extract samples of the main building rooftop and samples of the podium building rooftop using S_{bwd} and S_{fwd}.
- Utilize clustering algorithms to classify pixels in EI_{fwd} and obtain the main building rooftop B_m using the extracted samples.
- Podium building rooftop $B_p = B_r - B_m$.
- Apply Algorithm 2 to B_p. Classify B_p as podium building or building side.

In contour matching, a matched building edge has a long enough parallel line in the image. We proposed a method to identify matched edges. We divide the building contour template into multiple subsets based on the edges in the building rooftop contours. For each subset, the total number of G_i is denoted as num_e^{total}. For each G_i within the subset, the distance between Pt_i^{max} and Pt_i is calculated. To distinguish points inside the building contour from points outside the building contour, the distance of the point inside the building contour is set to a negative value. Considering that the lines in the image have dimensions, the distance intervals $[-D_{max}, -D_{max}+k]$, $[-D_{max}+1, -D_{max}+k+1]$, ..., $[D_{max}-k, D_{max}]$ are used to represent the parallel lines. The k represents the width of the parallel line and is set to 2 pixels. If the distance between Pt_i^{max} and Pt_i belongs to any internal, Pt_i^{max} belongs to this parallel line. The parallel line with the most contour points is the longest, denoting this contour point number as num_e. When $\frac{num_e}{num_e^{total}} > 0.5$, the edge is considered as a matched edge. Set S_{bwd} to represent matched edges set in the backward image, and S_{fwd} to represent matched edges set in the forward image. Figure 7 shows two building rooftop contours and the corresponding S_{bwd}, S_{fwd}.

As shown in Figure 7, the matched edges are different in the forward and backward images. Due to the tilt angles, the building sides in the backward image are occluded in the forward images. Additionally, the relative location between the podium building and the main building has changed. The differences between S_{bwd} and S_{fwd} provide samples for building contour segmentation. Define the set of edges $S_{me} = S_{bwd} \cap S_{fwd}$, where the edges in S_{me} belong to the main building rooftop. Define the set of edges $S_{pe} = S_{bwd} - S_{fwd}$, where the edges in S_{pe} belong to the podium building rooftop. By buffering S_{me} and intersecting it with the building contours, the samples of the main building are obtained. Similarly, applying these operations to S_{pe} provides samples of the podium building. In Figure 8a, the red edges represent S_{me}, and the blue edges represent S_{pe}. Meanwhile, Figure 8b shows samples of the main building rooftop, and Figure 8c shows samples of the podium building.

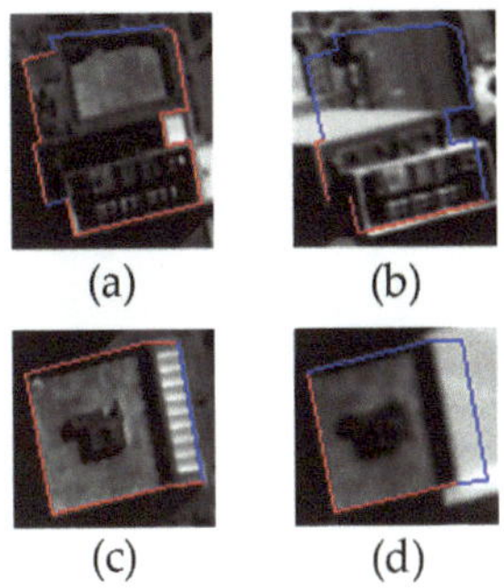

Figure 7. The building rooftop contours with multiple elevations and their matched edges. The red edges in the image indicate matched edges. (**a**) The building rooftop contour encompassing the podium building. (**c**) The building rooftop contour encompassing the building side. (**b**,**d**) The contour matching results in the forward image.

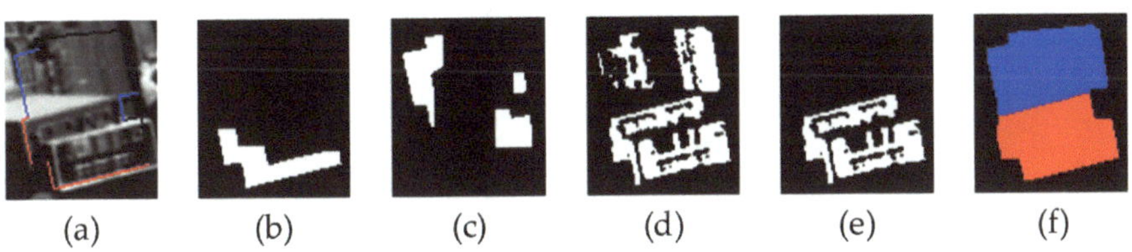

Figure 8. Building contour segmentation process. (**a**) The S_{me} and S_{pe}; (**b**) Samples of the main building rooftop; (**c**) Samples of the podium building; (**d**) Initial main building rooftop; (**e**) Main building rooftop; (**f**) Result of segmentation.

The pixels within the building rooftop in the forward images are classified into main building pixels and podium building pixels based on their grayscale. The K-means clustering algorithm is employed to group these pixels into eight clusters. For each cluster, the numbers of pixels in main building samples and podium building samples are counted, separately. If the number of pixels in the main building samples exceeds those within the podium building samples, this cluster is considered as a part of the main building rooftop. The resulting main building rooftop from this process is depicted in Figure 8d. Due to the limitations of panchromatic images, pixels with the same grayscale as the main building rooftop are misclassified. To address this issue, the parts overlapping with the samples of the main building are preserved, illustrated in Figure 8e. Thereafter, the longest edge in the original building contour is found to assist in gap filling. For each pixel outside the main building rooftop, parallel and perpendicular lines of the longest edge are drawn. If both ends of the parallel or perpendicular lines intersect with the main building rooftop, the pixel is considered part of the main building rooftop. We denote the main building rooftop as B_m, while the remaining building rooftop is a podium building, denoted as B_p. Figure 8f shows the classification result, where the red area represents B_m, and the blue area represents B_p.

For podium building rooftop B_p, the contour matching algorithm is executed. B_p is identified as a podium building when a building rooftop is matched in the forward image. Otherwise, it is considered as occluded building sides. Following Zhang's algorithm [26] as a reference, this paper conducted building contour segmentation experiments in Xi'an. Figure 9 shows the partial results of the building contour segmentation.

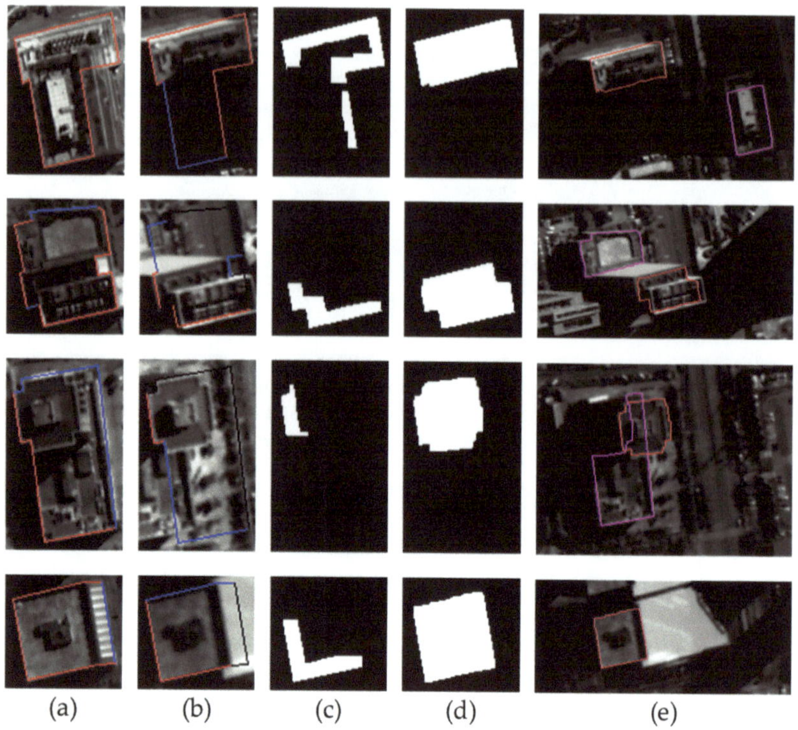

Figure 9. Building contour segmentation results. (**a**) The known building contours, where the red edges indicate matched edges. (**b**) The contour matching results, where the red edges indicate S_{me}. (**c**) The samples of the main building rooftop. (**d**) The main building rooftop extracted by our method. (**e**) The contour matching results after segmentation.

3. Results

3.1. Data Description and Experimental Area

This paper selected three regions—Yingde and Guangzhou in Guangdong Province, and Xi'an in Shaanxi Province—as experimental areas for the algorithm. Their basic details are as follows:

As for the Yingde experimental area, the GF-7 image was captured on 11 October 2020. The center coordinates of the backward image were 113.409°E and 24.326°N, with solar zenith and azimuth angles of 33.466° and 158.717°, respectively. A total of 841 building footprints within this experimental area were acquired. The images and the building footprints of the Yingde experimental area are shown in Figure 10. The DSM used in the experiments was computed using He et al.'s algorithm [32]. LiDAR data from the experimental area were collected as the reference for building heights. Figure 11 displays the DSM obtained from the LiDAR data and the DSM generated from the stereo images.

In the Guangzhou experimental area, the GF-7 image was captured on 14 March 2020. The center coordinates of the backward image were 113.329°E and 23.137°N, with solar zenith and azimuth angles of 32.013° and 140.211°, respectively, as shown in Figure 12. A total of 89,093 building rooftop contours were extracted from the backward image by a building extraction algorithm. The DSM utilized in the experiments was derived using He et al.'s algorithm [32]. LiDAR data from this region served as the reference for building heights. Figure 13 illustrates a portion of the extracted building rooftop contours, the DSM obtained from LiDAR data, and the DSM generated from stereo images.

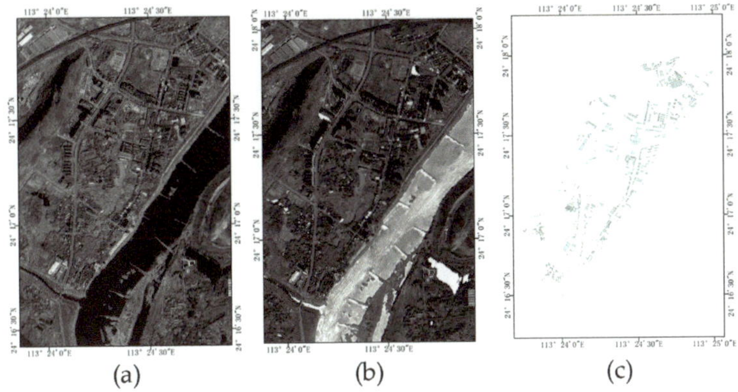

Figure 10. The images and building footprints of the Yingde experimental area. (**a**) The backward image, (**b**) the forward image, and (**c**) the building footprints.

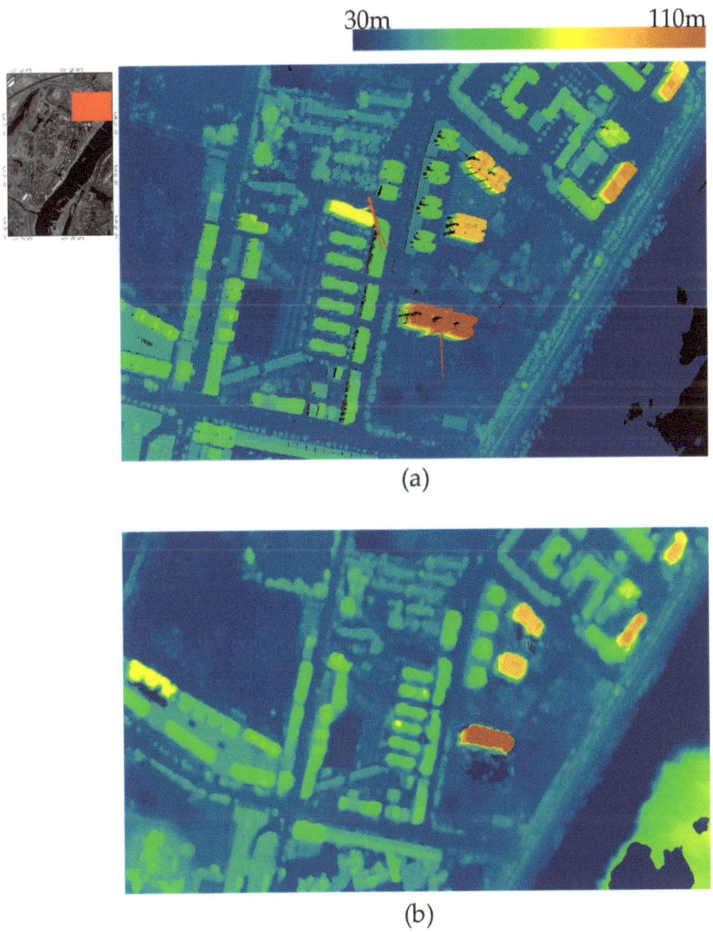

Figure 11. The DSM of Yingde experimental area. (**a**) The DSM obtained from the LiDAR data, with a spatial resolution of 1 m; (**b**) the DSM generated from the GF-7 stereo images.

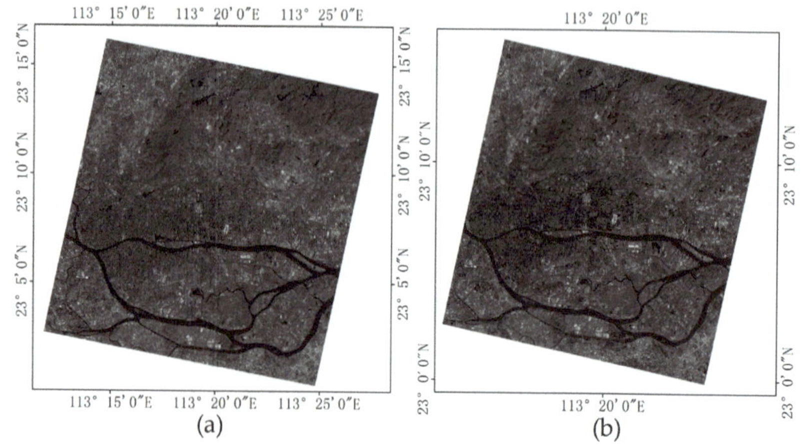

Figure 12. The GF-7 image of the Guangzhou experimental area. (**a**) The backward image; (**b**) the forward image.

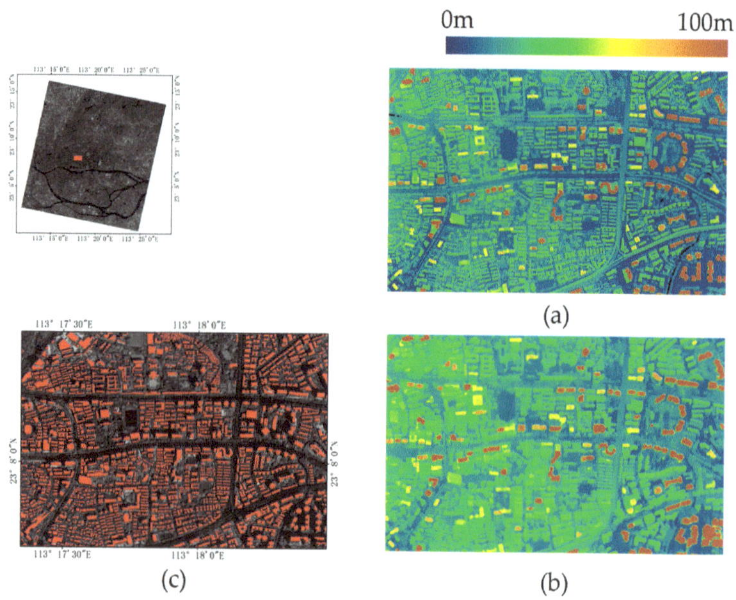

Figure 13. A portion of building rooftop contours and the DSM in the Guangzhou experimental area. (**a**) The DSM obtained from the LiDAR data, with a spatial resolution of 1 m; (**b**) the DSM generated from the GF-7 stereo images; (**c**) the building rooftop contours.

In the Xi'an experimental area, we utilized the dataset provided by Zhang et al. [26] The GF-7 image was captured on 17 February 2020, with the center coordinates of the backward image at 108.951°E and 34.255°N, having solar zenith and azimuth angles of 50.029° and 154.657°, respectively. The Xi'an experimental area encompasses the tallest building in Xi'an (350 m) and its surrounding areas. A total of 34 building rooftop contours were manually marked in the backward image, and reference building heights were obtained through manual marking of corresponding points. The DSM used in the experiments was calculated using He et al.'s algorithm [32]. Figure 14 illustrates the images, building rooftop contours, and the DSM generated from stereo images.

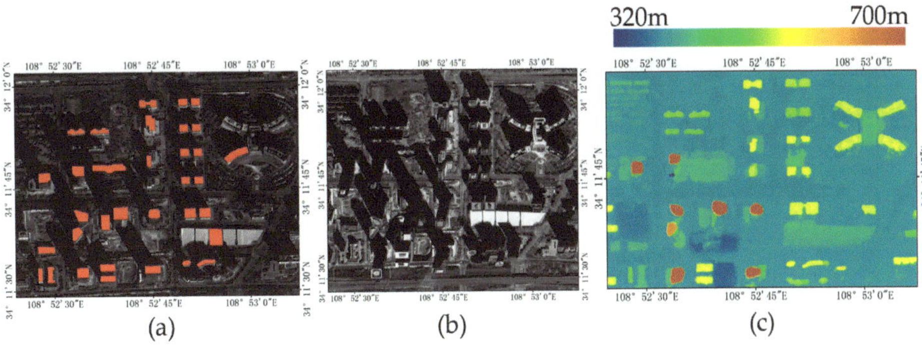

Figure 14. The Xi'an experimental area. (**a**) The backward image and building rooftop contours; (**b**) the forward image; (**c**) the DSM generated from the GF-7 stereo images.

In the Yingde and Guangzhou experimental areas, the reference building heights were calculated according to the vertical distance of ground around the building to the rooftop surface using LiDAR data. However, the production times of the LiDAR data and the GF-7 image were different, which led to different buildings in these data. To ensure the accuracy of the reference building heights in precision assessment, hundreds of buildings were randomly selected and manually removed the building that had discrepancies between the GF-7 images and the LiDAR data. In the Yingde and Guangzhou experimental areas, 343 and 506 buildings were obtained for precision assessment, respectively.

The buildings in the three experimental areas exhibit distinct characteristics that can validate our algorithm in different cases. Figure 15 illustrates the distribution of reference building heights: most buildings in Yingde are below 20 m, while in Guangzhou, the majority of building heights fall within the range of 20 to 100 m, and in Xi'an, half of the buildings are over 100 m. Additionally, the challenges related to contour matching differ across these study areas. In Xi'an, accurate building contours marked by humans are easy to match. Conversely, in Yingde, the building rooftops of adjacent footprints may overlap in images, as depicted in Figure 16a. In Guangzhou, the contour matching suffers from unclear edges, as depicted in Figure 16b.

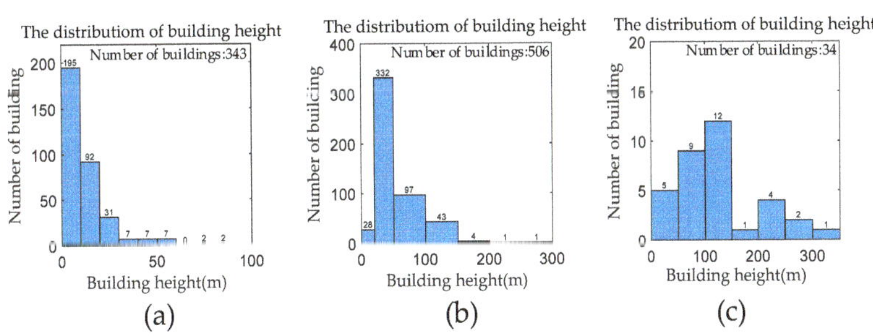

Figure 15. The distribution of reference building height. (**a**) Yingde; (**b**) Guangzhou; (**c**) Xi'an.

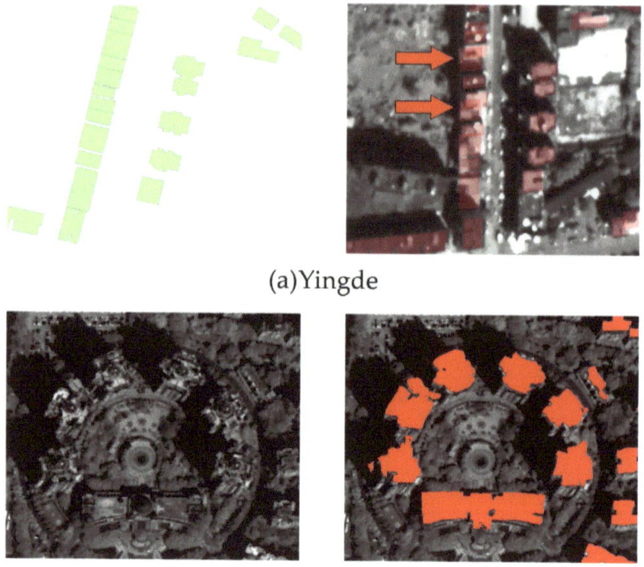

(a)Yingde

(b)Guangzhou

Figure 16. The factors influencing contour matching in the Yingde and Guangzhou experimental areas.

3.2. Evaluation Metrics

This paper evaluates the algorithm's accuracy by comparing the extracted building heights with the reference building heights. Mean error (ME), mean absolute error (MAE), and root mean square error (RMSE) were chosen as the evaluation metrics in this paper. They are calculated as follows:

$$\mathrm{ME} = \frac{1}{N}\sum_{i=1}^{N}\left(h_i - \overline{h_i}\right) \tag{7}$$

$$\mathrm{MAE} = \frac{1}{N}\sum_{i=1}^{N}\left|h_i - \overline{h_i}\right| \tag{8}$$

$$\mathrm{RMSE} = \sqrt{\frac{1}{N}\sum_{i=1}^{N}\left(h_i - \overline{h_i}\right)^2} \tag{9}$$

In the equations, h_i represents the extracted building height, while $\overline{h_i}$ denotes the reference building height.

Due to the building samples used in the experiments, significance testing is conducted to assess whether differences in experiment results are statistically meaningful or could have occurred by chance alone. The t-test was employed to compare the MAEs of two experimental groups. The null hypothesis and alternative hypothesis of the t-test are detailed in the notes following the table.

3.3. Performance of Building Height Extraction

The evaluation result is shown in Figure 17. The MAE and RMSE for each group are calculated and presented in Table 1 below. The right-tailed, two-sample t-test was conducted to compare the MAEs. The results of the t-test are summarized in Table 2. Additionally, Figure 18 displays the 3D reconstruction models of buildings. According to the statistical results and significance testing, our algorithm performed worst in Guangzhou and best in Xi'an.

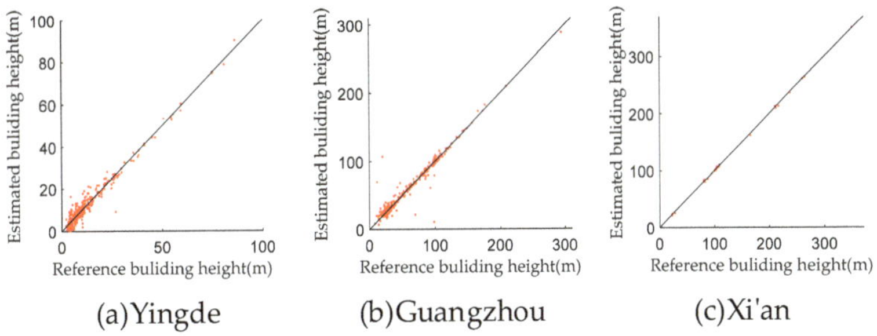

Figure 17. The building height extraction results of our algorithm.

Table 1. Accuracy statistics of our method.

	MAE (m)	RMSE (m)
Yingde	1.96	2.68
Guangzhou	3.76	7.60
Xi'an	1.55	1.93

Table 2. Results of right-tailed, two-sample *t*-test for the proposed algorithm.

Test Case		t	p
Guangzhou	Yingde	3.5637	0.0002
Guangzhou	Xi'an	4.2637	0.0000
Yingde	Xi'an	1.8538	0.0348

Note: 1. For the first row, the null hypothesis states that there is no difference in MAE between the two groups, while the alternative hypothesis suggests that the MAE of the below 20 m group is greater than the MAE of the 20–100 m group. 2. The significance level for all tests was set at 5%.

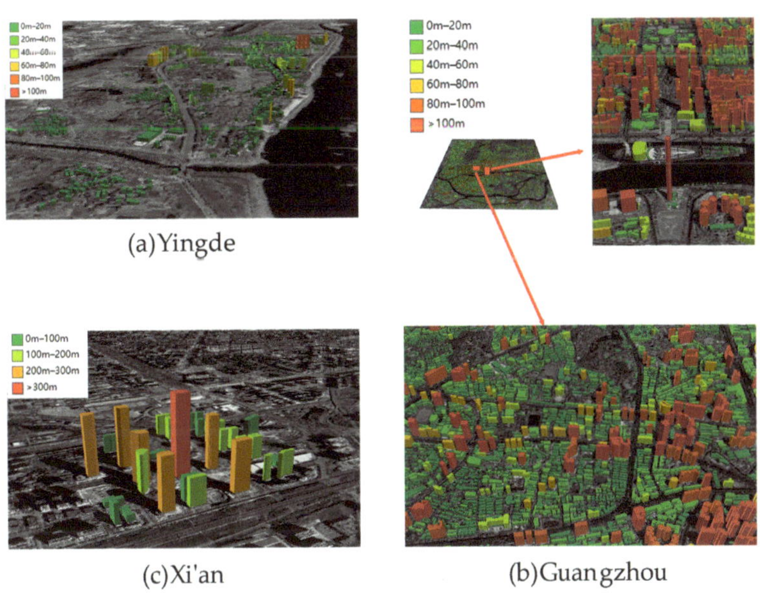

Figure 18. Three-dimensional reconstruction models of buildings.

Our algorithm was implemented in C++ and ran on a desktop computer with an Intel Core i5-6500 processor clocked at 3.20 GHz, featuring four cores and four threads. The algorithm utilized OpenMP for parallelization to leverage multi-core processing capabilities In Guangzhou's experimental areas, contour matching processed 89,093 buildings in a total time of 11,191 s, while ground filtering processed the DSM with dimensions of 34,613 × 38,824 in a total time of 14,041 s.

3.4. Comparative Experiment

The building height extraction methods based on the GF-7 satellite image chosen for the comparison experiments are as follows:

(1) The first comparison calculates building heights using the maximum and minimum elevations within the DSM within the building buffer zone [35], hereafter referred to as the 'DSM method'.
(2) In the second comparison, the ground elevation around the building is extracted by our algorithm, and the building rooftop elevation is extracted using the maximum elevations within DSM elevations within the building buffer zone, hereafter referred to as the 'DSM + DEM method'.
(3) Wang et al.'s method [30] was chosen as the third comparison, hereafter referred to as the 'nDSM method'.
(4) Zhang et al.'s method [26] was compared with ours, hereafter referred to as 'Zhang's method'.

Table 3 summarizes the accuracy of the comparative experiment. As Zhang's algorithm cannot use building footprints as input data, we cite their experimental results in Xi'an [26] for comparison with ours. The right-tailed, two-sample t-test was conducted to compare the MAE of these methods. The results are summarized in Table 4. ME was used to reflect the distribution of errors in this comparative experiment, and the one-sample t-test was conducted to test whether errors followed a normal distribution with a mean of zero. Table 5 shows the result of the one-sample t-test. Figure 19 showcases the distribution of errors in building height extraction. The statistical analysis demonstrated that the building height extraction accuracy achieved by our algorithm outperformed comparative methods across all three study areas. The significance testing in Table 5 shows that the error distribution of the DSM method and DSM + DEM method did not have a mean equal to zero. This means that the building height extracted by these methods was higher than it actually was.

Table 3. Accuracy statistics of building height extraction in the comparative experiment.

	Yingde			Guangzhou			Xi'an		
	ME (m)	MAE (m)	RMSE (m)	ME (m)	MAE (m)	RMSE (m)	ME (m)	MAE (m)	RMSE (m)
DSM	4.48	4.84	7.52	6.19	6.69	10.92	6.74	7.00	8.56
DSM + DEM	4.01	4.35	6.70	4.84	5.40	9.78	3.85	4.85	5.24
nDSM	3.99	4.33	5.47	0.35	4.32	8.65	0.86	4.40	6.17
Zhang	-	-	-	-	-	-	-	1.69	2.23
Ours	−0.32	1.96	2.68	0.22	3.76	7.60	−0.15	1.55	1.93

Table 4. Results of right-tailed, two-sample t-test for the comparative experiment.

Test Case			t	p
Yingde	DSM	Ours	8.8137	0.0000
	DSM + DEM	Ours	8.1771	0.0000
	nDSM	Ours	11.4755	0.0000
	DSM	DSM + DEM	1.1702	0.1212

Table 4. *Cont.*

	Test Case		t	p
Guangzhou	DSM	Ours	7.2432	0.0000
	DSM + DEM	Ours	4.8014	0.0000
	nDSM	Ours	2.6618	0.0039
	DSM	DSM + DEM	2.4266	0.0077
Xi'an	DSM	Ours	6.1901	0.0000
	DSM + DEM	Ours	5.1362	0.0000
	nDSM	Ours	3.6606	0.0004
	DSM	DSM + DEM	2.6572	0.0052

Note: 1. For the first row, the null hypothesis states that there is no difference in MAE between the two groups, while the alternative hypothesis suggests that the MAE of the DSM method is greater than the MAE of our method. 2. The significance level for all tests was set at 5%.

Table 5. Results of one-sample t-test for error distributions.

	Test Case	t	p
Yingde	DSM	13.7416	0.0000
	DSM + DEM	13.8075	0.0000
	nDSM	19.7254	0.0000
	Ours	−2.2283	0.0265
Guangzhou	DSM	15.4601	0.0000
	DSM + DEM	12.7919	0.0000
	nDSM	0.9139	0.3612
	Ours	0.6516	0.5150
Xi'an	DSM	7.3354	0.0000
	DSM + DEM	6.2139	0.0000
	nDSM	0.8068	0.4255
	Ours	−0.4548	0.6515

Note: 1. For the first row, the null hypothesis states that the errors of the DSM method come from a normal distribution with a mean equal to zero and unknown variance, while the alternative hypothesis suggests that the error distribution does not have a mean equal to zero. 2. The significance level for all tests was set at 5%.

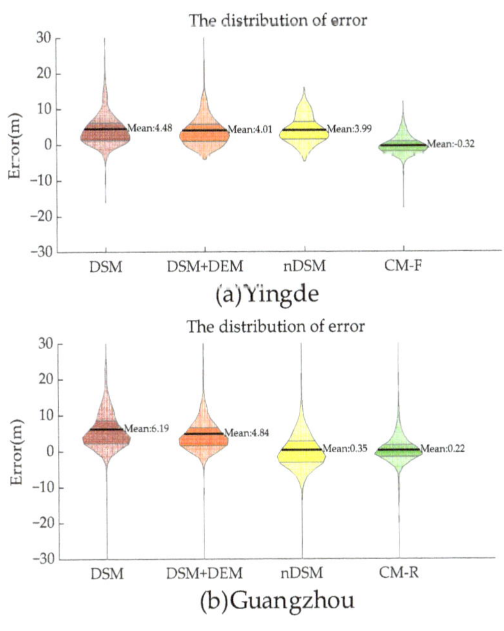

Figure 19. *Cont.*

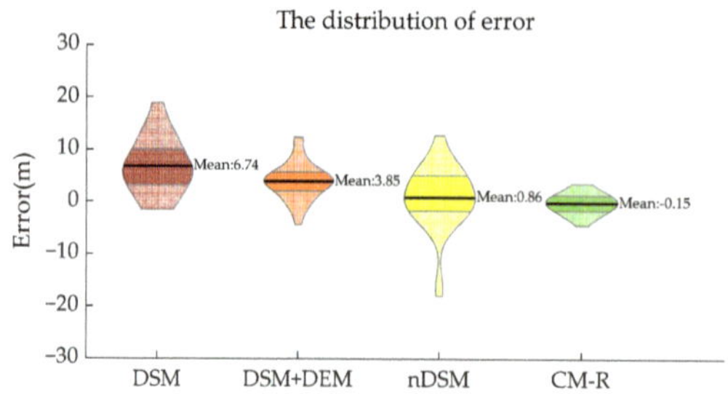

(c)Xi'an

Figure 19. Distribution of building height extraction errors in comparative experiments. (The contour matching algorithm for the building footprint is referred to as the 'CM-F'; the contour matching algorithm for building rooftop contour is referred to as the 'CM-R').

To make a comparison with Zhang's method, the *t*-test for a hypothesized mean was conducted. The null hypothesis states that the absolute errors of our method came from a distribution with a mean of 1.69 m. The *t*-test yielded a *t*-value of −0.6928 with a corresponding *p*-value of 0.4933. This means that Zhang's method demonstrated comparable accuracy to our algorithm in the Xi'an experimental area. However, our method can utilize building footprints as input data, making it more versatile in its application.

3.5. Ablation Experiment

To improve the performance of contour matching, this paper proposes two improvements: contour template correction based on the edges extracted on a backward image and local maximum values filtering by the DSM. The effect of improvements was examined in the ablation experiment. The following algorithms were used in ablation experiments:

(1) Conventional contour matching algorithm [39], hereafter referred to as the 'CM-C'.
(2) Contour matching algorithm with contour template correction based on the edges extracted on backward image, hereafter referred to as the 'CM-I'.
(3) Contour matching algorithm with local maximum values filtering by the DSM, hereafter referred to as the 'CM-D'.

In Yingde, the contour matching algorithm for the building footprint only includes the module that local maximum values filtering. Therefore, CM-C was performed for the ablation experiment. In Guangzhou, all methods were used for the ablation experiment. In Xi'an, due to the high precision of the building rooftop, there was no mismatch in the conventional contour matching method. Therefore, no ablation experiment was conducted.

According to the three-sigma rule of thumb, the thresholds for identifying mismatches were computed using the errors of our method. Table 6 presents the thresholds and the counts of matched buildings and mismatch. Figure 20 illustrates the distribution of absolute error in building heights. The experimental results demonstrate that our improvement can effectively reduce mismatches.

Table 6. Accuracy statistics of building height extraction in the ablation experiment.

	Yingde (343 Buildings)			Guangzhou (506 Buildings)		
	3σ	Matched Buildings (AE < 3σ)	Mismatch	3σ	Matched Buildings (AE < 3σ)	Mismatch
CM-C		242	101		368	138
CM-I		-	-		425	81
CM-D	6.70	-	-	7.93	467	39
CM-F		336	7		-	-
CM-R		-	-		476	30

Note: The contour matching algorithm for the building footprint is referred to as the 'CM-F'. The contour matching algorithm for the building rooftop contour is referred to as the 'CM-R'.

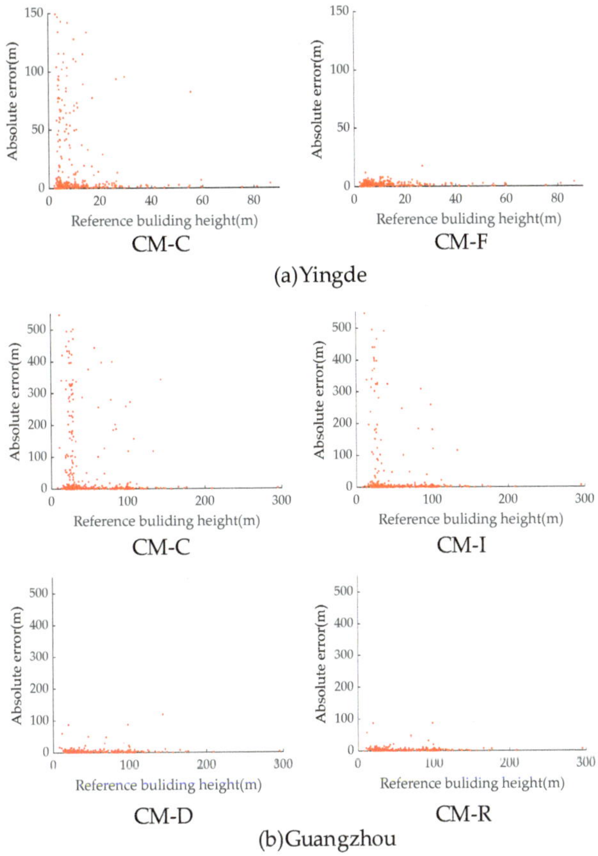

Figure 20. Distribution of building height extraction errors in the ablation experiment.

4. Discussion

4.1. Buildings of Different Heights

The building height extraction methods were compared on the buildings of different heights. The buildings were divided into three groups according to the reference height: below 20 m, between 20 and 100 m, and taller than 100 m. Table 7 provides a summary of the accuracy metrics. In Yingde, conventional methods exhibited poorer performance on buildings between 20 and 100 m compared to those below 20 m. Similarly, they fared worse on buildings taller than 100 m compared to those between 20 and 100 m in Guangzhou. This can be attributed to the disparity discontinuity issue encountered by dense matching algorithms.

In contrast, our proposed algorithm demonstrated superior performance on high-rise buildings relative to low-rise buildings. This was because low-rise buildings are more susceptible to occlusion, whereas high-rise buildings tend to have larger and more distinct features. As shown in Figure 21, there were instances of building loss in the DSM from stereo images for high-rise buildings exceeding 300 m. Nevertheless, our proposed algorithm is capable of matching building rooftops in such cases.

Table 7. Accuracy statistics of buildings with different heights.

		Number		MAE (m)	RMSE (m)
Yingde	<20 m	287	ours	2.03	2.63
			DSM	4.60	7.35
			DSM + DEM	4.33	6.74
			nDSM	4.27	5.49
	20–100 m	56	ours	1.63	2.94
			DSM	6.08	8.31
			DSM + DEM	4.49	6.49
			nDSM	4.62	5.40
	≥100 m	0		-	-
Guangzhou	<20 m	28	ours	5.98	12.07
			DSM	14.72	22.71
			DSM + DEM	13.42	21.67
			nDSM	9.29	17.63
	20–100 m	429	ours	3.01	7.57
			DSM	5.72	9.55
			DSM + DEM	4.74	8.67
			nDSM	3.86	7.94
	≥100 m	49	ours	2.45	3.31
			DSM	10.56	11.77
			DSM + DEM	6.65	7.82
			nDSM	5.57	6.51
Xi'an	<20 m	0		-	-
	20–100 m	14	ours	1.57	1.95
			DSM	7.21	8.49
			DSM + DEM	4.18	4.45
			nDSM	5.80	8.06
	≥100 m	20	ours	1.54	1.92
			DSM	6.85	8.60
			DSM + DEM	3.37	5.73
			nDSM	3.42	4.37

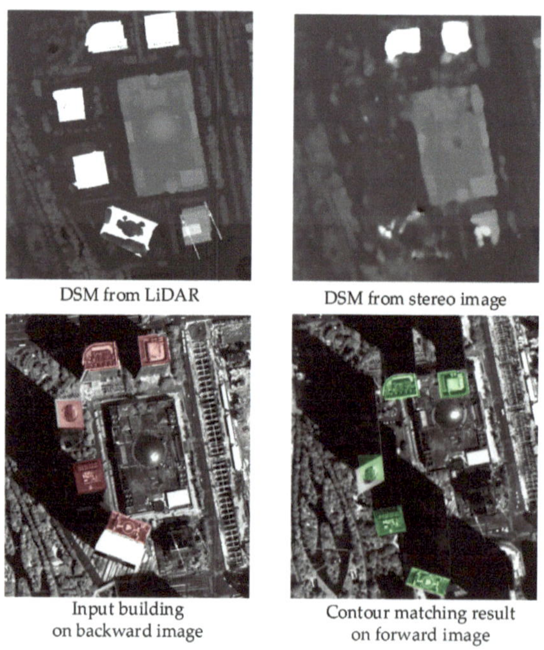

Figure 21. Examples of buildings above 300 m in the Guangzhou experimental area.

4.2. Building in Different Environments

In this paper, the error sources of the algorithm were analyzed in three zones with different environments. Figure 22 shows the three zones. 'Zone 1' is situated in the Xi'an experimental area, characterized by flat terrain and minimal vegetation. 'Zone 2' is located in the Guangzhou experiment area, featuring flat terrain but substantial occlusion by trees. 'Zone 3', also located in Guangzhou, exhibits occluded undulating terrain.

(a) Zone 1 (b) Zone 2 (c) Zone 3

Figure 22. The zones with different environments.

Table 8 summarizes the accuracy of rooftop elevation and ground elevation. According to the ME of rooftop elevation in the three zones, the roof elevation obtained from the DSM was higher than the actual value. The appendages on the rooftop, such as elevator rooms, stairwells, and water tanks, contributed to this discrepancy, as they were higher than the rooftop plane. This primarily accounts for the higher building height extracted by the DSM method in the comparison experiment. In contrast, our method extracts the elevation of the rooftop plane by matching the building rooftop. In applications such as per capita housing area estimation, considering the structural height as the building height becomes necessary. Our method is more suitable for addressing these cases.

Table 8. Accuracy statistics of buildings with different environments.

	Number		Rooftop Elevation			Ground Elevation		
			ME (m)	MAE (m)	RMSE (m)	ME (m)	MAE (m)	RMSE (m)
Zone 1	34	ours	−0.07	1.15	1.42	0.08	1.46	1.86
		DSM	3.93	4.11	5.07	−2.81	3.53	4.83
		nDSM	2.42	3.58	4.67	1.56	3.52	4.84
Zone 2	42	ours	−0.33	0.74	1.16	−0.40	1.49	2.09
		DSM	6.84	6.84	9.00	2.15	3.41	4.73
		nDSM	3.31	3.46	4.16	6.54	6.54	7.23
Zone 3	30	ours	−1.82	1.99	2.73	−3.26	3.80	5.60
		DSM	2.96	3.04	3.28	−0.90	1.18	1.58
		nDSM	2.24	2.71	3.06	2.75	2.75	3.08

According to the ME of ground elevation, the ground elevation extracted by the nDSM method is higher than the actual value. As shown in Figure 6a, 3D breaklines were modeled as smooth transitions from the ground level to the building level. The smooth transitions were easily classified as ground points by the CSF algorithm, resulting in the DEM corresponding to the building location being higher than the surrounding ground. We eliminated inaccurate ground points around buildings, resulting in a more accurate ground elevation, as shown in Figure 23.

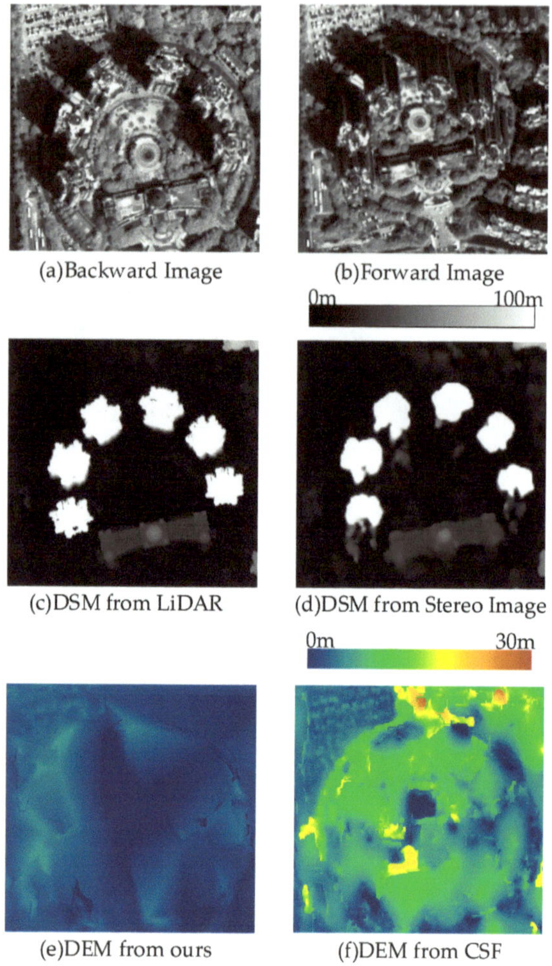

Figure 23. Comparison of ground filtering algorithms.

For traditional methods, the MEs of ground elevation in Zone 2 were larger than those in Zone 1. This indicates that the traditional method faces difficulty in extracting ground elevation in areas with high vegetation coverage. The proposed method uses image classification to ensure that the elevation of occluded ground is estimated from slightly distant ground points. It performs well on flat terrain such as Zone 2.

4.3. Limitation

Unclear building edges. Despite implementing two improvements, namely, contour template correction based on the edges extracted on the backward image and local maximum values filtering by the DSM, mismatches caused by unclear building edges remained in the Guangzhou experiment. To address this issue in future research, semantic segmentation can be used to exclude the edges that do not belong to building rooftops, and better edge extraction methods can also be used to extract more complete building edges for more accurate matching.

Occluded undulating terrain. As observed in the experimental results of Zone 3 in Section 4.2, obtaining ground elevation from the DSM in areas where undulating terrain is occluded by trees poses a significant challenge. To overcome this limitation, integrating additional data sources such as ground measurement data or other satellite images could offer a solution.

Pitched roof. As discussed in Section 4.2, the contour matching algorithm performed poorly in Zone 3, attributed to the presence of buildings with pitched roofs. Identification of pitched roofs is still a challenging task due to the limitations of image resolution. We aim to address this challenge in the future by leveraging higher resolution images.

5. Conclusions

This paper proposes a method for extracting building heights from high-resolution GF-7 stereo imagery. The method employs contour matching techniques to enhance building rooftop elevation extraction. Within the contour matching process, the method filters local maximum values by a DSM to resolve the mismatch issue. Moreover, the contour template correction is used to ensure higher precision in cases of unclear building edges. To improve the accuracy of the ground elevation extraction around the building, this method utilized image classification from the GF-7 multispectral imagery to identify and remove error-prone regions within the DSM, aiming to enhance the accuracy of ground filtering. The proposed method was validated in Yingde, Guangzhou, and Xi'an, showcasing its performance against comparative algorithms. The proposed method has more advantages for high-rise buildings. In the rooftop elevation extraction, the proposed algorithm takes the rooftop as an object, unaffected by issues such as smooth transitions in the DSM and rooftop appendages affecting the rooftop, resulting in more accurate results. In the ground elevation extraction, the proposed method effectively removes non-ground points and inaccurate ground points from the DSM, yielding accurate results in flat terrain.

However, problems such as unclear building edges and occluded undulating terrain are still challenges in building height extraction. In future research, semantic segmentation for identifying building edges and other data sources for ground elevation estimation can be considered to improve the accuracies of the elevation of rooftop and the ground elevation. Additionally, different satellite images from different cities, different countries, and even climate zones can be used to validate and improve the proposed methods.

Author Contributions: Conceptualization, Y.C., S.Z. and W.J.; methodology, Y.C. and W.J.; software, Y.C.; validation, Y.C.; formal analysis, Y.C.; investigation, Y.C.; resources, W.J.; data curation, Y.C.; writing—original draft preparation, Y.C.; writing—review and editing, G.Y.; visualization, Y.C. and S.Z.; supervision, S.Z.; project administration, S.Z.; funding acquisition, S.Z. All authors have read and agreed to the published version of the manuscript.

Funding: This work was supported by the High-Resolution Remote Sensing Application Demonstration System for Urban Fine Management under grant 06-Y30F04-9001-20/22.

Data Availability Statement: Data available on request due to restrictions. Our method is based on the original stereo images, which is restricted to be accessed on web according to the data policy of China, we are sorry that we cannot share our research data.

Conflicts of Interest: The authors declare no conflicts of interest.

References

1. Mahtta, R.; Mahendra, A.; Seto, K.C. Building up or Spreading out? Typologies of Urban Growth across 478 Cities of 1 Million+. *Environ. Res. Lett.* **2019**, *14*, 124077. [CrossRef]
2. Du, Y.; Mak, C.M.; Tang, B. Effects of Building Height and Porosity on Pedestrian Level Wind Comfort in a High-Density Urban Built Environment. *Build. Simul.* **2018**, *11*, 1215–1228. [CrossRef]
3. Li, X.; Yang, B.; Liang, F.; Zhang, H.; Xu, Y.; Dong, Z. Modeling Urban Canopy Air Temperature at City-Block Scale Based on Urban 3D Morphology Parameters—A Study in Tianjin, North China. *Build. Environ.* **2023**, *230*, 110000. [CrossRef]
4. Xu, S.; Li, G.; Zhang, H.; Xie, M.; Mendis, T.; Du, H. Effect of Block Morphology on Building Energy Consumption of Office Blocks: A Case of Wuhan, China. *Buildings* **2023**, *13*, 768. [CrossRef]
5. Zhou, X.; Huang, Z.; Scheuer, B.; Wang, H.; Zhou, G.; Liu, Y. High-Resolution Estimation of Building Energy Consumption at the City Level. *Energy* **2023**, *275*, 127476. [CrossRef]
6. Hang, J.; Li, Y.; Sandberg, M.; Buccolieri, R.; Di Sabatino, S. The Influence of Building Height Variability on Pollutant Dispersion and Pedestrian Ventilation in Idealized High-Rise Urban Areas. *Build. Environ.* **2012**, *56*, 346–360. [CrossRef]

7. Kim, J.-W.; Baik, J.-J.; Park, S.-B.; Han, B.-S. Impacts of Building-Height Variability on Turbulent Coherent Structures and Pollutant Dispersion: Large-Eddy Simulations. *Atmos. Pollut. Res.* **2023**, *14*, 101736. [CrossRef]
8. Zhang, X.; Liao, Q.; Yin, X.; Yin, Z.; Cao, Q. Spatial Characteristics and Influencing Factors of Multi-Scale Urban Living Space (ULS) Carbon Emissions in Tianjin, China. *Buildings* **2023**, *13*, 2393. [CrossRef]
9. Lian, H.; Zhang, J.; Li, G.; Ren, R. The Relationship between Residential Block Forms and Building Carbon Emissions to Achieve Carbon Neutrality Goals: A Case Study of Wuhan, China. *Sustainability* **2023**, *15*, 15751. [CrossRef]
10. Tosi, P.; De Rubeis, V.; Sbarra, P. Earthquake Perception Data Highlight Natural Frequency Details of Italian Buildings. *Earthq Spectra* **2023**, *39*, 1240–1254. [CrossRef]
11. Gui, S.; Qin, R. Automated LoD-2 Model Reconstruction from Very-High-Resolution Satellite-Derived Digital Surface Model and Orthophoto. *ISPRS J. Photogramm. Remote Sens.* **2021**, *181*, 1–19. [CrossRef]
12. Baltsavias, E.P. A Comparison between Photogrammetry and Laser Scanning. *ISPRS J. Photogramm. Remote Sens.* **1999**, *54*, 83–94 [CrossRef]
13. Sun, S.; Salvaggio, C. Aerial 3D Building Detection and Modeling From Airborne LiDAR Point Clouds. *IEEE J. Sel. Top. Appl. Earth Obs. Remote Sens.* **2013**, *6*, 1440–1449. [CrossRef]
14. Lao, J.; Wang, C.; Zhu, X.; Xi, X.; Nie, S.; Wang, J.; Cheng, F.; Zhou, G. Retrieving Building Height in Urban Areas Using ICESat-2 Photon-Counting LiDAR Data. *Int. J. Appl. Earth Obs. Geoinf.* **2021**, *104*, 102596. [CrossRef]
15. Zheng, Y.; Weng, Q. Model-Driven Reconstruction of 3-D Buildings Using LiDAR Data. *IEEE Geosci. Remote Sens. Lett.* **2015**, *12*, 1541–1545. [CrossRef]
16. Li, X.; Zhou, Y.; Gong, P.; Seto, K.C.; Clinton, N. Developing a Method to Estimate Building Height from Sentinel-1 Data. *Remote Sens. Environ.* **2020**, *240*, 111705. [CrossRef]
17. Esch, T.; Brzoska, E.; Dech, S.; Leutner, B.; Palacios-Lopez, D.; Metz-Marconcini, A.; Marconcini, M.; Roth, A.; Zeidler, J. World Settlement Footprint 3D—A First Three-Dimensional Survey of the Global Building Stock. *Remote Sens. Environ.* **2022**, *270*, 112877. [CrossRef]
18. Dong, B.; Zheng, Q.; Lin, Y.; Chen, B.; Ye, Z.; Huang, C.; Tong, C.; Li, S.; Deng, J.; Wang, K. Integrating Physical Model-Based Features and Spatial Contextual Information to Estimate Building Height in Complex Urban Areas. *Int. J. Appl. Earth Obs. Geoinf.* **2024**, *126*, 103625. [CrossRef]
19. Zhuang, D.; Zhang, L.; Zou, B. Interferometry Modeling and Model-Based Height Estimation for Buildings in Urban DSM Reconstruction Based on Interferometric Synthetic Aperture Radar Technology. *J. Appl. Remote Sens.* **2023**, *17*, 034508. [CrossRef]
20. Sun, Y.; Hua, Y.; Mou, L.; Zhu, X.X. Large-Scale Building Height Estimation from Single VHR SAR Image Using Fully Convolutional Network and GIS Building Footprints. In Proceedings of the 2019 Joint Urban Remote Sensing Event (JURSE), Vannes, France, 22–24 May 2019; IEEE: Vannes, France, 2019; pp. 1–4.
21. Izadi, M.; Saeedi, P. Three-Dimensional Polygonal Building Model Estimation From Single Satellite Images. *IEEE Trans. Geosci. Remote Sens.* **2012**, *50*, 2254–2272. [CrossRef]
22. Lee, T.; Kim, T. Automatic Building Height Extraction by Volumetric Shadow Analysis of Monoscopic Imagery. *Int. J. Remote Sens.* **2013**, *34*, 5834–5850. [CrossRef]
23. Qi, F.; Zhai, J.Z.; Dang, G. Building Height Estimation Using Google Earth. *Energy Build.* **2016**, *118*, 123–132. [CrossRef]
24. Zhao, Y.; Wu, B.; Li, Q.; Yang, L.; Fan, H.; Wu, J.; Yu, B. Combining ICESat-2 Photons and Google Earth Satellite Images for Building Height Extraction. *Int. J. Appl. Earth Obs. Geoinf.* **2023**, *117*, 103213. [CrossRef]
25. Zhang, H.; Xu, C.; Fan, Z.; Li, W.; Sun, K.; Li, D. Detection and Classification of Buildings by Height from Single Urban High-Resolution Remote Sensing Images. *Appl. Sci.* **2023**, *13*, 10729. [CrossRef]
26. Zhang, C.; Cui, Y.; Zhu, Z.; Jiang, S.; Jiang, W. Building Height Extraction from GF-7 Satellite Images Based on Roof Contour Constrained Stereo Matching. *Remote Sens.* **2022**, *14*, 1566. [CrossRef]
27. Liu, C.; Huang, X.; Wen, D.; Chen, H.; Gong, J. Assessing the Quality of Building Height Extraction from ZiYuan-3 Multi-View Imagery. *Remote Sens. Lett.* **2017**, *8*, 907–916. [CrossRef]
28. Hirschmuller, H. Stereo Processing by Semiglobal Matching and Mutual Information. *IEEE Trans. Pattern Anal. Mach. Intell.* **2008**, *30*, 328–341. [CrossRef] [PubMed]
29. Zhang, K.; Chen, S.-C.; Whitman, D.; Shyu, M.-L.; Yan, J.; Zhang, C. A Progressive Morphological Filter for Removing Nonground Measurements from Airborne LIDAR Data. *IEEE Trans. Geosci. Remote Sens.* **2003**, *41*, 872–882. [CrossRef]
30. Wang, J.; Hu, X.; Meng, Q.; Zhang, L.; Wang, C.; Liu, X.; Zhao, M. Developing a Method to Extract Building 3D Information from GF-7 Data. *Remote Sens.* **2021**, *13*, 4532. [CrossRef]
31. Zhang, W.; Qi, J.; Wan, P.; Wang, H.; Xie, D.; Wang, X.; Yan, G. An Easy-to-Use Airborne LiDAR Data Filtering Method Based on Cloth Simulation. *Remote Sens.* **2016**, *8*, 501. [CrossRef]
32. He, S.; Zhou, R.; Li, S.; Jiang, S.; Jiang, W. Disparity Estimation of High-Resolution Remote Sensing Images with Dual-Scale Matching Network. *Remote Sens.* **2021**, *13*, 5050. [CrossRef]
33. Laga, H.; Jospin, L.V.; Boussaid, F.; Bennamoun, M. A Survey on Deep Learning Techniques for Stereo-Based Depth Estimation. *IEEE Trans. Pattern Anal. Mach. Intell.* **2022**, *44*, 1738–1764. [CrossRef] [PubMed]
34. Li, S.; He, S.; Jiang, S.; Jiang, W.; Zhang, L. WHU-Stereo: A Challenging Benchmark for Stereo Matching of High-Resolution Satellite Images. *IEEE Trans. Geosci. Remote Sens.* **2023**, *61*, 5603914. [CrossRef]

35. Chen, P.; Huang, H.; Liu, J.; Wang, J.; Liu, C.; Zhang, N.; Su, M.; Zhang, D. Leveraging Chinese GaoFen-7 Imagery for High-Resolution Building Height Estimation in Multiple Cities. *Remote Sens. Environ.* **2023**, *298*, 113802. [CrossRef]
36. Cao, Y.; Huang, X. A Deep Learning Method for Building Height Estimation Using High-Resolution Multi-View Imagery over Urban Areas: A Case Study of 42 Chinese Cities. *Remote Sens. Environ.* **2021**, *264*, 112590. [CrossRef]
37. Perko, R.; Raggam, H.; Gutjahr, K.H.; Schardt, M. Advanced Dtm Generation from Very High Resolution Satellite Stereo Images. *ISPRS Ann. Photogramm. Remote Sens. Spat. Inf. Sci.* **2015**, *II-3/W4*, 165–172. [CrossRef]
38. Canny, J. A Computational Approach to Edge Detection. *IEEE Trans. Pattern Anal. Mach. Intell.* **1986**, *PAMI-8*, 679–698. [CrossRef]
39. Gong, J.; Hu, X.; Pang, S.; Wei, Y. Roof-Cut Guided Localization for Building Change Detection from Imagery and Footprint Map. *Photogramm. Eng. Remote Sens.* **2019**, *85*, 543–558. [CrossRef]
40. Ebisch, K. A Correction to the Douglas–Peucker Line Generalization Algorithm. *Comput. Geosci.* **2002**, *28*, 995–997. [CrossRef]
41. Huang, X.; Zhang, L. Morphological Building/Shadow Index for Building Extraction From High-Resolution Imagery Over Urban Areas. *IEEE J. Sel. Top. Appl. Earth Obs. Remote Sens.* **2012**, *5*, 12. [CrossRef]
42. Zhang, J.; Lin, X. Filtering Airborne LiDAR Data by Embedding Smoothness-Constrained Segmentation in Progressive TIN Densification. *ISPRS J. Photogramm. Remote Sens.* **2013**, *81*, 44–59. [CrossRef]

remote sensing

MDPI

Article

LVI-Fusion: A Robust Lidar-Visual-Inertial SLAM Scheme

Zhenbin Liu, Zengke Li *, Ao Liu, Kefan Shao, Qiang Guo and Chuanhao Wang

School of Environment Science and Spatial Informatics, China University of Mining and Technology, Xuzhou 221116, China; lzb@cumt.edu.cn (Z.L.); liuao@cumt.edu.cn (A.L.); guo_qiang934@cumt.edu.cn (Q.G.); ts23160184p31@cumt.edu.cn (C.W.)

* Correspondence: zengkeli@cumt.edu.cn

Abstract: With the development of simultaneous positioning and mapping technology in the field of automatic driving, the current simultaneous localization and mapping scheme is no longer limited to a single sensor and is developing in the direction of multi-sensor fusion to enhance the robustness and accuracy. In this study, a localization and mapping scheme named LVI-fusion based on multi-sensor fusion of camera, lidar and IMU is proposed. Different sensors have different data acquisition frequencies. To solve the problem of time inconsistency in heterogeneous sensor data tight coupling, the time alignment module is used to align the time stamp between the lidar, camera and IMU. The image segmentation algorithm is used to segment the dynamic target of the image and extract the static key points. At the same time, the optical flow tracking based on the static key points are carried out and a robust feature point depth recovery model is proposed to realize the robust estimation of feature point depth. Finally, lidar constraint factor, IMU pre-integral constraint factor and visual constraint factor together construct the error equation that is processed with a sliding window-based optimization module. Experimental results show that the proposed algorithm has competitive accuracy and robustness.

Keywords: IMU; monocular camera; lidar; SLAM; sensor fusion

1. Introduction

Surveying and mapping technology based on lidar and photogrammetry has developed rapidly. With the rapid rise of automatic driving, UAV (unmanned aerial vehicle) and other fields, surveying and mapping technology based on mobile platforms has also been further developed and has experienced a process from static to dynamic surveying and mapping. SLAM (simultaneous localization and mapping) technology as the basic module in these fields, has also been rapidly developed [1].

SLAM technology is roughly divided into two categories according to the different forms of sensors. Vision SLAM technology mainly uses sensors in the form of monocular, binocular, and RGB-D forms [2]. Lidar (Light Detection and Ranging)-based SLAM technology is dominated by 2D (two dimensions) lidar and 3D (three dimensions) lidar [3]. Among them, 2D lidar is mainly used in indoor plane positioning and mapping, and 3D lidar is used in outdoor 3D localization and mapping. SLAM technology based on vision sensor relies more on environmental texture information and lighting conditions, while SLAM technology based on lidar is prone to motion degradation in a structured environment. As an environment-independent sensor, the IMU (Inertial Measurement Unit) can measure the acceleration and angular velocity of the carrier at high frequencies. However, its positioning accuracy has been poor for a long time because the IMU contains the influence of zero bias and noise. At present, with the increasing complexity of mobile robot application scenarios, a single sensor can no longer meet the demand, and SLAM technology is gradually developing in the direction of multi-sensor fusion. SLAM technology that fuses vision with IMU sensors is also known as VI-SLAM. The fusion of lidar and IMU sensors is also known as LI-SLAM. The IMU can effectively assist the lidar sensor in point cloud undistortion and

Citation: Liu, Z.; Li, Z.; Liu, A.; Shao, K.; Guo, Q.; Wang, C. LVI-Fusion: A Robust Lidar-Visual-Inertial SLAM Scheme. *Remote Sens.* **2024**, *16*, 1524. https://doi.org/10.3390/rs16091524

Academic Editor: Joaquín Martínez-Sánchez

Received: 21 March 2024
Revised: 16 April 2024
Accepted: 24 April 2024
Published: 25 April 2024

provide pose constraints for vision and lidar sensors and high-frequency pose initial values to speed up optimization [4]. With the development of SLAM research, researchers realized that vision and lidar also have good complementary properties, and SLAM schemes for the fusion of lidar and vision sensors have gradually emerged [5].

The SLAM scheme based on multi-sensor fusion is mainly based on filtering and optimization [6]. The above two processing methods are essentially solving the maximum a posteriori estimation of the state variable. Experiments show that under the same computational complexity, the fusion method with graph optimization has better effect and more heterogeneous sensors are more easily fused [7]. Therefore, this study chooses the graph optimization method for fusion vision, lidar and IMU. LVI-SAM [8] (state-of-the-art SLAM algorithms) proposes a factorial graph framework to fuse vision, lidar, and IMU sensors. The LVI-SAM uses a lidar inertial odometer to assist in the initialization of the visual inertial odometer, which provides initial pose estimation for lidar matching. This approach is more like two separate systems running separately, without simultaneous data processing. Different from existing multi-sensor fusion SLAM schemes, LVI-fusion proposed in this paper is a tight-coupled system in a true sense. The contributions of this article are as follows:

1. Based on the issue of inconsistent time frequency between the camera and lidar, we proposed a time alignment module, which divide and merge point clouds according to visual time. This method can effectively solve the problem of time asynchrony in the tight coupling process between vision and lidar sensors.
2. The image segmentation algorithm is used to segment the dynamic target of the image, eliminating the influence of dynamic objects, and the static key points can be achieved.
3. A robust feature point depth estimation scheme is proposed. The sub-map is used to assign depth to each image key frame feature point, and the 3D world point coordinates are calculated for the same feature point under different camera positions and poses. When the depth estimation is accurate, the world coordinates recovered by the same feature point under different keyframe pose should be consistent. In this way, the depth of feature points can be estimated robustly.
4. In order to ensure the real-time performance of the back-end optimization, a sliding window optimization method is adopted for pose calculation, and we implemented a complete multi-sensor fusion SLAM scheme. To extensively validate the positioning accuracy performance of the proposed method, extensive experiments are carried out in the M2DGR dataset [9] and a campus dataset collected by our equipment. The results illustrate that the proposed approach outperforms the state-of-the-art SLAM schemes.

The rest of this article is organized as follows. The related work of vision SLAM, lidar SLAM and SLAM of vision and lidar fusion are presented in Section 2. Section 3 describes the factor graph framework proposed in this paper in detail. The experimental setup and precision evaluation is discussed in Section 4, and we draw our conclusions in Section 5

2. Related Works

At present, there are many excellent SLAM schemes based on vision and lidar [3]. However, providing a detailed overview of the existing SLAM technology is impossible due to the length restriction of the paper. Hence, this paper attempts to summarize representative SLAM schemes. According to the category of sensors, this study divides SLAM technology into three categories, visual SLAM, lidar SLAM, and SLAM of vision and lidar fusion [10]. The following is an overview of the three types of SLAM technologies.

2.1. Visual SLAM

Visual SLAM has been around for nearly 30 years. In the past decade, with the popularity of autonomous driving, drones, various service robots, VR/AR and other industries, SLAM has been widely studied by researchers. MonoSLAM is the first monocular SLAM scheme that can run in real time [11]. This algorithm uses Harris corner points for tracking at the front end, constant velocity model for forecasting, and EKF (Extended KalmanFilter) for pose estimation at the back end, which is of milestone significance. PTAM innovatively proposed the concepts of front-end and back-end of V-SLAM based on the monocular camera, where in the front-end was responsible for the extraction and tracking of feature points, and the back-end used BA (bundle adjustment) for the pose optimization update and map construction [12]. The ORB-SLAM [13] scheme builds an image pyramid for incoming images and uses ORB features [14] for feature extraction and matching. Compared with PTAM, ORB-SLAM scheme has better scale and rotation invariance and can achieve more stable tracking. Based on the ORB-SLAM foundation, the ORB-SLAM2 supports SLAM implementation of multiple camera models and adds map reuse and relocation functions [15]. The localization accuracy of the above visual SLAM scheme based on image feature point matching depends heavily on the accuracy of feature point matching, and often has poor effect in a texture environment. The direct method, another important branch of visual SLAM, builds a mathematical model based on the assumption of constant luminosity between adjacent frames, avoiding the process of key point extraction and feature description. The SVO scheme uses key point pixel blocks to construct pixel error recovery pose motion information, which is divided into two steps [16]. The pixels of key points between adjacent frames are compared to obtain a rough pose estimation. On this basis, key points of current frames are further matched with key points of map, and pose optimization is further carried out. The LSD-SLAM scheme [17] is a direct algorithm for semi-intensive reconstruction, which consists of three main steps: tracking, depth map estimation, and map optimization. Based on LSD-SLAM, DSO considers the exposure time and lens distortion, and puts the calibration results into the back-end optimization process and uses the sliding window optimization method to perform real-time motion estimation [18]. The above direct method can make full use of image information and build dense or semi-dense maps, but this modeling method has a poor positioning effect on scenes with large lighting changes. Visual-based SLAM schemes are prone to environmental problems, especially when the vision sensor is a monocular camera, there is also a problem of scale ambiguity, so the fusion of vision and IMU has been widely studied. Visual inertial fusion positioning systems can generally be divided into optimization methods and Filter methods, in which the multi-state constraint Kalman filter (MSCKF) is a typical representative of filter-based methods [19]. OKVIS-implemented binocular inertial odometer with an optimization method, which constructed the visual reprojection error and IMU constraints, is optimized by using a fixed sliding window of key frame [20]. The VINS series is one of the perfect examples of visual-IMU fusion SLAM systems based on optical flow tracking [21,22]. Based on ORB-SLAM2, ORB-SLAM3 proposes a fast and robust visual IMU initialization method, which is a representative scheme of visual IMU fusion based on feature method [23]. The above schemes are summarized in Table 1.

Table 1. Representational SLAM scheme based on visual and IMU.

Scheme	Release Time	Sensor Form	Characteristic
MonoSLAM [11]	2007	a	EKF + Feature point method
PTAM [12]	2007	a	Feature point method
ORB-SLAM [13]	2011	a	ORB feature point method
ORB-SLAM2 [15]	2015	b	ORB feature + multiple mode cameras
SVO [16]	2014	a	Semi-direct method
LSD-SLAM [17]	2014	a	Direct method + semi-dense reconstruction
DSO [18]	2020	b	Direct method+ Sparse reconstruction
MSCKF [19]	2020	c	IESKF (iterative error state Kalman filter)
OKVIS [20]	2020	c	Key frame + graph optimization
VINS-mono [21]	2017	c	Optical flow + graph optimization
VINS-fusion [22]	2019	c	Optical flow + multimode
ORB-SLAM3 [23]	2021	c	ORB feature+ multimode

a indicates support for monocular camera, b indicates support for multiple mode cameras, and c indicates support for visual integration with IMU.

2.2. Lidar SLAM

Lidar-based SLAM schemes can be divided into 2D SLAM and 3D SLAM, and since the emergence of Cartographer [24], 2D lidar SLAM indicates basic maturity. Compared with 2D lidar, 3D lidar can perceive more environmental information. At present, with the price of 3D lidar gradually decreasing, the size is getting smaller and smaller, and the SLAM scheme based on 3D lidar has gradually attracted the attention of researchers. The 3D lidar SLAM is represented by LOAM [25], and the scheme adopts point-to-line and point-to-surface matching, which has great enlightenment for subsequent 3D SLAM. Subsequent researchers have done a lot of work based on the LOAM framework. A-LOAM [26] uses the ceres-solve library to streamline the optimization code of Loam. LeGO-LOAM [27] segments ground points on the basis of LOAM and clusters point clouds to reduce the impact of noise on matching. SC-LeGO-LOAM [28] uses scan context [29] to add a loopback detection module on the basis of Lego-loam. Based on the 3D lidar SLAM, the researchers tried to combine the IMU with lidar, and the IMU assisted the de-distortion of the lidar point cloud. SLAM based on IMU and lidar fusion can be divided into categories based on filtering and optimization according to the backend. Table 2 summarizes some of the most representative lidar SLAM schemes from which most of the current research work begins.

Table 2. Visual-inertial navigation system (VINS) scheme for visual inertial measurement unit (IMU) fusion.

Scheme	Release Time	Sensor Form	Characteristic
LOAM [25]	2014	a	Milestone, based on feature matching
A-LOAM [26]	2018	a	Streamline LOAM code with optimization library
LeGO-LOAM [27]	2018	a	Ground point filtering, point cloud clustering
SC-LeGO-LOAM [28]	2020	a	Add loopback detection based on Scan Context
LIOM [30]	2020	b	CNN dynamic target elimination; ESKF filtering
LIO-Mapping [31]	2019	b	Graph optimization method
LIO-SAM [32]	2020	b	Factor graph optimization method
LINS [33]	2020	b	IESKF (iterative error state Kalman filter)
FAST-LIO [34]	2020	b	IEKF (Iterative Extended Kalman Filtering)
FAST-LIO2 [35]	2021	b	Incremental KD data structure (fast efficiency)
Faster-LIO [36]	2022	b	Use iVox data structure based on FAST-LIO2 to further improve efficiency

a indicates support for 3D lidar, b indicates support for 3D lidar and IMU.

2.3. SLAM of Vision and Lidar Fusion

Vision sensors can obtain rich environmental color information, lidar can obtain distance information to perceive the environment, and the two types of sensors have natural complementary properties. With the deepening of SLAM research, a number of excellent SLAM schemes based on the visual and lidar fusion have gradually emerged. LIMO [37] combines lidar and monocular vision, uses the depth measured by lidar to give depth information to visual feature points, and then predicts robot motion based on key frame BA. V-LOAM [38], a representative SLAM scheme of visual and lidar fusion, ranks second on the KITTI dataset [39]. For many years, V-LOAM adopts a positioning process from coarse to fine, obtains the initial pose according to the visual matching, and the lidar point cloud matches the frame to the local map according to the initial pose to obtain higher accuracy pose results. Lidar-based systems have proven to be superior compared to vision-based systems due to their accuracy and robustness. VIL-SLAM [40] combines tightly coupled stereo vision inertial odometer (VIO) with lidar mapping and lidar-enhanced visual loop closure to solve the problem of motion degradation of lidar in a structured environment. LIC_Fusion [41] is based on the efficient MSCKF framework, using the coefficient edge/suf feature points detected and tracked by the lidar, as well as sparse visual features and IMU readings, to complete the multimodal fusion. LIC_Fusion2 [42] is a lidar, camera and IMU fusion odometer based on sliding window optimization on the basis of LIC_Fusion, which has the function of online spatiotemporal calibration. ULVIO [43] constructs a factor graph that combines vision, lidar and inertial information for optimization. The point features extracted by vision, the line and surface features extracted by lidar, and the residual constructed by IMU pre-integration are put into the same factor graph for optimization. This method has high requirements for hardware time synchronization. R2live [44] estimates the state within the framework of the error-state iterated Kalman filter, and further improves the overall precision with factor graph optimization to guarantee real-time performance. R3live [45] based on R2live, utilizes measurements from solid-state lidar, inertial measurement units, and vision sensors to achieve robust and accurate state estimation. R3live contains two subsystems, namely lidar-Inertial Odometer (LIO) and Vision-Inertial Odometer (VIO). The LIO subsystem uses the measurements of lidar and inertial sensors to construct the geometry of a global map, which records the input lidar scans and estimates the state of the system by minimizing point-to-plane residuals. The VIO subsystem utilizes visual-inertial sensor data to render the texture of the map, render the RGB color of each point with the input image, and update the system state by minimizing the frame-to-frame PnP reprojection error and the frame-to-map photometric error. Based on LIO-SAM, LVI-SAM is coupled with a visual inertial odometer. The algorithm includes a lidar inertial odometer module and a visual inertial odometer module. The visual inertial odometer uses VINS-Mono. In the scenario of radar degradation, the visual odometer positioning results are used to replace the position and attitude of the lidar degradation direction, and the visual inertial odometer system is initialized with the results of the lidar inertial odometer. The visual word bag loopback detection results are also used in the radar inertial odometer to participate in the factor graph optimization. FAST-LIVO [46] integrates IMU vision and lidar using the iterative error Kalman filter to realize efficient and robust localization and mapping. Table 3 summarizes some of the most representative SLAM schemes based on visual and lidar fusion, based on which most of the current research work is carried out.

Table 3. Representative slam scheme based on visual and lidar fusion.

Scheme	Release Time	Sensor Form	Characteristic
LIMO [37]	2018	a	lidar-assisted visual recovery of feature point depth
V-LOAM [38]	2018	a	Match from high frequency to low frequency
VIL-SLAM [40]	2019	b	VIO assisted lidar positioning
LIC_Fusion [41]	2019	b	MSCKF filter (sensor online calibration)
LIC_Fusion2 [42]	2020	b	Sliding window filter
ULVIO [43]	2021	b	Factor Graph Optimization
R2live [44]	2021	b	ESKF filtering + factor graph optimization
R3live [45]	2021	b	Minimize the photometric error from frame to map
LVI-SAM [8]	2021	b	Factor Graph Optimization
FAST-LIVO [46]	2022	b	IESKF filtering

a indicates support for 3D lidar and camera, b indicates support for 3D lidar, camera and IMU.

3. System Overview

The LVI fusion framework designed in this paper consists of five parts as shown in Figure 1. Each module is described in detail below.

(1) Time alignment. Regardless of systems triggered by external clocks (such as GNSS), each sensor is collected at a different start time stamp, and different sensors have different data acquisition frequencies. State fusion estimation requires aligning data with different timestamps to the same time node. LVI-fusion takes the camera time stamp as the benchmark, splits the lidar point cloud data according to the camera time stamp, and merges the point cloud data between image frames into one frame point cloud. IMU data is interpolated according to the time stamp of the camera to obtain IMU data aligned with the camera time stamp. Through the above operations, the lidar data, IMU data and camera data can be time-stamped aligned.

(2) Data preprocessing. The state propagation of IMU data between adjacent image frames is carried out, and the point cloud data between two image frames is dedistorted according to the state prediction results, and the point cloud is unified to the end time of the point cloud of the frame. The YOLOv7 dynamic target recognition algorithm [47] is used to segment the dynamic target of the image, eliminate the influence of the dynamic target, obtain the static target image, construct the image pyramid of the deleted dynamic target image, extract the Harris key points from each layer of the image, and use the quadtree to homogenize the feature points to obtain uniformly distributed feature points. The tracked feature points are added to the image queue.

(3) Constraint construction. The IMU data between adjacent image key frames are pre-integrated, the pre-integral increment of adjacent image key frames is obtained, and the Jacobian matrix and covariance of the pre-integral error matrix are constructed. The local point cloud map near the current key frame is used to assign depth to the image feature points, and the image feature points with depth are obtained. The reprojection error constraints are constructed according to the 3D coordinates of the feature points and the image frames tracked to their coordinates. Due to the high frequency of the camera, the field of view Angle of the lidar data with the camera time is less than half of the original, and the key frame is selected according to the pose result obtained by the VIO odometer. When it is a key frame, the lidar data of the current frame and the lidar data of the previous two frames are combined into one frame point cloud data. Line features and surface features are extracted from key frame point cloud data and matched with a local map to construct pose constraint based on lidar. According to IMU pre-integral constraints, visual reprojection constraints and lidar matching constraints, the nonlinear optimization objective function can be constructed, and the real-time pose calculation can be performed by using the sliding window optimization method, and the optimization results are fed back to the visual inertial odometer.

(4) Closed loop detection. The closed-loop detection algorithm based on 3D lidar, and the visual-based bag of words model were used for closed-loop detection. When the constraints of both methods are met, the closed-loop constraints between visual and lidar are added to the global optimization.
(5) Global optimization. Opens a separate thread for global optimization of keyframe-based pose.

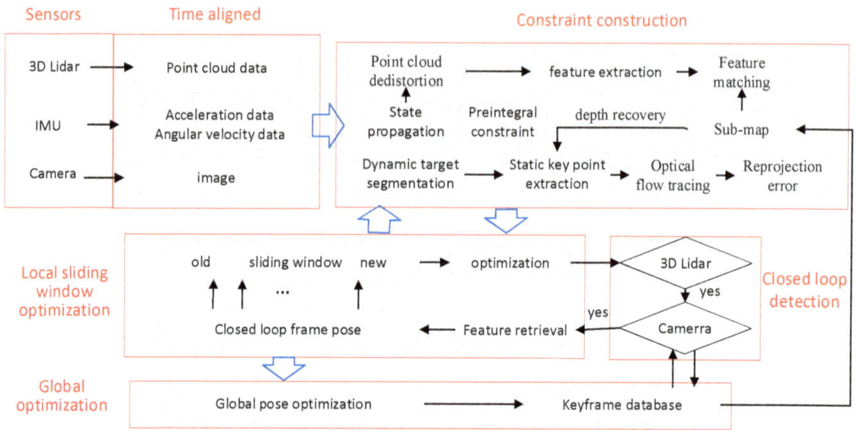

Figure 1. LVI-fusion framework based on visual, lidar and IMU fusion.

3.1. Symbolic Description

$(\cdot)^w$, $(\cdot)^b$, $(\cdot)^L$, and $(\cdot)^c$ represent the world coordinate system, IMU coordinate system, lidar coordinate system, and visual coordinate system, respectively. Define the body coordinate system to coincide with the IMU coordinate system. The variables χ represents all the variables in the sliding window. χ includes the state variables x of keyframes within the sliding window, as well as the inverse depth λ of all feature points within the sliding window. n represents the number of keyframes within the sliding window, and m represents the number of key points. x_k includes the position $p_{b_k}^w$, velocity $v_{b_k}^w$, attitude $q_{b_k}^w$, acceleration bias b_a, and angular velocity bias b_g. $q_{b_k}^w$ and $p_{b_k}^w$ depict the rotation and shift of the body coordinate system to the world coordinate systemwhen the k-th image is taken, where $q_{b_k}^w$ is a quaternion. $v_{b_k}^w$ depicts the velocity of the body coordinate system to the world coordinate system when the k-th image is taken. $p_{b_k}^w$. $v_{b_k}^w$, b_a, and b_g are all three-dimensional vectors.

$$\begin{aligned} \chi &= [x_0, x_1, \cdots x_n, \lambda_0, \lambda_1, \cdots, \lambda_m] \\ x_k &= \left[p_{b_k}^w, v_{b_k}^w, q_{b_k}^w, b_a, b_g\right], k \in [0, n] \end{aligned} \quad (1)$$

3.2. Time Aligned

There are two kinds of timestamps in ROS, one is the ROS system time stamp (ROS time), and the other is the time stamp of external hardware devices (such as cameras, lidar, etc.), also known as hardware time. The ROS timestamp is a floating-point number, measured in seconds, calculated from 00:00:00 UTC on 1 January 1970. The ROS timestamp is globally unique in the entire ROS system, that is, when nodes in the ROS system need to synchronize time, the ROS timestamp can be used as a standard, and each node can synchronize based on it. The hardware timestamp is provided by an external device and can be either a relative timestamp (the time difference between the device startup time or a fixed point in time) or an absolute timestamp (the time relative to a fixed point in time). Since the external device and the ROS system are different systems, their clocks may differ, so timestamp conversion is required to convert hardware timestamps to ROS timestamps, or ROS timestamps to hardware timestamps for operations such as time synchronization and data fusion. In this paper, we consider a system where lidar and camera are not triggered by an external clock (such as GNSS). In order to ensure the consistency of the time system, we assign time information to different sensor data according to the built-in time system of the robot operating system. The individual timestamp of the points can be obtained from the sensor's driver. If the timestamp for a point is not available, it also can be calculated by orientation difference. After the time reference of the lidar point cloud, the camera, and IMU are aligned to the ROS time system, the time alignment operation can be performed. Since the startup time and frequency of different sensors are different, this paper takes the frequency of the camera as the benchmark. The lidar data is split and merged, and the point cloud located between the time stamps is repackaged into a frame of point cloud data according to the time stamps between the adjacent images, as shown in Figure 2. Image acquisition can be considered instantaneous, and the data of a frame of lidar point cloud is continuous. In this paper, a frame of lidar point cloud data is dedistorted to the last moment of the frame. As shown in Figure 3. For specific operations, refer to FAST-LIO2, and then it can be considered that the frame of lidar point cloud data is acquired synchronously with the camera.

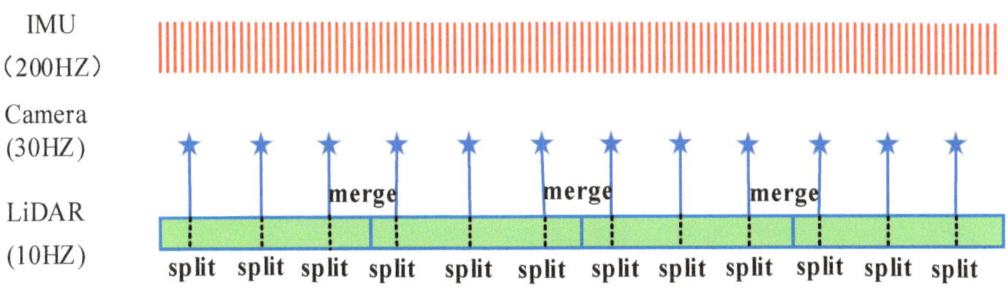

Figure 2. Time aligned: split-and-merge; dotted lines refer to the camera timestamp, and solid lines refer to the lidar timestamp. Lidar point cloud data is split and merged based on the camera timestamp (Stars represents the timestamp corresponding to the data collected by the camera).

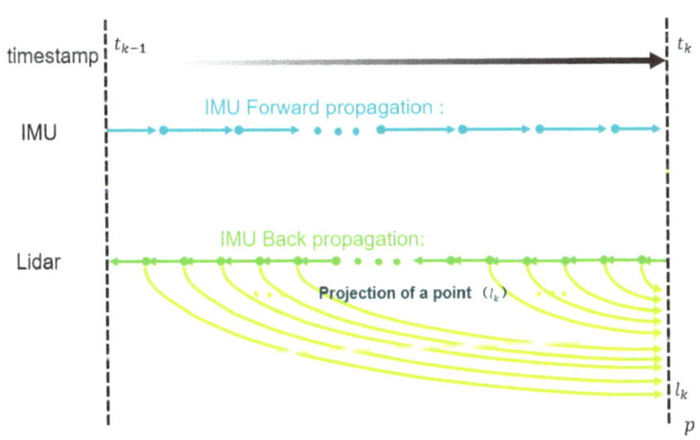

Figure 3. Align the point cloud to the end of the frame.

The IMU and camera time stamps are synchronized by timestamp interpolation, and the IMU data before and after the camera time stamps are interpolated to obtain the IMU data corresponding to the image time stamps, as can be seen from Figure 4.

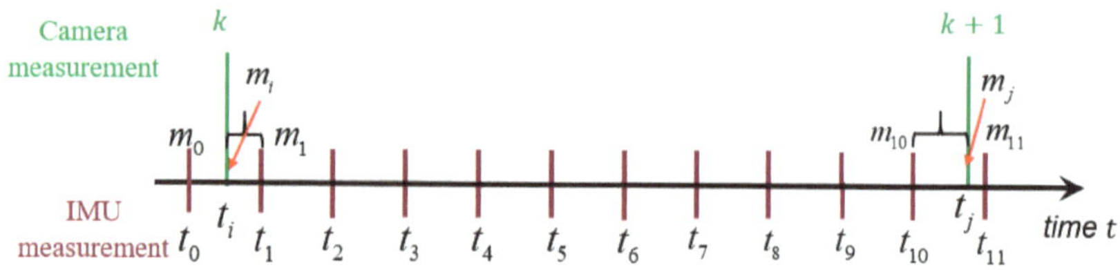

Figure 4. The IMU is aligned with the camera timestamp (Arrows correspond to the time of data collection).

The K-frame acquisition time of the camera is t_i. Due to the misalignment of time stamps, there is no IMU data at this time. The measured values of t_0 and t_1 before and after this time correspond to m_0 and m_1 respectively. The IMU data corresponding to t_i can be interpolated according to the Formula (2) to realize the alignment of time stamps. Similarly, the measured value at time t_j can be calculated according to Formula (3). Through the above processing, we do not need external timing equipment to complete the timestamp, which greatly increases the scene applicability of the system itself and provides a data basis for multi-sensor fusion based on graph optimization.

$$m_i = \frac{m_0(t_1 - t_i) + m_1(t_i - t_0)}{t_1 - t_0} \tag{2}$$

$$m_i = \frac{m_{10}(t_{11} - t_j) + m_{11}(t_j - t_{10})}{t_{11} - t_{10}} \tag{3}$$

3.3. *Key Point Depth Association*

Robust key point depth recovery is very important for positioning accuracy and robustness. The key point depth reply process of this paper is shown in Figure 5. Firstly, dynamic target segmentation is carried out on the image, and key points are extracted from the static target image. Secondly, the key point of the mask region boundary is eliminated to eliminate the error key point caused by the mask. Then, the local point cloud map is used to assign the depth value to the static key points, and the key points that are wrong in terms of depth recovery are checked, and the key points without depth are restored by triangulation.

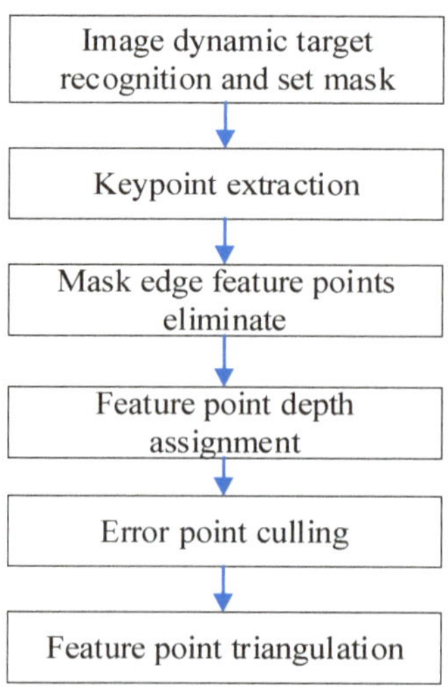

Figure 5. Key points depth recovery process.

3.3.1. Image Target Detection

In recent years, real-time target detection algorithms have been developed rapidly. For example, MCUNet [48] and NanoDet [49] worked to improve model inference speed on low-power edge CPU chips. The Yolox [50] algorithm focuses on improving the speed of model inference on various GPU devices. At present, the development of real-time object detection algorithms focuses on the design of the efficient backbone network modules of models. For real-time object detection algorithms used on cpus, backbone network design is mainly based on MobileNet [51], ShuffleNet [52] or GhostNet [53]. On the other hand, on the GPU, most of the mainstream real-time target detection algorithms use ResNet [54] or DLA [55], and then use the gradient strategy in CSPNet [56] to further optimize the module. In addition to the design of the model backbone network, the YOLOv7 algorithm also pays special attention to the optimization of the model training process. These modules and methods can enhance the training effect and improve the accuracy of target detection, but do not increase the inference cost. The YOLOv7 backbone network is mainly composed of extended efficient layer aggregation networks (E-ELAN) [55], and features of three scales are used to detect output targets, as shown in Figure 6. The YOLOv7 algorithm can achieve a good detection effect while maintaining the detection speed. Therefore, this paper selects the YOLOv7 algorithm to complete image-based dynamic target detection.

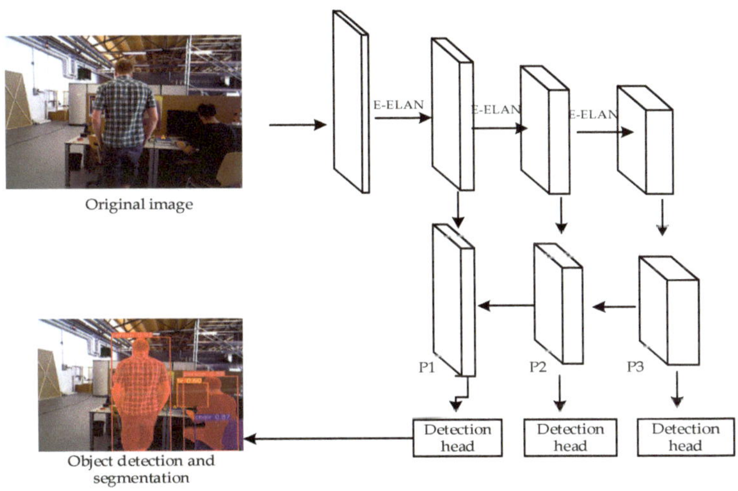

Figure 6. Network structure of YOLOV7 model.

Since YOLOv7 can detect a wide range of target categories, as shown in Figure 7b, in order to prevent static targets from being eliminated, this paper mainly identifies dynamic targets such as "people", "bicycles", "motorcycles" and "cars" according to realistic dynamic scenes, as shown in Figure 7c. In addition, in order to facilitate subsequent feature point extraction, the mask is set to pure white, as shown in Figure 7d.

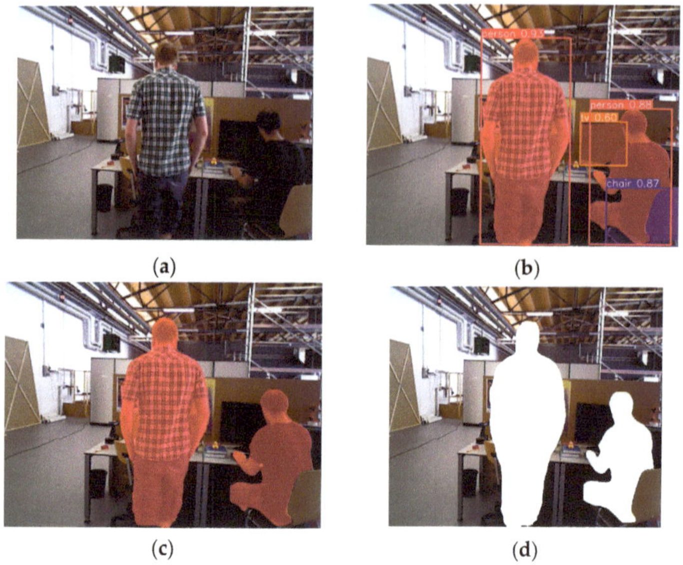

Figure 7. Schematic diagram of removing dynamic objects from YOLOv7. (**a**) Original image. (**b**) Original segmentation. (**c**) Specific dynamic target segmentation. (**d**) Dynamic target white mask.

Figure 8a shows the key points extraction results of images with deleted dynamic targets. Through Figure 8a, it is found that some key points have also been extracted on the contour of the dynamic target mask, which needs to be removed. Identify dynamic target edge feature points with a pixel value of 255 for surrounding pixels with a radius of 3 of key points and remove them. The elimination results are shown in Figure 8b. It can be found that the feature points on the edge of the dynamic feature are well eliminated and the key points under the static target image are obtained.

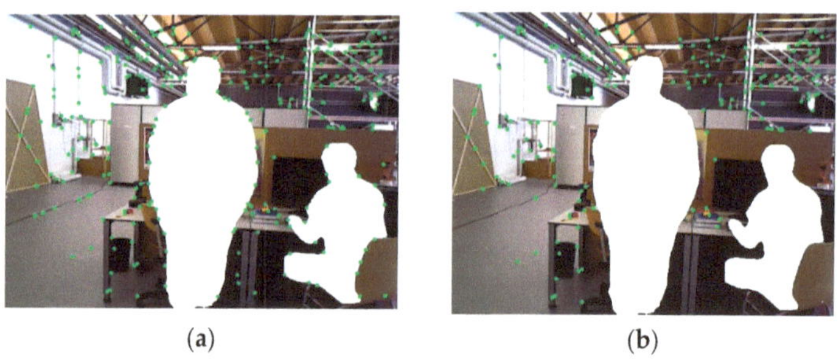

Figure 8. Static key points extraction results after tuning quadtree homogenization. (**a**) Quadtree equalization after removing dynamic targets. (**b**) Mask edge feature points are eliminated.

3.3.2. Key Point Depth Recovery

The sub-map is used to assign depth to each image key frame static key point. The feature points of the image are used to search for the three nearest-neighbor lidar points; the lidar points are fitted to the plane, and the distance between the feature points and the plane is calculated for depth assignment. When point p is in the same plane as the surrounding lidar point cloud, the following functional relationship is satisfied:

$$ {}^{w}p_1 = T_{c_i}^{w} \frac{{}^{c_i}w\left({}^{c_i}p_j - {}^{c_i}w\right) \cdot \vec{\eta}_i}{\|\vec{\eta}_i\|} \quad (4) $$

$$^{w}p_2 = T_{c_2}^{w} \frac{^{c_{i+1}}w\left(^{c_{i+1}}p_2 - {}^{c_{i+1}}w\right) \cdot \vec{\eta}_{i+1}}{\|\vec{\eta}_{i+1}\|} \tag{5}$$

$$^{w}p = {}^{w}p_1 = {}^{w}p_2 \tag{6}$$

^{w}p represents the coordinates of point p in the world coordinate system, $T_{c_i}^{w}$ and $T_{c_2}^{w}$ represent the conversion relationship between the camera coordinate system and the world coordinate system under different field angles, respectively. $\vec{\eta}_i$ and $\vec{\eta}_{i+1}$ represent plane normal vectors fitted by lidar point clouds at different viewing angles, respectively. ^{c_i}w and $^{c_{i+1}}w$ represent the normalized image plane coordinates of the same key point under different viewing angles, respectively, which can be calculated according to pixel coordinates and camera parameters.

Due to the complexity of the real environment, and the feature points are basically the positions where the image gradient changes greatly, such positions are often not in the same plane with the three nearest lidar point clouds, resulting in wrong depth estimation. The relationship of formula 6 is no longer satisfied, as can be seen from Figure 9. Therefore, we can use this feature to test the correctness of the correlation between feature points and depth values. When the modulus length of the coordinate difference between $^{w}p_i$ and $^{w}p_2$ is less than a certain threshold, it is considered that the correlation of the depth value is correct; otherwise, the correlation of depth value of the feature point is cancelled, and the coordinate of the triangle point is restored based on visual triangulation; the process is shown in Figure 10.

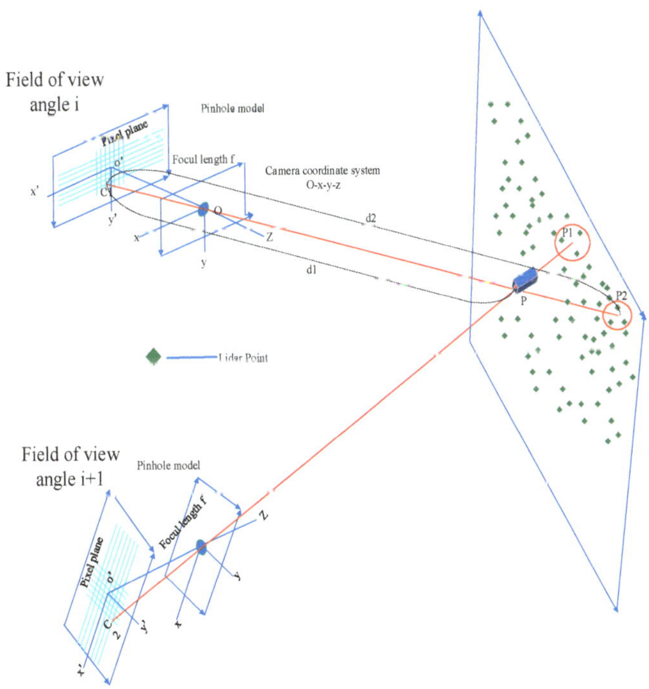

Figure 9. The projection relationship of the same key point under different keyframe perspectives.

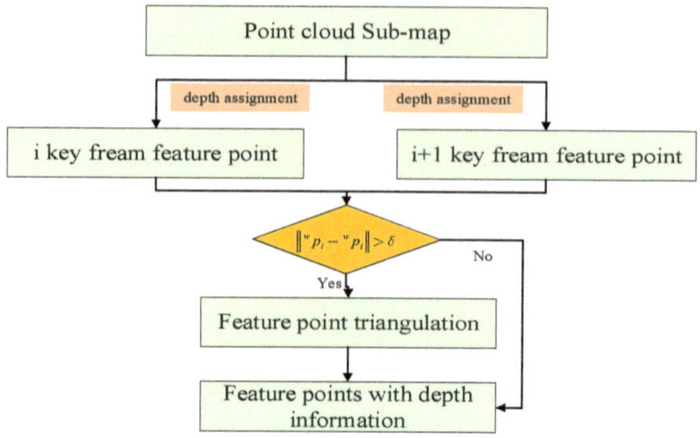

Figure 10. Process of determining the wrong depth assignment of feature points.

3.4. *Constraint Construction*

3.4.1. Pre-Integration Factor

The acceleration model of IMU is shown in Formula (7), and the angular velocity model of IMU is shown in Formula (8).

$$\hat{a}_t = a_t + b_{a_t} + R_w^I g^w + n_a \tag{7}$$

$$\hat{w}_t = w_t + b_{w_t} + n_w \tag{8}$$

$\hat{a}_t$ and $\hat{w}_t$ represent the raw measurements of the IMU sensor. The accelerometer noises n_a and n_w are assumed to obey white Gaussian noise, $n_a \sim \eta(0, \sigma_\alpha^2)$,$n_w \sim \eta(0, \sigma_w^2)$. R_w^I represents rotation from the world coordinate system to the carrier coordinate system. g^w represents the gravity vector in the world coordinate system, whose magnitude direction is known. The accelerometer bias and gyroscope bias follow random walks, as shown in Formulas (9) and (10).

$$\dot{b}_{a_t} = n_{b_a},\ n_{b_a} \sim \eta(0, \sigma_{b_a}^2) \tag{9}$$

$$\dot{b}_{w_t} = n_{b_w},\ n_{b_w} \sim \eta(0, \sigma_{b_w}^2) \tag{10}$$

There are multiple IMU data between the two image keyframes b_k and b_{k+1}. According to the dynamic equation of IMU, its integral form in continuous time is as follows:

$$\begin{aligned} \alpha_{b_{k+1}}^{b_k} &= \int\int_{t \in [t_k, t_{k+1}]} R_t^{b_k}(\hat{a}_t - b_{a_t})dt^2 \\ \beta_{b_{k+1}}^{b_k} &= \int_{t \in [t_k, t_{k+1}]} R_t^{b_k}(\hat{a}_t - b_{a_t})dt \\ \gamma_{b_{k+1}}^{b_k} &= \int_{t \in [t_k, t_{k+1}]} \frac{1}{2}\Omega(\hat{w}_t - b_{w_t})\gamma_t^{b_k}dt \end{aligned} \tag{11}$$

where

$$\Omega(w) = \begin{bmatrix} -\lfloor w \rfloor_\times & w \\ -w^T & 0 \end{bmatrix},\ \lfloor w \rfloor_\times = \begin{bmatrix} 0 & -w_z & w_y \\ w_z & 0 & -w_x \\ -w_y & w_x & 0 \end{bmatrix} \tag{12}$$

$\alpha_{b_{k+1}}^{b_k}$, $\beta_{b_{k+1}}^{b_k}$ and $\gamma_{b_{k+1}}^{b_k}$ represent the pose, velocity, and rotation angle corresponding to the pre-integral, respectively. The formula shows that the three quantities $\alpha_{b_{k+1}}^{b_k}$, $\beta_{b_{k+1}}^{b_k}$ and $\gamma_{b_{k+1}}^{b_k}$ have no relationship with the state of b_k. For two adjacent IMU observation data i

and $i+1$, whose time interval is Δt, Formula (11) can be written in a discretized form as Formula (13).

$$\begin{array}{l} \hat{a}_{i+1}^{b_k} = \hat{a}_i^{b_k} + \hat{\beta}_i^{b_k}\delta t + \frac{1}{2}R(\hat{\gamma}_i^{b_k})(\hat{\alpha}_i - b_{a_i})\delta t^2 \\ \hat{\beta}_{i+1}^{b_k} = \hat{\beta}_i^{b_k} + R(\hat{\gamma}_i^{b_k})(\hat{\alpha}_i - b_{a_i})\delta t \\ \hat{\gamma}_{i+1}^{b_k} \otimes \begin{bmatrix} 1 \\ \frac{1}{2}(\hat{w}_i - b_{w_i})\delta t \end{bmatrix} \end{array} \tag{13}$$

According to the equation of state and observation equation, the error equation of IMU can be obtained as Formula (14).

$$_B(\hat{z}_{b_{k+1}}^{b_k}, \chi) = \begin{bmatrix} \delta\alpha_{b_{k+1}}^{b_k} \\ \delta\beta_{b_{k+1}}^{b_k} \\ \delta\theta_{b_{k+1}}^{b_k} \\ \delta b_a \\ \delta b_g \end{bmatrix} = \begin{bmatrix} R_w^{b_k}(p_{b_{b+1}}^w - p_{b_k}^w + \frac{1}{2}g^w\Delta t_k^2 - v_{b_k}^w\Delta t_k) - \hat{\alpha}_{b_{k+1}}^{b_k} \\ R_w^{b_k}(v_{b_{k+1}}^w + g^w\Delta t_k - v_{b_k}^w) - \hat{\beta}_{b_{k+1}}^{b_k} \\ 2\left[q_{b_{k+1}}^{w^{-1}} \otimes q_{b_{k+1}}^w \otimes \left(\hat{\gamma}_{b_{k+1}}^{b_k}\right)^{-1}\right]_{xyz} \\ b_{ab_{k+1}} - b_{ab_k} \\ b_{wb_{k+1}} - b_{wb_k} \end{bmatrix} \tag{14}$$

According to Formula (14), the variation formula of the covariance equation corresponding to the error equation over time can be derived by using the error propagation theorem, and the specific derivation form is referred to [57].

3.4.2. Vision Factor

The visual part is shown in Figure 11. Thanks to the lidar sensor, ranging information from lidar point clouds can be used to provide depth information for visual features, and with depth information, it is easy to obtain 3D (three-dimensional) coordinates of key points. In addition, the threshold is used to judge and screen out the key points of the wrong depth information. The method of triangulation within the sliding window is used to recover the 3D coordinates of visual key points with incorrect depth information. The above process ensures that the 3D coordinates of the visual key points are more robust. Reprojection error constraints can be constructed in the sliding window to constrain the pose. In addition, the coordinates of triangulated key points are optimized to ensure the robustness of the coordinates of the triangulated feature points. The construction process of the error constraint for the reprojection error is described below.

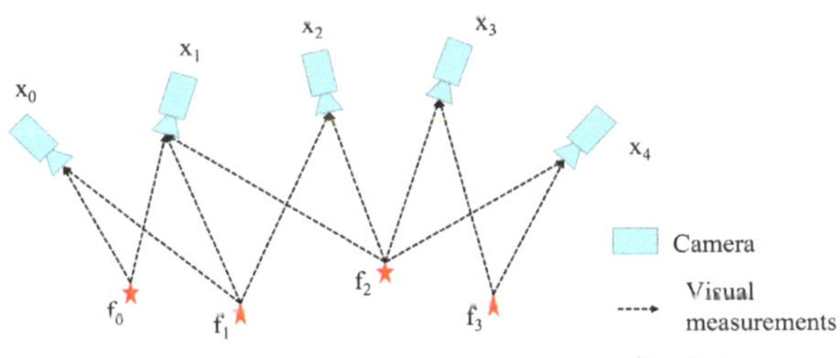

Figure 11. Structure diagram of visual restraint.

For the convenience of description, this study uses the pinhole camera model for modeling. The image observation value of the feature point f in the i-th frame is $(\hat{u}_f^{c_i}, \hat{v}_f^{c_i})$, and the point is projected to the j-th frame based on the result of the IMU status prediction. The coordinate of the feature point f in j-th frame can be calculated by Equation (15). The coordinates of feature point f in j-th $\hat{p}_f^{c_j}$ $(\hat{u}_f^{c_j}, \hat{v}_f^{c_j})$ can be traced according to optical flow. According to the predicted feature point coordinates and the feature point coordinates of optical flow tracking, the reprojection error equation can be constructed, as shown in Equation (16). K_c^{-1} represents the inverse of the camera internal parameter matrix.

Formula (17) is constructed by summing the square of all visual observation reprojection errors. C represents the feature points observed at least twice in the sliding window. The maximum likelihood estimation of state variables can be obtained by the nonlinear solution of Equation (17) using the L-M method.

$$p_f^{c_j} = R_b^c(R_w^{b_j}(R_{b_i}^w(R_c^b\frac{1}{\lambda_l}K_c{}^{-1}(\begin{bmatrix}\hat{u}_f^{c_i}\\ \hat{v}_f^{c_i}\end{bmatrix}) + p_c^b) + p_{b_i}^w - p_{b_j}^w) - p_c^b) \quad (15)$$

$$r_c(\hat{z}_l^{c_l}, \chi) = \hat{p}_l^{c_j} - \frac{p_l^{c_j}}{\|p_l^{c_j}\|} \quad (16)$$

$$\min_{\chi}\left\{\sum_{(l,i)\in C}\rho\|r_c(\hat{z}_l^{c_j}, \chi)\|_{P_l^{c_j}}^2\right\} \quad (17)$$

3.4.3. Lidar Factor

For time-stamped alignment with the vision sensor, the lidar data is divided and merged and the horizontal field of view angle of the repackaged point cloud data becomes one-third of the original when the frequency of the vision sensor is three times the frequency of the lidar sensor. Therefore, it is also difficult to match between frames based on the point cloud. It is necessary to build a local map to match the point cloud with the current frame. Therefore, this scheme needs to be stationary for a period of time, so that the lidar can fully scan and build a local map. When new time-aligned lidar data are received, we project the point cloud to the point cloud end time as FAST-LIO do. Surface features and line features are extracted from each frame of lidar and local point cloud map, and then matching constraints based on features are constructed. In order to maintain more efficient computing speed, ikd-Tree [35] is used for local map management and the nearest neighbor search of feature points. The construction process of feature extraction and lidar constraint is followed as LIO-SAM. Essentially, it minimizes the distance from the point to the line and the distance from the point to the surface, as shown in Formula (18).

$$\min_{\chi}\left\{\sum_{i=0}^{n_1} d_{e_i} + \sum_{j=0}^{n_2} d_{h_j}\right\} \quad (18)$$

d_{e_i} represents the distance from the i-th line feature point to the line feature, d_{h_j} represents the distance from the j-th plane feature point of to the corresponding plane. n_1 and n_1 represent the total number of line and surface features, respectively.

3.5. Local Sliding Window Optimization

The constraint factors constructed by different sensors all have common constraint variables, and the different constraint factors are combined, as shown in Formula (19). The first term $\{r_p - H_p\chi\}$ in Equation (19) represents the marginalized prior information. The Levenberg–Marquardt method is adopted in this paper to optimize the solution of the constraint equation. The size of the window can be adjusted according to the performance of the computer.

$$\min_{\chi}\left\{\|r_p - H_p\chi\|^2 + \sum_{k\in B}\|r_B(\hat{z}_{b_{k+1}}^{b_k}, \chi)\|^2 + \sum_{(l,i)\in C}\rho\|r_c(\hat{z}_l^{c_j}, \chi)\|_{P_l^{c_j}}^2 + \sum_{i=0}^{n_1} d_{e_i} + \sum_{j=0}^{n_2} d_{h_j}\right\} \quad (19)$$

3.6. Loopback Detection

This paper is based on the Scan Context (SC) [29] algorithm of 3D lidar for loop closure detection. In order to ensure that the lidar key frame can maintain a 360° field of view, the key frame is combined with the previous two frames of lidar data to form a frame of point cloud data and projected to the end of the key frame point cloud. Scan Context is calculated for each key frame and SC descriptors of different keyframes are matched to find point clouds in historical keyframes that are similar to current keyframes, so as to find loopback frames. It searches loopback frames by the similarity between the point clouds of individual keyframes, so it is not affected by geometric distance, and the loopback detection function can be completed even in large scenes. On this basis, the closed-loop detection of the visual word bag model is added. When the loopback detection of the above two methods is met at the same time, it is judged as a candidate frame. The sub-map is constructed with candidate frames for point cloud matching with the current frame. According to the matching situation, it is further determined whether it is a loop frame. When there is closed-loop detection, visual-based loop constraint and lidar loop constraint is added to the state estimation equation to minimize the cumulative error.

4. Experimental Setup and Evaluation

4.1. M2DGR Dataset

This paper uses the M2DGR dataset collected by Shanghai Jiao Tong University. M2DGR is the SLAM dataset collected by the ground robot navigation, which includes the look around RGB camera, infrared camera, event camera, 32-line lidar, IMU and original GNSS information, as shown in Figure 12. The dataset covers challenging scenes both indoor and outdoor, day and night, as shown in Figure 13. This paper selects 6 datasets from M2DGR for testing in different challenging scenarios and Table 4 summarizes the characteristics of different datasets.

Figure 12. Sensor integration platform.

Figure 13. M2DGR dataset partial scenarios.

Table 4. The characteristics of different datasets in M2DGR.

Sequence Name	Duration (s)	Features
hall_02	128	random walk, indoor, day
room_02	75	room, bright, indoor, day
door_02	127	outdoor to indoor, short-term, day
gate_03	283	outdoor, day
walk_01	291	back and forth, outdoor, day
street_05	469	straight line, outdoor, night,

4.1.1. Mapping Effect

All experiments in this paper were conducted in the Intel i7-107500H CPU test environment with 24 gb RAM. In this paper, the proposed algorithm LVI-fusion is used to test the mapping and positioning accuracy analysis. As shown in Figure 14, all scenes can establish accurate 3D point cloud maps. Next, we will further analyze the positioning track accuracy.

(**a**) hall_02

(**b**) room_02

(**c**) door_02

(**d**) gate_03

(**e**) walk_01

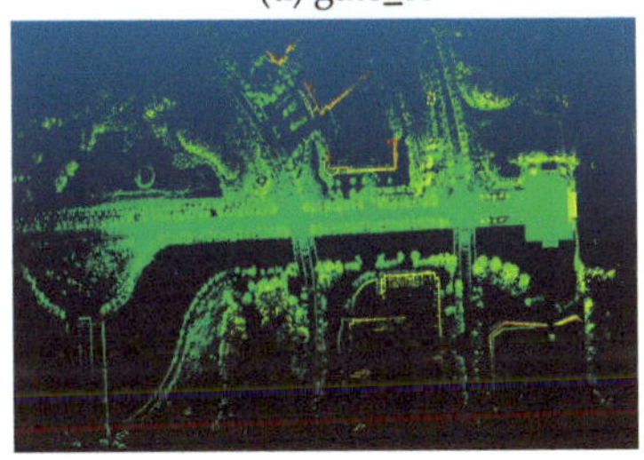

(**f**) street_05

Figure 14. The mapping results of the mode 3 proposed in this paper.

4.1.2. Precision Analysis

To show the positioning performance of LVI-fusion proposed in this paper, It can be seen intuitively from Figure 15 that there is no big deviation between LVI-fusion's positioning trajectory and the truth trajectory, and the trajectory shape is basically the same. The bar on the right of Figure 15 represents the error range and the unit is meters.

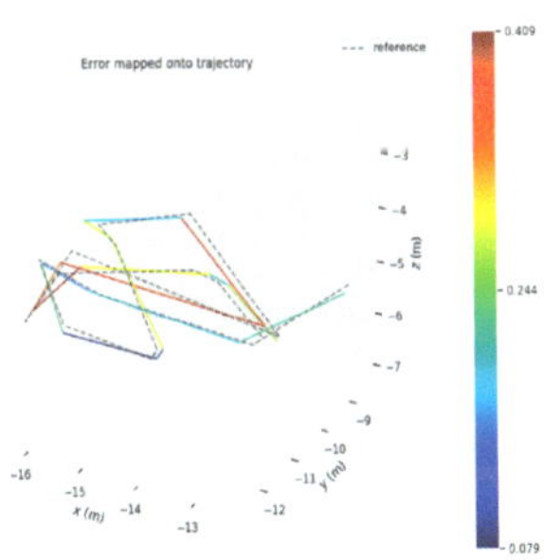

(**a**) hall_02

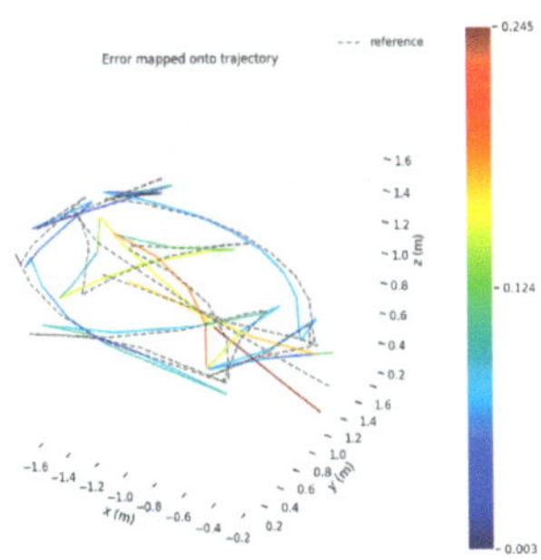

(**b**) room_02

Figure 15. *Cont.*

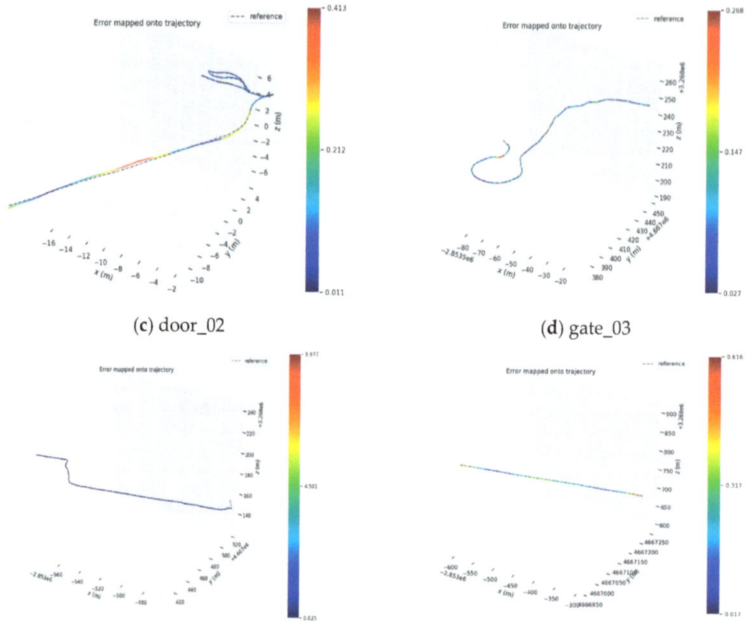

(**c**) door_02 (**d**) gate_03

(**e**) walk_01 (**f**) street_05

Figure 15. Track error of LVI-fusion proposed in this paper. Reference represents the truth value in the dataset, represented by a dotted line. The colored trajectory indicates the LVI-fusion running trajectory, and different colors indicate different degrees of error.

In order to further analyze the positioning performance of the LVI-fusion proposed in this paper, this paper uses the RMSE (Root Mean Square Error) index to calculate the positioning accuracy of the LVI-fusion. In addition, in order to better demonstrate the competitiveness of the algorithm proposed in this paper, this paper uses the current outstanding and representative SLAM schemes including VINS-Mono, A-LOAM, LIO-SAM, and LVI-SAM to test the above 6 scenarios, respectively. Based on the above 10 scenarios, the RMSE indicators of different SLAM schemes are shown in Table 5. The RMSE calculation formula of the estimated trajectory based on different SLAM schemes is as follows:

$$RMSE = \sqrt{\frac{1}{N}\sum_{i=1}^{N} \| \log\left(T_{gt,i}^{-1} T_{esti,i}\right)^{\vee} \|_2^2} \tag{20}$$

where $T_{esti,i}$ and $T_{gt,i}$ respectively represent the estimated pose and truth pose at time i, respectively, where $i = 1, \cdots, N$. Tables 5 and 6 show the positioning accuracy of different SLAM schemes. Table 4 shows the odometer accuracy of different SLAM schemes without Loop closure detection, and Table 6 contains the positioning accuracy after loopback detection. Among them, A-LOAM does not have the loopback closure detection module, so this paper only tests the accuracy of the front-end odometer for these two SLAM schemes. It can be found from Table 5, that the positioning accuracy of the lidar-based positioning scheme is generally better than that of the VINS-Mono scheme based on visual inertia fusion. Among them, the positioning results of the proposed algorithm in this paper are generally better than the A-LOAM, VINS-Mono, and LIO-SAM schemes except for individual scenarios, showing the advantages of multi-source sensor fusion. Through the positioning accuracy in different scenarios, it is found that the LVI-fusion can achieve better positioning accuracy than LVI-SAM. This is because the coupling degree of the proposed algorithm is higher, and all variables are optimized and solved at the same time. As can be seen from Table 5, the accuracy of LVI-fusion proposed in this paper has a significant advantage compared with the existing representative SLAM scheme, and the accuracy is increased by more than 20% compared with the LVI-SAM scheme.

Table 5. Comparing the (rmse)/m results of VINS-Mono, A-LOAM, LIO-SAM, LVI-SAM and our method based on M2DGR datasets (without loop closure).

Approach	Hall_02	Room_02	Door_02	Gate_03	Walk_01	Street_05
VINS-Mono	fail	0.462	1.653	5.838	9.976	fail
A-LOAM	0.208	0.121	0.168	0.246	3.303	0.657
LIO-SAM	0.399	0.125	0.124	0.111	0.891	0.407
LVI-SAM	0.279	0.123	0.186	0.113	0.885	0.394
LVI-fusion	0.214	0.103	0.117	0.104	0.627	0.371

Table 6. Comparing the (rmse)/m results of VINS-Mono, A-LOAM, LIO-SAM, LVI-SAM and our method based on M2DGR datasets (with loop clousure).

Approach	Hall_02	Room_02	Door_02	Gate_03	Walk_01	Street_05
VINS-Mono	fail	0.311	1.522	5.838	9.976	fail
LIO-SAM	0.291	0.125	0.106	0.111	0.830	0.405
LVI-SAM	0.270	0.120	0.171	0.114	0.888	0.395
LVI-fusion	0.181	0.101	0.099	0.105	0.631	0.370

4.2. Low-Dynamic Environment

In this paper, data acquisition is carried out based on the tracked robot, which is equipped with a camera, IMU, multi-line lidar (Robosense 16), and RTK (real-time kinematic) module for obtaining the truth value [58], as shown in Figure 16. The parameter indicators of Lidar and IMU are shown in Tables 7 and 8. The left eye of the MYNT EYE camera standard version is used as the image acquisition device, with an acquisition frequency of 25 Hz and a resolution of 752 × 480. Since the IMU is built into the tracked robot, it cannot be seen in Figure 16. We selected two representative scenes on the campus of China University of Mining and Technology, namely the square scene, the road scene, as shown in Figure 17, where the red trajectory is the positioning trajectory based on RTK.

Table 7. RS-LiDAR-16 parameters.

Parameter		RS-LiDAR-16
Ranging range		20 cm~150 m
Distance measurement accuracy		±2 cm
Field of view angle	horizontal	360°
	Vertical	+15°~−15°
Angle resolution	horizontal	0.2°
	Vertical	2°
Collect points per second		28,800
scan period		0.1 s

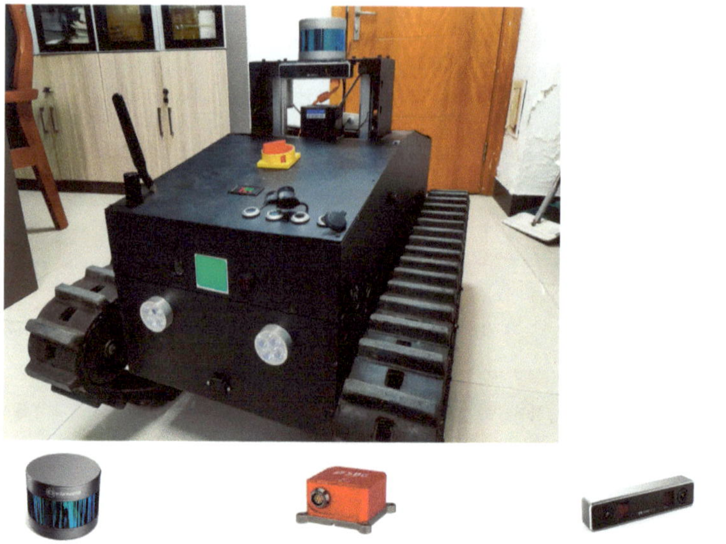

(**a**) robosense (**b**) IMU (**c**) MYNTEYE-S1030

Figure 16. Mobile measurement platform based on tracked robot.

Table 8. IMU parameters.

	Parameter	Index
Accelerometer	Speed Random Walk $(ug/\sqrt{hz})$	57
	Zero bias instability (ug)	14
gyroscope	angle random walk$(°/\sqrt{hr})$	0.18
	Zero bias instability $(°/hr)$	8
magnetometer	Noise (m Gauss)	3
	Zero bias stability (m Gauss)	1

(**a**) Road scene (**b**) Square scene

Figure 17. Data acquisition environment satellite map.

4.2.1. Mapping Effect

LVI-SAM is a multi- sensor fusion SLAM representative scheme based on graph optimization. In this paper, mapping experiments based on LVI-fusion and LVI-SAM are carried out based on the above three data, as shown in Figures 18 and 19. It can be seen from Figures 18 and 19 that LVI-fusion's drawing effect is clearer and more accurate.

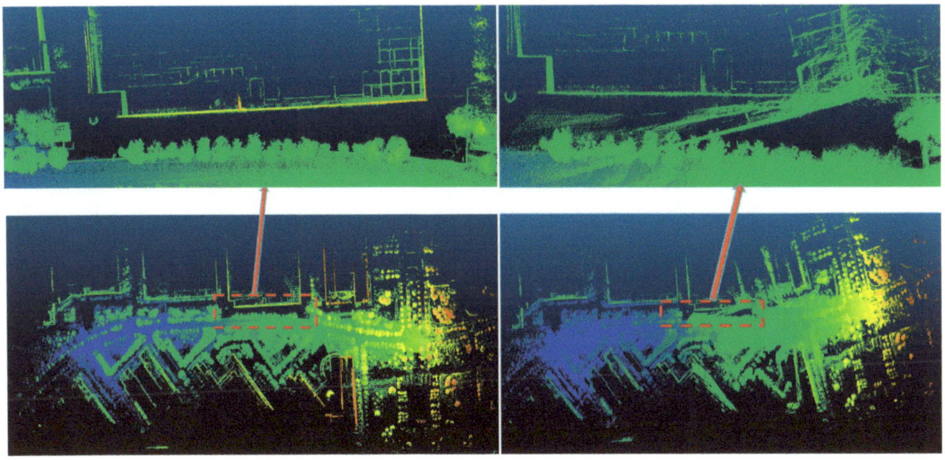

(**a**) LVI-fusion (**b**) LVI-SAM

Figure 18. Comparison of map construction effect based on LVI-SAM and LVI-fusion in the road scene (the picture the arrow points to is a detailed enlarged photo of the box).

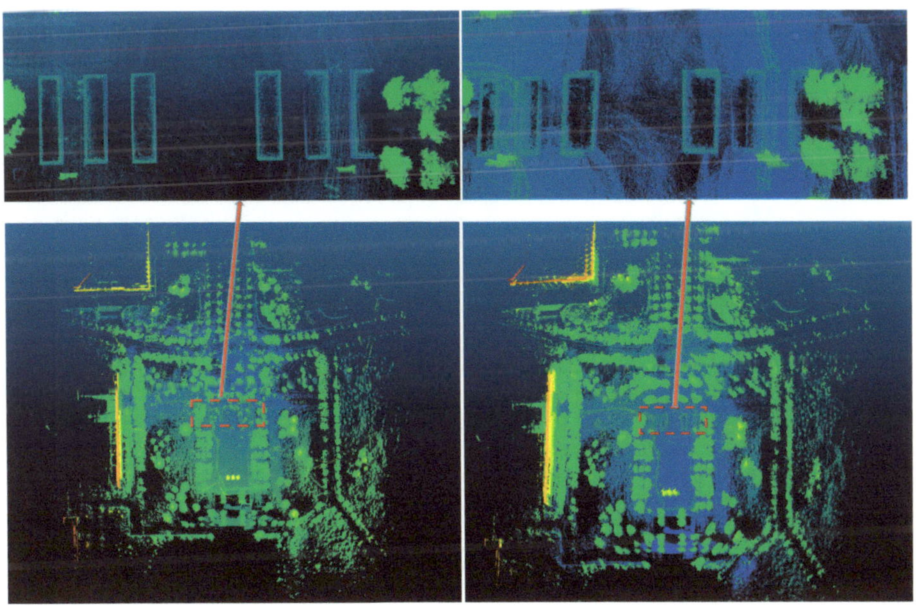

(**a**) LVI-fusion (**b**) LVI-SAM

Figure 19. Comparison of map construction effect based on LVI-SAM and LVI-fusion in the square scene (the picture the arrow points to is a detailed enlarged photo of the box).

4.2.2. Precision Analysis

In this paper, the EVO tool is used to draw the trajectory error graph of LVI-fusion, as shown in Figure 20. Table 9 shows the RMSE of LIO-SAM, LVI-SAM and LVI-fusion. It can be seen that the scheme proposed in this paper has the best precision and is more stable. In low-dynamic scenes, it was found that LVI-fusion has the highest accuracy and LIO-SAM has the lowest accuracy, and the multiple sensor fusion has more positioning advantages Due to the higher coupling degree of LVI-fusion, it can achieve better positioning accuracy compared to LVI-SAM.

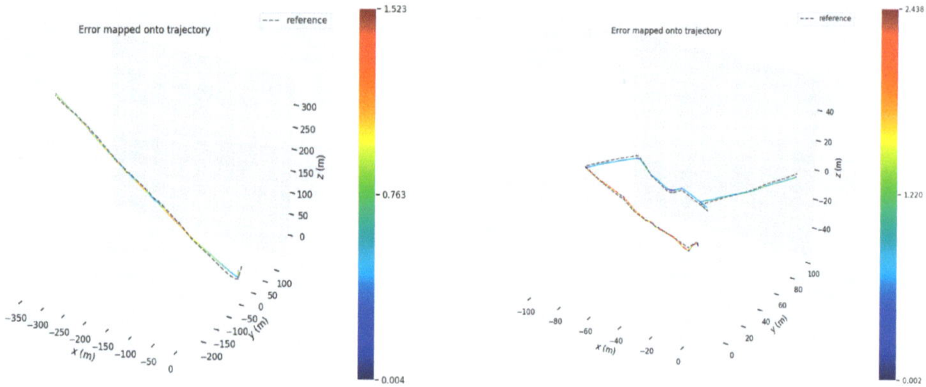

Figure 20. The colored trajectory indicates the LVI-fusion running trajectory, and different colors indicate different degrees of error.

Table 9. Comparing the (rmse)/m results of LIO-SAM, LVI-SAM and LVI-fusion.

Approach	Road Scene (m)	Square Scene (m)
LIO-SAM	1.16	1.09
LVI-SAM	1.04	0.98
LVI-fusion	0.80	0.79

4.3. *High-Dynamic Environment*

In order to verify the robustness of LVI-fusion in a dynamic environment, this paper selects the East gate of China University of Mining and Technology, a scene with abundant dynamic targets, as shown in Figure 21. The red trajectory in Figure 21a is the motion trajectory collected by the RTK positioning module. Figure 21b shows part of the data acquisition scenario. As can be seen from Figure 21b, the East gate of China University of Mining and Technology contains a large number of people, bicycles, electric vehicles, and taxis and other dynamic targets around the mobile measurement platform. Figure 22a shows the results of dynamic target segmentation based on the YOLOv7 dynamic target detection algorithm. Figure 22b shows the effect of the static key point extraction.

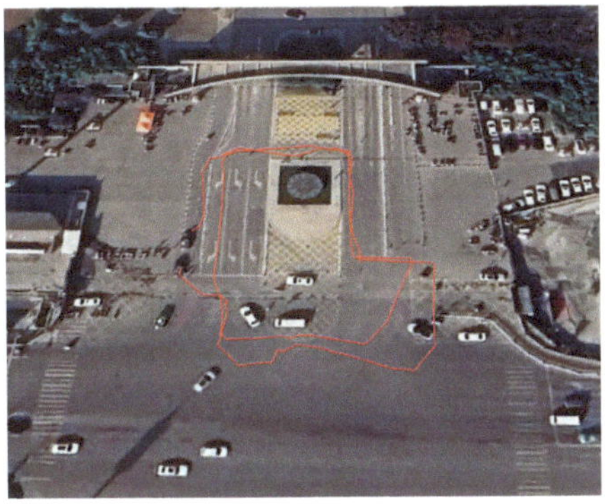

(a) Google satellite map

(b) Data collection scenario

Figure 21. Data acquisition of the 3D lidar/Vision/IMU dynamic scene.

(a) Dynamic target segmentation based on YOLOv7

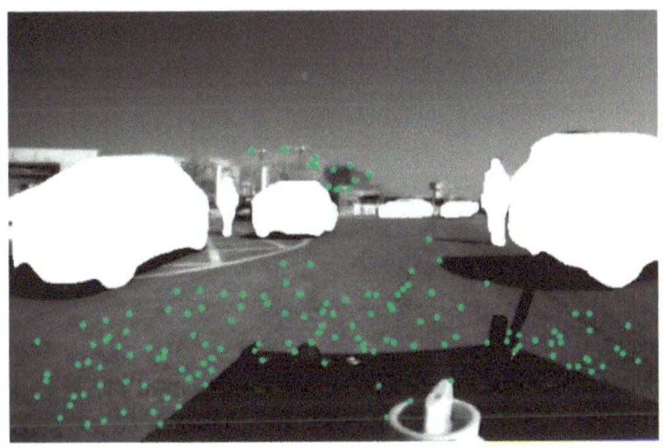

(b) Static feature point extraction

Figure 22. Static feature point extraction (The green dots represent the extracted key points).

LIO-SAM and LVI-fusion are representatives of multi-source fusion SLAM schemes based on optimization. Figure 23 shows the positioning trajectory diagram of LIO-SAM, LVI-SAM and LVI-fusion proposed in this paper. It can be seen that LVI-SAM has the worst positioning effect in a dynamic environment. Due to the presence of a large number of dynamic targets in the environment, incorrect point cloud information assigns values to visual dynamic key points, further leading to incorrect matching of visual key points. Therefore, the combination of the two is not as effective in positioning in high-dynamic environments as the LIO-SAM scheme. Due to the removal of dynamic key points and the use of only static key points for visual constraints, as well as the use of depth information for judgment, LVI-fusion removes key points with incorrect assignment, resulting in better localization performance compared to LVI-SAM and LIO-SAM. From Table 10, it can be seen that LVI-fusion has the highest positioning accuracy, with a 26% improvement compared to LIO-SAM and a 40% improvement compared to LVI-SAM. Figure 24 shows the mapping results of LVI-SAM and LVI-fusion. It can be seen that LVI-fusion has higher mapping quality, and no significant point cloud overlap appears.

Table 10. LIO-SAM, LVI-SAM, and LVI-fusion positioning accuracy.

Representative SLAM Scheme	LIO-SAM	LVI-SAM	LVI-Fusion
RMSE(m)	1.201	1.548	0.890

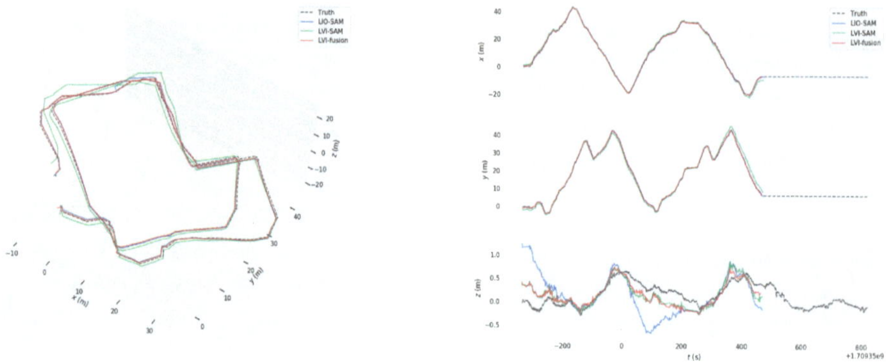

Figure 23. Positioning track of LIO-SAM, LVI-SAM and LVI-fusion.

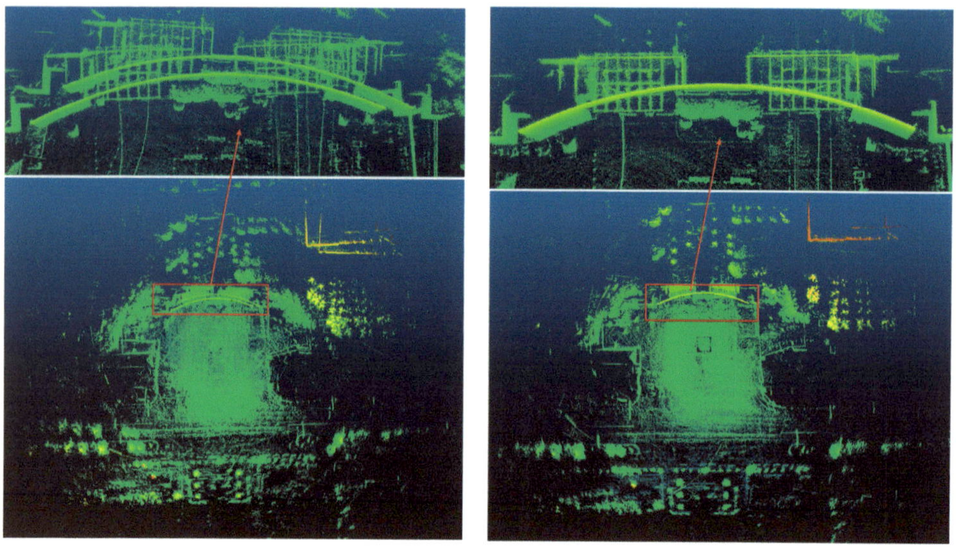

(**a**) LVI-SAM (**b**) LVI-fusion

Figure 24. Mapping effect of LVI-SAM and LVI-fusion.

5. Conclusions

This article proposes a robust SLAM scheme LVI-fusion for lidar/vision/IMU fusion. This scheme proposes a sensor soft synchronization time alignment method and utilizes lidar cloud depth assignment and triangulation to achieve a maximum number of image key point depth recovery. In addition, this scheme utilizes the YOLOV7 object recognition algorithm to eliminate the erroneous effects caused by matching key points in dynamic environments, achieving robust multi-source fusion localization and mapping. The positioning accuracy on the M2DGR dataset indicates that LVI-fusion can achieve better positioning accuracy compared to the current representative SLAM scheme. In addition, data collection is carried out in low-dynamic and high-dynamic environments through the built mobile measurement platform. Compared with the LVI-SAM scheme, LVI-fusion improves positioning accuracy by about 20% in low-dynamic scenes and by about 40% in high-dynamic scenes. The above results indicate that the LVI-fusion proposed in this article has better positioning accuracy in both low-dynamic and high-dynamic environments. And in dynamic environments, LVI-fusion has better robustness.

Although the LVI-fusion proposed in this paper can be robustly positioned and mapping, offline calibration is needed to transplant the algorithm to different hardware platforms, which brings great inconvenience to cross-platform applications. Therefore, it is an urgent problem to realize the high-precision and robust online calibration of external parameters between each sensor based on LVI-fusion.

Author Contributions: Conceptualization, Z.L. (Zhenbin Liu) and Z.L. (Zengke Li); methodology, Z.L. (Zhenbin Liu) and Z.L. (Zengke Li); software, Z.L. (Zhenbin Liu), A.L., K.S., Q.G. and C.W.; validation, Z.L. (Zhenbin Liu), A.L., K.S., Q.G. and C.W.; formal analysis, Z.L. (Zhenbin Liu); investigation, Z.L. (Zhenbin Liu); resources, Z.L. (Zhenbin Liu); data curation, Z.L. (Zhenbin Liu), A.L. and C.W.; writing—original draft preparation, Z.L. (Zhenbin Liu); writing—review and editing, Z.L. (Zhenbin Liu); visualization, Z.L. (Zhenbin Liu); supervision, Z.L. (Zengke Li); project administration, Z.L. (Zengke Li); funding acquisition, Z.L. (Zengke Li). All authors have read and agreed to the published version of the manuscript.

Funding: This work was supported by the National Natural Science Foundation of China (No. 42274020), Science and Technology Planning Project of Jiangsu Province (BE2023692) and National Natural Science Foundation of China (No. 41874006) (Corresponding author: Zengke Li).

Data Availability Statement: No new data were created or analyzed in this study. Data sharing is not applicable to this article.

Conflicts of Interest: The authors declare that they have no conflicts of interest.

References

1. Cadena, C.; Carlone, L.; Carrillo, H.; Latif, Y.; Scaramuzza, D.; Neira, J.; Reid, I.; Leonard, J.J. Past, Present, and Future of Simultaneous Localization and Mapping: Toward the Robust-Perception Age. *IEEE Trans. Robot.* **2016**, *32*, 1309–1332. [CrossRef]
2. Li, J.; Pei, L.; Zou, D. Attention-SLAM: A Visual Monocular SLAM Learning From Human Gaze. *IEEE Sens. J.* **2021**, *21*, 6408–6420. [CrossRef]
3. Debeunne, C.; Vivet, D. A review of visual-Lidar fusion based simultaneous localization and mapping. *Sensors* **2020**, *20*, 2068. [CrossRef] [PubMed]
4. Forster, C.; Carlone, L.; Dellaert, F. On-Manifold Preintegration for Real-Time Visual—Inertial Odometry. *IEEE Trans. Robot.* **2017**, *33*, 1–21. [CrossRef]
5. Tao, Y.; He, Y.; Ma, X. SLAM Method Based on Multi-Sensor Information Fusion. In Proceedings of the 2021 International Conference on Computer Network, Electronic and Automation (ICCNEA), Xi'an, China, 24–26 September 2021; pp. 289–293. [CrossRef]
6. Yu, H.; Wang, Q.; Yan, C.; Feng, Y.; Sun, Y.; Li, L. DLD-SLAM: RGB-D Visual Simultaneous Localisation and Mapping in Indoor Dynamic Environments Based on Deep Learning. *Remote Sens.* **2024**, *16*, 246. [CrossRef]
7. Huletski, A.; Kartashov, D.; Krinkin, K. Evaluation of the modern visual SLAM methods. In Proceedings of the 2015 Artificial Intelligence and Natural Language and Information Extraction, Social Media and Web Search FRUCT Conference (AINL-ISMW FRUCT), St. Petersburg, Russia, 9–14 November 2015; pp. 19–25. [CrossRef]
8. Shan, T.; Englot, B.; Ratti, C. LVI-SAM: Tightly-Coupled Lidar-Visual-Inertial Odometry via Smoothing and Mapping. In Proceedings of the 2021 IEEE International Conference on Robotics and Automation (ICRA), Xi'an, China, 30 May 2021–5 June 2021; pp. 5692–5698. [CrossRef]
9. Yin, J.; Li, A.; Li, T. M2DGR: A Multi-Sensor and Multi-Scenario SLAM Dataset for Ground Robots. *IEEE Robot. Auto Let.* **2022**, *7*, 2266–2273. [CrossRef]
10. Chghaf, M.; Rodriguez, S.; Ouardi, A.E. Camera, LiDAR and Multi-modal SLAM Systems for Autonomous Ground Vehicles: A Survey. *J. Intell. Robot. Syst.* **2022**, *105*, 2. [CrossRef]
11. Davison, A.J.; Reid, I.D.; Molton, N.D. MonoSLAM: Real-Time Single Camera SLAM. *IEEE Trans. Pattern Anal. Mach. Intell.* **2007**, *29*, 1052–1067. [CrossRef]
12. Klein, G.; Murray, D. Parallel Tracking and Mapping for Small AR Workspaces. In Proceedings of the 2007 6th IEEE and ACM International Symposium on Mixed and Augmented Reality, Nara, Japan, 13–16 November 2007; pp. 225–234. [CrossRef]
13. Mur-Artal, R.; Montiel, J.M. ORB-SLAM: A Versatile and Accurate Monocular SLAM System. *IEEE Trans. Robot.* **2017**, *31*, 1147–1163. [CrossRef]
14. Rublee, E.; Rabaud, V.; Konolige, K. ORB: An efficient alternative to SIFT or SURF. In Proceedings of the IEEE International Conference on Computer Vision, ICCV 2011, Barcelona, Spain, 6–13 November 2011. [CrossRef]
15. Mur-Artal, R.; Tardós, J.D. ORB-SLAM2: An Open-Source SLAM System for Monocular, Stereo, and RGB-D Cameras. *IEEE Trans. Robot.* **2017**, *33*, 1255–1262. [CrossRef]
16. Forster, C.; Pizzoli, M.; Scaramuzza, D. SVO: Fast semi-direct monocular visual odometry. In Proceedings of the 2014 IEEE International Conference on Robotics and Automation (ICRA), Hong Kong, China, 31 May–7 June 2014; pp. 15–22. [CrossRef]

17. Engel, J.; Thomas, S.; Cremers, D. Lsd-Salm: Large-Scale Direct Monocular Salm. In Proceedings of the European Conference on Computer Vision, Cham, Switzerland, 6–12 September 2014; pp. 834–849. [CrossRef]
18. Engel, J.; Koltun, V.; Cremers, D. Direct Sparse Odometry. *IEEE Trans. Pattern Anal. Mach. Intell.* **2018**, *40*, 611–625. [CrossRef] [PubMed]
19. Mourikis, A.I.; Roumeliotis, S.I. A Multi-State Constraint Kalman Filter for Vision-aided Inertial Navigation. In Proceedings of the 2007 IEEE International Conference on Robotics and Automation, Rome, Italy, 10–14 April 2007; pp. 3565–3572. [CrossRef]
20. Leutenegger, S.; Lynen, S.; Bosse, M. Keyframe-based visual–inertial odometry using nonlinear optimization. *Int. J. Robot. Res.* **2015**, *34*, 314–334. [CrossRef]
21. Qin, T.; Li, P.; Shen, S.T. VINS-Mono: A Robust and Versatile Monocular Visual-Inertial State Estimator. *IEEE Trans. Robot.* **2018**, *34*, 1004–1020. [CrossRef]
22. Qin, T.; Pan, J.; Cao, S. A general optimization-based framework for local odometry estimation with multiple sensors. *arXiv* **2019**, arXiv:1901.03638. [CrossRef]
23. Campos, C.; Elvira, R.; Rodríguez, J.J.G. Orb-slam3: An accurate open-source library for visual, visual–inertial, and multimap slam. *IEEE Trans. Robot.* **2021**, *37*, 1874–1890. [CrossRef]
24. Hess, W.; Kohler, D.; Rapp, H. Real-time loop closure in 2D LIDAR SLAM. In Proceedings of the 2016 IEEE International Conference on Robotics and Automation (ICRA), Stockholm, Sweden, 16–21 May 2016; pp. 1271–1278. [CrossRef]
25. Zhang, J.; Singh, S. LOAM: Lidar Odometry and Mapping in Real-time. *Robot. Sci. Syst.* **2014**, *2*, 1–9.
26. Qin, T.; Cao, S. A-LOAM. 2018. Available online: https://github.com/HKUST-Aerial-Robotics/A-LOAM (accessed on 23 April 2024).
27. Shan, T.; Englot, B. LeGO-LOAM: Lightweight and Ground-Optimized Lidar Odometry and Mapping on Variable Terrain. In Proceedings of the 2018 IEEE/RSJ International Conference on Intelligent Robots and Systems (IROS), Madrid, Spain, 1–5 October 2018; pp. 4758–4765. [CrossRef]
28. Kimm, G. SC-LeGO-LOAM. 2020. Available online: https://gitee.com/zhankun3280/lslidar_c16_lego_loam (accessed on 23 April 2024).
29. Kim, G.; Kim, A. Scan Context: Egocentric Spatial Descriptor for Place Recognition within 3D Point Cloud Map. In Proceedings of the 2018 IEEE/RSJ International Conference on Intelligent Robots and Systems (IROS), Madrid, Spain, 1–5 October 2018; pp. 4802–4809. [CrossRef]
30. Zhao, S.; Fang, Z.; Li, H. A Robust Laser-Inertial Odometry and Mapping Method for Large-Scale Highway Environments. In Proceedings of the 2019 IEEE/RSJ International Conference on Intelligent Robots and Systems (IROS), Macau, China, 3–8 November 2019; pp. 1285–1292. [CrossRef]
31. Ye, H.; Chen, Y.; Liu, M. Tightly Coupled 3D Lidar Inertial Odometry and Mapping. In Proceedings of the 2019 International Conference on Robotics and Automation (ICRA), Montreal, QC, Canada, 20–24 May 2019; pp. 3144–3150. [CrossRef]
32. Shan, T.; Englot, B.; Meyers, D. LIO-SAM: Tightly-coupled Lidar Inertial Odometry via Smoothing and Mapping. In Proceedings of the 2020 IEEE/RSJ International Conference on Intelligent Robots and Systems (IROS), Las Vegas, NV, USA, 24 October 2020–24 January 2021; pp. 5135–5142. [CrossRef]
33. Qin, C.; Ye, H.; Pranata, C.E. LINS: A Lidar-Inertial State Estimator for Robust and Efficient Navigation. In Proceedings of the 2020 IEEE International Conference on Robotics and Automation (ICRA), Paris, France, 31 May–31 August 2020; pp. 8899–8906. [CrossRef]
34. Xu, W.; Zhang, F. FAST-LIO: A Fast, Robust Lidar-Inertial Odometry Package by Tightly-Coupled Iterated Kalman Filter. *IEEE Robot. Autom. Let.* **2020**, *6*, 3317–3324. [CrossRef]
35. Xu, W.; Cai, Y.; He, D. FAST-LIO2: Fast Direct Lidar-Inertial Odometry. *IEEE Trans. Robot.* **2022**, *38*, 2053–2073. [CrossRef]
36. Bai, C.; Xiao, T.; Chen, Y. Faster-LIO: Lightweight Tightly Coupled Lidar-Inertial Odometry Using Parallel Sparse Incremental Voxels. *IEEE Robot. Autom. Let.* **2022**, *7*, 4861–4868. [CrossRef]
37. Graeter, J.; Wilczynski, A.; Lauer, M. LIMO: Lidar-Monocular Visual Odometry. In Proceedings of the 2018 IEEE/RSJ International Conference on Intelligent Robots and Systems (IROS), Madrid, Spain, 1–5 October 2018; pp. 7872–7879. [CrossRef]
38. Zhang, J.; Singh, S. Visual-Lidar odometry and mapping: Low-drift, robust, and fast. In Proceedings of the 2015 IEEE International Conference on Robotics and Automation (ICRA), Seattle, WA, USA, 26–30 May 2015; pp. 2174–2181. [CrossRef]
39. Geiger, A.; Lenz, P.; Stiller, C. Vision meets robotics: The kitti dataset. *Int. J. Robot. Res.* **2013**, *32*, 1231–1237.
40. Shao, W.; Vijayarangan, S.; Li, C. Stereo Visual Inertial Lidar Simultaneous Localization and Mapping. In Proceedings of the 2019 IEEE/RSJ International Conference on Intelligent Robots and Systems (IROS), Macau, China, 3–8 November 2019; pp. 370–377. [CrossRef]
41. Zuo, X.; Geneva, P.; Lee, W. LIC-Fusion: Lidar-Inertial-Camera Odometry. In Proceedings of the 2019 IEEE/RSJ International Conference on Intelligent Robots and Systems (IROS), Macau, China, 3–8 November 2019; pp. 5848–5854. [CrossRef]
42. Zuo, X. LIC-Fusion 2.0: Lidar-Inertial-Camera Odometry with Sliding-Window Plane-Feature Tracking. In Proceedings of the 2020 IEEE/RSJ International Conference on Intelligent Robots and Systems (IROS), Las Vegas, NV, USA, 24 October 2020–24 January 2021; pp. 5112–5119. [CrossRef]
43. Wisth, D.; Camurri, M.; Das, S. Unified Multi-Modal Landmark Tracking for Tightly Coupled Lidar-Visual-Inertial Odometry *IEEE Robot. Autom. Let.* **2021**, *6*, 1004–1011. [CrossRef]

44. Lin, J.; Zheng, C.; Xu, W. R^2 LIVE: A Robust, Real-Time, Lidar-Inertial-Visual Tightly-Coupled State Estimator and Mapping. *IEEE Robot. Autom. Let.* **2021**, *6*, 7469–7476. [CrossRef]
45. Lin, J.; Zheng, C. R^3LIVE: A Robust, Real-time, RGB-colored, Lidar-Inertial-Visual tightly-coupled state Estimation and mapping package. In Proceedings of the 2022 International Conference on Robotics and Automation (ICRA), Philadelphia, PA, USA, 23–27 May 2022; pp. 10672–10678. [CrossRef]
46. Zheng, C. FAST-LIVO: Fast and Tightly-coupled Sparse-Direct Lidar-Inertial-Visual Odometry. In Proceedings of the 2022 IEEE/RSJ International Conference on Intelligent Robots and Systems (IROS), Kyoto, Japan, 23–27 October 2022. [CrossRef]
47. Wang, C.Y.; Bochkovskiy, A.; Liao, H.Y.M. YOLOv7: Trainable bag-of-freebies sets new state-of-the-art for real-time object detectors. In Proceedings of the 2023 IEEE/CVF Conference on Computer Vision and Pattern Recognition (CVPR), Vancouver, BC, Canada, 17–24 June 2023. [CrossRef]
48. Lin, J.; Chen, W.M.; Lin, Y.; Cohn, J.; Han, S. MCUNet: Tiny Deep Learning on IoT Devices. *arXiv* **2007**, arXiv:2007.10319. [CrossRef]
49. Lyu, R. Nanodet-Plus: Super Fast and High Accuracy Lightweight Anchor-Free Object Detection Model. 2021. Available online: https://github.com/RangiLyu/nanodet (accessed on 23 April 2024).
50. Ge, Z.; Liu, S.; Wang, F. Yolox: Exceeding yolo series in 2021. *arXiv* **2021**, arXiv:2107.08430. [CrossRef]
51. Michele, A.; Colin, V.; Santika, D.D. Mobilenet convolutional neural networks and support vector machines for palmprint recognition. *Procedia Comput. Sci.* **2019**, *157*, 110–117. [CrossRef]
52. Zhang, X.; Zhou, X.; Lin, M. Shufflenet: An extremely efficient convolutional neural network for mobile devices. In Proceedings of the IEEE Conference on Computer Vision and Pattern Recognition, Salt Lake City, UT, USA, 18–22 June 2018; pp. 6848–6856. [CrossRef]
53. Han, K.; Wang, Y.; Tian, Q. Ghostnet: More features from cheap operations. In Proceedings of the IEEE/CVF Conference on Computer Vision and Pattern Recognition, Seattle, WA, USA, 13–19 June 2020; pp. 1580–1589. [CrossRef]
54. Targ, S.; Almeida, D.; Lyman, K. Resnet in resnet: Generalizing residual architectures. *arXiv* **2016**, arXiv:1603.08029. [CrossRef]
55. Yu, F.; Wang, D.; Shelhamer, E. Deep layer aggregation. In Proceedings of the IEEE Conference on Computer Vision and Pattern Recognition, Salt Lake City, UT, USA, 18–23 June 2018; pp. 2403–2412. [CrossRef]
56. Wang, C.Y.; Liao, H.Y.M.; Wu, Y.H. CSPNet: A new backbone that can enhance learning capability of CNN. In Proceedings of the IEEE/CVF Conference on Computer Vision and Pattern Recognition Workshops, Seattle, WA, USA, 14–19 June 2020; pp. 390–391. [CrossRef]
57. Sol'a, J. Quaternion kinematics for the error-state Kalman filter. *arXiv* **2017**, arXiv:1711.02508. [CrossRef]
58. Teunissen, P.J.G.; Khodabandeh, A. Review and principles of PPP-RTK methods. *J. Geod.* **2015**, *89*, 217–240. [CrossRef]

remote sensing

MDPI

Article

Enhanced Strapdown Inertial Navigation System (SINS)/LiDAR Tightly Integrated Simultaneous Localization and Mapping (SLAM) for Urban Structural Feature Weaken Occasions in Vehicular Platform

Xu Xu, Lianwu Guan *, Yanbin Gao, Yufei Chen and Zhejun Liu

College of Intelligent Systems Science and Engineering, Harbin Engineering University, Harbin 150001, China; xuxu66@hrbeu.edu.cn (X.X.)

* Correspondence: guanlianwu@hrbeu.edu.cn

Abstract: LiDAR-based simultaneous localization and mapping (SLAM) offer robustness against illumination changes, but the inherent sparsity of LiDAR point clouds poses challenges for continuous tracking and navigation, especially in feature-deprived scenarios. This paper proposes a novel LiDAR/SINS tightly integrated SLAM algorithm designed to address the localization challenges in urban environments characterized in sparse structural features. Firstly, the method extracts edge points from the LiDAR point cloud using a traditional segmentation method and clusters them to form distinctive edge lines. Then, a rotation-invariant feature—line distance—is calculated based on the edge line properties that were inspired by the traditional tightly integrated navigation system. This line distance is utilized as the observation in a Kalman filter that is integrated into a tightly coupled LiDAR/SINS system. This system tracks the same edge lines across multiple frames for filtering and correction instead of tracking points or LiDAR odometry results. Meanwhile, for loop closure, the method modifies the common SCANCONTEXT algorithm by designating all bins that do not reach the maximum height as special loop keys, which reduce false matches. Finally, the experimental validation conducted in urban environments with sparse structural features demonstrated a 17% improvement in positioning accuracy when compared to the conventional point-based methods.

Keywords: 3D LiDAR navigation; SLAM; tightly integrated navigation; LiDAR odometry and mapping; urban structural feature weaken occasions

Citation: Xu, X.; Guan, L.; Gao, Y.; Chen, Y.; Liu, Z. Enhanced Strapdown Inertial Navigation System (SINS)/LiDAR Tightly Integrated Simultaneous Localization and Mapping (SLAM) for Urban Structural Feature Weaken Occasions in Vehicular Platform. *Remote Sens.* **2024**, *16*, 2527. https://doi.org/10.3390/rs16142527

Academic Editors: Wanshou Jiang, San Jiang, Duojie Weng and Jianchen Liu

Received: 29 April 2024
Revised: 5 July 2024
Accepted: 7 July 2024
Published: 10 July 2024

1. Introduction

In recent years, the domain of simultaneous localization and mapping (SLAM) [1] has been an integral part of autonomous navigation, especially in environments where the reception of global navigation satellite system (GNSS) signals is unreliable or absent and where dynamic environmental conditions are the norm. SLAM aims to determine a robot's pose while simultaneously generating a map of its environment using onboard sensors. This process occurs in environments that may be unknown or partially known. The diversity of applicable sensors in use has naturally led to the bifurcation of SLAM into two primary SLAM categories: LiDAR-based SLAM and visual SLAM. Visual SLAM encompasses various subtypes, including monocular, stereo, and RGB-D [2]. LiDAR-based approaches demonstrate superior accuracy in pose estimation and maintain robust performance across varying environmental conditions, such as time of day and weather. In contrast, visual SLAM, as illustrated in Figure 1, is highly susceptible to factors like lighting and the availability of distinctive features, thus potentially limiting its effectiveness in certain settings [3]. Therefore, this paper concentrates on navigation systems leveraging LiDAR technology.

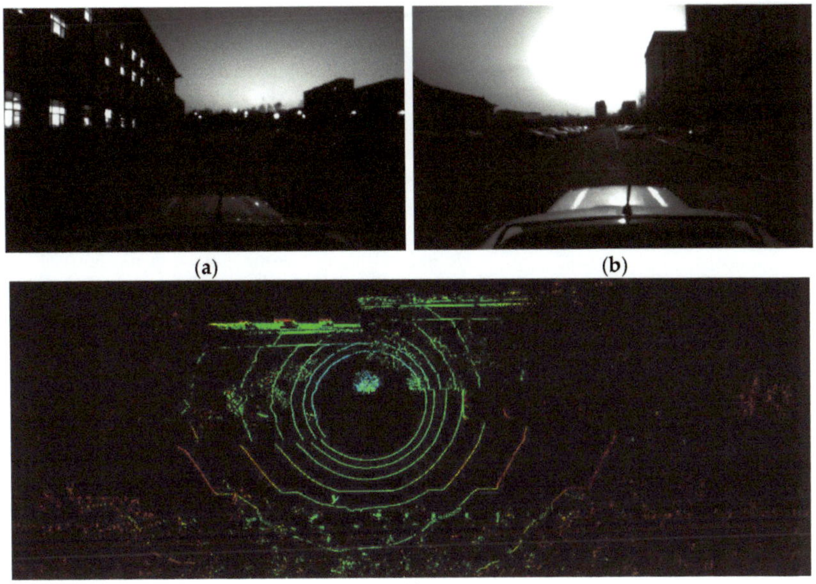

Figure 1. The figures illustrate the impact of the environment on different sensors; (**a**,**b**) show the effect of darkness and illumination on the visual sensor, respectively, while (**c**) indicates that LiDAR can work normally under such conditions.

The last two decades have witnessed significant strides in LiDAR-based SLAM methodologies, fueled by advancements in computer processing power and optimization algorithms. While machine learning-based LiDAR SLAM methods are constrained by the scope and quality of training data, two primary categories dominate the current landscape: normal distributions transform (NDT) [4] and iterative closest point (ICP) [5] algorithm. The NDT approach, which has seen extensive application in 2D LiDAR SLAM scenarios, entails the discretization of the point cloud data into a grid-like structure, the computation of Gaussian distributions for each cell, and subsequent alignment-based matching and fitting procedures [4]. However, the transition from 2D to 3D applications has exponentially increased computational demands, posing a significant challenge in meeting the stringent requirements for real-time processing [6]. To put it simply, when classified solely based on the quantity of point clouds and meshes, the computational data of the most basic 3D LiDAR are at least 16 times that of a 2D LiDAR, as they have at least 16 scanning projection planes. Although efforts to mitigate this issue have been made through algorithms such as SEO-NDT [7] and KD2D-NDT [8], they have occasionally resulted in trade-offs concerning accuracy and processing time in certain scenarios. ICP-based methods face similar challenges. However, the advent of LiDAR odometry and mapping (LOAM) [9] has marked a pivotal shift in focus; LOAM focuses the iteration process towards feature-rich points rather than the entire point cloud. This paradigm shift has propelled the widespread implementation of LOAM-inspired ICP techniques in addressing LiDAR SLAM challenges over the past decades.

LOAM distinguishes itself from conventional ICP techniques by classifying points in the point cloud based on their smoothness. This involves identifying and extracting "edge points", which are characterized by coarse texture, and "planar points", which exhibit fine texture. Subsequently, the derived feature points are systematically selected through a sector-based averaging technique. The system leverages these refined point clouds; the system performs odometry calculations at a frequency of 10 Hz using LiDAR data. Following the odometry computation, the aggregated point clouds are then employed for mapping at a reduced frequency of 1 Hz, thereby achieving a more accurate and efficient representation of the environment. To address LOAM's limitations in computational demands and loop closure, Shan and Englot proposed LeGO-LOAM [10]. This method, which stands for lightweight and ground-optimized LiDAR odometry and mapping, is specifically tailored for real-time six-degrees-of-freedom pose estimation with ground vehicles. However, further experiments have shown that the strategy of entirely segregating ground points from the surrounding point cloud environment for separate matching can result in a notable vertical drift. Furthermore, the methodologies for loop closure still face certain challenges.

Li He and colleagues investigated the application of Multiview 2D projection (M2DP) [11] to describe 3D points to achieve loop closure, but their findings showed limited scope and efficacy. Scan Context [12] and its advanced iteration, Scan Context++ [13], were introduced by Giseop Kim in 2018 and 2021, respectively. These innovative approaches have rapidly gained recognition as leading solutions for loop closure in 3D LiDAR-based SLAM. This is a non-histogram-based global descriptor that directly captures egocentric structural information from the sensor's field of view without relying on prior training. However, the aforementioned methods and their derivatives, such as F-LOAM [14], do not utilize strapdown inertial navigation systems (SINSs) or only use them for the rectification of LiDAR point clouds.

Compared to the mature field of vision-aided SINS, the integration of SINS and LiDAR within LiDAR-based SLAM algorithms remains largely unexplored. A study [15] employed a loosely coupled extended Kalman filter (EKF) to fuse IMU and LiDAR data within a two-dimensional framework. However, this approach lacked the robustness to handle the complexities of three-dimensional or multifaceted environments. Furthermore, a scholarly review published in 2022 [16] emphasized that within the majority of current systems employing the SINS/LiDAR integration systems, the SINS primarily functions to smooth trajectories and mitigate distortions. IMU data are often optionally integrated to predict platform motion and enhance registration accuracy during abrupt maneuvers. However, only gyroscopic measurements between consecutive LiDAR scans are utilized. Although these studies and related works often self-identify as "loose integration" based on the data fusion strategies outlined in this article, a more accurate designation would be "pseudo integration".

As illustrated in Figure 2, the concept of loose integration in LiDAR/SINS systems can be redefined from the GNSS/SINS loose integration navigation system. This approach involves combining position and other navigation data obtained from different sensors. In this process, none of the sensors involved in the integration have undergone in-depth data integration, but only a simple integration of the navigation results. By applying this redefined concept of loose integration, it becomes evident that studies such as [17,18], while claiming to employ tight integration, actually align more closely with the characteristics of loose integration. Specifically, these studies treat the individual systems as black boxes, focusing solely on integrating their outputs to generate the final navigation solution rather than performing in-depth data extraction and analysis.

To achieve a deeper level of sensor fusion than loose integration, the integration process should occur before the generation of individual navigation solutions. For instance, in a GNSS/SINS system, this translates to integrating data during the pseudorange measurement stage, prior to GNSS position determination. A key advantage of tightly coupled GNSS/SINS integration [19] over the loosely coupled approach is its reduced reliance on a high number of visible satellites. This integration scheme can function even with a single observable satellite, unlike loose integration, which typically requires at least four. Investigating tight integration within SLAM systems necessitates understanding the nature of the data employed for navigation. In LiDAR-based systems, these data comprise point clouds, while vision-based systems utilize feature point information. Some studies [20,21] have demonstrated that within the SLAM framework, the concept of lines exhibits greater stability than points, particularly during data transformations (rotation and translation) across multiple frames. Similarly, research on multi-frame feature tracking within multi-state constrained Kalman filters for vision-aided inertial navigation [22] has validated the enhanced accuracy and robustness of this approach compared to traditional methods. This has led to the development of a prototype tightly integrated LiDAR/SINS navigation system that utilizes line features extracted from the LiDAR point cloud as observations. The system employs continuous, multi-frame tracking of these line features to refine the SINS data. However, due to the sparse nature of the LiDAR point cloud, it is difficult to accurately track the same line. Consequently, revisiting the concept of distance as a measurement, akin to its application in GNSS/SINS systems, becomes crucial. Notably, distance, being a scalar quantity, offers a significant advantage—rotational invariance. This property can substantially reduce the computational burden of the integration process. The algorithm's core principle centers on leveraging shared features, specifically line distances, across multiple frames to enhance Kalman filter accuracy. In summary, this paper presents the following contributions:

1. This paper refines the edge point extraction process of the LOAM algorithm by implementing a more granular clustering approach. By classifying clustered edge points as either convex or concave, the mapping precision is enhanced. Leveraging the rotational invariance of line distances, a Kalman filter is developed that employs line distance error as its primary observation metric. This approach improves the system's robustness and accuracy.
2. This paper presents structural modifications to the LOAM algorithm that are predicated on the Scan Context framework to optimize its performance and ensure the data processing occurs more efficiently. The experiments have proven that the situation of incorrect loop closures in LiDAR SLAM has been mitigated effectively.
3. Extensive experiments conducted in various on-campus and off-campus environments validate the proposed algorithm and offer comparisons with traditional methods. These experiments highlight the superior performance of the proposed algorithm.

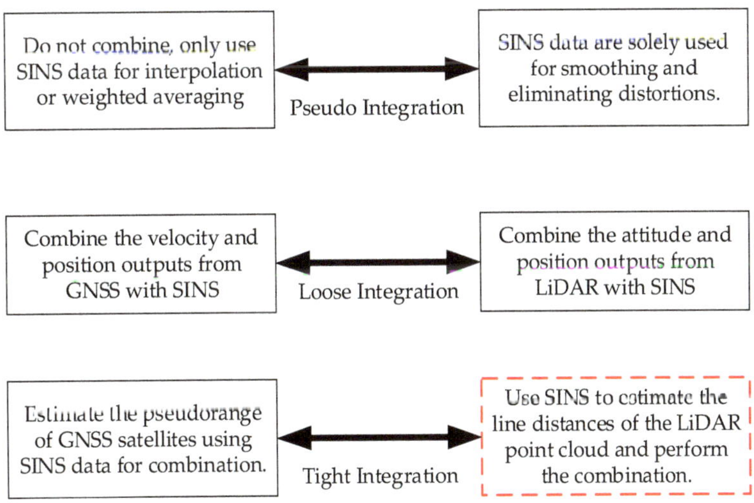

Figure 2. Use the concept of traditional SINS/GNSS integrated navigation systems to redefine the LiDAR integrated navigation system.

2. Method

This section outlines the workflow of our algorithm. This includes an insightful overview of the foundational principles that govern the pertinent hardware components, coupled with a thorough elucidation of the methodologies employed for the preprocessing of data. Section 3 will then delve into the system error model and measurement model, providing a comprehensive analysis of these crucial framework elements. Additionally, to elucidate the algorithmic details, the subsequent sections of this paper will operate under the assumption that the LiDAR point cloud was sourced from a 16-line LiDAR system by default. This is representative of commonly used systems such as Velodyne's VLP-16 (Velodyne Acoustics GmbH, Hamburg, Germany) and the LeiShen MS-C16 [23] (Leishen Intelligent System Co., Ltd., located in Shenzhen, China) employed in the experiments of this paper. These systems, with a horizontal angular resolution of 0.2° and a vertical resolution of 2°, generate a range image of 1800 by 16 pixels [23]. This translates to a point cloud with 16 projection planes, each containing 1800 points.

2.1. Algorithm Overview

Figure 3 provides the overview of a tightly integrated LiDAR/SINS SLAM algorithm. Let P be the original points received in a laser scan. However, because scanning occurs over a timeframe t (typically exceeding 0.1 s), the resulting point cloud represents the environment over this duration rather than instantaneously. Consequently, in dynamic environments, the recorded point cloud may exhibit distortions caused by movements, particularly pronounced during significant angular variations. To mitigate this, it is essential to utilize the high-frequency motion data provided by the SINS to project all points onto the reference timestamps, either the beginning of the period t_{k-1} or the end t_k.

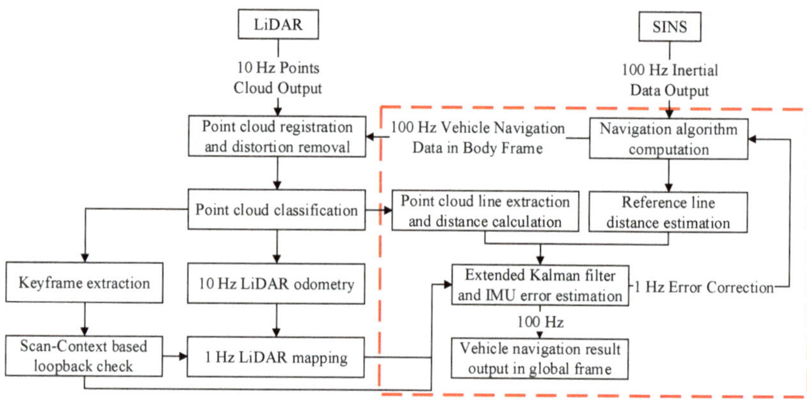

Figure 3. The algorithm overview of SINS-based 3D LiDAR tightly integrated SLAM.

After processing, the point cloud is denoted as $\hat{P}$ and proceeds to the next stage. Here, each point undergoes a meticulous classification into four categories based on its properties: (a) ground points, representing the surface on which the vehicle travels; (b) planar points, indicative of flat surfaces except the ground; (c) edge points, which demarcate boundaries or the perimeter of objects; and (d) the others, encompassing all points that do not fit into the previous categories. The subsequent section will elaborate on the point cloud classification technique, ensuring a thorough understanding. All points outside the ground in a key frame are compressed into scan context descriptors, and the key frames are set based on distance and the structure of the point clouds. Concurrently, after the clustering process, edge points are re-extracted to form edge lines. These edge lines will then serve as a basis for further computation of the reference line distances and facilitate tightly coupled filtering.

The LiDAR/SINS odometry primarily relies on the SINS navigation results, and the outputs further processed by LiDAR mapping, which matches and registers the undistorted point cloud onto a map at a frequency of 1 Hz. The Scan Context system performs loop closure detection based on both time and the distance traveled. When the similarity measure in the loop closure detection reaches a certain threshold, it is considered that the vehicle has returned to a previously visited location. Subsequently, the system optimizes the overall trajectory using this information. Successful loop closure detections will also contribute to the refinement of the SINS navigation and Kalman filtering processes.

2.2. Point Cloud Classification and Point Cloud Lines Extraction

Figure 4 shows the undistorted raw point cloud, ground points, edge points, and planar points, as well as edge line points, respectively. The following will detail the extraction methods for each point type.

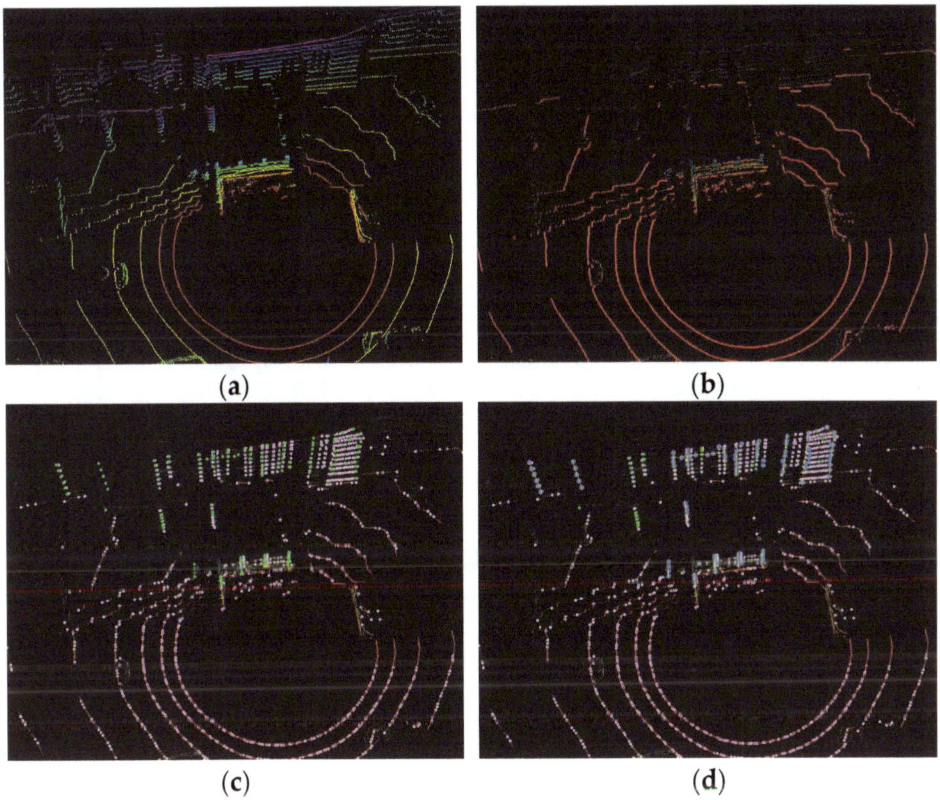

Figure 4. Feature extraction process for a scan in noisy environment. The original point cloud is shown in (**a**). (**b**) The red points are labeled as ground points. The rest of the points are the points that remain after segmentation. This method will be shown in Section 2.2.1. (**c**) Green and pink points indicate edge and planar features, which are mentioned in Section 2.2.2. (**d**) The blue points represent edge line points. The specific extraction method is explained in Section 2.2.3.

2.2.1. Ground Points

LeGO-LOAM employs a straightforward and efficient ground point extraction method, which involves specifically examining the 8 lines out of the total 16 that are positioned below 0° for detection [23].

In point cloud $\hat{P}$, point clouds are labeled with rings and scan sequence; let $p_{i,j} \in \hat{P}$, $i = 1,2,3 \cdots 16$, *and* $j = 1,2,3 \cdots 1800$. As shown in Figure 5, to calculate the angle between adjacent points $p_{i,j}$ and $p_{i+1,j}$, this paper assumes their coordinate differences are denoted as $diff\text{x}$, $diff\text{y}$ and $diff\text{z}$. The angle θ could be set as:

$$\theta = \tan^{-1}\left(diff\text{z}, \sqrt{diff\text{x}^2 + diff\text{y}^2}\right) \quad (1)$$

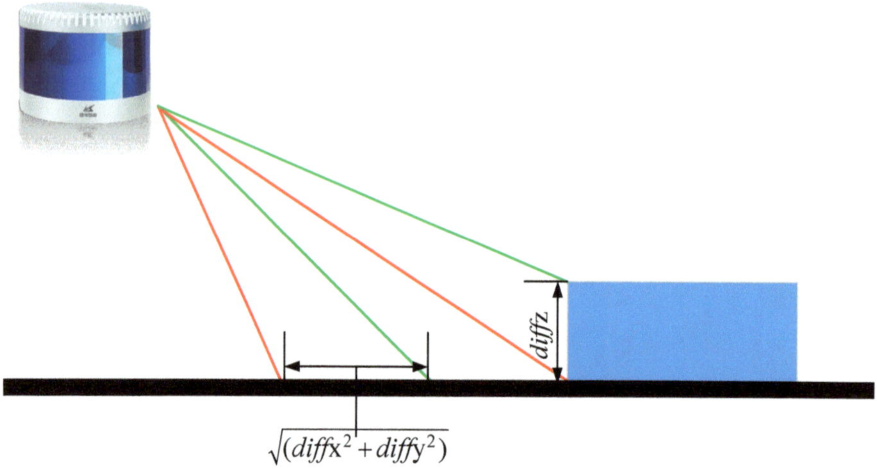

Figure 5. Ground points extracted from the point cloud $\hat{P}$.

Once $\theta < 10°$, points are marked as candidate ground points. Furthermore, an advanced point cloud sieving process [24] will utilize the RANSAC (random sample consensus) [25] technique to confirm the identification of ground points. This step is critical to avoid the misclassification of non-ground points as ground points, thereby ensuring the accuracy and reliability of the ground detection process. The fitted ground equation is as follows:

$$Ax + By + Cz + d = 0 \tag{2}$$

Then, an image-based segmentation method [26] is applied to the range image to group points into many clusters. Points from the same cluster are assigned a unique label.

2.2.2. Edge and Planar Points

The feature extraction process is similar to the method used in [9]; but, instead of extracting from the raw point cloud $\hat{P}$, we exclusively utilize the portion of the point cloud that remains unmarked as ground points. Let S be the set of points of p_i from the same ring of the point clouds. Half of the points are on either side of p_i. The set for this paper is presented in Table A1. Using the range values computed during segmentation, we can evaluate the roughness of point p_i in S,

$$c = \frac{1}{|S| \cdot \|r_i\|} \left\| \sum_{j \in S, j \neq i} (r_j - r_i) \right\| \tag{3}$$

where r_j means the range from p_i to the center of LiDAR.

Similar to LOAM, we use a threshold c_{th} to distinguish different types of features. We call the points with c larger than c_{th} edge points, and the points with c smaller than c_{th} planar points. Then, we sort the edge and planar points from minimum to maximum. The point cloud is segmented into several distinct parts, and a specific number of feature points are extracted from within each segment.

Following the extraction of feature points, another attribute will be computed, specifically, the concavity or convexity of the edge points. Figure 6 shows the difference between the concave points and convex points. Compare the distances between a specific point p_i and the remaining points within set S. If the majority of these points have distances greater than that of p_i, then p_i is considered a convex point. Otherwise, it is a concave point.

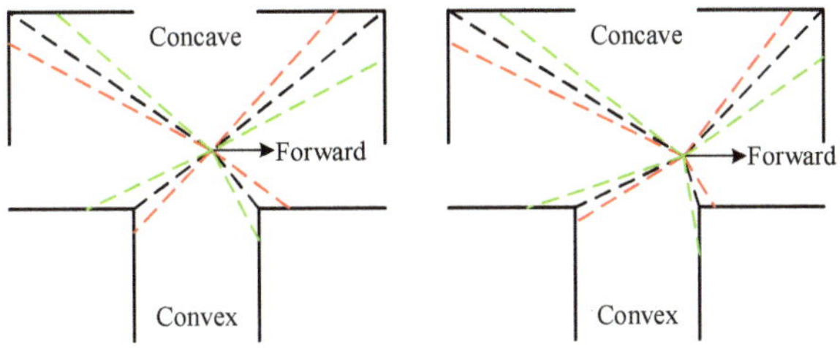

Figure 6. The concavity or convexity of the edge points.

2.2.3. Edge Lines

After classifying edge points as concave or convex, this paper employs K-means clustering [27] to group them into lines. Similarly, these lines will inherit concavity or convexity from the points that constitute them. This choice of using edge lines instead of individual edge points for subsequent computations stems from the inherent sparsity of LiDAR point clouds. Ensuring the capture of the exact same point across consecutive scans is a challenging proposition. In contrast, lines, when considered as collective entities, offer a higher degree of continuity and are much more amenable to persistent tracking. This approach enhances the reliability and robustness of the subsequent processing steps.

Section 3 will elaborate on the method for calculating point-to-line distances and the line selection criteria.

2.3. Scan Context

Scan Context was inspired by Shape Context [28], proposed by Belongie et al.; it is an algorithm for place recognition using 3D LiDAR scans. It works by:

1. Partitioning the point cloud into bins based on azimuthal and radial directions.
2. Encoding the point cloud into a matrix where each bin's value is the maximum height of points within it.
3. Calculating similarity between scan contexts using a column-wise distance measure.
4. Employing a two-phase search for loop detection that is invariant to viewpoint changes.

Figure 7 shows the bin division along azimuthal and radial directions. Using the top view of a point cloud from a 3D scan, the paper [10] partitioned ground areas into bins, which were split according to both azimuthal (from 0 to 2π within a LiDAR frame) and radial (from center to maximum sensing range) directions. They referred to the yellow area as a ring, the cyan area as a sector, and the black-filled area as a bin.

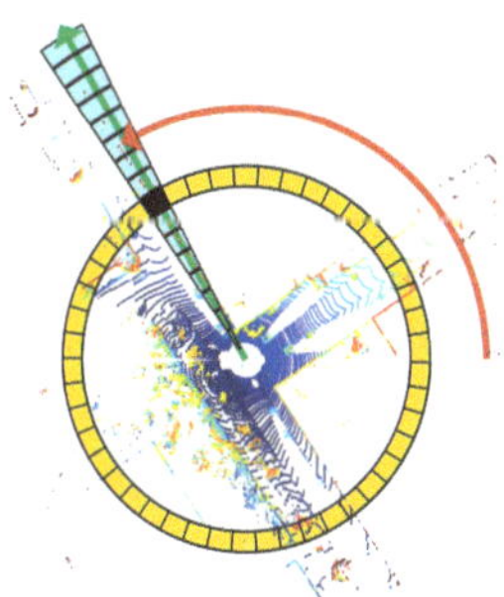

Figure 7. The scan context bins.

However, assigning the maximum height of points within a bin a value in the scan context can be problematic in certain situations. As depicted in Figure 8, due to the formation principle of LiDAR point clouds, the point cloud does not fully unfold at close ranges, which may result in the highest point not accurately representing the actual environmental point cloud.

Figure 8. Close-range point cloud scanning scenario in Cloud-Compare software (2.13).

A straightforward and effective solution is to perform a ring-based search for the highest point. If the highest point lies within the outermost ring of the 3D LiDAR and is lower than the adjacent bins, an additional annotation is made to record that the highest point has not been detected. The marked bin can then serve as a ring-key in scan context for the initial match.

Simultaneously, because the point cloud distribution is dense near and sparse far, for each point cloud P selected as a key frame, we can first calculate its centroid:

$$\hat{P}(O) = \frac{1}{n}\sum p_{i,j},\ \overline{p_{i,j}} = p_{i,j} - \hat{P}(O) \tag{4}$$

where n is the total number of the point cloud and $\overline{p_{i,j}}$ is the point cloud $p_{i,j}$ transformed back to the center of the original point cloud.

As is shown in Figure 9, the transformed point cloud will have a common center, which will save a significant amount of time in subsequent scan context description and matching processes.

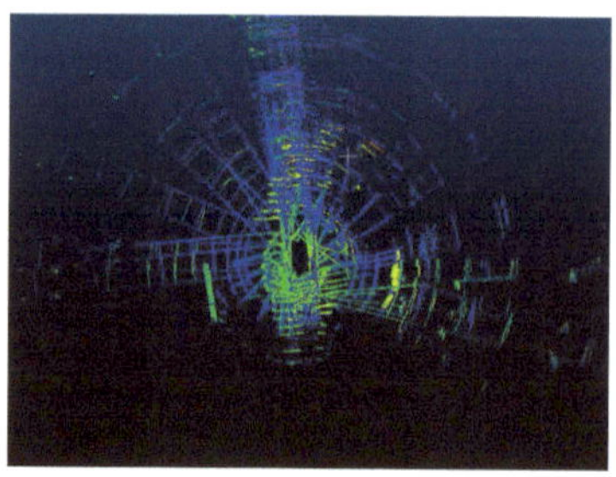

Figure 9. Transformed point clouds stacked in Cloud-Compare software (2.13).

3. LiDAR/SINS System Model

3.1. System Error Model

The SINS integrated navigation system error model was designed following the list in [29]:

$$\delta\dot{x} = F\delta x + Gw \tag{5}$$

$$F = \begin{bmatrix} F_{11} & F_{12} & 0_{3*3} & 0_{3*3} & 0_{3*3} \\ F_{21} & F_{22} & F_{23} & 0_{3*3} & R_b^n \\ F_{31} & F_{32} & F_{33} & R_b^n & 0_{3*3} \\ 0_{3*3} & 0_{3*3} & 0_{3*3} & F_{44} & 0_{3*3} \\ 0_{3*3} & 0_{3*3} & 0_{3*3} & 0_{3*3} & F_{55} \end{bmatrix} \tag{6}$$

The specific parameters in Equation (6) can be referred to in Equation (A1). And ϕ and λ are local latitude and longitude. R_M and R_N are meridian radius and normal radius. The h is the geodetic height. R_b^n is the transformation matrix from body frame to navigation frame.

For system state variables, $\delta x = \left[\delta p^n, \delta v^n, \delta a^n, b_\omega^n, b_f^n\right]^T$, δp^n denotes the positional errors in longitude, latitude, and height. δv^n presents the velocity errors related to the three directions above. δa^n is the error attitude. b_ω^n and b_f^n are the sensor noise errors of the gyroscopes and accelerometers, respectively.

The w is system noise, and the corresponding system noise matrix is given by:

$$G = \left[0_{9*1} \quad \sqrt{2\beta_{\omega x}\sigma_{\omega x}^2} \quad \sqrt{2\beta_{\omega y}\sigma_{\omega y}^2} \quad \sqrt{2\beta_{\omega z}\sigma_{\omega z}^2} \quad \sqrt{2\beta_{fx}\sigma_{fx}^2} \quad \sqrt{2\beta_{fy}\sigma_{fy}^2} \quad \sqrt{2\beta_{fz}\sigma_{fz}^2}\right]^T \tag{7}$$

where $\beta_{\omega x}$, $\beta_{\omega y}$, and $\beta_{\omega z}$ denote reciprocals of the correlation times of the autocorrelation sequence of b_ω^n while β_{fx}, β_{fy}, and β_{fz} are related to b_f^n. The $\sigma_{\omega x}^2$, $\sigma_{\omega y}^2$, $\sigma_{\omega z}^2$, σ_{fx}^2, σ_{fy}^2, and σ_{fz}^2 are variance associated with gyroscope and accelerometer errors.

3.2. The Observation Model

Traditionally, LiDAR-IMU integration has followed a loosely coupled approach. The observation variables of the model defined as the estimated position errors are:

$$Z = \begin{bmatrix} \phi_{Lidar} - \phi_{SINS} \\ \lambda_{Lidar} - \lambda_{SINS} \\ h_{Lidar} - h_{SINS} \end{bmatrix} \tag{8}$$

However, this integration approach merely treats LiDAR and SINS as two black boxes, simply combining their output results without any deeper level of mutual correction. This paper draws on the concept of tight integration between GNSS and SINS, selecting the error of the reference line distance d as the observation variable for filtering.

The selection of the error of d as the observation variable is based on the following considerations:

1. Frame invariance: In navigation systems, the relative orientation between the body frame and the navigation frame continuously changes. Distance, however, remains consistent across different frames.
2. Robustness to data loss: Compared to vision data, LiDAR point clouds are inherently sparse. Using feature points and their associated information directly as observation variables increases the susceptibility to data loss.

In three-dimensional space, the distance from a point L_0 to a straight line that was built by points L_1 and L_2 can be calculated as follow:

$$d = \frac{|(L_0 - L_1) \times (L_2 - L_1)|}{|L_2 - L_1|} \tag{9}$$

Imagine the points L_1 and L_2 with coordinates (x_{l1}, y_{l1}, z_{l1}) and (x_{l2}, y_{l2}, z_{l2}), respectively, and the origin of the vehicle or robot in time t_k could be estimated as $\tilde{p}_k = \left(\tilde{x}_k, \tilde{y}_k, \tilde{z}_k\right)$ while the ground truth is $p_k = (x_k, y_k, z_k)$; the relationship between them is:

$$\tilde{p}_k = \begin{Bmatrix} \tilde{x}_k \\ \tilde{y}_k \\ \tilde{z}_k \end{Bmatrix} = \begin{Bmatrix} x_k + \Delta x_k \\ y_k + \Delta y_k \\ z_k + \Delta z_k \end{Bmatrix} \tag{10}$$

where Δx_k, Δy_k, and Δz_k are the position errors in the ENU directions and $\delta p = (\Delta x_k/(R_{N+h})cos\phi, \Delta y_k/(R_M + h), \Delta z_k)\prime$. The figure above represents the relationship between the vehicle and the reference line in three-dimensional space, where L_0' denotes the actual position of the vehicle and L_0 signifies the vehicle's estimated position.

The measurement parameter in the EKF system is:

$$Z = \Delta d = d' - d \tag{11}$$

As is shown in Figure 10, Δd is much smaller than d or $d\prime$; it can be approximately considered that the normal vector to the line L_1L_2 connecting L_0' and L_0 is consistent. Then, the relationship between Δd and δp could be set as follows:

$$\Delta d = \vec{n}.\delta p \tag{12}$$

where $\vec{n} = (n_x, n_y, n_z)$ is the normalized vector from L_0L_d. Meanwhile, the corresponding system design matrix H is:

$$H = \begin{bmatrix} n_{x1}(R_N + h)cos\phi, n_{y1}(R_M + h), n_{z1}, 0_{1*12} \\ n_{x2}(R_N + h)cos\phi, n_{y2}(R_M + h), n_{z2}, 0_{1*12} \\ n_{x3}(R_N + h)cos\phi, n_{y3}(R_M + h), n_{z3}, 0_{1*12} \\ \cdots \\ n_{xi}(R_N + h)cos\phi, n_{yi}(R_M + h), n_{zi}, 0_{1*12} \end{bmatrix} \tag{13}$$

where i is the number of reference lines selected in the integrated system. It will always be changed with the changing of point clouds.

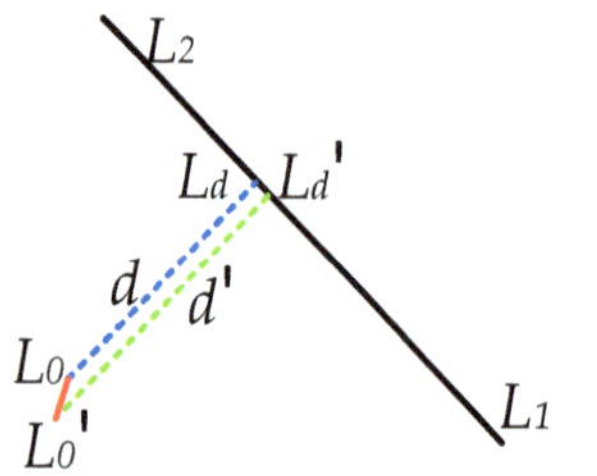

Figure 10. The distance from a point to a line.

For the LiDAR/IMU EKF system, the formulation linking the system observation variables and state variables is:

$$Z = H\delta x + v \tag{14}$$

where Z, H, and δx were defined previously and v represents the observation noise. As mentioned above, the LiDAR point cloud is the covariance value of v. LiDAR point clouds exhibit sparsity, and the uncertainty of a single point affects its entire surrounding space. The surrounding space is a truncated segment of a torus. For ease of calculation, this paper assumes it to be a rectangular prism, and its volume represents the covariance value of the measurement noise v.

For example, imagine a special distance of the reference line d_0. The covariance value of the noise of d_0 is $d_0^2(\sin^2\theta_h + \sin^2\theta_v)$ and θ_h is the horizontal separation angle of the LiDAR device while θ_v is the vertical separation angle.

As mentioned above, i refers to the number of reference lines; in this stated equation, this paper proposes that the following principles should be adhered to in controlling the reference lines involved in the filtering process.

1. Region division: The point cloud is segmented into multiple regions based on the scanning direction. Each region is characterized by the edge lines exhibiting distinct convexity/concavity properties.
2. Reference line tracking: The position of each reference line is tracked across multiple frames using SINS transformations. This ensures the consistent matching of the same reference line over an extended period.

3. Dynamic reference line management: Due to the inherent sparsity of point clouds, reference lines exceeding a predefined distance threshold are discarded. New reference lines are introduced to maintain robust matching.

3.3. Tracking of Reference Lines

1. The selection rules of the reference lines are mentioned in Section 3.2. Here, the tracking of these lines will be revealed with details.
2. As mentioned in Section 2.3, scan context built a series of point bins to extract the point cloud information. Imagine the attitude transformation matrix during the tracking period is R_{3*3} while the t_{3*1} represents the displacement provided by SINS. The projections L_1', L_2' of points L_1 and L_2 in the new point cloud could be calculated in Equation (12):

$$\begin{bmatrix} L_1' \\ 1 \end{bmatrix} = \begin{bmatrix} R_{3*3}^T & -t_{3*1} \\ 0_{1*3} & 1 \end{bmatrix} \begin{bmatrix} L_1 \\ 1 \end{bmatrix} \quad (15)$$

3. Connect L_1 and L_2 to obtain the scan context bins they pass through. By statistically analyzing these bins with their adjacent bins, select the edge line L_{new} that is also located in the same bin area. To further determine whether it is derived from a change in the original edge line, in addition to judging its concavity and convexity as well as the concavity and convexity and distance of the closest edge line, a further similarity analysis can be conducted on the line vectors.

$$s = 1 - \frac{\overrightarrow{L_1' L_2'} . \overrightarrow{L_{new}}}{\|\overrightarrow{L_1' L_2'}\| \|\overrightarrow{L_{new}}\|} \quad (16)$$

4. The smaller the value of s is, the higher the similarity between the two lines is. Based on the above conditions, it can be determined whether the new edge line is the target that needs to be tracked. The final threshold selection for this paper can be referred to in the data presented in Table 1.

Table 1. Parameters of method in this paper.

Parameters	Value
Separated Point Cloud Region	8
Reference Line Distance Threshold	80 m
Set of Points	10
Edge Line Similarity Detection Threshold	0.01

4. Experiment

4.1. Algorithm Parameter Settings

As mentioned in the above text, the algorithm parameter settings for the relevant experiments of this paper are all listed in the Table 1.

4.2. Sensors System

All the sensors were mounted on a sport utility vehicle (SUV) for data collection. The LiDAR unit and the GNSS antennas were installed on the roof, while the SINS equipment and power supply were secured within the SUV.

Figure 11 depicts the experimental setup, which utilized a high-performance fiber-optic gyroscope navigation system (Self-developed experimental equipment of Harbin Engineering University, Heilongjiang, China.) integrated with a GNSS receiver (K823 GNSS receiver, ComNav Technology Ltd., Shanghai, China) to provide ground truth data. Table 2 summarizes the specifications of this system. The precise alignment of sensor positions is crucial to minimize navigation errors arising from lever arm effects. Therefore, the central positions of all sensors were carefully measured and aligned with the azimuth axis. Detailed measurement data are available in Table A1. Note that this table omits sensors with less stringent relative positioning requirements, such as the magnetometer used for initial SLAM orientation.

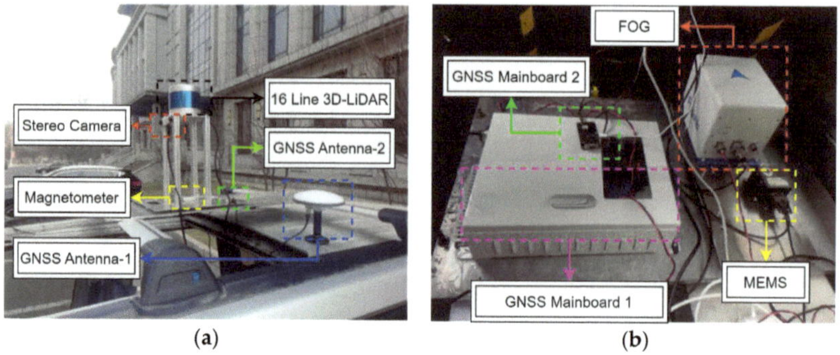

Figure 11. The experimental hardware: (**a**) shows the 3D LiDAR and GNSS antenna while (**b**) shows the GNSS processing, fiber optic gyroscope, and MEMS.

Table 2. The specifications of reference integrated navigation system.

Reference Accuracy	Specifications
Pitch	<0.02°
Yaw	<0.02°
Heading	<0.05°
Velocity (Integrated)	<0.1 m/s
Positioning (Integrated)	<1 m
Output Rate	100 Hz

The LiDAR sensor employed in this study was the Leishen 16 Line 3D-LiDAR, with performance parameters detailed in Table 3.

Table 3. The performance parameters of 3D-LiDAR.

Performance	Parameters
Detection Range	200 m
Point Rate	320,000 pts/s (single echo)
Distance Measurement Accuracy	±3 cm
Laser Wavelength	905 nm
Maximum Echo Count for Reception	2
Scanning Channels	16
Field-of-View Angle	360° × −15°~15°
Scanning Frequency	5~20 Hz
Angular Resolution	5 Hz: 0.09°/10 Hz: 0.18°
Power Supply Range	9 V~36 V DC
Operating Temperature	−20 °C~55 °C

Table 4 lists the performance parameters of the ADIS16445 (Analog Devices, Inc., Wilmington, MA, USA) Micro-Electro-Mechanical System (MEMS), a complete inertial system comprising a tri-axial gyroscope and a tri-axial accelerometer. The UM6 (Clearpath Robotics, Kitchener, ON, Canada) magnetometer provided a static heading with an accuracy of better than 2°, serving as the initial heading for the system. Similarly, the GNSS receiver provided the initial longitude, latitude, and altitude.

Data processing was performed on a laptop equipped with an Intel i7-6700 CPU, a GT960m graphics card, and 12 GB of RAM. The operating system was Ubuntu 16.04, running the ROS (robot operating system) kinetic distribution. This software environment supports both sophisticated data simulation and advanced graphical rendering.

Table 4. The performance parameters of ADIS16445.

Performance	Parameters
Gyroscope Dynamic Range	$\pm 250°/s$
Gyroscope Sensitivity	$0.01°/s$
Gyroscope Nonlinearity	$\pm 0.1\%$
Gyroscope Bias Stability	$12°/h$
Angular Random Walk	$0.56°/\sqrt{h}$
Accelerometer Dynamic Range	± 5 g
Accelerometer Sensitivity	0.25 mg
Accelerometer Nonlinearity	$\pm 0.2\%$
Accelerometer Bias Stability	0.075 mg
Velocity Random Walk	$0.0735\ m/s/\sqrt{h}$
Bandwidth	330 Hz
Output Rate	100 Hz

4.3. Experimental Area

All data in the paper were collected in April 2023 at Harbin Engineering University and its surrounding areas, with geographic coordinates approximately at 126.68° longitude and 45.77° latitude and an elevation of about 130 m. Based on the actual driving environment, the driving speed of the SUV in different experiments was controlled between 15 km/h and 30 km/h.

4.4. Result and Analysis

As shown in Figure 12, this work compared two currently popular LiDAR SLAM methods with the algorithm proposed in the text. When the information is relatively rich, both LOAM and the algorithm presented in this paper achieved satisfactory results. However, due to the incorrect loop closure judgment at the end, SC-Lego-LOAM resulted in a certain deviation in the overall outcome. When the scene information was not sufficiently rich, such as in Figure 13, the LOAM algorithm exhibited attitude deviations at the end, which led to errors in the navigation results. In contrast, SC-Lego-LOAM encountered more severe errors in loop closure, rendering it entirely inoperative.

Data_1 was collected at 6 PM on 4 April 2023, near Building 61 of Harbin Engineering University, with a total traveled distance of 1460 m. In this scenario, the SUV's route was to circle around the building for two laps, and the main purpose of the scene setup was to verify the effectiveness of the algorithm in the paper under general environmental conditions. In the initial 1000 m, SC-Lego-LOAM maintained relatively good performance. However, after the final incorrect loop closure, the total distance was re optimized, which led to a significant misalignment between the final distance and the azimuth angle. The method presented in this paper performed similarly to LOAM in the early stages, but because the original LOAM lacked loop closure detection functionality, its errors were bound to increase over time. Table 5 provides an overall summary of that experiment.

Data_2 was collected at 5 PM on 5 April 2023, near the parking lot of Harbin Engineering University, with a total traveled distance of 1403 m. As is shown in Figure 14 and Table 6, the structural feature weakened near the parking lot; the lack of structural features caused matching issues with the algorithm that relied on points. The experiment was mainly designed to demonstrate the stability of the algorithm in this paper relative to the comparative algorithms under the preset conditions of this paper. Judging from the comparison of results, the incorrect loop closure (red circle) by SC-Lego-LOAM led to severe issues once again, causing the method to fail entirely in this set of experiments. After the first loop closure, LOAM began to accumulate heading errors, which resulted in the continuous amplification of positioning errors in the subsequent SLAM due to the heading deviation.

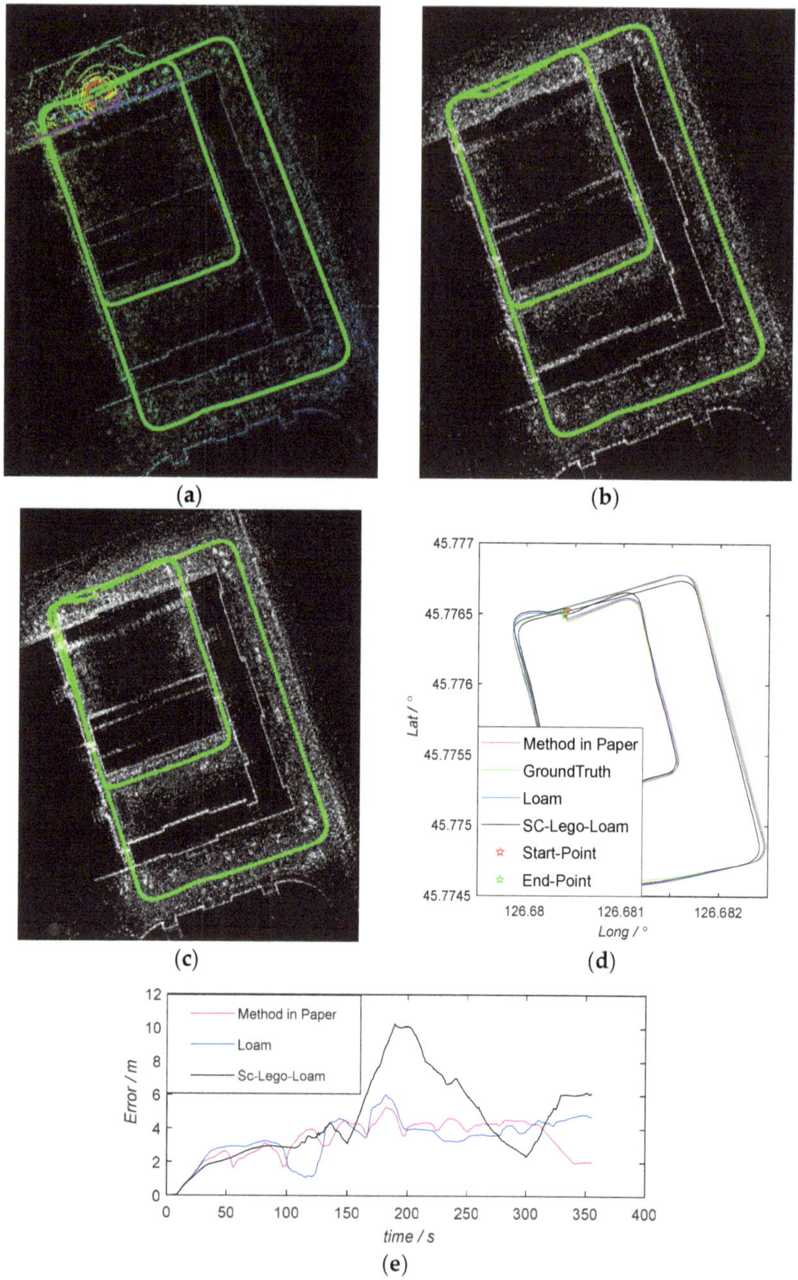

Figure 12. The result of LiDAR SLAM in Data_1: (**a**) was built by the LOAM; (**b**) was built by SC-Lego-LOAM; (**c**) shows the mapping result of the method from this paper; (**d**) depicts a direct comparison between various algorithms; (**e**) represents the positioning errors of the algorithms measured in meters.

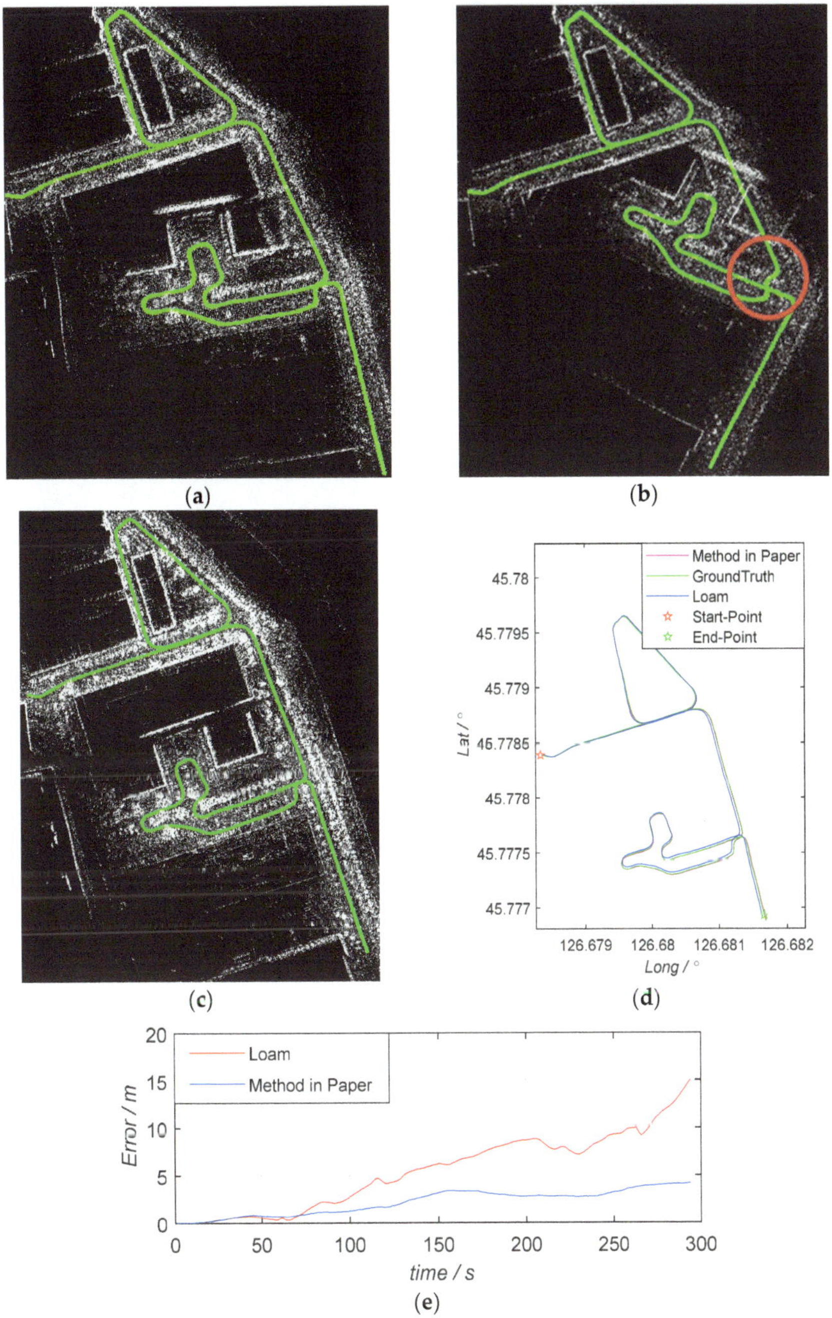

Figure 13. The result of LiDAR SLAM in Data_0405: (**a**) was built by the LOAM; (**b**) was built by SC-Lego-LOAM; (**c**) shows the mapping result of the method from this paper; (**d**) depicts a direct comparison between various algorithms; (**e**) represents the positioning errors of the algorithms measured in meters.

Table 5. The performance of SLAM methods for Data_1.

Method	LOAM	SC-Lego-LOAM	Method in Paper
Avg. Positioning Error (m)	3.89	4.14	3.33
Final Heading Error (°)	2.17	5.43	2.32
Max Positioning Error (m)	6.17	10.12	5.32
Travel Distance Error (m)	17	18	13

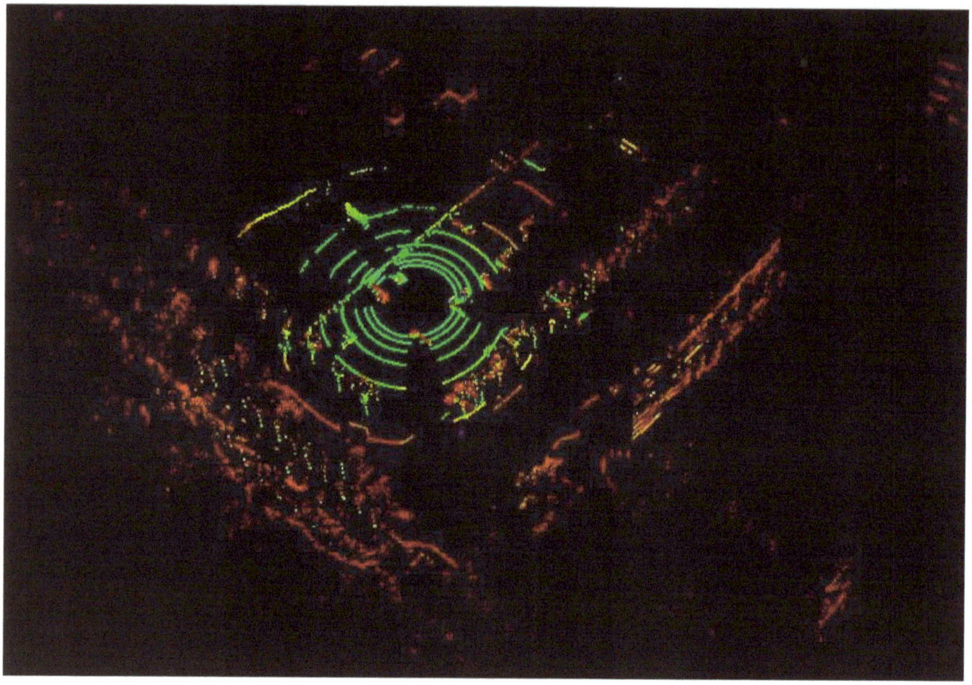

Figure 14. The point clouds near the parking lot.

Table 6. The performance of SLAM methods for Data_2.

Method	LOAM	SC-Lego-LOAM	Method in Paper
Avg. Positioning Error (m)	5.79	Failed	2.25
Final Heading Error (°)	6.22	Failed	1.02
Max Positioning Error (m)	15.17	Failed	4.12
Travel Distance Error (m)	22	Failed	11

The remaining data were also collected from 3–5 April 2023, at Harbin Engineering University and its surrounding areas. Data_3 had a loop closure point available shortly after the start, which could be used to eliminate accumulated errors. Data_4 featured a longer segment of nearly straight-line travel. These two sets of experiments were not designed intentionally with specific scenarios. They were standard test experiments; hence, they are not elaborately compared in detail within the text but are listed as supplementary experimental data in the Table 7.

Table 7. Comparison of positioning errors for other data.

Data	LOAM	SC-Lego-LOAM	Method in Paper	Travel Distance
Data_3	28.79 m	18.12 m	14.17 m	2205 m
Data_4	16.73 m	9.25 m	10.34 m	2234 m

5. Discussion

5.1. Results' Interpretation and Contribution

It redefines the fundamental computational unit in LiDAR SLAM by shifting the focus from LiDAR regression to an INS, rather than treating it as merely an accessory to LiDAR. The high-frequency output from the SINS navigation significantly reduced the computational load on the odometry component of LiDAR SLAM, thereby enhancing the accuracy of its positioning results.

In conventional scenarios, the performance of the algorithm proposed herein was comparable to that of LOAM. However, in scenarios where structural features were sparse or lacking, the algorithm demonstrated superior performance. The experimental results indicate that, while the algorithm achieved results similar to LOAM under typical conditions, it excelled in environments with limited structural features. In experiments that satisfied loop closure conditions, its relative advantage was even more pronounced. The relative accuracy improved by approximately 17%. From Figure 12, it can be observed that, in a general scenario, although the algorithm in the paper achieved good results, its basic performance was consistent with the other two algorithms. This scenario was only to verify the universal applicability of the algorithm in the paper, so there was no significant improvement in the specific comparative data. Figure 13 (DATA_2) is a preset scenario for the paper. Excluding the SC-Lego-LOAM algorithm, which was eliminated due to loop closure failure, from the error in Figure 13e, it can be seen that, after entering the parking lot environment, the error of the LOAM algorithm began to gradually increase, while the algorithm in the paper maintained a stable trajectory tracking. This fully demonstrated that the algorithm proposed in the paper achieved optimization for special scenarios while maintaining universality.

At the same time, this paper reorganizes and extracts the inherent characteristics of the point cloud. It goes a step further in the use of points, focusing the application of LiDAR point clouds on the edges that are less susceptible to the sparsity of point clouds and frequent changes in attitude matrices. This virtual edge composed of edge points is inherently a form of clustering. As long as points that meet the clustering criteria can be scanned, they can be continuously tracked in LiDAR SLAM and used to correct the positioning results of SINS. Of course, considering that the density of the point cloud has an attenuation characteristic with distance, in actual selection, further screening will only be carried out when a cluster contains at least four consecutive points and the line length exceeds 1 m. The selection of line distance rather than points, lines, or surfaces as the observation variable effectively reduces the computational load of the filter and increases the feasibility of real-time computation on low-performance devices.

For LiDAR SLAM loop closure based on scan context, the paper also makes certain rules changes. Experiments have proven that it can effectively reduce the occurrence of incorrect loop closure points and thereby enhance the overall accuracy of SLAM.

Although LiDAR SLAM algorithms based on machine learning have achieved excellent results, for in-vehicle processors, due to limitations in size and power, it is still difficult for their core to meet the real-time requirements in navigation. This work provides another feasible path within traditional algorithms.

5.2. Further Research

The experimental results presented in this paper demonstrate that the algorithm proposed within the text has superiority over traditional algorithms in both the odometry and mapping components of LiDAR SLAM. However, the results for Data_5 also indicate that in complex long-distance environments, relying solely on LiDAR and SINS for navigation still cannot achieve long-term precise positioning. This suggests that the algorithm presented in the paper should only be used as a supplementary method to maintain the original navigation accuracy when GNSS signals are lost, rather than a complete substitute, in urban environments. In future research, exploring how to integrate GNSS-related data to further enhance the performance of LiDAR SLAM will be investigated.

Additionally, another point that requires attention is that, similar to other LiDAR algorithms, the divergence issue in the height channel of the algorithm presented in the paper has not been significantly improved. When the LiDAR is scanning in open areas (such as forest trails), it becomes extremely difficult to obtain lateral edge lines, and at this point, 3D LiDAR SLAM can degrade to a performance like that of 2D LiDAR SLAM. Furthermore, how to further subdivide and utilize ground points will also be one of the key research projects' focuses in the future.

6. Conclusions

This paper presents a novel LiDAR/SINS tightly integrated SLAM algorithm designed to address the localization challenges in urban environments characterized by sparse structural features. Building upon the LOAM framework, the algorithm introduces further processing of LiDAR point cloud classification to extract edge lines through clustering. Leveraging the rotational invariance of distance, the algorithm constructs a Kalman filter system based on the distance variation in edge lines. This approach contributes to enhanced robustness and positioning accuracy.

Experimental results obtained in local urban scenarios demonstrated a 17% enhancement in positioning accuracy when compared to traditional point-based methods, particularly in environments characterized by sparse features. By proposing a line distance-based observation model and detailing the associated EKF framework and parameter settings, the proposed method redefines the concepts of loosely and tightly coupled integration within LiDAR/SINS systems.

Future research will explore the integration of GNSS data to further enhance the performance of the proposed LiDAR SLAM system, particularly in complex and long-distance navigation scenarios. Additionally, key areas of focus for future work include improving performance in open areas, particularly in the vertical channel, and optimizing ground point utilization.

This study not only achieves significant algorithmic improvements over existing methods but also paves a new technological pathway for autonomous driving and robotic navigation applications.

Author Contributions: Conceptualization, X.X. and L.G.; methodology, X.X.; software, X.X.; validation, X.X., Y.C. and Z.L.; formal analysis, X.X. and L.G; investigation, X.X.; resources, X.X. and Z.L.; data curation, X.X.; writing—original draft preparation, X.X.; writing—review and editing, X.X. and L.G.; visualization, X.X. and Y.C.; supervision, L.G. and Y.G.; project administration, Y.G.; funding acquisition, L.G. All authors have read and agreed to the published version of the manuscript.

Funding: This work was funded by the Department of Science and Technology of Heilongjiang Province (2023ZX01A21) and the National Natural Science Foundation of China (NSFC. 61803118).

Data Availability Statement: The datasets presented in this article are not readily available because they are part of an ongoing study. Requests to access the datasets should be directed to the corresponding author.

Conflicts of Interest: The authors declare no conflicts of interest.

Appendix A

The parameter settings for F in Equation (6) are as follows:

$$F_{11} = \begin{bmatrix} 0 & 0 & -\frac{\dot{\phi}}{R_M+h} \\ \dot{\lambda} tan\phi & 0 & -\frac{\dot{\lambda}}{R_M+h} \\ 0 & 0 & 0 \end{bmatrix}, F_{23} = \begin{bmatrix} 0 & f_u & -f_n \\ -f_u & 0 & f_e \\ f_n & -f_e & 0 \end{bmatrix},$$
$$F_{12} = \begin{bmatrix} 0 & \frac{1}{R_M+h} & 0 \\ \frac{1}{(R_N+h)cos\phi} & 0 & 0 \\ 0 & 0 & 1 \end{bmatrix}, F_{32} = \begin{bmatrix} 0 & \frac{1}{R_M+h} & 0 \\ \frac{-1}{R_N+h} & 0 & 0 \\ \frac{-tan\phi}{R_N+h} & 0 & 0 \end{bmatrix},$$
$$F_{21} = \begin{bmatrix} 2\omega_e(v_u sin\phi + v_n cos\phi) + \dot{\lambda} v_n / cos\phi & 0 & 0 \\ -2\omega_e v_e cos\phi - \dot{\lambda} v_e / cos\phi & 0 & 0 \\ -2\omega_e v_e sin\phi & 0 & 2g/R_N \end{bmatrix},$$
$$F_{22} = \begin{bmatrix} (v_n tan\phi - v_u)/(R_N + h) & \left(2\omega_e + \dot{\lambda}\right) sin\phi & -\left(2\omega_e + \dot{\lambda}\right) cos\phi \\ -2\omega_e v_e cos\phi - \dot{\lambda} v_e / cos\phi & -v_u/(R_M + h) & -\dot{\phi} \\ -2\omega_e v_e sin\phi & 2\dot{\phi} & 0 \end{bmatrix}, \quad \text{(A1)}$$
$$F_{31} = \begin{bmatrix} 0 & 0 & -\dot{\phi}/(R_M + h) \\ \omega_e sin\phi & 0 & \dot{\lambda} cos\phi/(R_N + h) \\ -\omega_e cos\phi - \dot{\lambda}/(R_N + h)cos\phi & 0 & \dot{\lambda} sin\phi/(R_N + h) \end{bmatrix},$$
$$F_{44} = \begin{bmatrix} -\beta_{\omega x} & 0 & 0 \\ 0 & -\beta_{\omega y} & 0 \\ 0 & 0 & -\beta_{\omega z} \end{bmatrix}, F_{55} = \begin{bmatrix} -\beta_{fx} & 0 & 0 \\ 0 & -\beta_{fy} & 0 \\ 0 & 0 & -\beta_{fz} \end{bmatrix},$$
$$F_{33} = \begin{bmatrix} 0 & \left(\omega_e + \dot{\lambda}\right) sin\phi & -\left(\omega_e + \dot{\lambda}\right) cos\phi \\ -\left(\omega_e + \dot{\lambda}\right) sin\phi & 0 & -\dot{\phi} \\ \left(\omega_e + \dot{\lambda}\right) cos\phi & \dot{\phi} & 0 \end{bmatrix}$$

Table A1. The positions of different sensors relative to the FOG center.

Sensors	Right	Forward	Up
MEMS	5.3 cm	−22.5 cm	−4.3 cm
LiDAR	−32.5 cm	78 cm	113.5 cm
GNSS Antenna	21.5 cm	68.5 cm	83 cm

References

1. Montemerlo, M.; Thrun, S.; Koller, D.; Wegbreit, B. Fast SLAM: A factored solution to the simultaneous localization and mapping problem. In Proceedings of the AAAI-02: Eighteenth National Conference on Artificial Intelligence, Edmonton, AL, Canada, 28 July–1 August 2002; Volume 593598.
2. Huang, L. Review on LiDAR-based SLAM techniques. In Proceedings of the 2021 International Conference on Signal Processing and Machine Learning (CONF-SPML), Stanford, CA, USA, 14 November 2021; pp. 163–168.
3. Cadena, C.; Carlone, L.; Carrillo, H.; Latif, Y.; Scaramuzza, D.; Neira, J.; Reid, I.; Leonard, J.J. Past, present, and future of simultaneous localization and mapping: Toward the robust-perception age. *IEEE Trans. Robot.* **2016**, *32*, 1309–1332. [CrossRef]
4. Biber, P.; Straßer, W. The normal distributions transform: A new approach to laser scan matching. In Proceedings of the 2003 IEEE/RSJ International Conference on Intelligent Robots and Systems (IROS 2003) (Cat. No. 03CH37453), Las Vegas, NV, USA, 27–31 October 2003; Volume 3, pp. 2743–2748.
5. Rusinkiewicz, S.; Levoy, M. Efficient Variants of the ICP Algorithm. In Proceedings of the of the Third International Conference on 3-D Digital Imaging and Modeling, Quebec City, QC, Canada, 28 May–1 June 2001; pp. 145–152.
6. Hung, Y.-W.; Chen, Y.-C.; Lo, C.; So, A.G.; Chang, S.-C. Dynamic workload allocation for edge computing. *IEEE Trans. Very Large-Scale Integr. (VLSI) Syst.* **2021**, *29*, 519–529. [CrossRef]
7. Deng, Q.; Sun, H.; Chen, F.; Shu, Y.; Wang, H.; Ha, Y. An Optimized FPGA-Based Real-Time NDT for 3D-LiDAR Localization in Smart Vehicles. *IEEE Trans. Circuits Syst. II: Express Briefs* **2021**, *68*, 3167–3171. [CrossRef]
8. Jiang, M.; Song, S.; Li, Y.; Liu, J.; Feng, X. Scan registration for mechanical scanning imaging sonar using kD2D-NDT. In Proceedings of the 2018 Chinese Control and Decision Conference (CCDC), Shenyang, China, 9–11 June 2018; pp. 6425–6430.

9. Zhang, J.; Singh, S. LOAM: Lidar odometry and mapping in real-time. *Robot. Sci. Syst.* **2014**, *2*, 1–9.
10. Shan, T.; Englot, B. LeGO-LOAM: Lightweight and Ground-Optimized Lidar Odometry and Mapping on Variable Terrain In Proceedings of the 2018 IEEE/RSJ International Conference on Intelligent Robots and Systems (IROS), Madrid, Spain, 1–5 October 2018. [CrossRef]
11. He, L.; Wang, X.; Zhang, H. M2DP: A novel 3D point cloud descriptor and its application in loop closure detection. In Proceedings of the 2016 IEEE/RSJ International Conference on Intelligent Robots and Systems (IROS), Daejeon, Republic of Korea, 9–14 October 2016. [CrossRef]
12. Kim, G.; Kim, A. Scan Context: Egocentric Spatial Descriptor for Place Recognition within 3D Point Cloud Map. In Proceedings of the 2018 IEEE/RSJ International Conference on Intelligent Robots and Systems (IROS), Madrid, Spain, 1–5 October 2018; pp. 4802–4809. [CrossRef]
13. Kim, G.; Choi, S.; Kim, A. Scan Context++: Structural Place Recognition Robust to Rotation and Lateral Variations in Urban Environments. *IEEE Trans. Robot.* **2022**, *38*, 1856–1874. [CrossRef]
14. Wang, H.; Wang, C.; Chen, C.L.; Xie, L. F-loam: Fast lidar odometry and mapping. In Proceedings of the 2021 IEEE/RSJ International Conference on Intelligent Robots and Systems (IROS), Prague, Czech Republic, 27 September–1 October 2021; pp. 4390–4396.
15. Tang, J.; Chen, Y.; Niu, X.; Wang, L.; Chen, L.; Liu, J.; Shi, C.; Hyyppa, J. Lidar scan matching aided inertial navigation system in gnss-denied environments. *Sensors* **2015**, *15*, 16710–16728. [CrossRef] [PubMed]
16. Xu, X.; Zhang, L.; Yang, J.; Cao, C.; Wang, W.; Ran, Y.; Tan, Z.; Luo, M. A Review of Multi-Sensor Fusion SLAM Systems Based on 3D LIDAR. *Remote Sens.* **2022**, *14*, 2835. [CrossRef]
17. Koide, K.; Yokozuka, M.; Oishi, S.; Banno, A. Globally consistent and tightly coupled 3D LiDAR inertial mapping. In Proceedings of the 2022 International Conference on Robotics and Automation (ICRA), Philadelphia, PA, USA, 23–27 May 2022; pp. 5622–5628.
18. Ye, H.; Chen, Y.; Liu, M. Tightly coupled 3d lidar inertial odometry and mapping. In Proceedings of the 2019 International Conference on Robotics and Automation (ICRA), Montreal, QC, Canada, 20–24 May 2019; pp. 3144–3150.
19. Rabbou, M.A.; El-Rabbany, A. Tightly coupled integration of GPS precise point positioning and MEMS-based inertial systems. *GPS Solut.* **2015**, *19*, 601–609. [CrossRef]
20. Rong, H.; Gao, Y.; Guan, L.; Ramirez-Serrano, A.; Xu, X.; Zhu, Y. Point-Line Visual Stereo SLAM Using EDlines and PL-BoW. *Remote Sens.* **2021**, *13*, 3591. [CrossRef]
21. Wang, H.; Guan, L.; Yu, X.; Zhang, Z. PL-ISLAM: An Accurate Monocular Visual-Inertial SLAM with Point and Line Features. In Proceedings of the 2022 IEEE International Conference on Mechatronics and Automation (ICMA), Guilin, China, 7–10 August 2022; pp. 1141–1146. [CrossRef]
22. Mourikis, A.I.; Roumeliotis, S.I. A multi-state constraint Kalman filter for vision-aided inertial navigation. In Proceedings of the 2007 IEEE International Conference on Robotics and Automation, Roma, Italy, 10–14 April 2007; pp. 3565–3572.
23. 32/16-Line Mechanical Line Mechanical LiDAR | Leishen Intelligent System. Available online: https://www.lslidar.com/product/c32-16-mechanical-lidar/ (accessed on 6 July 2024).
24. Himmelsbach, M.; Hundelshausen, F.V.; Wuensche, H.-J. Fast Segmentation of 3D Point Clouds for Ground Vehicles. In Proceedings of the IEEE Intelligent Vehicles Symposium, La Jolla, CA, USA, 21–24 June 2010; pp. 560–565.
25. Fischler, M.A.; Bolles, R.C. Random sample consensus: A paradigm for model fitting with applications to image analysis and automated cartography. *Commun. ACM* **1981**, *24*, 381–395. [CrossRef]
26. Bogoslavskyi, I.; Stachniss, C. Fast Range Image-based Segmentation of Sparse 3D Laser Scans for Online Operation. In Proceedings of the IEEE/RSJ International Conference on Intelligent Robots and Systems, Daejeon, Republic of Korea, 9–14 October 2016; pp. 163–169.
27. Arthur, D.; Vassilvitskii, S. k-means++: The advantages of careful seeding. In Proceedings of the 18th Annual ACM-SIAM Symposium on Discrete algorithms (SODA), New Orleans, LA, USA, 7–9 January 2007; Volume 7, pp. 1027–1035.
28. Belongie, S.; Malik, J.; Puzicha, J. Shape matching and object recognition using shape contexts. *IEEE Trans. Pattern Anal. Mach. Intell.* **2002**, *24*, 509–522. [CrossRef]
29. Guan, L.; Cong, X.; Sun, Y.; Gao, Y.; Iqbal, U.; Noureldin, A. Enhanced MEMS SINS aided pipeline surveying system by pipeline junction detection in small diameter pipeline. *IFAC-Pap.* **2017**, *50*, 3560–3565. [CrossRef]

remote sensing

Article

Fast Robust Point Cloud Registration Based on Compatibility Graph and Accelerated Guided Sampling

Chengjun Wang [1,2,†], Zhen Zheng [1,2,†], Bingting Zha [1,2,*] and Haojie Li [1,2]

1 ZNDY of Ministerial Key Laboratory, Nanjing University of Science and Technology, Nanjing 210094, China; wangchengjun@njust.edu.cn (C.W.); zhengzhen@njust.edu.cn (Z.Z.); haojieli@njust.edu.cn (H.L.)

2 School of Mechanical Engineering, Nanjing University of Science and Technology, Nanjing 210094, China

* Correspondence: zhabingting@njust.edu.cn

† These authors contributed equally to this work.

Abstract: Point cloud registration is a crucial technique in photogrammetry, remote sensing, etc. A generalized 3D point cloud registration framework has been developed to estimate the optimal rigid transformation between two point clouds using 3D key point correspondences. However, challenges arise due to the uncertainty in 3D key point detection techniques and the similarity of local surface features. These factors often lead to feature descriptors establishing correspondences containing significant outliers. Current point cloud registration algorithms are typically hindered by these outliers, affecting both their efficiency and accuracy. In this paper, we propose a fast and robust point cloud registration method based on a compatibility graph and accelerated guided sampling. By constructing a compatible graph with correspondences, a minimum subset sampling method combining compatible edge sampling and compatible vertex sampling is proposed to reduce the influence of outliers on the estimation of the registration parameters. Additionally, an accelerated guided sampling strategy based on preference scores is presented, which effectively utilizes model parameters generated during the iterative process to guide the sampling toward inliers, thereby enhancing computational efficiency and the probability of estimating optimal parameters. Experiments are carried out on both synthetic and real-world data. The experimental results demonstrate that our proposed algorithm achieves a significant balance between registration accuracy and efficiency compared to state-of-the-art registration algorithms such as RANSIC and GROR. Even with up to 2000 initial correspondences and an outlier ratio of 99%, our algorithm achieves a minimum rotation error of 0.737° and a minimum translation error of 0.0201 m, completing the registration process within 1 s.

Keywords: point cloud; registration; compatibility graph; accelerated guided sampling; correspondence

Citation: Wang, C.; Zheng, Z.; Zha, B.; Li, H. Fast Robust Point Cloud Registration Based on Compatibility Graph and Accelerated Guided Sampling. *Remote Sens.* **2024**, *16*, 2789. https://doi.org/10.3390/rs16152789

Academic Editors: Wanshou Jiang, San Jiang, Duojie Weng and Jianchen Liu

Received: 18 June 2024
Revised: 26 July 2024
Accepted: 27 July 2024
Published: 30 July 2024

1. Introduction

Point cloud registration is a fundamental task in remote sensing [1,2], robot perception [3,4], photogrammetry [5], and other fields, and has been applied to a variety of technologies such as 3D reconstruction [6], structural health monitoring [7,8], target recognition and localization [9], simultaneous localization, and mapping [10]. Due to the fixed limitations of the laser scanner in terms of field of view and viewpoints, a single data acquisition with fixed viewpoints can only capture part of the point cloud of a scene. In order to obtain a complete 3D representation of the scene, it is necessary to fuse and splice the point clouds with different viewpoints. The goal of point cloud registration lies in estimating the optimal rigid transformation between the two point clouds in order to accurately align the point clouds under different viewpoints.

The feature-based global registration method is the mainstream method for point cloud registration [11]. It generally consists of two stages: the feature extraction stage and the robust transform estimation stage. The feature extraction stage extracts key points and generates feature descriptors, and establishes the correspondence between two point clouds based on the similarity of the feature descriptors between the points. The robust transform estimation stage estimates the transformation parameters between two point clouds based on the correspondence. Many well-differentiated point cloud description methods have been proposed, such as FPFH [12], RoPS [13], SDASS [14], etc. However, noise is still unavoidable, mainly due to (1) Most of the point clouds partially overlapping, so the established correspondences may be inliers only if they are located in overlapping regions, while correspondences in non-overlapping regions introduce a large number of outliers. (2) The presence of many similar local surfaces in the point cloud, resulting in very similar corresponding feature descriptors, and forming false correspondences. Since the correspondences established in the feature extraction stage usually have a large number of outliers, the reliability of the robust transform estimation is seriously affected. Therefore, one of the difficulties in feature-based point cloud registration is how to select inliers from the correspondence containing a large number of outliers and then accurately estimate the transformation parameters.

In recent years, a large number of robust transform estimation methods have been proposed. Random sampling consistency (RANSAC) [15] is the most commonly used method in robust transform estimation. RANSAC solves the transform parameters by iteratively sampling the minimum subset, and then selects the hypothesis of the maximum number of inliers as the estimation parameter. RANSAC requires a large number of iterations and does not guarantee obtaining the optimal solution, and, moreover, it cannot deal with the situation where the outlier ratio is very high. In order to cope with the problem of a very high outlier ratio in the correspondence, many methods choose to use the geometric properties corresponding to the inliers to identify the inliers. GORE [16] utilizes geometric consistency to exclude outliers. VODRAC [17] and RANSIC [18] establish the minimum subset by judging the compatibility between the sampled points, and use the compatible subset to generate the hypothesis transformation matrix. However, pairwise consistency is not sufficient since outliers are equally likely to occasionally satisfy length consistency. SC2-PCR [19] is further used to distinguish between inliers and outliers by computing second-order spatial compatibility. These methods have been shown to be effective in improving the parameter estimation problem in the case of a high outlier ratio, but there are still some limitations, such as the extremely low computational efficiency of RANSIC when the outlier ratio is too high, which limits the practical application of the algorithm.

Despite the great progress made in current research, it is still a challenging task to determine the inliers from correspondences containing a large number of outliers. Currently, feature-based point cloud registration algorithms suffer from the following problems: (1) Due to the diversity of scenes, the outlier ratio varies in different scenes, which limits the robustness and adaptability of the algorithms. (2) When the number of correspondences is high or the outlier ratio is too high, the parameter estimation process becomes very time-consuming and inefficient. (3) How to select the inliers from a large number of outliers and estimate the transformation parameters accurately is very difficult. To address the problem of fast robust point cloud registration containing a large number of outliers, we propose a fast robust point cloud registration algorithm based on a compatibility graph and accelerated guided sampling, which can realize the accurate registration of the corresponding point cloud that is seriously contaminated by the outliers, and, at the same time, has a high computational efficiency. The contribution of this paper is mainly:

- Constructing a compatibility graph based on the compatibility between inliers and proposing a minimum subset sampling method combining graph edge sampling and graph vertex sampling to reduce the influence of outliers on the registration results.
- Introducing a preference-based accelerated guided sampling strategy that utilizes the hypothetical model generated during the iterative process to guide the subsequent samples to be biased toward the inliers, achieving efficient and robust point cloud registration.

- Compared to many existing state-of-the-art methods, the proposed algorithm is able to cope with a very high outlier ratio (outlier ratio > 99%) and strikes a remarkable balance between registration accuracy and efficiency.

2. Related Works

A key step in feature-based point cloud registration algorithms is to establish the correspondences between the source and target point clouds based on local feature descriptors. Feature descriptors have been widely studied in the past decades, and traditional descriptors such as PFH [20], FPFH [12], SHOT [21], and RoPS [13] describe the local geometric structure of the point cloud from different measurements. In order to further improve the descriptive performance of descriptors, TOLDI [22], SDASS [14], and KDD [23], introducing additional information such as local reference frame or point density features can more effectively describe the local features of the point cloud and generate more reliable correspondences. With the rapid development of deep learning technology, learning-based descriptors have received more attention due to their excellent differentiation and robustness. Learning-based local feature extraction modules usually use frameworks such as point-pair features [24,25], local reference frame [26], and rotationally invariant networks [27]. These learning-based feature description methods have good generalization but are usually computationally inefficient. Recently, Transformer [28] has also been successfully applied to 3D feature matching with promising results. Predator [29] introduces an overlap-aware module based on self-crossing and self-attention. CoFiNet [30] utilizes an attentional mechanism to aggregate the contextual information between two piece point clouds. GeoTrans [31] employs a geometric Transformer module to encode rotationally invariant geometric features of point clouds, which generates model assumptions using local correspondences and performs model validation using global fitness, thus accomplishing local-to-global alignment. These algorithms are effective in detecting overlapping regions and are shown to have the potential to solve the low overlap rate registration problem. Although current feature-matching methods can establish robust correspondences, a large number of outliers in the constructed correspondence set still inevitably exist. Therefore, it is necessary to rely on model-fitting methods for robust rigid transformation estimation.

The main robust estimation methods that have been used to solve the point cloud registration problem include M-estimation [32], truncated least squares [33], Lp-paradigm [34], and RANSAC family [15]. Since a large number of outliers will inevitably exist in the initial correspondence, how to estimate the accurate model parameters from the data containing a large number of outliers is the difficulty of robust transformation. In order to solve this problem, many researchers have proposed registration algorithms based on outlier filtering, and the core of these methods lies in removing the wrong matches in the correspondences, so as to avoid dealing with outliers in the registration process. Fast global registration (FGR) [35] is one of the typical algorithms that removes outlier points by geometric tests, then uses Geman McClure as the objective function and proposes a global method that combines a line process with robust estimation to optimize the model parameter estimation process. Similarly, Li et al. [36] constructed a topological graph based on correspondences, then proposed a side-voting strategy to remove outliers, and proposed a Cauchy-weighted Lq-paradigm as the cost function to achieve robust registration with a 90% outlier rate. A guaranteed outlier removal strategy was introduced in GORE [16], which removes outliers from correspondences by computing a simple geometric consistency test. A cleaner set of correspondences is obtained, which guarantees a globally optimal solution, but its high computational complexity leads to very low efficiency. CLIPPER [37] incorporated the concept of geometric consistency into the graph theoretic framework by finding dense subgraphs to determine the inliers. In order to improve the registration efficiency, Yang et al. [38] introduced a truncated least squares cost that is insensitive to the outliers to deal with the transformation parameter estimation problem, and rewrote the problem as a quadratically constrained quadratic programming problem. They proposed a convex semidefinite programming relaxation for the optimal solution, which can achieve the computation of the verifiable optimal solution under the condition of 95% outliers while guaranteeing efficiency. Zhang et al. [39] proposed a point cloud registration approach based on a Maximal cluster (MAC). MAC first constructs the initial correspondence compatibility graph, then searches for the largest clusters in the graph, and finally selects the largest clusters with large weights to calculate the transformation assumptions in combination with the SVD algorithm. While this approach accurately obtains the optimal transformation parameters, it suffers from low computational efficiency. Li et al. [40] use the correspondence matrix and the generalized correspondence matrix to seek the corresponding tight upper bounds and lower bounds, and then combine them with an adaptive Cauchy's estimator for optimal parameter estimation. Yao et al. [41] proposed a global-to-local registration method and introduced a hypergraph consistency module to learn the high-order consistency of guided sampling to obtain more reliable clusters of inliers. Second-order spatial compatibility was proposed in SC2-PCR++ [19] to distinguish the inliers from the outliers at an early stage. GROR [42] introduced the concepts of graph node reliability and graph edge reliability by constructing a correspondence graph to quickly and accurately remove the inliers from the outliers. Li et al. [43] proposed a maximum group correspondence selection strategy based on reliable edges, which combines the adaptive Maxwell–Boltzmann (AMB) algorithm and confidence intervals to estimate the rotation and translation parameters.

The RANSAC algorithm is another pipeline widely used in correspondence-based point cloud registration, but the randomness of the algorithm itself leads to its low accuracy and the need for a large number of samples in order to find a relatively correct solution, which is likely to fail on the data with serious contamination of outliers. Many improved algorithms have been proposed to address the problems of RANSAC [44–46]. Maximum Likelihood Estimated Sample Consistency (MLESAC) [47] improves the robustness of RANSAC by replacing the cost function from the size of the consistent samples to maximize the likelihood. Locally Optimized RANSAC (LO-RANSAC) [45] performs local optimization by deriving solutions from random samples, which improves speed by two to three times compared to standard RANSAC. Wu et al. [48] introduced a particle swarm optimization algorithm in RANSAC to directly sample the model parameters and achieved good results in image alignment. GESAC [49] introduced a graph to enhance the sample consistency and achieved effective registration even if there are outliers in the smallest subset of the sampled points. ICOS [50] accelerated the search for inliers by constructing a compatibility structure. One-Point RANSAC [32] introduced a scale-annealing bi-weighted estimator to stepwise optimize the estimation of the transform parameters. Invariant and compatible random selection of minimum subsets are introduced in RANSIC [18]. Hu et al. [17] proposed a fast robust point cloud registration algorithm based on election-compatible weighted two-point random sampling (VODRAC), which combines scale-invariant constraints with a two-point random sampling framework, and can achieve fast candidate inliers search. Cheng et al. [51] proposed a point cloud registration algorithm based on local sampling and global hypothesis generation. Gentner et al. [52] proposed a graph-based maximum consistency alignment algorithm (GMCR), in which a novel consistency function was introduced specifically to translate the consistency maximizing objective into the graph domain. The algorithm is robust to various types of outliers. C-RANSAC [53] introduces a scale histogram-based outlier filtering method and involves a master–slave handshake mechanism for optimal parameter estimation, which achieves high-accuracy registration and fast convergence.

3. Methods

In this Section, we propose a novel fast robust point cloud registration based on a compatibility graph and accelerated guided sampling. We first introduce the problem formulation of registration and describe the framework of the proposed method. Then, we introduce in detail the key processes, including the correspondence compatibility graph construction, the minimum compatible subset sampling, the preference-based guided sampling strategy, and the complete registration algorithm.

3.1. Problem Formulation

The procedure of the feature-based point cloud registration algorithm is to establish the correspondences between the source and target point clouds based on the local feature descriptors, and then estimate the registration parameters based on the correspondences. We first give the method of the correspondence establishment. Assume that the two point clouds to be aligned are called source point cloud P_s and target point cloud P_t. (1) Due to the excessive number of points in the initial point cloud, which contains a large amount of redundant information, the key point estimation technique is first used to estimate the key points $P_{sf} = \{x_i | 1 \leq i \leq N\}$ and $P_{tf} = \{y_j | 1 \leq j \leq N\}$ of the source and target point clouds, respectively. (2) Generate feature description vectors for key points using feature descriptors, e.g., classical FPFH, learning-based GeoTrans. (3) For each key point x_i in P_{sf}, the nearest neighbor y_i corresponding to x_i in P_{tf} is obtained based on the feature description vectors using a KD-Tree, so that the initial correspondence set $C = \{(x_i, y_i) | 1 \leq i \leq N\}$ of P_s and P_t can be established.

Since P_s and P_t are usually partially overlapping, and the feature descriptors cannot completely and accurately distinguish each point in P_s and P_t, a large number of incorrect correspondences inevitably exist in C. The purpose of the feature-based point cloud registration method is to estimate the transformation parameters of the source and target point clouds based on the correspondence set. The objective function is denoted as

$$\underset{\mathbf{R},\mathbf{t}}{minimize} \sum_{i=1}^{N} ||y_i - (\mathbf{R}x_i + \mathbf{t})||^2 \tag{1}$$

where $\mathbf{R} \in SO(3)$ is an orthogonal rotation matrix, $\mathbf{t}$ is a 3×1 translation vector, (x_i, y_i) is a correspondence in the correspondence set C, $|| \cdot ||$ denotes L2-norm.

Due to the large number of wrong correspondences in the initial correspondences, the above objective function can be further expressed as a maximizing consensus problem, denoted as

$$\begin{aligned} &\underset{\mathbf{R},\mathbf{t},I \subset C}{maximize} |I| \\ &Subject \;\; to \; ||y_i - (\mathbf{R}x_i + \mathbf{t})|| < \varepsilon, \forall (x_i, y_i) \in I \end{aligned} \tag{2}$$

where I is called the consensus set, $|I|$ denotes the size of the consensus set, ε is an inliers threshold, and $(\mathbf{R}, \mathbf{t})$ corresponding to the consensus set is considered to be the optimal transformation parameter. In order to search for the maximum consensus set in the initial correspondences, the commonly adopted approach is to sample a series of minimum subsets (a subset consisting of three points) from the initial correspondences for estimating $(\mathbf{R}, \mathbf{t})$, and then to compute the correspondences in the initial correspondences that are consistent with the minimum subset, i.e., correspondences that satisfy $||y_i - (\mathbf{R}x_i + \mathbf{t})|| < \varepsilon$. Finally, the set with the most consensus correspondences is selected as the maximum consensus set. To address the above question, we propose a sampling consistency algorithm that combines a compatibility graph and accelerated guided sampling. The overall framework of the algorithm is shown in Figure 1, by constructing the compatibility graph structure of the initial correspondences, combining graph edge sampling and graph vertex sampling to obtain the minimum compatible subset, and introducing a preference-based accelerated guided sampling strategy to search for the optimal minimum subset so as to determine the maximum consensus set and estimate the transformation parameters.

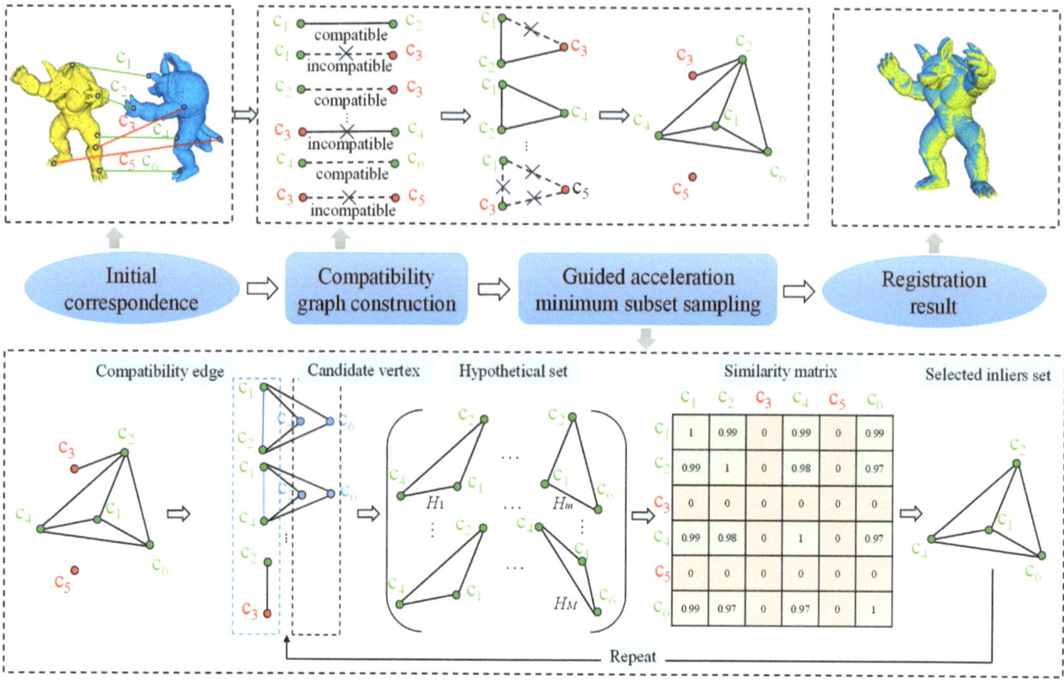

Figure 1. Overview of the proposed method. First, taking the initial correspondences as inputs, the compatibility graph is constructed by calculating the compatibility of each correspondence with other correspondences. Then the minimum compatible subset is constructed by combining compatible edge sampling and candidate vertex sampling, model hypotheses are generated, preference scores for model hypotheses are computed for each correspondence, and similarity matrices are further constructed to select the set of possible inliers to participate in the subsequent iterations. Finally, the transformation parameters are calculated based on the maximum consensus set obtained from the iterations, and the registration is completed using the transformation parameters.

3.2. *Correspondence Compatibility Graph Construction*

In our approach, the selection of the inliers of the correspondences will be performed on a graph structure, which is a better representation of the compatibility degree between correspondences than the Euclidean distance space. Therefore, it is first necessary to construct an undirected graph of the initial correspondences, where each correspondence is represented as a graph vertex, and geometrically compatible nodes are connected by graph edges.

For the initial correspondence set C, suppose that two elements in C are $c_i = (p_{si}, p_{ti})$ and $c_j = (p_{sj}, p_{tj})$, where p_{si}, p_{sj} denote two points in the source point cloud and p_{ti}, p_{tj} denote two points in the target point cloud corresponding to p_{si}, p_{sj}. The compatibility between c_i and c_j can be quantitatively measured as

$$d_{cmp}(c_i, c_j) = \left| \, ||p_{si} - p_{sj}|| - ||p_{ti} - p_{tj}|| \, \right|. \tag{3}$$

When c_i and c_j are ideal inliers, $d_{cmp}(c_i, c_j) = 0$. Noise inevitably exists in the point cloud, and $d_{cmp}(c_i, c_j)$ cannot be strictly 0. Therefore, when $d_{cmp}(c_i, c_j) < \varepsilon$, it indicates that c_i and c_j are compatible and considered to be inliers.

Construct a compatibility graph based on the compatibility between any vertices, given an initial set of correspondences $C = \{c_i | 1 \leq i \leq N\}$, the graph formed by them denoted as $G = (V, E)$, with V being the vertices of the graph and $V = \{c_i | 1 \leq i \leq N\}$, E

being the edges of the graph and $E = \{e_{ij} | 1 \leq i \leq N, 1 \leq i \leq N\}$, where $e_{ij} = (c_i, c_j)$. In the process of constructing the graph, for two correspondences c_i and c_j, they are considered compatible so that e_{ij} is in E only when $d_{cmp}(c_i, c_j) < \varepsilon$, and in this way the compatibility graph of C is constructed. At the same time, we build an $N \times N$ compatibility matrix $\mathbf{M}_C$, and when $d_{cmp}(c_i, c_j) < \varepsilon$, the corresponding positional element of $\mathbf{M}_C$, $\mathbf{M}_C(i, j) = 1$, which indicates that the correspondences c_i and c_j are compatible, and $\mathbf{M}_C(i, j) = 0$, otherwise denoted as:

$$\mathbf{M}_C(i,j) = \begin{cases} 1, \ if \ d_{cmp}(c_i, c_j) < \varepsilon \\ 0, \ otherwise \end{cases} \tag{4}$$

3.3. Minimum Compatible Subset Sampling

In the traditional RANSAC algorithm, point cloud registration requires randomly selecting three correspondences to form a minimum subset, and then combining them with Horn's triad-based method [54] to estimate the transformation parameters between two point clouds. Due to the presence of a large number of outliers, the probability of RANSAC sampling to a minimum subset of all inliers is extremely low. According to [17], when the outlier ratio in the correspondences is certain, the number of iterations required for RANSAC to sample a subset of all inliers grows exponentially with the size of the minimum subset, and a large number of iterations are often required to obtain a more optimal solution. In order to reduce the influence of the outliers, this paper introduces a compatible minimum subset sampling method based on the constructed compatibility graph, and the method consists of two layers, the edge sampling layer, and the vertex sampling layer.

In the edge sampling layer, we first randomly select an edge $e_{ij} = (c_i, c_j)$ in the compatibility graph, and search the vertices connected to this edge to form a triangle as the candidate correspondence set $\Phi = \{c_k \mid 0 \leq k \leq K\}$. Then, enter the vertex sampling layer and randomly select a point c_k in Φ, with e_{ij} forming a minimal subset $s_k = \{c_i, c_j, c_k\}$. Next, we use Horn's method to compute the rotation and translation parameters, and compute the consensus set L_k corresponding to the smallest subset s_k of the candidate correspondence set Φ from the estimated transformation parameters. Repeat sampling in Φ until reaching the set maximum iteration number of vertex sampling MI_v to obtain a series of consensus sets $L_V = \{L_k \mid 0 \leq k \leq MI_v\}$, and always retain the largest consensus set in L_V as the best consensus set for the vertex sampling layer, i.e., $L_{best} = \underset{L_k \in L_v}{argmax}(|L_k|)$. To avoid too much redundant computation, each time we obtain a new L_{best}, we update MI_v according to L_{best}. After the vertex sampling layer is completed, return to the edge sampling layer and use L_{best} to compute the transformation parameters, and calculate the consensus set G_n corresponding to the currently sampled edge in the initial correspondence C. Repeat the edge sampling until reaching the set maximum iteration number MI_e of edge sampling, and the iterative process generates a series of consensus sets $G_E = \{G_n \mid 0 \leq n \leq MI_e\}$. Always retain the maximum consensus set $G_{best} = \underset{G_n \in G_E}{argmax}(|G_n|)$ during the iterative process. Similarly, each time we obtain a new G_{best}, we update MI_e according to G_{best}. Finally, estimate the registration parameters using SVD [55] based on G_{best}. We dynamically adjust the maximum iteration number MI_e of edge sampling and the maximum iteration number MI_v of vertex sampling according to the consensus set size. Similarly to RANSAC [15], the maximum iteration number is updated by the following rule.

$$MI_v = \frac{\log(1 - P_1)}{\log\left(1 - \frac{|L_{best}|}{|\Phi|}\right)} \tag{5}$$

$$MI_e = \frac{\log(1 - P_2)}{\log\left(1 - \left(\frac{|G_{best}|}{N}\right)^2\right)} \tag{6}$$

where $|\cdot|$ denotes the set size. P_1 and P_2 denote the probability of sampling at least one all-inlier subset for vertex sampling and edge sampling, respectively, and we set $P_1 = P_2 = 0.99$.

3.4. Preference-Based Guided Sampling Strategy

For an established compatibility graph, given a vertex $c_n \in G$ in the graph and c_n denotes a correspondence, define the set of vertices in the compatibility graph that are compatible with c_n as

$$N_{c_n} = \{c_{n'} | \mathbf{M}_C(n, n') = 1\}. \tag{7}$$

Based on the constructed compatibility graph G, according to the introduced minimum compatible subset sampling method, compatible edges are sampled in the graph and combined with compatible vertices to estimate the model, and a locally optimal model hypothesis can be obtained for each compatible edge sampled. Assuming that M edges are initially sampled through iterations, M model hypotheses are generated accordingly, denoted as $H = \{h_m | 1 \le m \le M\}$, where $h_m = (\mathbf{R}_m, \mathbf{t}_m)$, and the M+1th model is generated by the guided sampling strategy. Specifically, for each data $c_n = (x_n, y_n)$, we compute the residual distance $r(c_n, h_m) = ||y_n - (\mathbf{R}_m x_n + \mathbf{t}_m)||$ of c_n with respect to the mth model hypothesis based on the Euclidean distance. We then introduce the preference function, which represents the degree of preference of a correspondence c_n over a model hypothesis h_m, as follows:

$$f_m^n = \begin{cases} e^{-r^2(c_n, h_m)/\delta^2}, & if\ r(c_n, h_m) < \tau_m \\ 0, & otherwise \end{cases} \tag{8}$$

where τ_m is an inlier threshold and δ is a regularization constant. Thus, the preferences of a correspondence c_n for M model hypotheses in the set of model hypotheses H can be expressed as a set $\mathbf{f}^n = [f_1^n, f_2^n, \cdots, f_M^n]$. For any two correspondences c_n and $c_{n'}$, whose preference vectors are computed as $\mathbf{f}^n$ and $\mathbf{f}^{n'}$, respectively. We use cosine similarity to compute the residual correlation between the two correspondences, denoted as

$$\varphi(c_n, c_{n'}) = \frac{\left\langle \mathbf{f}^n, \mathbf{f}^{n'} \right\rangle}{\left|\left|\mathbf{f}^n\right|\right| \times \left|\left|\mathbf{f}^{n'}\right|\right|} \tag{9}$$

where $\langle \cdot, \cdot \rangle$ and $|| \cdot ||$ denote the inner product and L2 norm, respectively. It is intuitive that inliers should be compatible with each other, and inliers should share many of the same modeling assumptions with each other. Thus, if two correspondences c_n and $c_{n'}$ are inliers, the corresponding similarity scores of them are high. Otherwise, if c_n and $c_{n'}$ are outliers, they do not have similar preferences for different modeling assumptions, so the corresponding similarity scores will be low.

Based on the mutual compatibility between inliers, high similarity scores of inliers should be accompanied by the existence of edge connections, so the similarity between c_n and $c_{n'}$ is further defined as

$$w(c_n, c_{n'}) = \begin{cases} \varphi(c_n, c_{n'}), & c_{n'} \in N_{c_n} \\ 0, & c_{n'} \notin N_{c_n} \end{cases} \tag{10}$$

where N_{c_n} donates the set of vertices in the compatibility graph that are compatible with c_n, and satisfies Equation (7). Thus, the similarity scores of the spatially incompatible correspondences are set to 0, in which case the similarity matrix $\mathbf{M}_s$ between the correspondences can be obtained and $\mathbf{M}_s(n, n') = w(c_n, c_{n'})$. Assuming that c_n is fixed, the similarity score between c_n and the jth corresponding c_j in C is $D_{nj} = \mathbf{M}_s(n, j)$. Finally, the similarity between c_n and the rest of the data in C constitutes an association vector $\mathbf{D}_n$

$$\mathbf{D}_n = [D_{n1}, D_{n2}, \ldots, D_{nj}, \ldots, D_{nN}]. \tag{11}$$

As in [56], define the gap γ_j as the difference between the maximum value of $\mathbf{D}_n$ and D_{nj}

$$\gamma_j = \max(\mathbf{D}_n) - D_{nj} \tag{12}$$

γ_j is not less than 0, the smaller γ_j is, the more similar c_n is to c_j. Define the probability of γ_j as

$$\eta(\gamma_j) = \gamma_j \Big/ \sum_{k=1}^{N} \gamma_k \tag{13}$$

According to [57], the information provided by the jth correspondence is denoted as

$$e_j = -\log(\eta(\gamma_j) + \xi) \tag{14}$$

where ξ is a small positive value, and the sum of the information entropies of the remaining points in C on c_n is

$$EP_n = \sum_{j=1}^{N} \eta(\gamma_j) e_j. \tag{15}$$

The information entropy is computed for each vertex in G to form the vector $\mathbf{EP} = [EP_n]_{n=1}^{N}$. The smaller EP_n indicates that c_n is more likely to be an inlier, so the vertex with smaller information entropy is selected according to $\mathbf{EP}$ as the set of vertices participating in the subsequent sampling of the compatible edges for the next model estimation. The vertex selection strategy is denoted as follows.

$$\chi = \{c_n | EP_n < mean(\mathbf{EP})\}. \tag{16}$$

Using this method to select significant vertices that are more likely to be inliers, and sampling compatible edges in the set of significant vertices in the next iteration, effectively increases the probability of sampling the smallest subset of all inliers and speeds up the estimation of the optimal model.

3.5. Complete Registration Algorithm

Based on the compatibility graph and preference guided sampling, we further propose a complete correspondence-based point cloud registration algorithm for clouds with a high outlier ratio. In order to control the selection process of significant vertices, we define a batch size b as well as a maximum inlier update time *max_up*. b controls the frequency of vertex information entropy computation; i.e., we perform the vertex information entropy computation only for every b model hypothesis generated. And *max_up* is used as the end condition of the algorithm; i.e., after significant vertices have been selected *max_up* times, it is considered that the inliers have been involved in enough iterations to have obtained the exact transformation parameters; i.e., it is considered that the optimal solution has been obtained and the iteration is ended. The flow of the algorithm is shown in Algorithm 1.

Algorithm 1. Proposed Method

Input: Initial correspondences C: $\{c_i = (x_i, y_i) | 1 \le i \le N\}$; inlier threshold ε, τ_m; regularization constant δ; batch size b; maximum inliers update times max_up; positive constant ξ;
Output: optimal (**R**, **t**); maximum consensus set G_{best};
$G_{best} = \varnothing$, $MI_e = 10^5$, edge sample iteration number $I_e = 0$;
inliers update times $t\_up = 0$; preference calculations times $t\_pr = 0$; $C_{in} = C$;
Construct compatibility graph G, obtain $V = \{c_i | 1 \le i \le N\}$, $E = \{e_{ij} | 1 \le i \le N, 1 \le i \le N\}$, get $\mathbf{M}_C = \{(i,j) | 1 \le i \le N, 1 \le i \le N\}$ with Equation (4);
while $I_e \le MI_e$ **do**
  $I_e = I_e + 1$;
  Randomly select 2 points (c_i, c_j) from C_{in};
  if $\mathbf{M}_C(i,j) = 1$ **then**
    Search for candidate set Φ according to $\mathbf{M}_C$, set $L_{best} = \varnothing$, vertex sample iteration number $I_v = 0$, $MI_v = 10^5$;
    while $I_v \le MI_v$ **do**
      $I_v = I_v + 1$;
      Randomly select 1 vertex c_k from Φ;
      Use Horn's minimal method to estimate **R**, **t** with $\{c_i, c_j, c_k\}$;
      Find consensus set $L_{temp} \in \Phi$, using **R**, **t**;
      if $|L_{temp}| \ge |L_{best}|$ **then**
        $L_{best} = L_{temp}$, and update MI_v with Equation (5);
      end
      if $I_v \ge MI_v$ **then**
        break
      end
    end
    Use SVD to estimate **R**, **t** with L_{best};
    Find consensus set $G_{temp} \in C$, using **R**, **t**;
    if $|G_{temp}| \ge |G_{best}|$ **then**
      $G_{best} = G_{temp}$, and update MI_e with Equation (6);
    end
    Calculate the degree of preference f_m^n of C for **R**, **t**;
    $t\_pr = t\_pr + 1$;
    if mod$(t_pr, b) = 0$ **then**
      $t\_up = t\_up + 1$;
      Calculate the information entropy **EP** of C with Equation (15);
      Obtain possible inliers set, update C_{in} according to **EP** with Equation (16);
    end
  end
  if $I_e \ge MI_e$ or $t\_up > max\_up$ **then**
    break
  end
end
Use SVD to estimate **R**, **t** with G_{best};
return **R**, **t**, G_{best};

4. Experimental Results

To validate the effectiveness of the proposed algorithm, we conducted a series of experiments on several datasets, including the synthetic dataset Stanford 3D Scanning Repository dataset, the indoor dataset 3DMatch, the low-overlap indoor dataset 3DLoMatch, and the outdoor dataset KITTI. The addresses of all datasets can be seen in the *Data Availability Statement*.

The Stanford 3D Scanning Repository dataset contains several mesh models, which were obtained by scanning with a range scanner, followed by registration and surface reconstruction techniques. In order to verify the basic performance of our algorithm, we constructed test data pairs by randomly generating rotation matrices and translation vectors as ground truth transformations. 3Dmatch [58] is a point cloud dataset of eight indoor scenes obtained from RGBD sequences, containing a total of 1623 test pairs, each with a real camera pose and an overlap of more than 30%. 3DLoMatch [29] is a dataset of the same eight scenes with an overlap of between 10% and 30%, containing a total of 1781 pairs. KITTI [59] is a large-scale outdoor LIDAR dataset, which provides 11 sequences with pose annotations. Its ground truth transformations are obtained by GPS with refinement by the standard iterative closest point (ICP) algorithm [60]. This dataset contains several thousand frames of data in each sequence, and data in the same sequence have a high overlap rate.

4.1. Synthetic Data Experiment

We use the *armadillo* [61] point cloud model from the Stanford 3D Scanning Repository dataset for basic performance validation of the algorithm. First, 1000 points are sampled in the initial point cloud model as key points P_s, then its scale is changed so that the point cloud is inside a 1m×1m×1m enclosing box. And then rigid transformation $\mathbf{R}\in SO(3)$ and $\mathbf{t}\in R^3$ are randomly generated and the initial point cloud model is transformed to obtain the transformed point cloud, where the transformed key points are P_t. To make the experiment closer to the real situation, we add Gaussian noise with a mean value of 0 and a standard deviation $\sigma = 0.01$ to the transformed key points P_t to simulate the noise present in the actual collected data, and obtain an inlier set $C_{in} = (P_s, P_t)$. To obtain the outliers, we generate N_{out} random points Q_{out} in a spherical space with the center of gravity of P_t as the spherical center and the length of the diagonal of the bounding box of P_t as the radius. Then, randomly select N_{out} points P_{out} in P_s and release P_{out} from matching with the corresponding points in P_t. Next, establish the correspondence between P_{out} and Q_{out} to form the outlier set $C_{out} = (P_{out}, Q_{out})$, and the corresponding set C containing outliers is obtained by replacing the positions in C_{in} with the same index as C_{out}. In order to simulate the case of different outlier ratios, by changing the value of N_{out}, set the outlier ratio at {20%, 40%, 60%, 80%, 90%, 92%, 94%, 96%, 98%, 99%}. Figure 2 shows the key points obtained by subsampling the point cloud and the initial correspondences of different outlier ratios, respectively. Due to the randomness of the noise distribution, each experiment is repeated 50 times to ensure the stability of the results.

In order to quantitatively assess the performance of the registration algorithms, the widely used rotation error (E_R) and translation error (E_t) are used as evaluation criteria [62], which are respectively

$$\begin{cases} E_R = |\arccos\frac{tr(\mathbf{R}_{GT}^{\mathrm{T}}\mathbf{R}_e)-1}{2}| \cdot \frac{180^\circ}{\pi} \\ E_t = ||\mathbf{t}_{GT} - \mathbf{t}_e|| \end{cases} \tag{17}$$

where $\mathbf{R}_{GT}$ and $\mathbf{t}_{GT}$ denote the true values of the rotation and translation matrices, respectively. $\mathbf{R}_e$ and $\mathbf{t}_e$ denote the estimated values of the rotation and translation matrices, respectively, computed by the registration algorithm. tr(·) denotes the trace of the matrix. E_R is used to measure the angular difference between $\mathbf{R}_{GT}$ and $\mathbf{R}_e$, and E_t is used to measure the Euclidean distance between $\mathbf{t}_{GT}$ and $\mathbf{t}_e$. In addition, we evaluate the efficiency of the algorithm by comparing the running time (T_c) required for the registration.

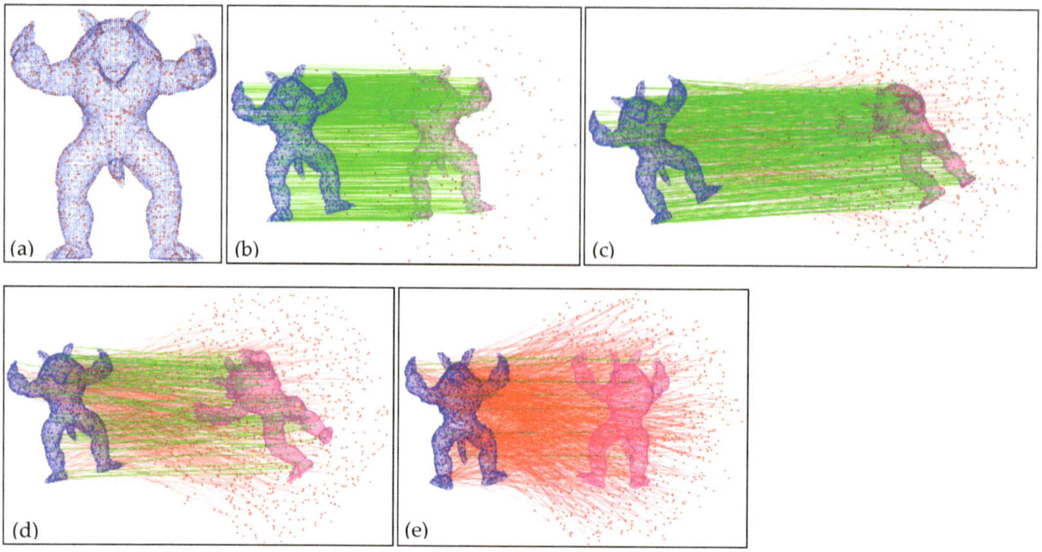

Figure 2. The generation of simulation data, where green lines denote the inliers while red lines denote the outliers, and the bolded points indicate key points: (**a**) Key points, (**b**) 20% outliers are added, (**c**) 60% outliers are added, (**d**) 90% outliers are added, and (**e**) 99% outliers are added.

In order to test the influence of the algorithm parameters on the experimental results, parameter analysis experiments are carried out. The main parameters involved in the proposed algorithm are the batch size b and the maximum number of updates of the inliers set max_up. For b, we first set $max\_up = 3$, and increase b from 10 to 50 in steps of 10. Then we fix b to 20, and increase max_up from 2 to 6 in steps of 1. Experiments were carried out on data with different outlier ratios, and each parameter condition was run 50 times to record the mean rotation error, mean translation error, and mean time cost. The results of the experiments are shown in Figure 3.

According to the results, it can be seen that when $b = 10$, the rotation error and translation error are large; this is because at this time it is not possible to fully sample the inliers, resulting in the results having a larger error. When $b = 20$, the rotation and translation errors are relatively small, while the computational efficiency is high, and the accuracy and efficiency are in good balance. The time cost will increase significantly if b continues to increase. For max_up, when the outlier ratio is less than 98%, max_up has less influence on the experimental results. When the outlier ratio is 99%, $max\_up = 3$ corresponds to a small mean rotation error and mean translation error. At the same time, the time cost is very little, which achieves a good balance in terms of accuracy and efficiency. Therefore, in this paper, b and max_up are set to 20 and 3, respectively.

In order to verify the performance of our algorithm equivalent to advanced robust point cloud registration algorithms, we compare the proposed method with six state-of-the-art algorithms, namely, RANSAC [15], GORE [16], One-Point RANSAC [32], GROR [42], RANSIC [18], and VODRAC [17]. Among these algorithms, RANSAC is the widely used initial registration algorithm, GORE and GROR are the most recently proposed provable and have good outlier filtering performance. One-Point RANSAC, RANSIC, and VODRAC are recently proposed state-of-the-art algorithms and show excellent performance in point cloud registration tasks heavily contaminated by outliers. Specifically, we set the maximum number of iterations to 10^5 for all RANSAC-type algorithms and set the inlier threshold to $6pr$ for all algorithms, where pr denotes the resolution of the input point cloud [63]. pr is obtained by summing and averaging the distances between each point and its nearest neighbor. The parameters of the different algorithms are shown in Table 1 and are the same for the rest of the experiments.

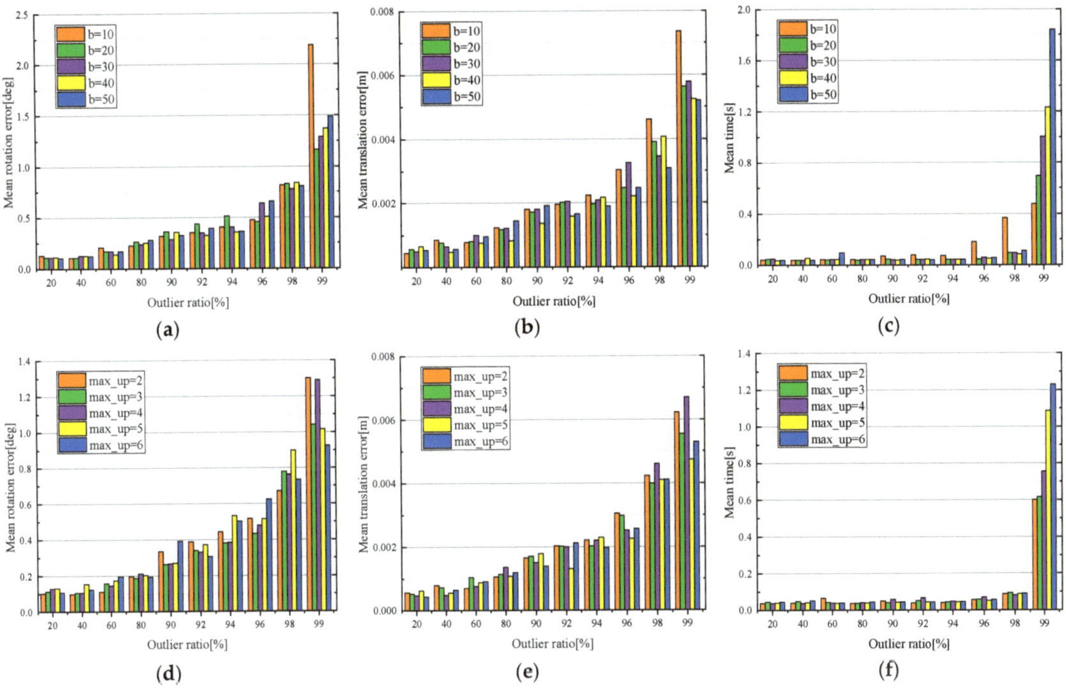

Figure 3. Influence of the parameters b and max_up on the performance of proposed method: (**a**) Rotation error for sensitivity test of b, (**b**) translation error for sensitivity test of b, (**c**) running time for sensitivity test of b, (**d**) rotation error for sensitivity test of max_up, (**e**) translation error for sensitivity test of max_up, (**f**) running time for sensitivity test of max_up.

Table 1. Detailed Settings of the Compared Algorithms.

Method	Parameters
RANSAC	Maximum number of iterations: 10^5; inlier threshold: $6pr$
GORE	Lower bound: 0; repeat: true; consistent threshold: $6pr$
One-Point RANSAC	Confidence: 0.99; subset size: 1; Maximum number of iterations: 10^5; step size: 1.3
GROR	reliable set size: 800; inlier threshold: $6pr$
RANSIC	Maximum number of iterations: 10^5; Confidence: 0.99
VODRAC	Maximum number of iterations: 10^5; Confidence: 0.99; inlier threshold: $6pr$
Ours	Maximum number of iterations: 10^5; inlier threshold: $6pr$ $P_1 = P_2 = 0.99$; $b = 20$; $max\_up = 3$; $\delta = 10pr$; $\xi = 10^{-6}$

The registration results of different algorithms are shown in Figure 4 and some visualization results are shown in Figure 5. From the results, it can be seen that RANSAC can be useful when the outlier ratio is lower than 80%. The algorithm fails when the outlier ratio continues to increase, and the rotation and translation errors of the parameters estimated by RANSAC are large. GORE maintains a stable performance under different outlier ratios due to its ability to reliably remove the outliers and its robustness to noise. However, it exhibits limited registration accuracy, and the computational complexity of GORE is high. The time it takes to complete registration is usually several orders of magnitude higher compared to the other algorithms. When the outlier ratio is lower than 98%, One-Point RANSAC shows competitive performance in terms of registration accuracy and registration efficiency, but when the outlier ratio is 99%, the registration accuracy of the algorithm decreases rapidly, and the algorithm usually fails to estimate the correct registration parameters. GROR maintains good registration accuracy at different outlier ratios. While the registration efficiency decreases with the increase in the outlier ratio, the algorithm takes a long time to complete the registration. Both RANSIC and VODRAC have good robustness to outliers, and maintain high registration accuracy even when the outlier ratio is very high When the outlier ratio is as high as 99%, the rotation and translation errors of the parameter estimated by RANSIC are about 1.221° and 0.0061 m, respectively, and the rotation and translation errors of the parameter estimated by VODRAC are about 1.655° and 0.0075 m. The registration efficiency of VODRAC increases first and then decreases with the increase in the outlier ratio. When the outlier ratio is 80%, the registration efficiency is highest, and it takes about 0.085 s to complete the registration. When the outlier ratio is lower than 96%, the registration efficiency of RANSIC is very high, and is higher than that of VODRAC. When the outlier ratio is higher than 96%, the time required for RANSIC to complete the registration increases significantly, and when the outlier ratio is 99%, RANSIC takes about 20.46 s to complete the registration.

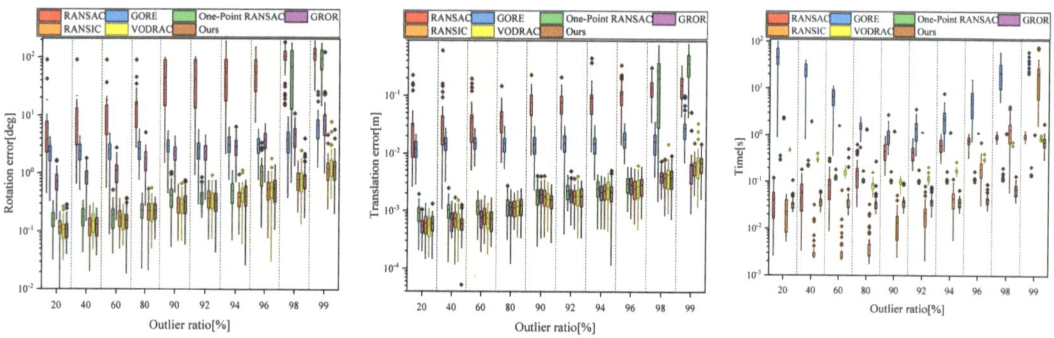

(a) (b) (c)

Figure 4. Registration performance on simulated data. In the figure, ▮ indicates data between 25% and 75% of all data in the result in descending order of magnitude; I indicates maximum and minimum values; - indicates average value; ◆ denotes outliers: (**a**) Box-plot of rotation error. (**b**) Box-plot of translation error. (**c**) Box-plot of time cost.

(**a**) (**b**) (**c**) (**d**)

Figure 5. Visualization results on synthetic dataset, where blue points indicate source point cloud. green lines and green points denote the inliers while red lines and red points denote the outliers: (**a**) Correspondences with 60% outliers, (**b**) Registration result of (**a**), (**c**) Correspondences with 99% outliers, (**d**) Registration result of (**c**).

As can be seen from the results, the proposed algorithm has excellent performance in terms of rotation error, translation error, and time cost. When the outlier ratio is lower than

96%, compared with other algorithms, the proposed algorithm exhibits remarkably low levels of both rotation and translation errors, and its time cost remains consistently stable at 0.035 s. When the outlier ratio is higher than 96%, the proposed algorithm still maintains a very high registration accuracy. When the outlier ratio is 96% and 98%, the proposed algorithm efficiency is significantly higher than the RANSIC and VODRAC algorithms. When the outlier ratio is 99%, One-Point RANSAC is no longer able to accurately estimate the registration parameters despite its low time cost, while the proposed algorithm still has a high accuracy. The excellent performance of the algorithm proposed in this paper can be attributed to the following factors: (1) The compatibility graph is constructed with full consideration of the compatibility relationship between the inliers, which can avoid the influence of the outliers on the registration results and ensure the accuracy of the algorithm. (2) The minimum subset sampling is split from three-point random sampling into compatible edge sampling and candidate subset sampling, which effectively reduces the computational complexity. (3) A guided accelerated sampling strategy is introduced, by calculating the preference between the correspondence and the estimated parameter to determine the inliers faster, which effectively improves the speed of convergence of the parameter estimation.

4.2. *Challenging Real-World Data Experiments*

To evaluate the registration performance of the proposed algorithm on real-world data, we conduct registration experiments using the 3DMatch dataset [58], which contains a total of 8 scenes, namely, *Kitchen*, *Home1*, *Home2*, *Hotel1*, *Hotel2*, *Hotel3*, *Studyroom*, and *Lab*. In each scene, we select 20 data pairs that overlap as test data. For each data pair, we adopt the Harris3D key point detection algorithm [64] to sample about 2000 key points in the source and target point clouds, respectively. Then, we use the FPFH [12] descriptor to obtain the feature vectors of the key points, and then further establish the initial correspondences between the two point clouds based on the feature descriptors. A pair of data is selected from each scene, and the initial correspondences are shown in Figure 6, where the red lines indicate the wrong correspondences, i.e., the outliers, and the green lines indicate the correct correspondences, i.e., the inliers. It can be seen that the initial correspondence set is contaminated by a large number of outliers, which makes it extremely challenging to align accurately.

The registration experiments on the 3DMatch dataset also compare six registration algorithms, including RANSAC, GORE, GROR, One-Point RANSAC, RANSIC, and VODRAC. We compare the rotation error, translation error, and time cost of the different algorithms. In order to qualitatively demonstrate the performance of the different algorithms, we select a pair of data pairs with low overlap between the source and the target point clouds from each scene for visualization, and the results of the different algorithms are shown in Figure 7. Visually, RANSAC can only roughly align the *Lab* scene, and similarly, GORE performs poorly, One-Point RANSAC performs slightly better and can effectively align two scenes, and GROR has a large improvement in performance, effectively aligning six scenes, but the algorithm fails for *Home2* and *Hotel3*. Both RANSIC and VODRAC can complete the registration of all scenes, but RANSIC takes a lot of time to align each scene, and VODRAC is more efficient but still less efficient when the initial number of correspondences is very large. Our algorithm efficiently completes the registration of all the scenes, and the registration efficiency is very high in all cases, which proves the robustness and efficiency of the proposed algorithm.

Since FPFH is a manually designed feature descriptor, the correspondences established by it usually contain a large number of outliers with an outlier ratio of up to 99%. We record the rotation error, translation error, and time consumption of different algorithms. The experimental results are shown in Figure 8. The average outlier ratios and registration results of the experimental data for different scenarios are shown in Table 2.

Registration Accuracy Analysis: As shown in Table 2, the initial correspondences contain a large number of outliers, and the average outlier ratio of each scene is close to

99%. RANSAC can only achieve approximate registration for a few scenes, and the rotation and translation errors of most of the scenes are very large. This limitation is due to the fact that RANSAC needs to sample randomly in a large number of initial correspondences, which results in its inability to achieve effective registration within the set number of iterations. Its rotation and translation errors reach a maximum of 86.584° and 2.116 m. According to Figure 8a,b, under this condition of the number of correspondence sets and the outlier ratio, GORE and One-point RANSAC are also ineffective for most of the data pairs. They can only achieve accurate registration for a small portion of pairs, and the robustness of the algorithms needs to be further improved. In contrast, the performance of GROR is greatly improved. For most of the correspondences that are heavily contaminated by outliers, GROR can achieve accurate registration. As can be seen, many of the GROR registration results have a rotation error of less than 1°, and a translation error of less than 0.05 m. However, GROR still faces failures for individual data pairs. We speculate that this is due to the fact that an excessive number of outlier points affect the reliability of the algorithms in terms of graph node reliability and graph edge reliability, which leads to inaccurate final registration results. Under the condition that the correspondence set is heavily contaminated, RANSIC and VODRAC show excellent performance. Both of them can achieve accurate registration for each scene, in which the mean rotation error of RANSIC reaches a minimum of 0.984°, and the mean translation error reaches a minimum of 0.041 m. The mean rotation error of VODRAC reaches a minimum of 0.842° and the mean translation error reaches a minimum of 0.0302 m, which are significantly better than those of RANSAC, GORE, and One-Point RANSAC. According to the experimental results, the registration algorithm proposed in this paper reaches the advanced level in terms of registration accuracy, and can complete the accurate registration of all data pairs. The algorithm has good robustness to outliers, the mean rotation error of the registration results reaches 0.737° at the lowest level, and the mean translation error reaches 0.0201 m at the lowest level, which has a very high accuracy.

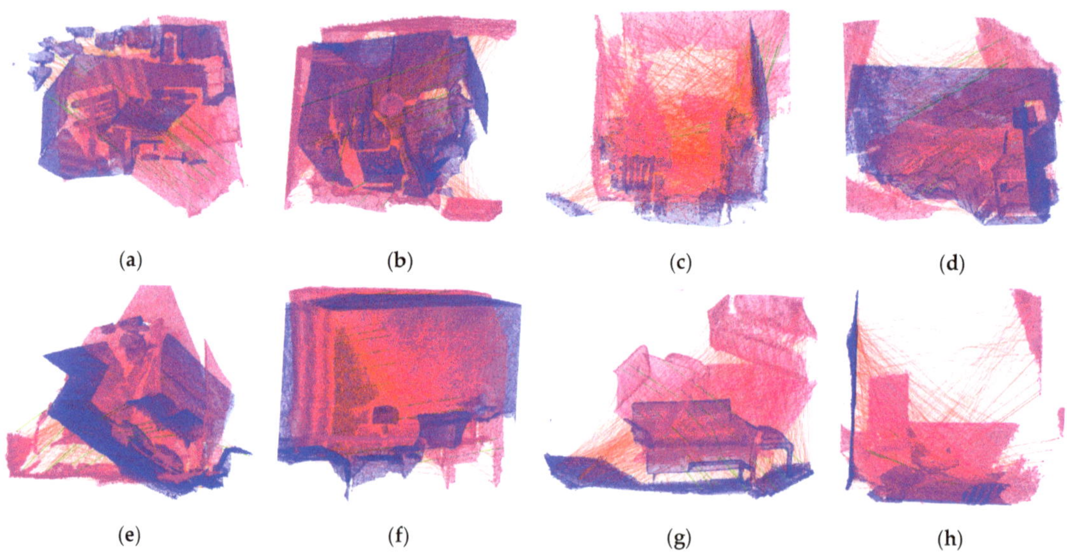

Figure 6. Initial correspondences for different scenarios, where green lines denote the inliers while red lines denote the outliers, blue points and red points indicate the source and target point clouds, and the bolded points indicate key points: (**a**) *Kitchen*, inliers/totals: 21/1876. (**b**) *Home1*, inliers/totals: 23/2682. (**c**) *Home2*, inliers/totals: 20/2494. (**d**) *Hotel1*, inliers/totals: 27/1947. (**e**) *Hotel2*, inliers/totals: 21/2167. (**f**) *Hotel3*, inliers/totals: 20/3044. (**g**) *Studyroom*, inliers/totals: 20/1500. (**h**) *Lab*, inliers/totals: 20/1359.

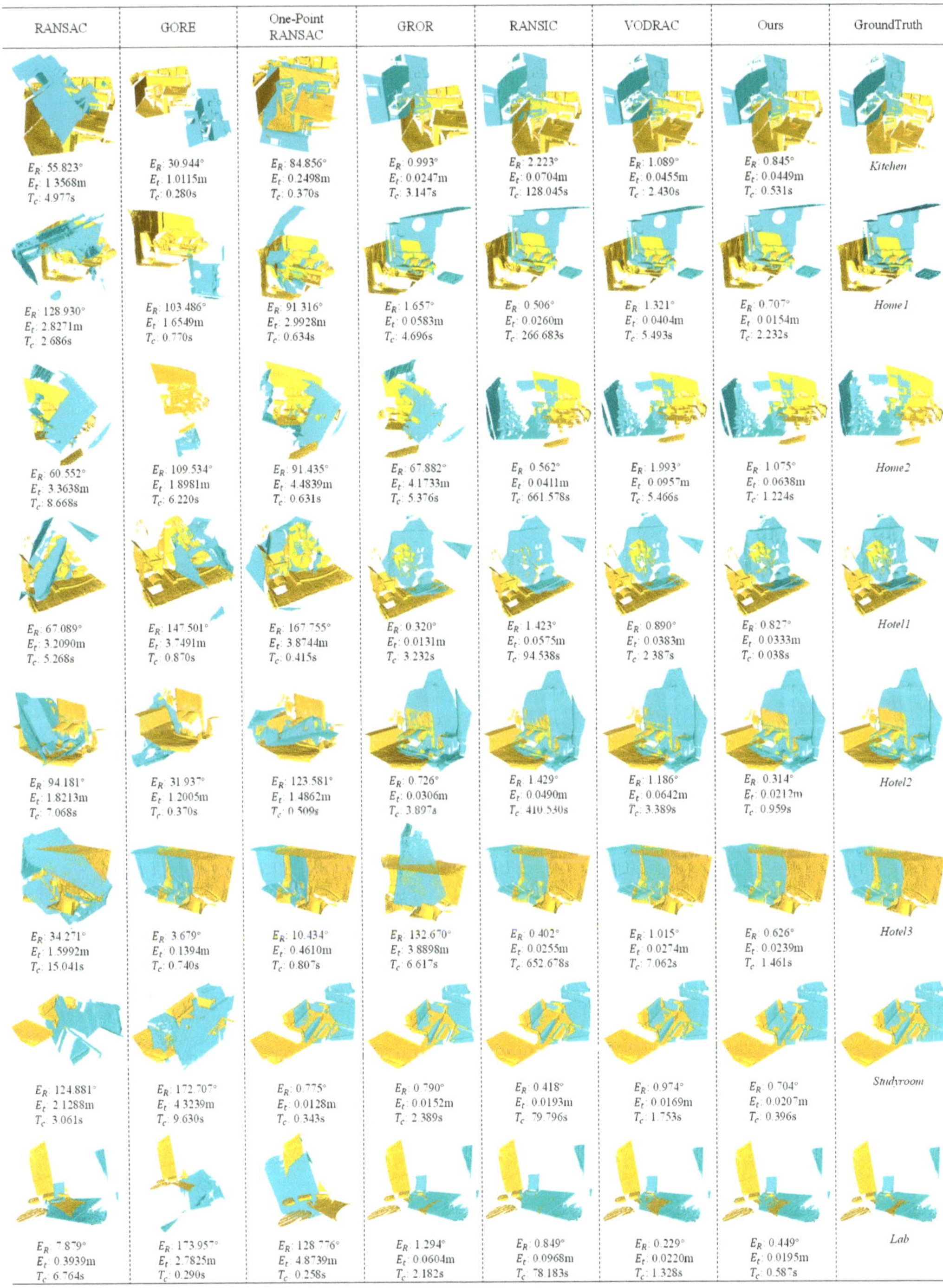

Figure 7. The visual performance of real-world data experiment of algorithms, where yellow and cyan indicate the source and target data.

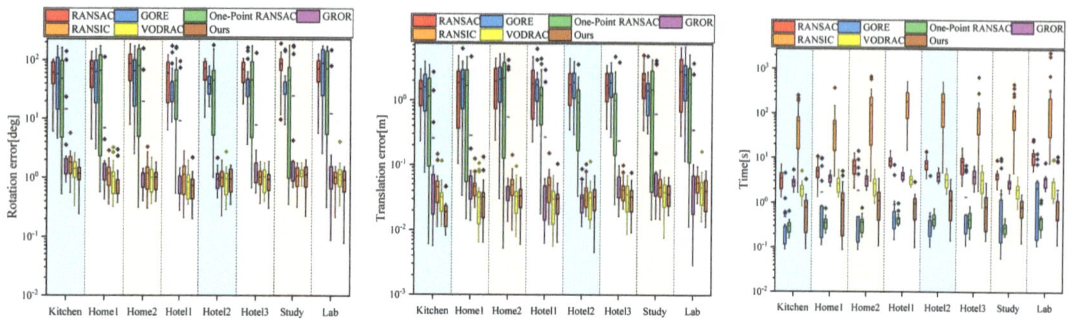

(a) (b) (c)

Figure 8. Registration performance on 3DMatch dataset. In the figure, ▌ indicates data between 25% and 75% of all data in the result in descending order of magnitude; I indicates maximum and minimum values; - indicates average value; ◆ denotes outliers: (**a**) Box-plot of rotation error. (**b**) Box-plot of translation error. (**c**) Box-plot of time cost.

Table 2. Quantitative results on 3DMatch dataset.

	Kitchen	*Home1*	*Home2*	*Hotel1*	*Hotel2*	*Hotel3*	*Studyroom*	*Lab*
Mean outlier ratio	98.55%	98.74%	98.70%	98.96%	98.93%	98.83%	98.69%	98.74%
				Mean Rotation Error (°)				
RANSAC	60.854	70.984	86.465	60.237	72.172	70.028	86.584	71.358
GORE	56.389	62.745	65.342	42.885	38.998	47.645	42.732	89.329
One-Point RANSAC	50.128	65.939	79.379	44.533	45.151	45.930	45.533	72.256
GROR	14.382	6.920	19.685	9.135	0.943	7.842	24.886	12.356
RANSIC	1.794	1.173	1.189	1.133	0.984	1.029	1.079	1.022
VODRAC	1.395	1.040	1.004	0.842	0.949	1.047	1.194	1.142
Ours	1.147	0.909	0.999	0.737	0.933	0.931	1.122	0.921
				Mean Translation Error (m)				
RANSAC	1.4804	1.8729	1.9635	1.8119	1.7625	1.7233	2.1161	2.8324
GORE	1.5556	2.1000	2.1068	1.7491	1.8869	1.8829	1.4261	2.5204
One-Point RANSAC	1.1090	1.6728	2.1970	1.1921	0.9329	1.0032	1.4640	1.8214
GROR	0.2562	0.2826	0.5506	0.2687	0.0316	0.2386	0.6235	0.3536
RANSIC	0.0472	0.0469	0.0494	0.0490	0.0406	0.0417	0.0463	0.0540
VODRAC	0.0327	0.0321	0.0333	0.0302	0.0315	0.0357	0.0386	0.0461
Ours	0.0201	0.0327	0.0339	0.0304	0.0326	0.0327	0.0362	0.0407
				Mean Time Cost (s)				
RANSAC	3.398	4.967	6.426	8.389	7.457	8.276	4.134	11.439
GORE	0.469	1.697	1.867	0.921	0.494	1.370	1.618	2.154
One-Point RANSAC	0.299	0.354	0.364	0.434	0.432	0.430	0.283	0.446
GROR	2.778	3.554	3.332	4.069	3.828	3.857	2.829	3.494
RANSIC	69.001	57.093	151.787	182.039	183.542	119.140	116.414	326.279
VODRAC	1.983	2.643	2.773	3.300	3.246	3.260	2.013	2.989
Ours	0.759	1.129	1.121	0.948	1.155	0.826	0.804	2.001

Registration efficiency analysis: Figure 8c shows the registration time distribution of different algorithms on eight scenes, and Table 2 records the average registration time of different algorithms for each scene. According to the results, it can be seen that the registration efficiency of RANSAC is low, and its running time is mainly related to the preset number of iterations, which requires a large number of iterative calculations to obtain relatively better results. The running time of GORE is very short because there are too few inliers in the correspondence. The algorithm cannot efficiently compute the upper and lower bounds, and it skips the computation of the parameter updating process. One-Point RANSAC has high running efficiency, which is due to the fact that the algorithm decomposes the registration problem, and the parameter space is drastically reduced, allowing it to quickly find what it considers to be the optimal solution. GROR has a high registration efficiency, generally taking 2~4 s to complete the registration. VODRAC has a slightly higher registration efficiency than that of GROR, which is due to the fact that it has the step of random sampling consistency decomposition. In the case of high outlier ratios, despite achieving accurate registration, the registration efficiency of RANSIC is very low, usually requiring tens or hundreds of seconds to complete the registration. In contrast, our algorithm has very high registration efficiency, and even if the initial number of correspondences reaches 2000 and is heavily contaminated by outliers, it can still compute very accurate registration results in less than 1 s in most cases.

4.3. Low-Overlap Point Cloud Registration Experiments

In order to verify the ability of the proposed algorithm to handle point cloud pairs with low overlap rate, we carried out experiments on 3DLoMatch, which contains 1781 test point cloud pairs and has a low overlap rate between point cloud pairs, with the overlap rate ranging from 10% to 30%. It is difficult to establish correspondences between these point cloud pairs by handcrafted descriptors. Recently, Transformer-based correspondence estimators have shown excellent performance on point clouds with low overlap rates to establish reliable correspondences between point cloud pairs. We use GeoTrans [31] to establish the correspondences of 3DLoMatch data and incorporate the proposed parameter estimation method to improve the registration performance. We evaluate the performance of the algorithm by using E_R, E_t and registration recall (RR) [65]. RR is the proportion of the results with E_R, E_t under the error threshold to the total number of test samples, i.e., the rate of successful registration, and we set the threshold to (15°, 0.3 m). As correspondences established using GeoTrans usually contain enough inliers, most data pairs can be successfully aligned. Following [65], since part of the failed registration can generate large rotation and translation errors, we only computed the mean rotation error ($\overline{E}_R$) and translation error ($\overline{E}_t$) of successfully registered point cloud pairs of each method to avoid unreliable metrics. A local-to-global (LGR) parameter estimation method is proposed in GeoTrans, and the experiments are compared with this method. The basic RANSAC algorithm and advanced algorithms including GROR and RANSAIC are also compared.

The experimental results obtained are shown in Table 3, and some qualitative results are shown in Figure 9. According to the experimental results, it can be seen that the proposed algorithm can effectively improve the registration recall by 3.41% compared to LGR, due to the more effective handling of the case of fewer inliers within the correspondences. As GROR and RANSIC are able to detect inliers in the correspondences, both of them provide some performance gains, while RANSAC has a poorer performance. The experiments illustrate that the proposed algorithm can effectively align point cloud pairs with very low overlap and achieve significant performance in conjunction with the learning-based descriptor.

Table 3. Registration results on 3DLoMatch with learning-based correspondences.

Method	$\overline{E}_R$ (°)	$\overline{E}_t$ (m)	RR (%)
LGR	2.992	0.0867	77.50
RANSAC	4.516	0.1385	61.93
GROR	3.186	0.1012	80.85
RANSIC	3.549	0.1143	79.79
Ours	2.967	0.0962	80.91

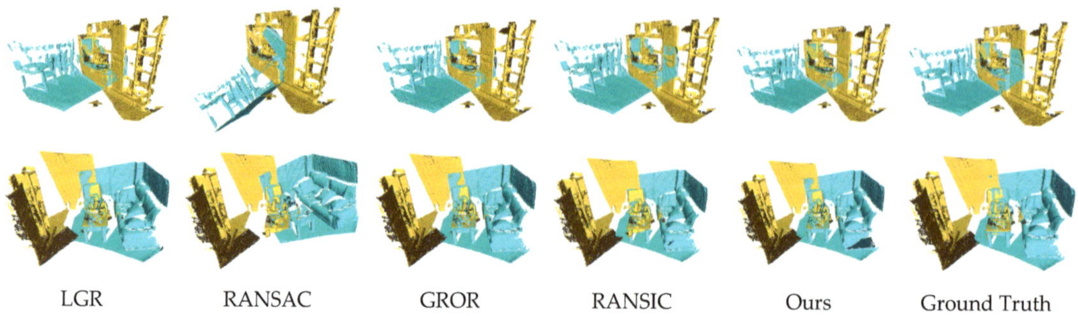

Figure 9. Visualization of two groups of point cloud registration results on the 3DLoMatch, where yellow and cyan indicate the source and target data. The upper group has an overlap rate of 11.14% and the lower group has an overlap rate of 11.98%. From left to right: results of LGR, RANSAC, GROR, RANSIC, ours, and ground truth. RANSAC fails for the data pair in the first row, while other algorithms successfully align these data pairs.

4.4. Outdoor Scene Registration Experiments

To further validate the ability of the proposed algorithm to handle more complex scenarios, we conducted experiments on the outdoor LIDAR dataset KITTI, where the data scale of the outdoor scene is much larger than that of the indoor scene. As in [66], we selected scenes 8 to 10 as the test dataset and obtained a total of 555 test data pairs. Again, we used GeoTrans to establish the correspondences between the point cloud pairs and then combined the parameter estimation methods to estimate the registration parameters between the point cloud pairs. $\overline{E}_R$, $\overline{E}_t$, and RR are used to evaluate the experimental results, and the error threshold of RR is set to (5°, 0.6 m). LGR, RANSAC, GROR, and RANSAIC are used as comparison algorithms.

The experimental results are shown in Table 4, and some visualized results are shown in Figure 10. From the experimental results, it can be seen that LGR, RANSAIC, and the proposed algorithm obtain the highest registration recall with high parameter estimation accuracy, and the proposed algorithm reaches the optimum in terms of rotation error. RANSAC achieves high registration recall, although the estimated parameters are usually sub-optimal but mostly within acceptable range. GROR performs slightly worse on the KITTI dataset compared to the other algorithms. It is experimentally verified that the proposed algorithm has excellent performance in estimating registration parameters and is common across different scenarios.

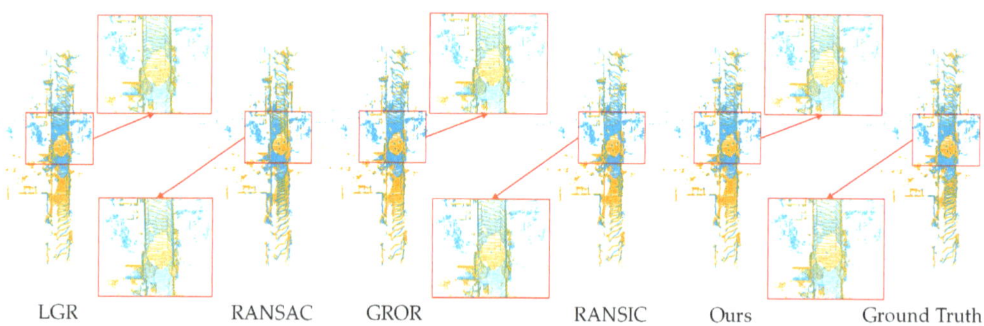

Figure 10. Visualization results on KITTI, where yellow and blue indicate the source and target data. From left to right: results of LGR, RANSAC, GROR, RANSIC, ours, and ground truth. All algorithms successfully align the data pairs.

Table 4. Registration results on KITTI with learning-based correspondences.

Method	$\overline{E}_R$ (°)	$\overline{E}_t$ (m)	RR (%)
LGR	0.378	0.0693	99.10
RANSAC	0.803	0.1861	98.38
GROR	0.505	0.1287	97.84
RANSIC	0.385	0.0872	99.10
Ours	0.341	0.0804	99.10

5. Discussion

In this paper, we propose a method for solving the problem of estimating transformation parameters in feature-based point cloud registration algorithms. For feature-based point cloud registration, high outlier ratios in correspondences established by feature descriptors are a common problem. The outlier ratio in correspondences established by classical handcrafted descriptors such as FPFH is usually higher than 90%. In this case, the proposed algorithm still obtains a high registration accuracy and maintains the optimal accuracy compared to algorithms such as RANSAC, GORE, and One-Point RANSAC. Meanwhile, the proposed algorithm has high registration efficiency, which is tens times faster than RANSIC and several times faster than GROR and VODRAC under the condition of a very high outlier ratio. In conclusion, our algorithm has superior robustness, accuracy, and computational efficiency compared with other state-of-the-art methods.

In terms of algorithm generality, the proposed algorithm takes correspondences as input and outputs the final registration parameters. Point cloud registration can be accomplished by combining any feature matching and correspondence establishment methods, such as handcrafted descriptors and learning-based descriptors. Due to the advanced feature description performance of the learning-based descriptors, combining them with the proposed method can be used for point cloud registration in low-overlap and complex scenarios, and a remarkable registration performance can be obtained. Combining the proposed algorithm with GeoTrans achieves a 3.41% improvement in registration recall on low-overlap point cloud datasets compared to LGR. The algorithm can also be applied to the registration of large-scale scenarios, and the proposed algorithm combined with GeoTrans for large-scale point cloud data registration also obtains the optimal performance.

Although the proposed algorithm is able to achieve fast and robust point cloud registration, it still has some limitations. Firstly, the proposed algorithm still relies on the initial correspondences. If the number of inliers in the correspondences is too small, the proposed algorithm may not be able to find enough correct inliers for parameter estimation, leading to the failure of the algorithm. Secondly, the algorithm relies on the Euclidean distance to determine the compatibility between the correspondences. However, the Euclidean distance has an inherent ambiguity in 3D space; i.e., the Euclidean distances from the surface of the sphere to the center of the sphere are all equal. This property may lead to a lack of stability in the compatibility calculation, thus affecting the performance of the algorithm.

6. Conclusions

In this paper, we present an efficient and robust point cloud registration method that directly outputs the final alignment registration based on correspondences and excels in terms of accuracy, efficiency, and robustness. Compared to many existing techniques, the algorithm in this paper operates efficiently under very high outlier conditions and strikes an excellent balance between efficiency and accuracy. In order to minimize the influence of the outliers, this paper introduces the concept of the compatibility graph, and proposes a minimum subset sampling method for the combination of compatible edges and compatible vertices, which effectively avoids the participation of a large number of outliers in the computation. A preference-guided accelerated sampling strategy is further proposed to effectively utilize the estimated transformation parameters at the initial stage, calculate the preference score of each vertex based on the transformation parameters, and then guide the execution of the sampling in the direction of more likely to be an inlier to improve the efficiency of registration. Finally, the transformation parameters are estimated based on the maximum set of compatible vertices to complete the accurate registration. Based on a synthetic and real dataset, the proposed registration algorithm is compared and analyzed with classical and advanced algorithms. Simulation experiments demonstrate the robustness and efficiency of the algorithm, which can still accomplish registration quickly when the outlier ratio is as high as 99%. Real data show that the algorithm can successfully perform point cloud registration even if the correspondence established by the feature description contains a large number of outliers. Compared with the state-of-the-art algorithms, the proposed algorithm is able to realize a point cloud registration several times faster while maintaining a comparable or higher registration accuracy. By combining the proposed algorithm with a learning-based feature description method, the registration accuracy can be further improved and can be applied to low overlap and large-scale point cloud registration tasks.

In follow-up work, as the proposed algorithm is still closely related to the quality of the initial correspondences, and more inliers can give more accurate results, designing more reliable correspondence establishment methods will be a priority. In addition, the proposed method relies on the Euclidean distance of the correspondence to compute the compatibility, and the compatibility results obtained are not stable enough, so exploring the compatibility of the correspondence with higher orders to further improve the parameter estimation performance and registration accuracy is another future research work.

Author Contributions: Conceptualization, C.W.; Formal analysis, Z.Z.; Funding acquisition, Z.Z., B.Z. and H.L.; Methodology, C.W.; Software, C.W. and Z.Z.; Supervision, B.Z. and H.L.; Validation, Z.Z.; Visualization, B.Z.; Writing—original draft, C.W.; Writing—review and editing, C.W., B.Z. and H.L. All authors have read and agreed to the published version of the manuscript.

Funding: The research in this article was funded by the National Natural Science Foundation of China, grant number 52201399; the foundation of JWKJW Field [grant number 2022-JCJQ-JJ-0394]; the Central University Special Funding for Basic Scientific Research, grant No. 30918012201; Postgraduate Research & Practice Innovation Program of Jiangsu Province.

Data Availability Statement: All datasets used in this study are publicly available. The Stanford 3D Scanning Repository dataset is available at Stanford 3D Scanning Repository (https://graphics.stanford.edu/data/3Dscanrep/, accessed on 6 November 2023). The 3DMatch dataset is available at 3DMatch (https://3dmatch.cs.princeton.edu/, accessed on 17 December 2023). The 3DLoMatch dataset is available at Predator (https://github.com/prs-eth/OverlapPredator, accessed on 7 June 2024). The KITTI is available at KITTI Vision Benchmark Suite (https://www.cvlibs.net/datasets/kitti/eval_odometry.php, accessed on 13 June 2024).

Conflicts of Interest: The authors declare that they have no known competing financial interests or personal relationships that could have appeared to influence the work reported in this paper.

References

1. Han, T.; Zhang, R.; Kan, J.; Dong, R.; Zhao, X.; Yao, S. A Point Cloud Registration Framework with Color Information Integration. *Remote Sens.* **2024**, *16*, 743. [CrossRef]
2. Chen, Y.; Mei, Y.; Yu, B.; Xu, W.; Wu, Y.; Zhang, D.; Yan, X. A Robust Multi-Local to Global with Outlier Filtering for Point Cloud Registration. *Remote Sens.* **2023**, *15*, 5641. [CrossRef]
3. Miao, Y.; Liu, Y.; Ma, H.; Jin, H. The Pose Estimation of Mobile Robot Based on Improved Point Cloud Registration. *Int. J. Adv. Robot. Syst.* **2016**, *13*, 52. [CrossRef]
4. Hu, K.; Chen, Z.; Kang, H.; Tang, Y. 3D Vision Technologies for a Self-Developed Structural External Crack Damage Recognition Robot. *Autom. Constr.* **2024**, *159*, 105262. [CrossRef]
5. Szabó, S.; Enyedi, P.; Horváth, M.; Kovács, Z.; Burai, P.; Csoknyai, T.; Szabó, G. Automated Registration of Potential Locations for Solar Energy Production with Light Detection And Ranging (LiDAR) and Small Format Photogrammetry. *J. Clean. Prod.* **2016**, *112*, 3820–3829. [CrossRef]
6. Choi, S.; Zhou, Q.-Y.; Koltun, V. Robust Reconstruction of Indoor Scenes. In Proceedings of the 2015 IEEE Conference on Computer Vision and Pattern Recognition (CVPR), Boston, MA, USA, 7–12 June 2015; pp. 5556–5565.

7. Wan, S.; Guan, S.; Tang, Y. Advancing bridge structural health monitoring: Insights into knowledge-driven and data-driven approaches. *J. Data Sci. Intell. Syst.* **2023**, *2*, 129–140. [CrossRef]
8. Zhang, H.; Zhu, Y.; Xiong, W.; Cai, C.S. Point Cloud Registration Methods for Long-Span Bridge Spatial Deformation Monitoring Using Terrestrial Laser Scanning. *Struct. Control Health Monit.* **2023**, *2023*, 2629418. [CrossRef]
9. Du, G.; Wang, K.; Lian, S.; Zhao, K. Vision-Based Robotic Grasping From Object Localization, Object Pose Estimation to Grasp Estimation for Parallel Grippers: A Review. *Artif. Intell. Rev.* **2021**, *54*, 1677–1734. [CrossRef]
10. Kim, P.; Chen, J.; Cho, Y.K. SLAM-Driven Robotic Mapping and Registration of 3D Point Clouds. *Autom. Constr.* **2018**, *89*, 38–48. [CrossRef]
11. Yang, J.; Li, H.; Campbell, D.; Jia, Y. Go-ICP: A Globally Optimal Solution to 3D ICP Point-Set Registration. *IEEE Trans. Pattern Anal. Mach. Intell.* **2016**, *38*, 2241–2254. [CrossRef]
12. Rusu, R.B.; Blodow, N.; Beetz, M. Fast Point Feature Histograms (FPFH) for 3D Registration. In Proceedings of the 2009 IEEE International Conference on Robotics and Automation, Kobe, Japan, 12–17 May 2009.
13. Guo, Y.; Sohel, F.; Bennamoun, M.; Lu, M.; Wan, J. Rotational Projection Statistics for 3D Local Surface Description and Object Recognition. *Int. J. Comput. Vis.* **2013**, *105*, 63–86. [CrossRef]
14. Zhao, B.; Le, X.; Xi, J. A Novel SDASS Descriptor for Fully Encoding the Information of a 3D Local Surface. *Inf. Sci.* **2019**, *483*, 363–382. [CrossRef]
15. Fischler, M.A.; Bolles, R.C. Random sample consensus: A paradigm for model fitting with applications to image analysis and automated cartography. *Commun. ACM* **1981**, *24*, 381–395. [CrossRef]
16. Bustos, A.P.; Chin, T.-J. Guaranteed Outlier Removal for Point Cloud Registration with Correspondences. *IEEE Trans. Pattern Anal. Mach. Intell.* **2018**, *40*, 2868–2882. [CrossRef] [PubMed]
17. Hu, E.; Sun, L. VODRAC: Efficient and Robust Correspondence-Based Point Cloud Registration with Extreme Outlier Ratios. *J. King Saud Univ.-Comput. Inf. Sci.* **2023**, *35*, 38–55. [CrossRef]
18. Sun, L. RANSIC: Fast and Highly Robust Estimation for Rotation Search and Point Cloud Registration Using Invariant Compatibility. *IEEE Robot. Autom. Lett.* **2022**, *7*, 143–150. [CrossRef]
19. Chen, Z.; Sun, K.; Yang, F.; Guo, L.; Tao, W. SC^2-PCR++: Rethinking the Generation and Selection for Efficient and Robust Point Cloud Registration. *IEEE Trans. Pattern Anal. Mach. Intell.* **2023**, *45*, 12358–12376. [CrossRef]
20. Rusu, R.B.; Marton, Z.C.; Blodow, N.; Beetz, M. Persistent Point Feature Histograms for 3D Point Clouds. In Proceedings of the 10th International Conference on Intelligent Autonomous Systems, Baden-Baden, Germany, 23–25 July 2008; pp. 119–128.
21. Salti, S.; Tombari, F.; Di Stefano, L. SHOT: Unique Signatures of Histograms for Surface and Texture Description. *Comput. Vis. Image Underst.* **2014**, *125*, 251–264. [CrossRef]
22. Yang, J.; Zhang, Q.; Xiao, Y.; Cao, Z. TOLDI: An Effective and Robust Approach for 3D Local Shape Description. *Pattern Recognit.* **2017**, *65*, 175–187. [CrossRef]
23. Zhang, Y.; Li, C.; Guo, B.; Guo, C.; Zhang, S. KDD: A Kernel Density Based Descriptor for 3D Point Clouds. *Pattern Recognit.* **2021**, *111*, 107691. [CrossRef]
24. Deng, H.; Birdal, T.; Ilic, S. PPFNet: Global Context Aware Local Features for Robust 3D Point Matching. In Proceedings of the 2018 IEEE/CVF Conference on Computer Vision and Pattern Recognition, Salt Lake City, UT, USA, 18–23 June 2018; pp. 195–205.
25. Deng, H.; Birdal, T.; Ilic, S. PPF-FoldNet: Unsupervised Learning of Rotation Invariant 3D Local Descriptors. In Proceedings of the European Conference on Computer Vision (ECCV), Munich, Germany, 8–14 September 2018; pp. 602–618.
26. Gojcic, Z.; Zhou, C.; Wegner, J.D.; Wieser, A. The Perfect Match: 3D Point Cloud Matching with Smoothed Densities. In Proceedings of the IEEE/CVF Conference on Computer Vision and Pattern Recognition, Long Beach, CA, USA, 15–20 June 2019; pp. 5545–5554.
27. Ao, S.; Hu, Q.; Yang, B.; Markham, A.; Guo, Y. SpinNet: Learning a General Surface Descriptor for 3D Point Cloud Registration. In Proceedings of the IEEE/CVF Conference on Computer Vision and Pattern Recognition, Nashville, TN, USA, 20–25 June 2021; pp. 11753–11762.
28. Vaswani, A.; Shazeer, N.; Parmar, N.; Uszkoreit, J.; Jones, L.; Gomez, A.N.; Kaiser, L.; Polosukhin, I. Attention Is All You Need. In Proceedings of the 31st Conference on Neural Information Processing Systems (NIPS 2017), Long Beach, CA, USA, 4–9 December 2017.
29. Huang, S.; Gojcic, Z.; Usvyatsov, M.; Wieser, A.; Schindler, K. PREDATOR: Registration of 3D Point Clouds with Low Overlap. In Proceedings of the IEEE/CVF Conference on Computer Vision and Pattern Recognition, Nashville, TN, USA, 20–25 June 2021; pp. 4267–4276.
30. Yu, H.; Li, F.; Saleh, M.; Busam, B.; Ilic, S. CoFiNet: Reliable Coarse-to-Fine Correspondences for Robust Point Cloud Registration. *Adv. Neural Inf. Process. Syst.* **2021**, *34*, 23872–23884.
31. Qin, Z.; Yu, H.; Wang, C.; Guo, Y.; Peng, Y.; Xu, K. Geometric Transformer for Fast and Robust Point Cloud Registration. In Proceedings of the IEEE/CVF Conference on Computer Vision and Pattern Recognition, New Orleans, LA, USA, 18–24 June 2022; pp. 11143–11152.
32. Li, J.; Hu, Q.; Ai, M. Point Cloud Registration Based on One-Point RANSAC and Scale-Annealing Biweight Estimation. *IEEE Trans. Geosci. Remote Sens.* **2021**, *59*, 9716–9729. [CrossRef]
33. Yang, H.; Shi, J.; Carlone, L. TEASER: Fast and Certifiable Point Cloud Registration. *IEEE Trans. Robot.* **2020**, *37*, 314–333. [CrossRef]

34. Li, J.; Zhong, R.; Hu, Q.; Ai, M. Feature-Based Laser Scan Matching and Its Application for Indoor Mapping. *Sensors* **2016**, *16*, 1265 [CrossRef]
35. Zhou, Q.-Y.; Park, J.; Koltun, V. Fast Global Registration. In *Computer Vision–ECCV 2016*; Leibe, B., Matas, J., Sebe, N., Welling, M., Eds.; Lecture Notes in Computer Science; Springer International Publishing: Cham, Switzerland, 2016; Volume 9906, pp. 766–782
36. Li, J.; Zhao, P.; Hu, Q.; Ai, M. Robust Point Cloud Registration Based on Topological Graph and Cauchy Weighted l q -Norm. *ISPRS J. Photogramm. Remote Sens.* **2020**, *160*, 244–259. [CrossRef]
37. Lusk, P.C.; Fathian, K.; How, J.P. CLIPPER: A Graph-Theoretic Framework for Robust Data Association. In Proceedings of the 2021 IEEE International Conference on Robotics and Automation (ICRA), Xi'an, China, 30 May–5 June 2021; pp. 13828–13834.
38. Yang, H.; Carlone, L. A Quaternion-Based Certifiably Optimal Solution to the Wahba Problem With Outliers. In Proceedings of the 2019 IEEE/CVF International Conference on Computer Vision (ICCV), Seoul, Republic of Korea, 27 October–2 November 2019; pp. 1665–1674.
39. Zhang, X.; Yang, J.; Zhang, S.; Zhang, Y. 3D Registration with Maximal Cliques. In Proceedings of the IEEE/CVF Conference on Computer Vision and Pattern Recognition, Vancouver, BC, Canada, 17–24 June 2023; pp. 17745–17754.
40. Li, J. A Practical O(N2) Outlier Removal Method for Point Cloud Registration. *IEEE Trans. Pattern Anal. Mach. Intell.* **2021**, *44*, 3926–3939.
41. Yao, R.; Du, S.; Cui, W.; Ye, A.; Wen, F.; Zhang, H.; Tian, Z.; Gao, Y. Hunter: Exploring High-Order Consistency for Point Cloud Registration With Severe Outliers. *IEEE Trans. Pattern Anal. Mach. Intell.* **2023**, *45*, 14760–14776. [CrossRef]
42. Yan, L.; Wei, P.; Xie, H.; Dai, J.; Wu, H.; Huang, M. A New Outlier Removal Strategy Based on Reliability of Correspondence Graph for Fast Point Cloud Registration. *IEEE Trans. Pattern Anal. Mach. Intell.* **2022**, *45*, 7986–8002.
43. Li, R.; Yuan, X.; Gan, S.; Bi, R.; Gao, S.; Luo, W.; Chen, C. An Effective Point Cloud Registration Method Based on Robust Removal of Outliers. *IEEE Trans. Geosci. Remote Sens.* **2024**, *62*, 1–16. [CrossRef]
44. Chum, O.; Matas, J.; Kittler, J. Locally Optimized RANSAC. In *Pattern Recognition*; Michaelis, B., Krell, G., Eds.; Lecture Notes in Computer Science; Springer: Berlin/Heidelberg, Germany, 2003; Volume 2781, pp. 236–243.
45. Barath, D.; Matas, J.; Noskova, J. MAGSAC: Marginalizing Sample Consensus. In Proceedings of the 2019 IEEE/CVF Conference on Computer Vision and Pattern Recognition (CVPR), Long Beach, CA, USA, 15–20 June 2019; pp. 10189–10197.
46. Cavalli, L.; Barath, D.; Pollefeys, M.; Larsson, V. Consensus-Adaptive RANSAC. *arXiv* **2023**, arXiv:2307.14030.
47. Torr, P.H.S.; Zisserman, A. MLESAC: A New Robust Estimator with Application to Estimating Image Geometry. *Comput. Vis. Image Underst.* **2000**, *78*, 138–156. [CrossRef]
48. Wu, Y.; Miao, Q.; Ma, W.; Gong, M.; Wang, S. PSOSAC: Particle Swarm Optimization Sample Consensus Algorithm for Remote Sensing Image Registration. *IEEE Geosci. Remote Sensing Lett.* **2018**, *15*, 242–246. [CrossRef]
49. Li, J.; Hu, Q.; Ai, M. GESAC: Robust Graph Enhanced Sample Consensus for Point Cloud Registration. *ISPRS J. Photogramm. Remote Sens.* **2020**, *167*, 363–374. [CrossRef]
50. Sun, L. ICOS: Efficient and Highly Robust Rotation Search and Point Cloud Registration with Correspondences. *arXiv* **2021**, arXiv:2104.14763.
51. Cheng, Y.; Huang, Z.; Quan, S.; Cao, X.; Zhang, S.; Yang, J. Sampling Locally, Hypothesis Globally: Accurate 3D Point Cloud Registration with a RANSAC Variant. *Vis. Intell.* **2023**, *1*, 20. [CrossRef]
52. Gentner, M.; Kumar Murali, P.; Kaboli, M. GMCR: Graph-Based Maximum Consensus Estimation for Point Cloud Registration. In Proceedings of the 2023 IEEE International Conference on Robotics and Automation (ICRA), London, UK, 29 May 2023; pp. 4967–4974.
53. Chung, K.-L.; Chang, W.-T. Centralized RANSAC-Based Point Cloud Registration With Fast Convergence and High Accuracy. *IEEE J. Sel. Top. Appl. Earth Obs. Remote Sens.* **2024**, *17*, 5431–5442. [CrossRef]
54. Horn, B.K.P. Closed-Form Solution of Absolute Orientation Using Unit Quaternions. *J. Opt. Soc. Am. A* **1987**, *4*, 629. [CrossRef]
55. Arun, K.S.; Huang, T.S.; Blostein, S.D. Least-squares fitting of two 3-d point sets. *IEEE Trans. Pattern Anal. Mach. Intell.* **1987**, *5*, 698–700. [CrossRef]
56. Han, W.; Tat-Jun, C.; Suter, D. Simultaneously Fitting and Segmenting Multiple-Structure Data with Outliers. *IEEE Trans. Pattern Anal. Mach. Intell.* **2012**, *34*, 1177–1192.
57. Lai, T.; Wang, H.; Yan, Y.; Chin, T.-J.; Zheng, J.; Li, B. Accelerated Guided Sampling for Multistructure Model Fitting. *IEEE Trans. Cybern.* **2020**, *50*, 4530–4543. [CrossRef] [PubMed]
58. Zeng, A.; Song, S.; Niessner, M.; Fisher, M.; Xiao, J.; Funkhouser, T. 3DMatch: Learning Local Geometric Descriptors from RGB-D Reconstructions. In Proceedings of the 2017 IEEE Conference on Computer Vision and Pattern Recognition (CVPR), Honolulu, HI, USA, 21–26 July 2017; pp. 199–208.
59. Geiger, A.; Lenz, P.; Urtasun, R. Are We Ready for Autonomous Driving? The KITTI Vision Benchmark Suite. In Proceedings of the 2012 IEEE Conference on Computer Vision and Pattern Recognition, Providence, RI, USA, 16–21 June 2012; pp. 3354–3361.
60. Besl, P.J.; McKay, N.D. A Method for Registration of 3-D Shapes. *IEEE Trans. Pattern Anal. Mach. Intell.* **1992**, *14*, 239–256. [CrossRef]
61. Krishnamurthy, V.; Levoy, M. Fitting smooth surfaces to dense polygon meshes. In Proceedings of the 23rd Annual Conference on Computer Graphics and Interactive Techniques, New Orleans, LA, USA, 4–9 August 1996; pp. 313–324.
62. Wei, P.; Yan, L.; Xie, H.; Huang, M. Automatic Coarse Registration of Point Clouds Using Plane Contour Shape Descriptor and Topological Graph Voting. *Autom. Constr.* **2022**, *134*, 104055. [CrossRef]

63. Yang, J.; Xiao, Y.; Cao, Z.; Yang, W. Ranking 3D Feature Correspondences via Consistency Voting. *Pattern Recognit. Lett.* **2019**, *117*, 1–8. [CrossRef]
64. Sipiran, I.; Bustos, B. Harris 3D: A Robust Extension of the Harris Operator for Interest Point Detection on 3D Meshes. *Vis. Comput.* **2011**, *27*, 963–976. [CrossRef]
65. Bai, X.; Luo, Z.; Zhou, L.; Chen, H.; Li, L.; Hu, Z.; Fu, H.; Tai, C.-L. PointDSC: Robust Point Cloud Registration Using Deep Spatial Consistency. In Proceedings of the 2021 IEEE/CVF Conference on Computer Vision and Pattern Recognition (CVPR), Nashville, TN, USA, 20–25 June 2021; pp. 15854–15864.
66. Choy, C.; Park, J.; Koltun, V. Fully Convolutional Geometric Features. In Proceedings of the 2019 IEEE/CVF International Conference on Computer Vision (ICCV), Seoul, Republic of Korea, 27 October–2 November 2019; pp. 8957–8965.

remote sensing

MDPI

Article

A Building Point Cloud Extraction Algorithm in Complex Scenes

Zhonghua Su [1,2], Jing Peng [3], Dajian Feng [3], Shihua Li [2], Yi Yuan [2] and Guiyun Zhou [2,*]

1 School of Computer and Software Engineering, Xihua University, Chengdu 610039, China; suzhonghua@mail.xhu.edu.cn
2 School of Resources and Environment, University of Electronic Science and Technology of China, Chengdu 611731, China; lishihua@uestc.edu.cn (S.L.); yanyee@std.uestc.edu.cn (Y.Y.)
3 Communications and Information Technology Headquarters, Sichuan Provincial Public Security Department, Chengdu 610041, China; pengjing855@foxmail.com (J.P.); fengdajian56599@foxmail.com (D.F.)
* Correspondence: zhouguiyun@uestc.edu.cn

Abstract: Buildings are significant components of digital cities, and their precise extraction is essential for the three-dimensional modeling of cities. However, it is difficult to accurately extract building features effectively in complex scenes, especially where trees and buildings are tightly adhered. This paper proposes a highly accurate building point cloud extraction method based solely on the geometric information of points in two stages. The coarsely extracted building point cloud in the first stage is iteratively refined with the help of mask polygons and the region growing algorithm in the second stage. To enhance accuracy, this paper combines the Alpha Shape algorithm with the neighborhood expansion method to generate mask polygons, which help fill in missing boundary points caused by the region growing algorithm. In addition, this paper performs mask extraction on the original points rather than non-ground points to solve the problem of incorrect identification of facade points near the ground using the cloth simulation filtering algorithm. The proposed method has shown excellent extraction accuracy on the Urban-LiDAR and Vaihingen datasets. Specifically, the proposed method outperforms the PointNet network by 20.73% in precision for roof extraction of the Vaihingen dataset and achieves comparable performance with the state-of-the-art HDL-JME-GGO network. Additionally, the proposed method demonstrated high accuracy in extracting building points, even in scenes where buildings were closely adjacent to trees.

Keywords: building; point cloud; geometric information

Citation: Su, Z.; Peng, J.; Feng, D.; Li, S.; Yuan, Y.; Zhou, G. A Building Point Cloud Extraction Algorithm in Complex Scenes. *Remote Sens.* **2024**, *16*, 1934. https://doi.org/10.3390/rs16111934

Academic Editor: Massimiliano Pepe

Received: 6 April 2024
Revised: 8 May 2024
Accepted: 16 May 2024
Published: 28 May 2024

1. Introduction

Building points are widely used in a variety of fields, including urban planning, cultural preservation, and disaster management, due to their capacity to capture detailed geometric features [1,2]. With the rapid development of cities, the surrounding environment of buildings has become complicated, making accurate building extraction a difficult task [3–6].

Building point cloud extraction methods can be classified into two categories based on data sources: single-source methods and multi-source methods. Single-source methods only use LiDAR data to extract building points. Zou et al. [7] proposed an adaptive strips approach for extracting buildings, which used adaptive-weight polynomials to classify each point and extract the edge points of buildings based on the regional clustering relationship among the points. This method only utilized the three-dimensional coordinate values of LiDAR data without the need for other auxiliary information to successfully identify buildings. Huang et al. [8] developed a top-down method based on the object entity to extract building points. Ground points were separated from non-ground points, and non-ground points were split to identify smooth zones. The building regions were then distinguished from smooth regions by top-level processing using their geometric and penetrating properties. Lastly, employing topological, geometric, and penetrating properties, the down-level processing was used to eliminate non-building points surrounding structures from each building region. The method produced good results in terms of area-based and object-based quality. Hui et al. [9] developed a multi-constraint graph segmentation method that converted point-based building extraction into object-based building extraction through multi-constraint graph segmentation and then utilized the spatial geometry information of objects and a multi-scale progressive growth algorithm to obtain building points. These methods perform well in extracting buildings in general urban environments and enable automated building recognition. However, when dealing with tree points closely attached to buildings, there is a possibility of misclassifying them as buildings.

The multi-source methods integrate LiDAR, aerial images, and ground planning maps into building point extraction, typically employing traditional and deep learning techniques. In the traditional technique, Qin and Fang [10] proposed a hierarchical building extraction method from high-resolution multispectral aerial images and Digital Surface Model (DSM) data. The method began with shadow detection using the morphological index, followed by the calculation of NDVI for correction. Subsequently, the top-hat reconstruction of DSM was combined with the NDVI to create the initial building mask data. Finally, the extracted building data was optimized using graph segmentation based on an improved super-pixel method. Acar et al. [11] introduced a building roof extraction algorithm that incorporated multiple data sources. Initially, the NDVI was calculated using spectral information, followed by applying a threshold to distinguish between vegetation and non-vegetation data. Subsequently, the Triangular Mesh Progressive Encoder Filter algorithm was employed to separate ground data. Lastly, the random sample consensus algorithm was utilized to extract the planar information of buildings. The algorithm achieved an average accuracy of 95%, completeness of 98%, and quality of 93%. Hron and Halounová [12] introduced a method for autonomously creating topologically correct roof-building models using building footprints and vertical aerial images. The method enabled the detection and categorization of roof edges in orthophotos by leveraging spatial relationships and height data from a digital surface model. This strategy enabled buildings with complicated designs to be divided into small portions that could be treated separately.

In the deep learning technique, Ghamisi et al. [13] proposed a fusion approach that combines extinction curves and convolutional neural networks for spectral-spatial classification of LiDAR and hyperspectral data. Firstly, extinction curves were extracted from different attributes to capture elevation and spatial information from both LiDAR and hyperspectral data. Afterwards, the extracted features were merged through either feature concatenation or graph feature fusion. Finally, the merged features were fed into a deep learning-based classifier for generating classification maps. Using optical imagery and unregistered airborne LiDAR data, Nguyen et al. [14] proposed an unsupervised and fully autonomous snake model without manual beginning points or training data to extract buildings. It was demonstrated that the method could recover buildings of different colors from intricate surroundings with a high degree of overall accuracy. Yuan et al. [15] proposed an end-to-end fully convolutional neural model based on residual networks for handling high-resolution aerial imagery and LiDAR data. The residual network effectively extracted high-level features, thus reducing the performance degradation associated with increasing network depth. The network demonstrated excellent performance, achieving an IoU of 93.19% and an OA of 97.56% on the WHU dataset and an IoU of 94.72% and an OA of 97.84% on the Boston dataset.

Combining LiDAR with aerial images and other data can significantly enhance the accuracy of building extraction. However, it is still challenging to combine data from different sources into the same reference coordinate system.

To improve building extraction accuracy, this paper proposes a highly accurate building point cloud extraction method based solely on the geometric information of the points. The method is divided into two stages: coarse extraction and fine extraction. In the coarse extraction stage, a coarsely extracted building point cloud is obtained using the cloth simulation filtering algorithm and the region growing algorithm. In the fine extraction stage, the coarsely extracted building point cloud is iteratively refined using mask polygons and the region growing algorithm. This step-by-step refinement process allows for a more accurate extraction of the building point cloud. The proposed method is evaluated on the Urban-LiDAR and Vaihingen datasets, demonstrating excellent extraction accuracy. The main contributions of this paper are summarized as follows:

1. This paper combines the Alpha Shape algorithm with the neighborhood expansion method to compensate for the shortcomings of the region growing algorithm in the coarse extraction stage, thereby obtaining more complete building points.
2. To address the issue of misidentifying facade points near the ground, we perform mask extraction on the original points instead of non-ground points. This approach allows us to obtain more comprehensive facade points within the mask polygons compared to the ones obtained using the cloth simulation filtering algorithm.
3. Even in cases where buildings are closely adjacent to trees, the proposed method can successfully separate and extract building points from tree points, thereby improving accuracy and reliability.

2. Methods

This section introduces the proposed method for building extraction in complex scenes in detail. Our method is divided into two stages, namely coarse extraction and fine extraction, to achieve accurate extraction of the building point cloud.

In the coarse extraction stage of the building point cloud, our proposed method identifies non-ground points in the point cloud using the cloth simulation filtering (CSF) algorithm and uses a region growing algorithm to obtain the coarse extraction of the building point cloud. At this stage, the region growing algorithm may fail to identify some building boundary points.

In the fine extraction stage of the building point cloud, our proposed method obtains mask polygons based on the coarsely extracted building points by applying the Alpha Shape algorithm and the neighborhood expansion method. The building point cloud is enlarged and replaced by non-ground points within mask polygons. Discrete tree points are removed from the building point cloud using the region growing algorithm and the Euclidean clustering algorithm. The building point cloud is then upgraded by merging with the facade point cloud near the ground. Noise points are removed using the radius filtering algorithm to obtain the final building point cloud. The detailed workflow and visualization flowchart for the building point cloud extraction are shown in Figures 1 and 2.

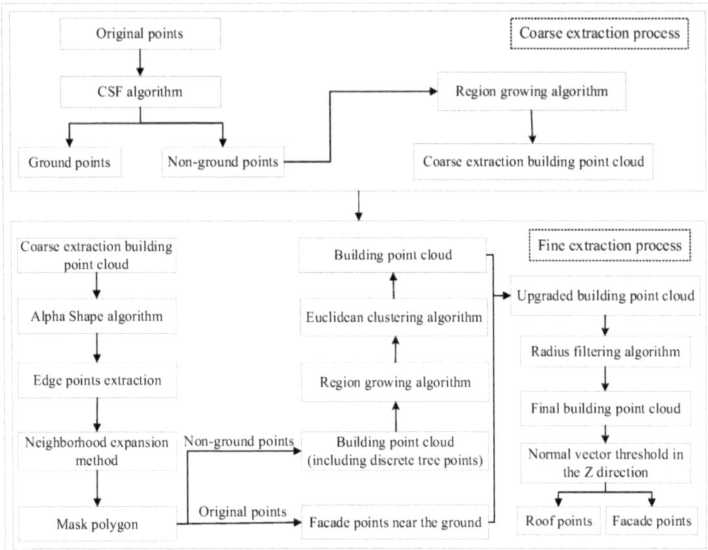

Figure 1. Workflow of the building point cloud extraction.

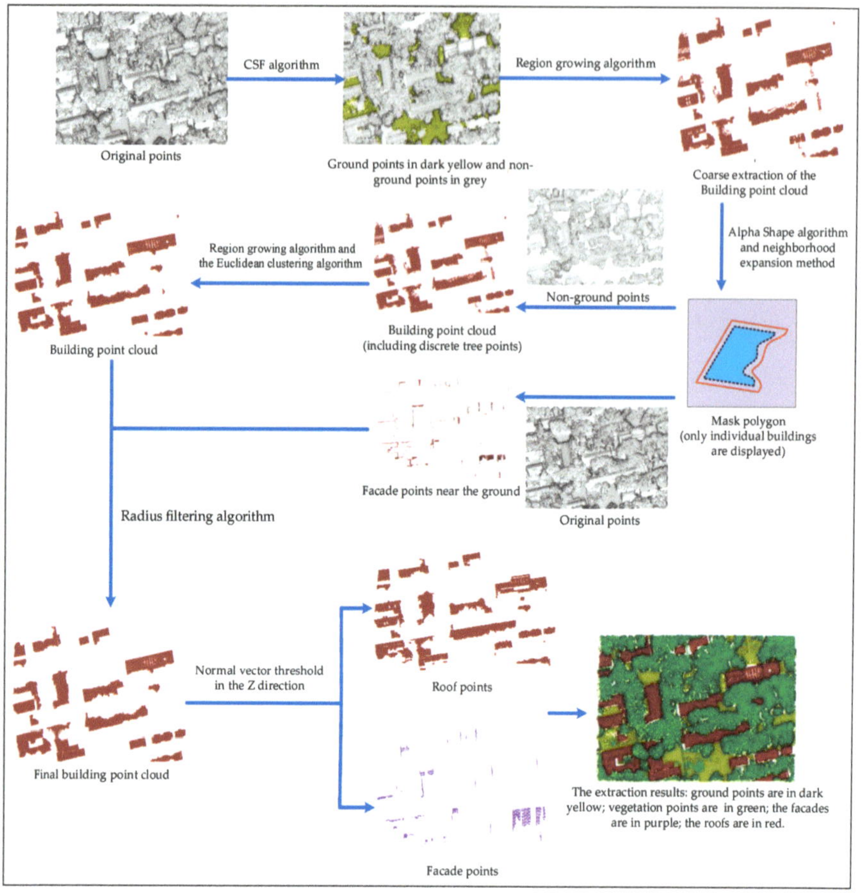

Figure 2. Visualization flowchart of the building point cloud extraction.

2.1. Coarse Extraction of the Building Point Cloud

Due to the large terrain undulations and uneven density distribution of points, traditional filtering algorithms have difficulty obtaining high-accuracy non-ground points. In order to remove ground points with high accuracy, this paper uses the CSF algorithm to separate non-ground points from ground points.

The basic idea of the CSF algorithm is to invert the original points and use a cloth model composed of spring-connected cloth particles to simulate the filtering process [16]. The position of particles on grid nodes in space determines the shape of the fabric [17]. According to Newton's Second Law, the relationship between particle position and force can be expressed as follows [18]:

$$m\frac{\partial X(t)}{\partial t^2} = F_e(X,t) + F_i(X,t), \tag{1}$$

where m is the mass of the particle. $X(t)$ is the position of the particle at time t. $F_e(X,t)$ is the external force on the particle. $F_i(X,t)$ is the internal force of the particle at position X at time t.

According to Equation (1), we first only calculate the influence of gravity on each particle, resulting in the position of each particle [18]:

$$X(t+\Delta t) = 2X(t) - X(t-\Delta t) + \frac{G}{m}\Delta t^2, \quad (2)$$

where G is the gravity. $X(t)$ is the position of the particle at time t, and Δt is the step length of time.

Next, consider the internal forces between particles to limit their displacement in the void area of the inverted points. The displacement of each particle is calculated as follows [18]:

$$\vec{d} = \frac{1}{2}b\left(\vec{p_k} - \vec{p_0}\right)\cdot\vec{n}, k = 1,2,3,\ldots \quad (3)$$

where $\vec{d}$ is the displacement vector of particles. b is a parameter that determines whether a particle can move ($b = 1$ indicates it can move; $b = 0$ indicates it cannot move); p_k is the position of adjacent particles of p_0. $\vec{n} = (0,0,1)^T$.

Finally, the relative position of particles is adjusted based on the internal forces between them and the fabric stiffness parameters. If the distance between the actual point and the simulated particles is less than the pre-set threshold, it is considered a ground point; otherwise, it is considered a non-ground point (Figure 3).

Figure 3. The point cloud is divided into ground points and non-ground points using the CSF filtering algorithm (ground points are displayed in dark yellow, and non-ground points are displayed in blue).

After identifying non-ground points in the point cloud, we use the region growing algorithm to obtain the coarse extraction of the building point cloud from non-ground points. The algorithm selects the point with the minimum curvature as the initial seed point. Given a neighboring point A of a seed point B, if the angle between the normal vector of A ($N_{neighbor}$) and that of B(N_{seed}) is less than a given threshold θ (Equation (4)) and the curvature value of A($\sigma_{neighbor}$) is less than a given threshold value σ (Equation (5)), point A is considered a new seed point. The region continues to grow until all points are processed (Figure 4) [19].

$$\mathrm{acos}\left(\frac{N_{seed}}{||N_{seed}||}\cdot\frac{N_{neighbor}}{||N_{neighbor}||}\right) < \theta, \quad (4)$$

$$\sigma_{neighbor} < \sigma. \quad (5)$$

Here, θ and σ are usually small enough to avoid incorrectly identifying non-building points that are approximately planar as building points. In this case, the region growing algorithm may fail to extract some building boundary points due to the large angles between the local normal vectors of adjacent points (Figure 5b).

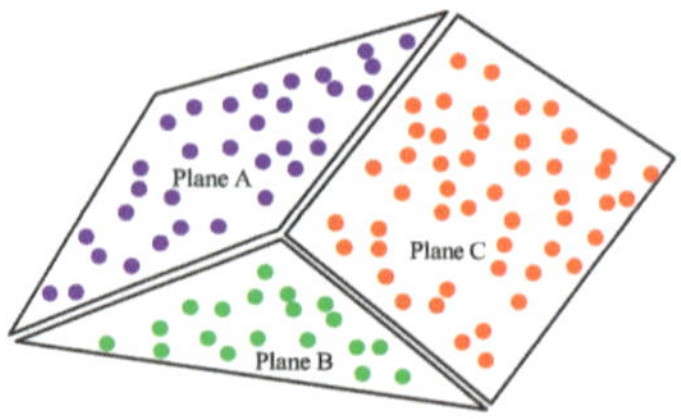

Figure 4. Plane segmentation results using the region growing algorithm.

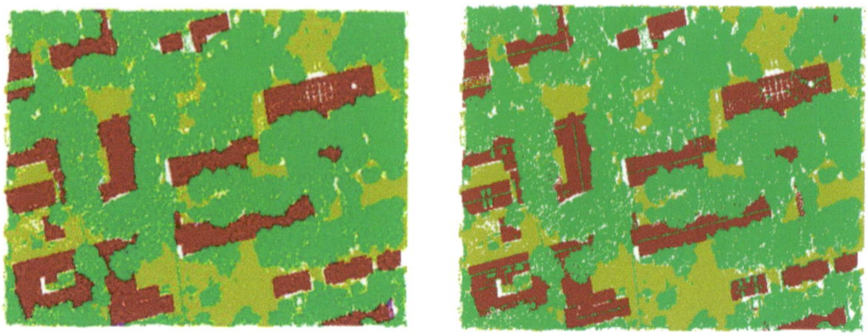

Figure 5. (**a**) Ground truth; (**b**) coarse extraction results using the region growing algorithm, with buildings in red, trees in green, and ground points in dark yellow.

2.2. Fine Extraction of the Building Point Cloud

Considering that the region growing algorithm may fail to include the boundary points of the buildings during the coarse extraction stage, the building point cloud is enlarged and replaced with the help of mask polygons.

In this paper, mask polygons are used to identify the points located within them. To obtain mask polygons, we first project the coarsely extracted building point cloud onto the XOY plane. Then, we use the Alpha Shape algorithm [20] to extract edge points from the projected points and finally extend the edge points through the neighborhood expansion method based on corresponding multipliers.

Mask polygons are extracted in the following steps (Figure 6):

(1) All possible pairs of projected points are processed in the same way. For any pair of points $P_1(x_1, y_1)$ and $P_2(x_2, y_2)$ from projected point cloud on the XOY plane of point cloud S, the center point $P_3(x_3, y_3)$ of the circle whose distance from is calculated and is equal to α based on the distance intersection method (Figure 7) [21]:

$$\begin{cases} x_3 = x_1 + \frac{1}{2}(x_2 - x_1) + H(y_2 - y_1) \\ y_3 = y_1 + \frac{1}{2}(y_2 - y_1) + H(x_2 - x_1) \end{cases}, \quad (6)$$

where

$$\begin{cases} H = \sqrt{\frac{\alpha^2}{S^2_{P_1P_2}} - \frac{1}{4}} \\ S^2_{p_1p_2} = (x_1 - x_2)^2 + (y_1 - y_2)^2 \end{cases}. \quad (7)$$

(2) The distance d between each point in S and P_3 is calculated. If d is less than α, the point is considered to be inside the circle; otherwise, it is deemed to be outside the circle. If there are P_1 and P_2 such that there are no other points inside the circle, then P_1 and P_2 are defined as edge points, and P_1P_2 is defined as a boundary line. The edge points are obtained until all point pairs in S have been processed.

(3) The centroid coordinates Cen_{point} of all edge points and the distance Dis_{point} from each edge point to the Cen_{point}, as well as the direction vector Dir_{point} from the Cen_{point} to

each edge point, are calculated. *Multi* refers to the corresponding multipliers. The expanded corresponding edge point Exp_{point} is as follows:

$$\text{Exp}_{point} = \text{Cen}_{point} + \text{Dir}_{point} \times Multi \times \text{Dis}_{point}. \tag{8}$$

(4) Edge points are sorted based on the polar angles between adjacent points and connect them to form a closed polygon for extracting points within the polygon.

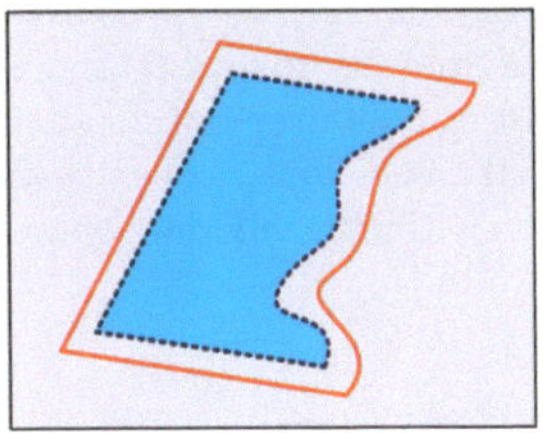

Figure 6. Mask polygon extraction using a combination of the Alpha Shape algorithm and neighborhood expansion method.

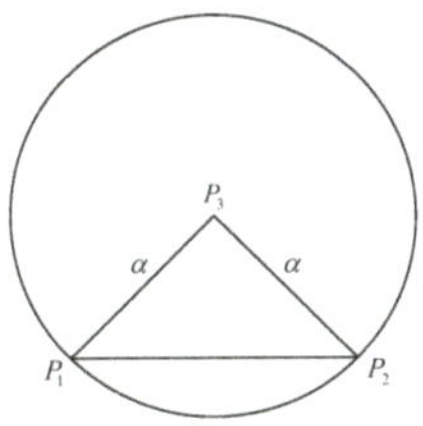

Figure 7. Calculation of the center coordinates of a circle based on the distance intersection method.

The steps for connecting edge points are as follows: First, the center point of all edge points is calculated. Then, the edge points are sorted based on their polar angles relative to the center point in a counterclockwise direction in ascending order. Finally, all edge points are connected in counterclockwise order to create a closed polygon (Figure 8).

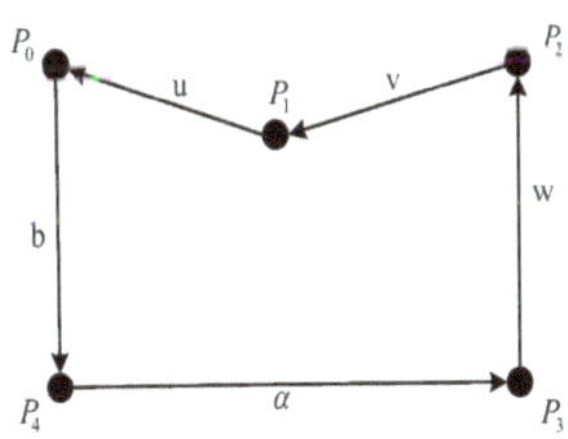

Figure 8. Polygonal connection based on the polar angles.

After the mask polygons are obtained based on the coarsely extracted building point cloud, the building point cloud is enlarged and replaced by all non-ground points within the mask polygons. Due to the possibility of adding certain tree points to the building point cloud, we use the region growing algorithm and the Euclidean clustering algorithm [22] to filter out some discrete tree points from the building point cloud.

The specific operation process of the Euclidean clustering algorithm is as follows:

(1) The K nearest neighbor points for any point P in space are found using the KD-Tree nearest neighbor search algorithm.

(2) For the K nearest neighbor points, the Euclidean distance between each point and P is calculated.

(3) If there are points within the K nearest neighbors that have a distance smaller than the set threshold, these points are clustered into a set Q.
(4) The above process is repeated until the number of elements in set Q no longer increases.

At this stage, the threshold values for the normal vector and curvature in the region growing algorithm are relatively large to include the boundary points of the buildings.

Subsequently, the building point cloud is upgraded by merging with the façade point cloud near the ground, which is obtained by conducting mask extraction on the original points instead of non-ground points and setting appropriate values for the Z-axis to adjust the height to a certain distance from the ground (Figure 9). Given that the façade point cloud may overlap with the existing building point cloud, the duplicate points are removed from the merged building point cloud. Finally, we use the radius filtering algorithm to remove the discrete noise points within the building point cloud.

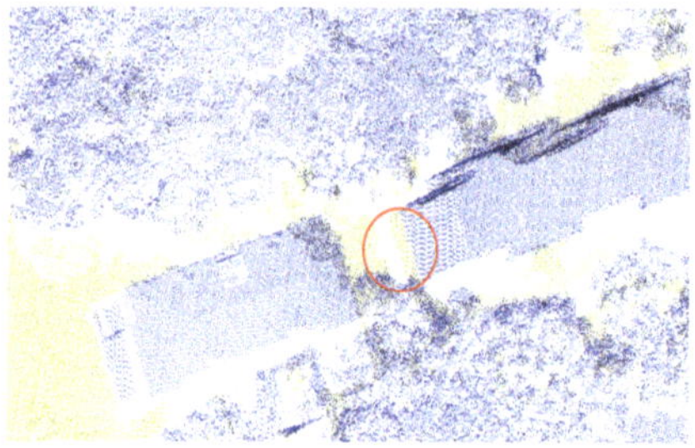

Figure 9. Misclassification of building points using the CSF algorithm within the red circle, with ground points in dark yellow and non-ground points in blue.

The main idea of the radius filtering algorithm is to assume that each point in the original points contains at least a certain number of neighboring points within a specified radius neighborhood [23]. When this assumption is satisfied, the point is considered a valid point and retained. On the contrary, if the conditions are not met, it will be identified as a noise point and removed. As an example, Figure 10 specifies a radius of d. If at least one adjacent point is specified within this radius, only the blue points in the figure will be removed from the point cloud. If at least two adjacent points are specified within the radius, both the purple and black points will be removed.

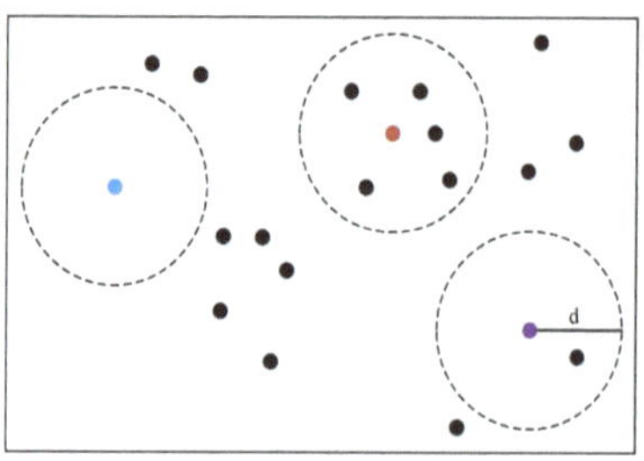

Figure 10. Radius filtering algorithm.

3. Experiment Settings

3.1. Study Areas

To evaluate the performance of our proposed method, we conducted experiments on two datasets: the Urban-LiDAR dataset (https://www.lidar360.com/ accessed on 2 May 2022) and the Vaihingen dataset (http://www2.isprs.org/ accessed on 7 April 2022). The Urban-LiDAR dataset consists of a total of 719,823 points. The dataset includes various types of objects, including buildings, trees, and ground points, as shown in Figure 11.

The terrain in this area has undergone significant changes, with dense vegetation and high buildings.

Figure 11. Urban-LiDAR dataset.

The Vaihingen dataset contains 411,722 points. The Vaihingen dataset is divided into two parts: Vaihi-1 and Vaihi-2, which have been processed separately in this paper, as shown in Figure 12 (displayed by elevation). In the Vaihingen dataset, non-ground points are composed of buildings, powerlines, low vegetation, cars, fences, hedges, shrubs, and trees; ground points are composed of impervious surfaces. The Vaihingen dataset is collected by the Leica ALS50 system with a point density of 4–8 m^{-2}. The terrain in this area is relatively flat, with sparse vegetation and low buildings.

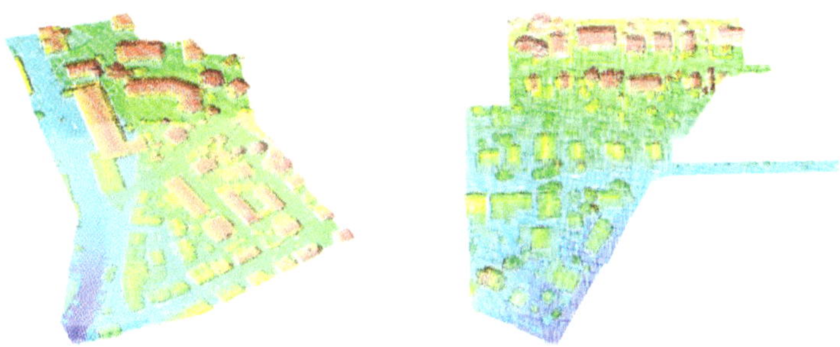

(**a**) (**b**)

Figure 12. Vaihingen dataset. (**a**) Vaihi-1 data; (**b**) Vaihi-2 data.

3.2. Parameter Settings

In the process of extracting the building point cloud, this paper involves some important algorithms, including the CSF algorithm, the region growing algorithm, and the Euclidean clustering algorithm. In this article, the parameters we set are mainly based on the density of points and terrain undulations. The specific parameter settings are shown in Table 1, where the parameter settings of the region growing algorithm are used for the coarse extraction stage of building points.

When using the CSF algorithm to separate ground and non-ground points, the following key parameters play an important role: (1) cloth_resolution represents the size of the terrain coverage grid, that is, the setting of the grid resolution, which affects the precision of generating a digital terrain model (DTM). A larger cloth resolution usually leads to a rougher DTM generated; (2) max_iterations represents the maximum number of iterations; (3) classification_threshold represents the distance threshold between the actual point and the simulated terrain, used to divide the point cloud into ground points and non-ground points.

Table 1. Parameter settings of some important algorithms.

Algorithm	Parameter	Urban-LiDAR	Vaihi-1	Vaihi-2
CSF algorithm	cloth_resolution	1.0	0.3	1.0
	max_iterations	500	500	500
	classification_thresold	2.0	1.5	2.2
Region growing algorithm	theta_threshold	5	30	10
	curvature_threshold	0.05	0.05	0.03
	neighbor_number	20	15	30
	min_pts_per_cluster	100	40	50
	max_pts_per_cluster	10,000	10,000	10,000
European clustering algorithm	tolerance	0.58	1.5	1.25
	min_cluster_size	80	180	180
	max_cluster_size	100,000	10,000	15,000

In the coarse extraction stage of the building point cloud, the region growing algorithm is used to extract building points from non-ground points. The region growing algorithm involves the following key parameters: (1) theta_threshold represents the smoothing threshold; (2) curvature_threshold represents the curvature threshold; (3) neighbor_number represents the number of neighborhood search points; (4) min_pts_per_cluster represents the minimum number of points for each cluster; and (5) max_pts_per_cluster represents the maximum number of points in each cluster.

When using the Euclidean clustering algorithm to filter discrete tree points and obtain building points, the Euclidean clustering algorithm involves several important parameters: (1) tolerance represents the search radius of nearest neighbor search, which is the minimum Euclidean distance between two different clusters; (2) min_cluster_size represents the minimum number of cluster points; (3) max_cluster_size represents the maximum number of cluster points.

3.3. Evaluation Indicators

This paper uses precision, recall, and the F1 score as evaluation indicators to verify the effectiveness of the proposed method in extracting building points.

Precision represents the proportion of correctly predicted building points to all predicted building points [24]:

$$\text{Precision} = \frac{\text{TP}}{\text{TP} + \text{FP}}, \tag{9}$$

Recall represents the proportion of correctly predicted building points to actual building points [24]:

$$\text{Recall} = \text{TP}/(\text{TP} + \text{FN}), \tag{10}$$

The F1 score is the weighted average of precision and recall, which is closer to the smaller value of precision and recall [24]:

$$\text{F1} = 2 \times \frac{\text{Precision} \times \text{Recall}}{\text{Precision} + \text{Recall}} \tag{11}$$

where TP represents the number of correctly predicted building points, FP represents that non-building points are incorrectly predicted as building points, and FN represents that building points are incorrectly predicted as non-building points.

3.4. Benchmark Algorithm

To verify the effectiveness of the proposed method, a manually interactive recognition of the building point cloud was used as a reference. In the Urban LiDAR dataset, this paper mainly analyzes the building point cloud obtained through manual interactive recognition. In the Vaihingen dataset, this paper compares the PointNet [25], PointNet++ [26], and

HDL-JME-GGO [27] networks with the proposed method. The PointNet, PointNet++, and HDL-JME-GGO networks estimate test data by learning from training data (Figure 13).

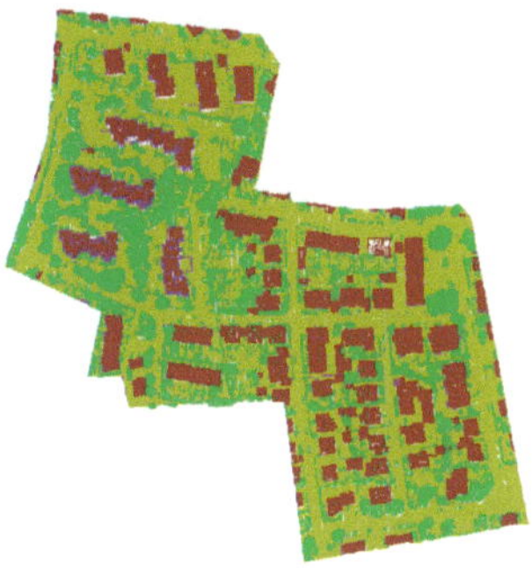

Figure 13. Training data. Ground points are in dark yellow; the facades are in purple; the roofs are in red; and other elements are in green.

The basic idea of the PointNet network is to utilize a multi-layer perceptron to capture the feature information of the point, followed by the use of maximum pooling to aggregate these point features into a global feature representation. The PointNet network is able to directly process unordered point cloud data without considering the order of points.

The PointNet++ network incorporates a hierarchical structure comprising a sampling layer, a grouping layer, and a feature extraction layer. This structure allows for the organization of each point and its surrounding neighborhood into local regions, which are then processed using the PointNet network to extract features from the corresponding point cloud. By employing this hierarchical structure, the network becomes capable of effectively learning local feature information as the context scale expands.

The HDL-JME-GGO network utilizes layered data to enhance deep feature learning using the PointNet++ network. It incorporates a joint learning method based on nonlinear manifolds to globally optimize and embed deep features into a low-dimensional space, taking into account the contextual information of spatial and deep features. It effectively addresses artifacts caused by partitioning and sampling in the processing of large-scale datasets. This network achieves global regularization by optimizing initial labels to ensure spatial regularity, resulting in locally continuous and globally optimal classification results.

4. Results

We evaluated the building extraction performance of the proposed method on the Urban-LiDAR dataset and the Vaihingen dataset. The building point cloud could be divided into two non-overlapping point clouds: the facade point cloud and the roof point cloud. The separation of facade points and roof points was achieved based on the normal vector threshold in the Z direction. The extraction results of the proposed method on Urban-LiDAR, Vaihi-1, and Vaihi-2 data are shown in Figures 14–16, respectively. It was evident from the figures that the proposed method achieved a high level of accuracy in extracting building points.

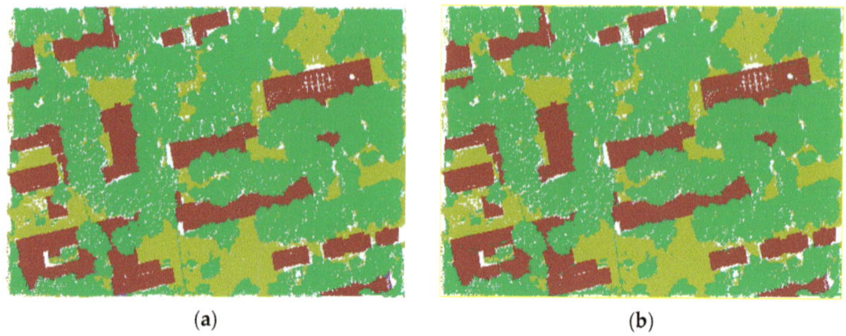

Figure 14. Urban-LiDAR's extraction results: ground points are in dark yellow; tree points are in green; the facades are in purple; the roofs are in red. (a) Ground truth; (b) the extraction results using the proposed method.

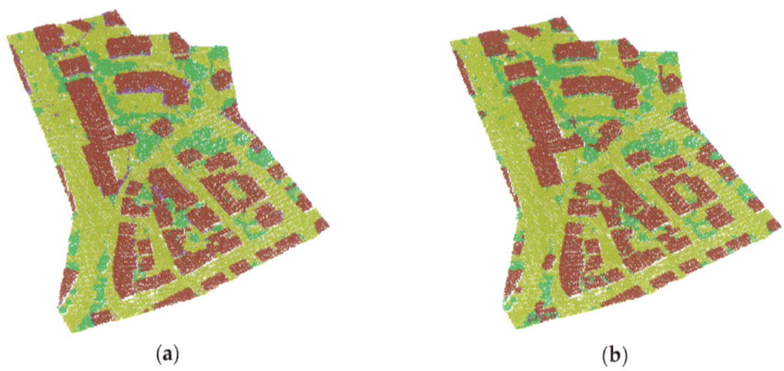

Figure 15. Vaihi-1's extraction results: ground points are in dark yellow; tree points are in green; the facades are in purple; the roofs are in red. (a) Ground truth; (b) the extraction results using the proposed method.

Figure 16. Vaihi-2's extraction results: ground points are in dark yellow; the facades are in purple; the roofs are in red. (a) Ground truth; (b) the extraction results using the proposed method.

5. Discussion

This paper evaluated the extraction results of the proposed method on Urban-LiDAR data, as shown in Table 2. For the roofs, the proposed method yielded a precision of 98.74%, a recall of 98.47%, and an F1 score of 98.60%. For the facades, the values were 97.98%, 70.94%, and 82.30%, respectively.

Table 2. Accuracy assessment of Urban-LiDAR's extraction (%).

	Precision	Recall	F1 Score
Roof	98.74	98.47	98.60
Façade	97.98	70.94	82.30

In addition, we analyzed the extraction accuracy of the roof in the Urban-LiDAR data. From Table 3, it can be seen that the highest precision, recall, and F1 scores all reached 100% (Roof 14 and Roof 29). The minimum accuracy rate of the roof was 79.57%, the recall was 89.13%, and the F1 score was 84.08% (Roof 28). The experimental results showed that the proposed method exhibited high accuracy and completeness in roof segmentation.

Table 3. Accuracy assessment of Urban-LiDAR's roof extraction (%).

ID	Precision	Recall	F1 Score
0	99.54	99.77	99.66
1	98.25	98.92	98.58
2	99.80	98.42	99.11
3	96.05	98.00	97.02
4	97.19	98.56	97.87
5	95.22	95.62	95.42
6	99.85	99.80	99.82
7	100	98.14	99.06
8	84.08	91.31	87.55
9	98.72	98.88	98.80
10	98.68	97.35	98.01
11	98.82	98.38	98.60
12	98.00	98.52	98.26
13	99.50	97.70	98.59
14	100	100	100
15	99.12	96.64	97.86
16	98.79	97.65	98.22
17	99.94	99.32	99.63
18	96.67	98.28	97.47
19	88.47	93.46	90.90
20	93.29	96.37	94.80
21	99.87	97.78	98.81
22	99.62	99.17	99.39
23	97.69	97.96	97.82
24	99.41	92.02	95.57
25	97.46	92.42	94.87
26	96.18	98.06	97.11
27	98.91	98.68	98.79
28	79.57	89.13	84.08
29	100	100	100
30	92.76	95.66	94.19

Although the CSF algorithm can effectively separate ground points from non-ground points, it may mistakenly identify façade points that are closer to the ground as ground points. To solve this difficult problem, this paper extracted masks based on original points rather than non-ground points and set appropriate values for the Z-axis to obtain the façade point cloud near the ground. Comparing Figure 17c, the façade points within mask polygons in the original points were more complete than those in the non-ground points acquired using the CSF algorithm.

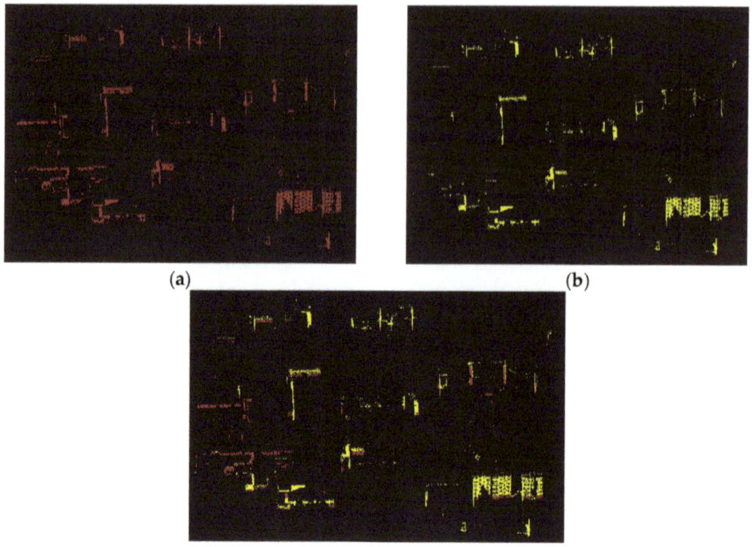

Figure 17. (**a**) Facade points within mask polygons in the original points; (**b**) the facade points within mask polygons in the non-ground points; (**c**) the overlay of (**a**,**b**).

In addition, we evaluated the effectiveness of building point extraction in two different scenes from the Urban-LiDAR dataset: a complex scene and a low-density scene. Figure 18c displayed the extracted building point cloud using the proposed method in the complex scene, and the precision, recall, and F1 score of the roof were 98.82%, 98.38%, and 98.60%, respectively. It demonstrated that the proposed method could extract building points accurately in the complex scene. Figure 19c shows the extraction results using the proposed method in the scene with low point density. The recall of the roof was only 92.02%, but the precision was 99.41%, and the F1 score was 95.57%. It could be seen that there were relatively dense points with significant fluctuations at the edges of the original points, and even if we used the region growing algorithm to process it, points at that location could still be lost.

Our proposed method is compared with three segmentation networks: PointNet, PointNet++, and HDL-JME-GGO on the Vaihingen dataset. The performance indicators are listed in Table 4. The proposed method performed outstandingly in roof extraction, achieving a precision 20.73% higher than that of the PointNet network. However, the F1 score of the proposed method was only lower by 0.28% compared to the HDL-JME-GGO network. For facade extraction, the precision of the proposed method was 49.63% higher than that of the PointNet network, 16.53% higher than that of the PointNet++ network, but only 3.87% lower than that of the HDL-JME-GGO network. While our proposed method achieved slightly lower accuracy than the HDL-JME-GGO network, it considerably outperformed the PointNet and PointNet++ networks in extracting building points based on geometric information.

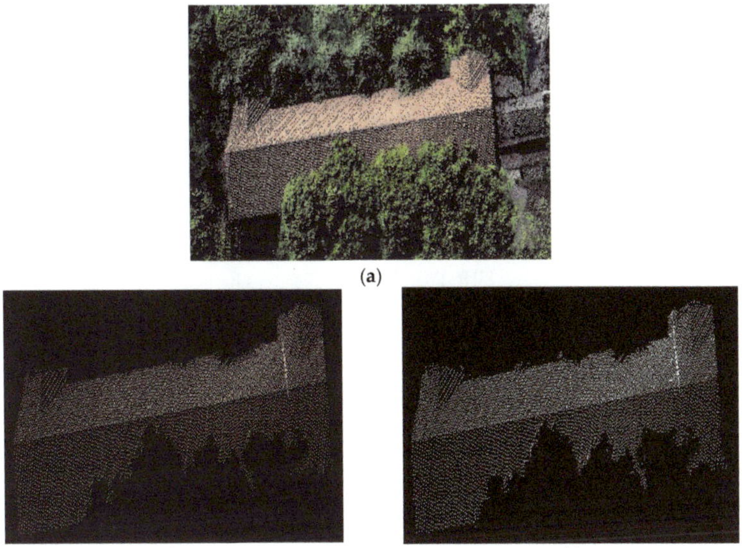

(a)

(b) (c)

Figure 18. Extraction of buildings results in complex scenes: (**a**) original data; (**b**) label data; (**c**) the extraction results of the building using the proposed method.

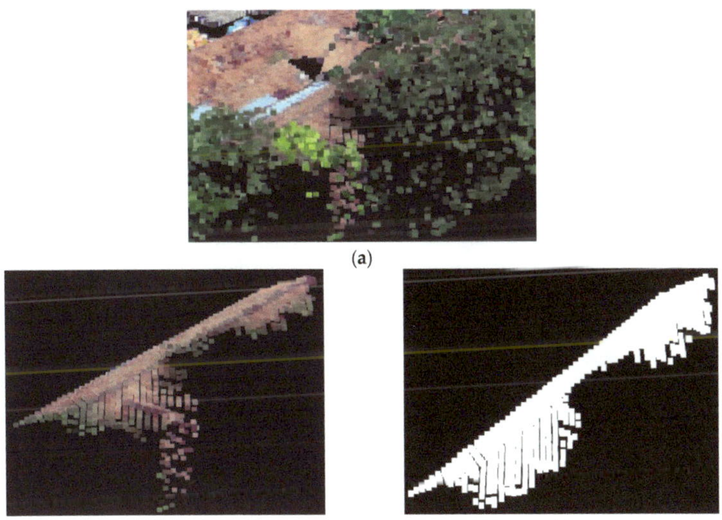

(a)

(b) (c)

Figure 19. Extraction of the buildings with low cloud density: (**a**) original point cloud; (**b**) manually delineated reference building points. The integration of texture information into data collected by unmanned aerial vehicles (UAVs) may introduce errors, as exemplified by the points highlighted in blue in the figure, which should ideally be categorized as building points; (**c**) the extracted building points using the proposed method.

Table 4. Accuracy assessment of Vaihingen's extraction (%).

Algorithm	Indicator	Roof	Facade
PointNet	Precision	73.0 (↑20.73)	10.7 (↑49.63)
	Recall	82.2	0.1
	F1 score	77.6	5.4
PointNet++	Precision	92.8	43.8 (↑16.53)
	Recall	81.0	38.3
	F1 score	86.9	41.0
HDL-JME-GGO	Precision	92.8	64.2 (↓3.87)
	Recall	89.3	24.2
	F1 score	91.1 (↓0.28)	44.2
The Proposed Method	Precision	93.73	60.33
	Recall	88.08	27.33
	F1 score	90.82	37.62

Because the Vaihingen dataset was composed of the Vaihi-1 point cloud and the Vaihi-2 point cloud, we conducted a detailed analysis of the extraction results on the two-point clouds. For roof extraction, the proposed method achieved precision, recall, and an F1 score of 91.49%, 92.32%, and 91.90% for the Vaihi-1 point cloud and 96.27%, 83.93%, and 89.68% for the Vaihi-2 point cloud, respectively (Table 5).

Table 5. Accuracy assessment of Vaihi-1 and Vaihi-2's extraction (%).

	Precision		Recall		F1 Score	
	Vaih-1	Vaih-2	Vaih-1	Vaih-2	Vaih-1	Vaih-2
Roof	91.49	96.27	92.32	83.93	91.90	89.68
Facade	58.33	61.45	17.77	38.36	27.24	47.23

Furthermore, we selected 21 buildings and analyzed the roof extraction accuracy for both the Vaihi-1 point cloud and the Vaihi-2 point cloud (Table 6). For the Vaihi-1 point cloud, the proposed method achieved the highest precision, recall, and F1 score, all reaching 100%. The proposed method yielded the lowest precision, recall, and F1 score at 71.91%, 81.51%, and 76.41%, respectively. Regarding the Vaihi-2 point cloud, the proposed method achieved the highest precision (99.90%), recall (98.39%), and F1 score (99.04%). Conversely, the proposed algorithm exhibited the lowest precision (86.80%), recall (55.14%), and F1 score (71.05%). These results indicate the proposed method's capability to achieve high-accuracy results in roof extraction.

Although the proposed method achieved high accuracy in extracting the Vaihi-1 point cloud and the Vaihi-2 point cloud, there were still some shortcomings. Due to the limitations of the CSF algorithm, it may have difficulty extracting certain roof points close to the ground, such as those points shown in the white circle in Figure 20b. In addition, it was difficult to extract building points solely based on geometric information for some roofs with significant undulations, as shown in the black circle of building points in Figures 20b and 21b.

Table 6. Accuracy assessment of Vaihi-1 and Vaihi-2's roof extraction (%).

ID	Precision		Recall		F1 Score	
Roof	Vaih-1	Vaih-2	Vaih-1	Vaih-2	Vaih-1	Vaih-2
0	100	86.80	100	91.83	100	89.24
1	88.94	98.04	93.87	90.77	91.34	94.27
2	100	92.65	99.80	92.45	99.90	92.55
3	97.83	97.91	99.77	95.02	98.79	96.44
4	99.45	99.90	99.73	94.43	99.59	97.09
5	97.75	99.88	97.61	78.13	97.68	87.68
6	99.39	94.78	95.46	84.80	97.39	89.51
7	100	99.88	95.58	55.14	97.74	71.05
8	99.02	99.41	99.18	68.67	99.10	81.23
9	71.91	99.72	81.51	94.67	76.41	97.13
10	98.52	99.29	98.89	94.40	98.70	96.78
11	100	97.21	100	93.43	100	95.28
12	98.14	99.70	86.17	98.38	91.77	99.04
13	98.18	96.57	96.83	97.70	97.50	97.13
14	98.96	99.55	95.65	98.31	97.28	98.93
15	99.43	99.84	99.15	87.41	99.29	93.21
16	100	99.07	99.76	98.05	99.88	98.56
17	99.29	99.23	99.29	90.44	99.29	94.63
18	100	99.16	89.25	98.39	94.32	98.77
19	96.92	97.68	98.43	91.75	97.67	94.62
20	97.35	96.92	96.89	97.55	97.12	97.23

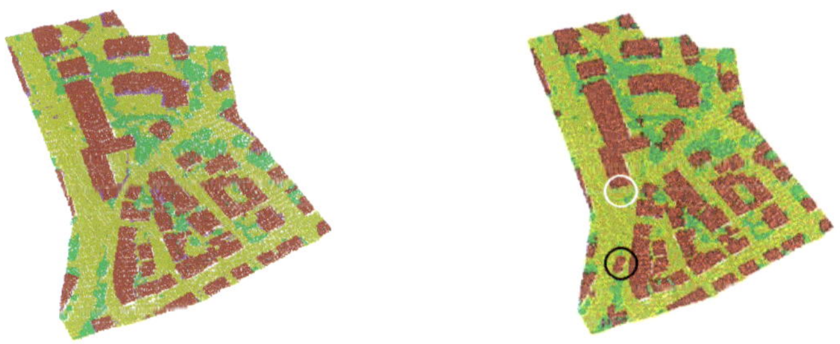

(a) (b)

Figure 20. Vaihi-1 data. (**a**) Label of Vaihi-1; (**b**) Vaihi-1's extraction results using the proposed method.

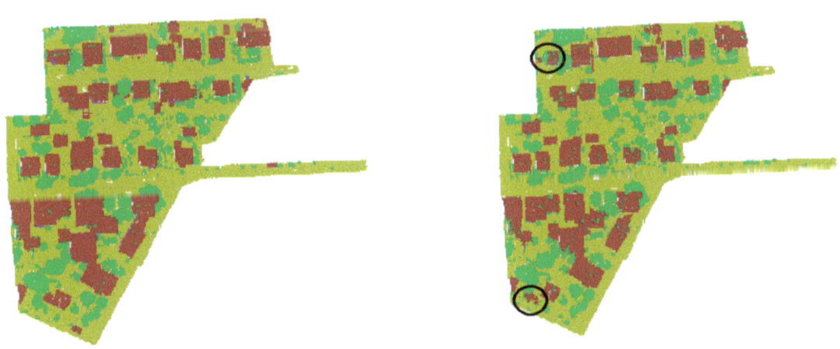

(a) (b)

Figure 21. Vaihi-2 data. (**a**) Label of Vaihi-2; (**b**) Vaihi-2's extraction results using the proposed method.

6. Conclusions

This paper proposes a highly accurate building point cloud extraction method based solely on the geometric information of points. The method is divided into two stages: coarse extraction and fine extraction. In the coarse extraction stage, a coarsely extracted building point cloud is obtained using the cloth simulation filtering algorithm and the region growing algorithm. In the fine extraction stage, the coarsely extracted building point cloud is iteratively refined using mask polygons and the region growing algorithm. The proposed method has shown excellent extraction accuracy on the Urban-LiDAR and Vaihingen datasets. On the Urban-LiDAR dataset, the method achieved a precision of 98.74%, a recall of 98.47%, and an F1 score of 98.60% for roof extraction. For facade extraction on the same dataset, the precision, recall, and F1 scores were 97.98%, 70.94%, and 82.30%, respectively. On the Vaihingen dataset, the proposed method outperformed the PointNet network by 20.73% in roof extraction precision and achieved comparable performance with the HDL-JME-GGO network. For facade extraction, the method surpassed the PointNet network by 49.63% in precision, the PointNet++ network by 16.53%, and fell slightly behind the HDL-JME-GGO network by only 3.87%. Additionally, the proposed method can still extract building points with high accuracy, even in cases where buildings are closely adjacent to trees. However, relying solely on geometric information for building extraction may face significant challenges for roofs with significant fluctuations or in situations where point density is low. We will introduce more feature information, such as color or texture, to enhance the ability to extract buildings, thereby achieving more accurate and complete building extraction in the future.

Author Contributions: Z.S., J.P., D.F. and G.Z. designed and performed the experiments. Z.S., J.P., D.F., S.L., Y.Y. and G.Z. contributed to the manuscript writing. All authors have read and agreed to the published version of the manuscript.

Funding: This research was funded by the National Natural Science Foundation of China (42271427), the Second Tibetan Plateau Scientific Expedition and Research (2022QZKK0101), and the Science and Technology Program of the Ministry of Public Security of China (2022JSZ09).

Data Availability Statement: Urban-LiDAR and Vaihingen data were obtained from https://www.lidar360.com/ (accessed on 2 May 2022), and Vaihingen data were acquired from http://www2.isprs.org/ (accessed on 7 April 2022).

Acknowledgments: The authors would like to thank the anonymous referees for constructive criticism and comments.

Conflicts of Interest: The authors declare no conflicts of interest.

References

1. Wang, X.; Li, P. Extraction of urban building damage using spectral, height and corner information from VHR satellite images and airborne LiDAR data. *ISPRS J. Photogramm. Remote Sens.* **2020**, *159*, 322–336. [CrossRef]
2. Adamopoulos, E.; Rinaudo, F.; Ardissono, L. A critical comparison of 3D digitization techniques for heritage objects. *ISPRS Int. J. Geo-Inf.* **2020**, *10*, 10. [CrossRef]
3. Xu, Y.; Stilla, U. Toward building and civil infrastructure reconstruction from point clouds: A review on data and key techniques. *IEEE J. Sel. Top. Appl. Earth Obs. Remote Sens.* **2021**, *14*, 2857–2885. [CrossRef]
4. Schrotter, G.; Hürzeler, C. The digital twin of the city of Zurich for urban planning. *PFG–J. Photogramm. Remote Sens. Geoinf. Sci.* **2020**, *88*, 99–112. [CrossRef]
5. Tarsha Kurdi, F.; Gharineiat, Z.; Campbell, G.; Awrangjeb, M.; Dey, E.K. Automatic filtering of lidar building point cloud in case of trees associated to building roof. *Remote Sens.* **2022**, *14*, 430. [CrossRef]
6. Martín-Jiménez, J.; Del Pozo, S.; Sánchez-Aparicio, M.; Lagüela, S. Multi-scale roof characterization from LiDAR data and aerial orthoimagery: Automatic computation of building photovoltaic capacity. *Autom. Constr.* **2020**, *109*, 102965. [CrossRef]
7. Zou, X.; Feng, Y.; Li, H.; Zhu, J. An Adaptive Strips Method for Extraction Buildings From Light Detection and Ranging Data. *IEEE Geosci. Remote Sens. Lett.* **2017**, *14*, 1651–1655. [CrossRef]
8. Huang, R.; Yang, B.; Liang, F.; Dai, W. A top-down strategy for buildings extraction from complex urban scenes using airborne LiDAR point clouds. *Infrared Phys. Technol.* **2018**, *92*, 203–218. [CrossRef]
9. Hui, Z.; Li, Z.; Cheng, P.; Ziggah, Y.Y.; Fan, J.L. Building extraction from airborne lidar data based on multi-constraints graph segmentation. *Remote Sens.* **2021**, *13*, 3766. [CrossRef]
10. Qin, R.; Fang, W. A hierarchical building detection method for very high resolution remotely sensed images combined with DSM using graph cut optimization. *Photogramm. Eng. Remote Sens.* **2014**, *80*, 37–47. [CrossRef]
11. Acar, H.; Karsli, F.; Ozturk, M.; Dihkan, M. Automatic detection of building roofs from point clouds produced by the dense image matching technique. *Int. J. Remote Sens.* **2018**, *40*, 138–155. [CrossRef]

12. Hron, V.; Halounová, L. Automatic reconstruction of roof models from building outlines and aerial image data. *Acta Polytech.* **2019**, *59*, 448–457. [CrossRef]
13. Ghamisi, P.; Höfle, B.; Zhu, X.X. Hyperspectral and lidar data fusion using extinction profiles and deep convolutional neural network. *IEEE J. Sel. Top. Appl. Earth Obs. Remote Sens.* **2017**, *10*, 3011–3024. [CrossRef]
14. Nguyen, T.H.; Daniel, S.; Gueriot, D.; Sintes, C.; Caillec, J.M.L. Unsupervised Automatic Building Extraction Using Active Contour Model on Unregistered Optical Imagery and Airborne LiDAR Data. In Proceedings of the PIA19+MRSS19—Photogrammetric Image Analysis & Munich Remote Sensing Symposium, Munich, Germany, 18–20 September 2019; Volume XLII-2/W16, pp. 181–188. [CrossRef]
15. Yuan, Q.; Shafri, H.Z.H.; Alias, A.H.; Hashim, S.J. Multiscale semantic feature optimization and fusion network for building extraction using high-resolution aerial images and LiDAR data. *Remote Sens.* **2021**, *13*, 2473. [CrossRef]
16. Li, F.; Zhu, H.; Luo, Z.; Shen, H.; Li, L. An adaptive surface interpolation filter using cloth simulation and relief amplitude for airborne laser scanning data. *Remote Sens.* **2021**, *13*, 2938. [CrossRef]
17. Provot, X. Deformation constraints in a mass-spring model to describe rigid cloth behaviour. In *Graphics Interface*; Canadian Information Processing Society: Mississauga, ON, Canada, 1995; p. 147. Available online: http://www-rocq.inria.fr/syntim/research/provot/ (accessed on 3 May 2022).
18. Zhang, W.; Qi, J.; Wan, P.; Wang, H.; Xie, D.; Wang, X.; Yan, G. An easy-to-use airborne LiDAR data filtering method based on cloth simulation. *Remote Sens.* **2016**, *8*, 501. [CrossRef]
19. Su, Z.; Gao, Z.; Zhou, G.; Li, S.; Song, L.; Lu, X.; Kang, N. Building Plane Segmentation Based on Point Clouds. *Remote Sens.* **2022**, *12*, 95. [CrossRef]
20. Dos Santos, R.C.; Galo, M.; Carrilho, A.C. Building boundary extraction from LiDAR data using a local estimated parameter for alpha shape algorithm. *Int. Arch. Photogramm. Remote Sens. Spat. Inf. Sci.* **2018**, *42*, 127–132. [CrossRef]
21. Shen, W.; Li, J.; Chen, Y.; Deng, L.; Peng, G. Algorithms study of building boundary extraction and normalization based on LiDAR data. *J. Remote Sens.* **2008**, *05*, 692–698. [CrossRef]
22. Sun, Z.; Li, Z.; Liu, Y. An improved lidar data segmentation algorithm based on euclidean clustering. In Proceedings of the 11th International Conference on Modelling, Identification and Control, Tianjin, China, 13–15 July 2019; Springer: Singapore, 2020; pp. 1119–1130. [CrossRef]
23. Xu, Z.; Yan, W. The Filter Algorithm Based on Lidar Point Cloud. *Inf. Commun.* **2018**, *3*, 80–82. [CrossRef]
24. Li, W.; Wang, F.; Xia, G. A geometry-attentional network for ALS point cloud classification. *ISPRS J. Photogramm. Remote Sens.* **2020**, *164*, 26–40. [CrossRef]
25. Charles, R.Q.; Su, H.; Kaichun, M.; Guibas, L.J. PointNet: Deep Learning on Point Sets for 3D Classification and Segmentation. In Proceedings of the 2017 IEEE Conference on Computer Vision and Pattern Recognition (CVPR), Honolulu, HI, USA, 21–26 July 2017; pp. 77–85. [CrossRef]
26. Charles, R.Q.; Yi, L.; Su, H.; Guibas, L.J. PointNet++: Deep hierarchical feature learning on point sets in a metric space. In Proceedings of the 31st Conference on Neural Information Processing Systems (NIPS 2017), Long Beach, CA, USA, 4–9 December 2017; Volume 30. [CrossRef]
27. Huang, R.; Xu, Y.; Hong, D.; Yao, W.; Ghamisi, P.; Stilla, U. Deep point embedding for urban classification using ALS point clouds: A new perspective from local to global. *ISPRS J. Photogramm. Remote Sens.* **2020**, *163*, 62–81. [CrossRef]

remote sensing

Article

Research on a Matching Method for Vehicle-Borne Laser Point Cloud and Panoramic Images Based on Occlusion Removal

Jiashu Ji [1], Weiwei Wang [1,*], Yipeng Ning [2], Hanwen Bo [1] and Yufei Ren [1]

[1] QiLu Aerospace Information Research Institute, Jinan 250132, China; jijs@aircas.ac.cn (J.J.); bohw@aircas.ac.cn (H.B.); renyf@aircas.ac.cn (Y.R.)
[2] School of Surveying and Geo-Informatics, Shandong Jianzhu University, Jinan 250102, China; ningyipeng19@sdjzu.edu.cn
* Correspondence: wangww@aircas.ac.cn

Abstract: Vehicle-borne mobile mapping systems (MMSs) have been proven as an efficient means of photogrammetry and remote sensing, as they simultaneously acquire panoramic images, point clouds, and positional information along the collection route from a ground-based perspective. Obtaining accurate matching results between point clouds and images is a key issue in data application from vehicle-borne MMSs. Traditional matching methods, such as point cloud projection, depth map generation, and point cloud coloring, are significantly affected by the processing methods of point clouds and matching logic. In this study, we propose a method for generating matching relationships based on panoramic images, utilizing the raw point cloud map, a series of trajectory points, and the corresponding panoramic images acquired using a vehicle-borne MMS as input data. Through a point-cloud-processing workflow, irrelevant points in the point cloud map are removed, and the point cloud scenes corresponding to the trajectory points are extracted. A collinear model based on spherical projection is employed during the matching process to project the point cloud scenes to the panoramic images. An algorithm for vectorial angle selection is also designed to address filtering out the occluded point cloud projections during the matching process, generating a series of matching results between point clouds and panoramic images corresponding to the trajectory points. Experimental verification indicates that the method generates matching results with an average pixel error of approximately 2.82 pixels, and an average positional error of approximately 4 cm, thus demonstrating efficient processing. This method is suitable for the data fusion of panoramic images and point clouds acquired using vehicle-borne MMSs in road scenes, provides support for various algorithms based on visual features, and has promising applications in fields such as navigation, positioning, surveying, and mapping.

Keywords: vehicle-borne mobile mapping system; laser point cloud; panoramic imaging; matching; occlusion removal

Citation: Ji, J.; Wang, W.; Ning, Y.; Bo, H.; Ren, Y. Research on a Matching Method for Vehicle-Borne Laser Point Cloud and Panoramic Images Based on Occlusion Removal. *Remote Sens.* **2024**, *16*, 2531. https://doi.org/10.3390/rs16142531

Academic Editors: Wanshou Jiang, San Jiang, Duojie Weng and Jianchen Liu

Received: 7 May 2024
Revised: 5 July 2024
Accepted: 7 July 2024
Published: 10 July 2024

1. Introduction

In recent years, the technology of vehicle-borne mobile mapping systems (MMSs) has seen rapid development. Vehicle-borne MMSs often integrate laser scanners, cameras, an inertial measurement unit (IMU), a global navigation satellite system (GNSS), and other equipment [1]; when the vehicle is traveling at a certain speed, the GNSS and IMU obtain, in real-time, the position and attitude data of the equipment, while LiDAR and a panoramic camera synchronously obtain a series of point clouds and panoramic images of the scene. This technology can efficiently collect ground information from streets, industrial parks, mining warehouses, forests and other scenes around the clock, and is widely used in fields such as road maintenance, 3D scene construction, topographic mapping, positioning, and navigation [2–6]. The high-precision fusion of multi-source data can bring the complementary advantages of different types of sensors; the point cloud data obtained using a vehicle-borne MMS have rich spatial location information with high accuracy [7]; and panoramic data can be integrated for imaging, with a 360-degree field of view, and contain a large amount of textural and color information [8]. Achieving the matching of the two types of data is a hot research direction in vehicle-borne MMS applications.

Matching between the point cloud and images often requires several steps, including coordinate transformation, registration, and fusion [9]. In terms of coordinate transformation, matching currently relies mainly on the collinear equation, direct linear, and pyramid methods [10] to establish the relationship between the point cloud and pixel data. The important factors directly affecting registration include the quality and synchronization rate of data collection from various sensors, the accuracy regarding the relative position between sensors, and the method of solving and correlating the various types of collected raw data [11]. With the iteration of the hardware system and calibration method, the errors generated in the first two aspects have reached a relatively low level. In the past few years, the focus of research has been on the use of algorithms to achieve further fine registration. Yao et al. used feature points to divide the point clouds generated with multiple laser scanners into blocks, and determined the accurate relationship between points and pixels by correlating the visual field of the panoramic camera with the point cloud blocks and the principle of the collinear equation [12]. Zhang et al. used the spherical epipolar line method and spherical absolute orientation model to achieve dense matching between images and point clouds based on Harris angle extraction, thereby obtaining more accurate registration parameters [13]. Wang et al. extracted the rods in the panoramic image and the corresponding point clouds, re-projected them to a virtual image, and obtained refined correspondence by maximizing the overlap area through particle swarm optimization [14]. Zhu et al. proposed a relative orientation model for panoramic images (PROM), which used feature point matching between adjacent panoramic images to calculate their relative poses, combined with the absolute pose of the starting panoramic image, to achieve registration under the condition that the image pose parameters are unknown [15]. Li et al. used Fast-RCNN to extract vehicles in panoramic images and matched them with possible corresponding vehicles in the point clouds through initial poses, thereby improving registration accuracy [16]. Wang et al. used the semantic segmentation method to remove the sky in panoramic images, projected the point clouds after ground removal to images to obtain registration primitives, and then achieved fine registration through the whale algorithm [17].

Fusion is often divided into point cloud coloring based on 3D point clouds and point cloud projection and depth maps based on 2D images [18]. Point cloud coloring is the process of assigning actual material colors to each laser point, which is commonly used in point cloud classification or 3D real scene modeling [19]; its implementation is based on using timestamps to find the optimal image correspondence of the point cloud, searching further for the pixel correspondence of the laser point, and assigning color attributes to the point clouds. Yuan et al. proposed an iterative strategy through which to construct color and textural information from point clouds using multi-view observations; for the color conflicts of point clouds generated from different viewpoints, they used differential image regions to assign values to the point clouds [20]. Shinohara et al. used PointNet++ to estimate the color of points in the point clouds and used a differentiable renderer to convert the colored point cloud into an image. The difference between the real image and the converted image was used as a loss function with which to train the network [21]. Depth maps are similar to point cloud projection; based on the coordinate transformation theory, point clouds are projected onto the image, and corresponding pixels are given depth of field or spatial position information, thereby achieving the measurement and positioning of ground objects through two-dimensional images [22]. The main difference between the two methods is that the depth map assigns an initial depth value to all pixels, replaces them with the depth values of the point clouds, and sets a scaling factor of panoramic image to increase the continuity of depth measurement [23]. Ku et al. used only basic computer vision operations (such as dilation, smoothing, etc.) to solve the problem of generating depth maps from point clouds, achieving good performance [24]. Xiang et al. proposed a deep learning network, 3dDepthNet, to generate accurate dense depth maps from pairs of point clouds and color images, then trained and validated it based on the KITTI dataset [25]. Bai et al. proposed a lightweight network with a significant reduction in the number of parameters, achieving a depth map generation algorithm applicable to embedded devices [26].

At present, for the data-matching method based on two-dimensional images, a vehicle-borne MMS often undergoes calibration after delivery, which can enable it to obtain highly accurate sensor external parameter data, thus reducing the workload for registration. However, there are still some unresolved issues in other aspects. Firstly, if data matching is performed based on deep learning, then the application effect is largely limited by the quality of the dataset and requires significant computational resources.

Secondly, due to the fact that point clouds represent discrete data, the projection generated with a point cloud representing an object on the image may be a series of discontinuous pixels. The point cloud map generated using a vehicle-borne MMS contains the 3D point clouds of most objects in the scene; meanwhile, a single panoramic image reflects the imaging from a certain perspective, and some objects in the scene are occluded. Therefore, when transforming point clouds into images, a point cloud projection in a region representing an occluded object might be generated incorrectly, in addition to the visible point cloud projections that should exist, which will cause a number of pixels with incorrect depth values to appear in the generated results of the point cloud map or depth map, and errors will be generated when querying the location coordinates based on the pixels. In addition, a generated point cloud map contains a large number of noise points and data with no measurement values (such as ground point clouds), which directly leads to accuracy interference in the generated results and reduces computational efficiency. In order to address the above issues, a method for matching point clouds with panoramic images based on occlusion removal, which assigns spatial position information to the main target on the image, is proposed. The algorithm removes invalid points contained in the generated point clouds and filters erroneous projection points from occluded objects, thus obtaining high-precision matching results.

2. Materials and Methods

In order to achieve accurate matching between point clouds and panoramic images, it is necessary to determine the fusion area, based on the coverage range of a single panoramic image, and search for the corresponding relationship between point clouds and images based on the time series. Specifically, the key technology can be divided into three parts—sequence point cloud scene generation, fusion coordinate conversion, and image matching—and the implementation method of each is described below.

2.1. Sequence Point Cloud Scene Generation

A vehicle-borne MMS can obtain the original point cloud data from the scanner, and the trajectory data from the IMU, during the acquisition process. This equipment outputs point cloud maps through a preprocessing system. Within the preprocessing system, the inertial navigation data are corrected based on the base station to obtain Position and Orientation System (POS) data that represent the trajectory of the inertial navigation center point. The system integrates the point cloud from each frame in the local coordinate system into a point cloud map in the world coordinate system, based on the scanner data and the POS data.

The point cloud map contains the point cloud data of the entire scanned scene, with a lot of data, but also useless information such as noise and ground points. In order to improve the efficiency and accuracy of point cloud projection into panoramic images, it is necessary to preprocess the point cloud. In this study, we designed a set of point-cloud-processing algorithms; the pseudocode is shown in Figure 1.

Workflow 1 Sequence Point Cloud Scene Generation

```
Input:  RouteP: The path of original point cloud map.
        RouteT: The path of the camera center trajectory file.
Output: Las_i: A point cloud file corresponding to a track point.
Initialize
every_k_points = 2: The interval of the uniform sampling algorithm.
mean_k = 50: The search range of the statistical filtering algorithm.
stddev_mul_thresh = 3: The threshold of the statistical filtering algorithm.
cloth_resolution = 1: The resolution of the CSF algorithm.
rigidness = 3: The number of times the particles moved.
class_threshold = 0.5: The threshold for classifying ground points.
OnePoint: Spatial information of a trajectory point.
filter_limits = [-120,120],[-100,100],[-3,40]:
Intercept range in the xyz axis of the region segmentation algorithm.
Main Program
P = ReadPointcloud(RouteP)
P_d = PcDownsample(P, every_k_points)
P_s = PcSorFiltering(P_d, mean_k, stddev_mul_thresh)
P_n = PcGroundFiltering(Ps, cloth_resolution, rigidness, class_threshold)
Csv = ReadTrajectory(RouteT)
while getline(Csv, OnePoint) do
    P_c = PassFilter(OnePoint, filter_limits)
    Las_i = LasWrite(P_c)
end
```

Figure 1. Flow of sequence point cloud scene generation.

During this process, the point cloud data were read, and then the uniform sampling algorithm was used to reduce the density of the point cloud data to an appropriate level. Secondly, the statistical filtering algorithm was used to remove irrelevant points, such as outliers in the point clouds. Thereafter, the cloth simulation filter (CSF) algorithm was used to separate non-ground point cloud data [27]. Furthermore, the trajectory data of the camera optical center were read, and, based on the region segmentation algorithm, a suitable scene range was set and the point cloud roughly corresponding to each trajectory point was segmented. Finally, the point cloud data were saved, and the point cloud file corresponding to each panoramic image frame was generated. It should be noted that the setting of various thresholds in this process primarily achieved good processing effects in the experimental scenario described in this paper; it can be adjusted according to the actual circumstances. The density of the point cloud, as the principal data feature, influences the selection of certain thresholds in this process. For sparser point clouds, 'every_k_points' in the uniform sampling step and 'mean_k' in the statistical filtering step should be reduced, while 'cloth_resolution' in the CSF algorithm step should be appropriately increased. Conversely, for denser point clouds, these parameters should be adjusted in the opposite direction.

2.2. Coordinate Transformation

The world geodetic coordinate system is affected by the positioning and pose determination method, as well as the initial coordinate system of the 3D point cloud generated from a vehicle-borne MMS [28]. In order to integrate panoramic images with point clouds, firstly, coordinate system transformation is required for the point cloud, and then collinear relationships and spherical projection are applied to complete the mapping of the point cloud to the panoramic image [29].

To complete coordinate transformation from the point cloud to the corresponding image, the point cloud coordinates need to be unified to the camera coordinate system, with the camera center as the origin at the time of image capture. Each camera shooting time corresponds to the timestamp of a POS trajectory point, and the pose information of the IMU at that time can be queried in the trajectory. By using this pose and the relative position relationship obtained from calibrating the IMU and camera, the pose of the camera optical center at the time of shooting can be determined.

Given the coordinates of the point cloud and the pose of the camera optical center in the world coordinate system at the time of shooting, the coordinates of the point cloud in the camera optical center coordinate system can be calculated using the following formula

$$\begin{bmatrix} x_c \\ y_c \\ z_c \end{bmatrix} = R\left(\begin{bmatrix} x_w \\ y_w \\ z_w \end{bmatrix} - T\right) \tag{1}$$

In Formula (1), $(x_{\mathrm{w}}, y_w, z_w)$ represents the coordinates of the point cloud in the world coordinate system, (x_c, y_c, z_c) represents the coordinates of the point cloud in the camera coordinate system, and R and T represent the rotation matrix and translation matrix of the camera optical center in the world coordinate system, corresponding to the pose of the camera optical center. The camera trajectory pose points are usually represented by a set of Euler angles. In order to convert Euler angles into a rotation matrix, different Euler angle definitions have different rotation matrix generation rules. In this experiment, the corresponding order and rules in the East–North–Up coordinate system are used to generate the rotation matrix.

After obtaining the coordinates of the point cloud in the camera coordinate system, with the camera coordinate system origin as the center of sphere, the coordinates from the camera coordinate system are converted to the spherical coordinate system, the calculation formula for which is as follows:

$$\begin{cases} \theta = \arctan\frac{x_c}{y_c} & x_c > 0,\ y_c > 0 \\ \theta = \pi + \arctan\frac{x_c}{y_c} & x_c < 0,\ y_c > 0 \\ \theta = \pi + \arctan\frac{x_c}{y_c} & x_c < 0,\ y_c < 0 \\ \theta = 2\pi + \arctan\frac{x_c}{y_c} & x_c > 0,\ y_c < 0 \end{cases} \\ \phi = \frac{\pi}{2} - \arctan\frac{z_c}{\sqrt{x_c{}^2 + y_c{}^2}} \tag{2}$$

In Formula (2), (θ, ϕ) is the coordinate of the point cloud in the spherical coordinate system, which can be understood as its longitude and latitude on the sphere.

The panoramic image can also be converted into a panoramic sphere, corresponding to the spherical coordinate system. To convert the point cloud from a spherical coordinate system to a panoramic pixel coordinate system, the calculation formula is as follows:

$$\begin{cases} m = r\theta \\ n = r\phi \\ r = w/2\pi \end{cases} \tag{3}$$

In Formula (3), (m, n) is the coordinate of the point cloud in the pixel coordinate system, and r is the spherical radius corresponding to the panoramic image. Using the above formula, the point cloud in each frame of the scene can be projected onto the corresponding region on the panoramic image.

2.3. Data Matching

As previously mentioned, scene data are processed and segmented from a point cloud map scanned in the field, containing the overall 3D information of the region. When converting this information to a two-dimensional image, there may be situations in which the point cloud information from objects occluded in the camera's perspective is projected onto the panoramic image, but these projection point data do not conform to physical reality and should be filtered out. In this study, we designed a filtering algorithm, based on vectorial angle, to divide the processing logic into the two following main parts:

Nearest point preservation: all point cloud data are traversed during the projection process; if a pixel corresponds to multiple data points, only the data points closest to the camera center are retained.

Filter out occluded points based on vectorial angle: to determine whether a point to be processed is an occluded point, we propose an algorithm based on angle filtering under spherical projection. Within the range of the spherical projection, a pair of points that do not have an occlusion relationship from a certain perspective form a larger angle with the viewpoint, and, similarly, a pair of points that are easily occluded tend to form a smaller angle.

As shown in Figure 2, the angle in (a) between the vector $M_{iw}M_{\text{in}}$ formed from the visible point M_{iw} is the starting point to adjacent points M_{in}, and the vector $M_{iw}O$, formed from the visible point to the spherical center O, is significantly greater than the angle in (b) between the vector $M_{ow}M_{\text{in}}$ and the vector $M_{ow}O$, both of which start with the occluded point M_{ow}.

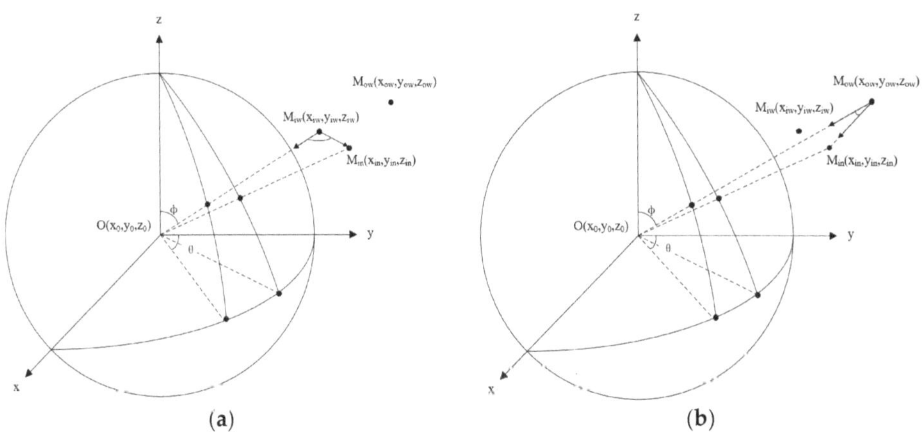

Figure 2. Different angles formed by visible and invisible points as vertices in spherical projection: (**a**) visible points; (**b**) invisible points.

By calculating the angle between vectors, it can be judged whether a point in 3D space is occluded by other points during centroid projection. Taking Figure (a) as an example, firstly, the unit vector between the point to be processed and the camera optical center can be constructed, the calculation formula for which is as follows:

$$\begin{cases} v_{ix} = \frac{x_0 - x_{iw}}{v_{is}} \\ v_{iy} = \frac{y_0 - y_{iw}}{v_{is}} \\ v_{iz} = \frac{z_0 - z_{iw}}{v_{is}} \\ v_{is} = \sqrt{(x_0 - x_{iw})^2 + (y_0 - y_{iw})^2 + (z_0 - z_{iw})^2} \end{cases} \quad (4)$$

In Formula (4), (x_{iw}, y_{iw}, z_{iw}) is the coordinate of the point to be processed in the world coordinate system, (x_0, y_0, z_0) is the world coordinate of the camera optical center, v_{ix}, v_{iy}, and v_{iz} are the normalized components of the unit vector from the point to the camera optical center in the three directions of the coordinate axis, and v_{is} is the modulus of the vector from the point to the camera optical center.

A radius threshold r_{filter} is set, and all pixels within the threshold range as nearby pixels are found. For these nearby pixels, their corresponding point cloud is queried and the unit vectors formed by these points with the point to be determined are calculated as follows:

$$\begin{cases} w_{ix} = \frac{x_{in}-x_{iw}}{w_{is}} \\ w_{iy} = \frac{y_{in}-y_{iw}}{w_{is}} \\ w_{iz} = \frac{z_{in}-z_{iw}}{w_{is}} \\ w_{is} = \sqrt{(x_{in}-x_{iw})^2+(y_{in}-y_{iw})^2+(z_{in}-z_{iw})^2} \end{cases} \quad (5)$$

In Formula (5), (x_{in}, y_{in}, z_{in}) is the coordinate of the point cloud corresponding to adjacent pixels in the world coordinate system, w_{ix}, w_{iy}, and w_{iz} are the normalized components of the unit vector from the point to be processed to adjacent points in the three directions of the coordinate axis, and $\vec{w_i}$ is the modulus of the vector from the point to adjacent points.

Finally, the angle between the two vectors is calculated using the following formula:

$$\alpha = \arccos\left(\frac{\vec{v_i}\cdot\vec{w_i}}{\left|\vec{v_i}\right|\left|\vec{w_i}\right|}\right) \quad (6)$$

Simplification leads to the following formula:

$$\alpha = \arccos[(x_{in}-x_{iw})(x_0-x_{iw})+(y_{in}-y_{iw})(y_0-y_{iw})+(z_{in}-z_{iw})(z_0-z_{iw})] \quad (7)$$

In Formula (7), α is the angle formed by the point to be processed and the camera optical center and the adjacent point. The adjacent points are traversed according to the threshold value, and the minimum angle generated between the point to be processed and the adjacent points within the radius threshold is recorded. The angle screening threshold th_{angle} is set. If the minimum angle is less than the screening threshold, the point is judged to be invisible, filtered as an occluded point, and is not reflected on the panoramic image projection.

Based on the research methods described above, we designed and implemented an algorithm for generating the matching results of point clouds and panoramic images, the pseudocode for which is shown in Figure 3:

Workflow 2 Matching Result Generation

```
Input: RouteP: The path of the point cloud files corresponding to the track points.
       RouteC: The path of the image files corresponding to track points.
       RouteT: The path of the camera center trajectory file.
Output: ProjectionVisFile_i: A visualization file of point clouds projection.
        MatchingVisFile_i: A visualization file of matching result.
        MatchingFile_i: A matching result file corresponding to a track point.
Initialize
item: The name and spatial information of a track point.
filter_radius: The search radius of the occlusion removal algorithm.
occlusion_threshold: The threshold of the occlusion removal algorithm.
Main Program
ParaList = ReadCSV(RouteT)
for item in ParaList do
    Img = ReadImg(RouteC + item['name'])
    Las = ReadLas(RouteP_i + item['name'])
    ImgPointMat = GetImgCoord4Pic(Las, item['position'])
    ProjectionVisFile_i = ImgWrite(Img, ImgPointMat)
    MatchingMat = PtOccludedFilter(ImgPointMat, item['position'],
                                   filter_radius, occlusion_threshold)
    MatchingVisFile_i = ImgWrite(Img, MatchingMat)
    MatchingFile_i = ResultWrite(MatchingMat)
end
```

Figure 3. Flow of matching result generation.

The algorithm first read the camera optical center trajectory; for each trajectory point, it read the corresponding point cloud scene and panoramic image. The projection result of the point cloud scene on the corresponding panoramic image was obtained through the coordinate transformation function in step 10. In step 12, the data-matching function took the point cloud projection result, the position of camera optical center, and the parameters of the occlusion removal algorithm as input data and traversed all points that had been processed through the nearest point preservation. The function was used to calculate the vector angles formed by these points, and their neighbors within the search radius, and determined whether any points were occluded based on the threshold value of the angle. For all points that completed occlusion determination, the corresponding relationship between the generated non-occluded points and pixels was retained, and a series of pixel-to-point correspondences was output in the form of a database file, which contains the coordinates of some pixels on the panoramic images and the corresponding spatial points in the world coordinate system. The algorithm traversed all trajectory points, generating a series of matching result files and corresponding visualization files in a specified folder, thus realizing batch processing of point cloud and image matching based on trajectory information.

3. Experimentation and Results

In order to verify the correctness and reliability of the matching method detailed in this article, a complete implementation process for matching point clouds and panoramic images was designed. The specific technical route is shown in Figure 4.

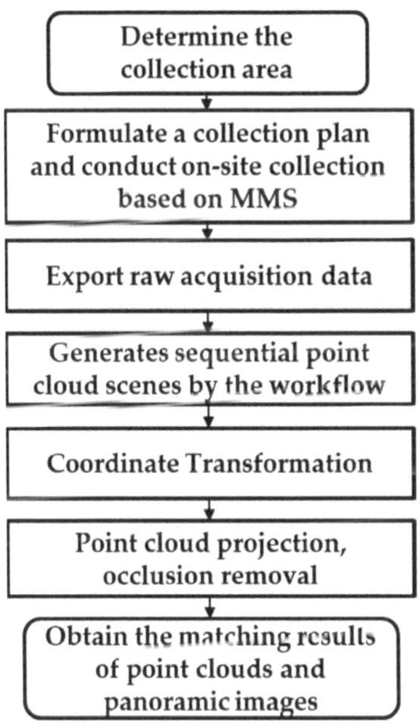

Figure 4. The implementation process of this study.

3.1. Preparation of Experiment

In this study, the experimental basic data were collected and generated using the ARS-1000L mobile mapping platform produced by the company Hi-target DigitalCloud. The system composition is shown in Figure 5.

Figure 5. Vehicle-borne mobile mapping system used in the experiment.

In the hardware system, the acquisition equipment mainly includes a Riegl VUX-1HA laser scanner, aSPAN-ISA-100C inertial navigation equipment, and a Ladybug5Plus panoramic camera. The equipment is fixed and connected via mechanical devices and calibrated with high precision so that original sensor data can be obtained. The main performance parameters are shown in Table 1.

Table 1. Main performance parameters of equipment.

<table>
<tr><th></th><th>Performance Parameters</th><th colspan="2">ARS-1000L</th></tr>
<tr><td rowspan="2">Basic parameters</td><td>Absolute accuracy</td><td colspan="2">±5 cm</td></tr>
<tr><td>Weight (main unit)</td><td colspan="2">4.6 kg</td></tr>
<tr><td rowspan="6">Laser scanning unit</td><td>Max. measuring distance</td><td colspan="2">1350 m@60%</td></tr>
<tr><td>Laser frequency</td><td colspan="2">820 K Hz</td></tr>
<tr><td>Scanning speed</td><td colspan="2">10–200 lines/s</td></tr>
<tr><td>Ranging accuracy</td><td colspan="2">10 mm@150 m</td></tr>
<tr><td>Scanning angle</td><td colspan="2">330°</td></tr>
<tr><td>Angle resolution</td><td colspan="2">0.001°</td></tr>
<tr><td rowspan="3">Panoramic camera unit</td><td>Panoramic image resolution</td><td colspan="2">8192 × 4096</td></tr>
<tr><td>Pixel size</td><td colspan="2">3.45 μm</td></tr>
<tr><td>Focal length</td><td colspan="2">4.4 mm</td></tr>
<tr><td rowspan="4">Inertial navigation unit</td><td rowspan="2">Positioning accuracy</td><td>Plane</td><td>0.01 m</td></tr>
<tr><td>Elevation</td><td>0.02 m</td></tr>
<tr><td>Directional accuracy</td><td colspan="2">0.010°</td></tr>
<tr><td>Attitude accuracy</td><td colspan="2">0.005°</td></tr>
</table>

The Shandong Xiaoya Group campus was selected as the site for experimental data acquisition in order to complete the verification of the proposed method. Located at

No. 44, Industrial North Road, Licheng District, Jinan City, Shandong Province, the area features expansive and level roads, along with numerous well-organized buildings, which facilitated accuracy verification. Data acquisition was carried out in the afternoon, when direct sunlight was not strong. The driving route of the experimental vehicle was planned, from the nearby road to the industrial park, the mobile scanning platform was started, and the vehicle was driven on the planned route at a speed of approximately 15 km/h. The original data of the park were synchronously acquired with each sensor, and then a point cloud map of the planned route, a trajectory corresponding to the camera optical center, and panoramic images corresponding to trajectory points were generated.

3.2. Experimental Evaluation of Matching Effect

To qualitatively analyze the matching effects, we utilized point cloud maps, camera optical center trajectories, and the panoramic images corresponding to the trajectory points as input data. Data processing was conducted sequentially through the workflows designed previously for generating sequential point cloud scenes and matching results. We investigated threshold selection for occlusion removal algorithms, examined the visual effects of the method, and assessed its processing efficiency. Furthermore, in order to facilitate a comparative analysis of the algorithms, the same experimental dataset was used to generate colored point clouds, using a point cloud coloring algorithm, and to generate point cloud projection, directly based on the original point cloud map. The visual effects produced using both algorithms, as well as their computational efficiencies, were obtained and compared.

3.2.1. Selection of Threshold Value for Occlusion Removal Algorithm

In the filtering algorithm based on vector angles, the setting of the threshold directly affects the result of the processing. The radius threshold represents the search range of the surrounding data points from the points to be evaluated in the image coordinate system. Sufficient surrounding points help to make more accurate judgments. The angle selection threshold determines the screening level of the algorithm for the points to be evaluated. The higher the threshold, the more points will be filtered out. In the matching process, the density and distribution characteristics of point clouds, the sizes of images, and other characteristics are all related to the selection of algorithm thresholds. The appropriate selection of algorithm thresholds should be based on reasonable empirical ranges, with multiple experimental evaluations conducted to select a good set of thresholds as the parameters of subsequent algorithms.

According to the characteristics of data acquisition utilized in this experiment, the range of r_{filter} was set to be within 10 pixels, and the range of th_{angle} was set to be within 0.1. Taking the scene corresponding to a trajectory point in the industrial park as the test dataset, different filtering thresholds were set within this range to compare the filtering effect produced using the algorithm. Table 2 and Figure 6, respectively, show the influence of different thresholds within the range on the number of filtered points, and the visualization of matching results at representative positions after completing the occlusion removal process. It can be seen that th_{angle} determines the overall filtering effect. The smaller the threshold value, the more filtered points, and the more obvious the distinction between occluded objects. The filtering effect on the occluded part at the junction of two surfaces of an object is significantly improved, but too small a threshold value can easily lead to erroneous filtering of visible points, especially those at the far end, resulting in sparse point cloud projection. The increase in r_{filter} improves the accuracy of occlusion judgment, which has a significant impact on the point cloud projection at the boundary between different objects, producing the effect of distinguishing the boundaries of objects. However, too large a value can cause a significantly loss of valuable point clouds at the boundary.

Table 2. The number of points filtered out using the algorithm at different thresholds.

r_{filter}	th_{angle}	Number of Projection Points Filtered Out	Percentage of Total
2	0.1	1,137,043	35.77%
5	0.1	1,311,833	41.27%
7	0.1	1,378,921	43.38%
10	0.1	1,461,944	45.99%
5	0.07	1,371,933	43.16%
5	0.05	1,427,822	44.91%
5	0.03	1,531,863	48.19%
5	0.01	1,897,458	59.69%

When threshold A is set to 0.1 and B is set to 5, the filtering result is more in line with the physical reality, effectively filtering out the points of the occluded object and preserving the points of the object within view, with a low loss of position information. Scanning devices of the same type with similar settings tend to have consistent image and point cloud generation characteristics; so, the current threshold value can be used as an empirical parameter for processing scenes of this type. For scenes with different data characteristics, accurate occlusion removal can be achieved by applying different threshold and search radius values.

It is evident from algorithmic principles that the density of the point clouds is the most significant data characteristic affecting processing outcomes. To achieve the desired accuracy in occlusion removal, the threshold of the algorithm should be adjusted according to the density of the point clouds. In certain scenarios, where the target object is situated at a greater distance from the acquisition device, or where the acquisition vehicle is traveling at a higher speed, the resulting density of the point clouds is lower; for these situations, the search radius for each point to be processed should be increased, correspondingly elevating the threshold r_{filter}. The sparseness of points also results in an increase in the angle formed by occluded points, necessitating a corresponding increase in the threshold th_{angle}. Conversely, when the density of the point clouds is increased, the threshold of the algorithm should be correspondingly reduced.

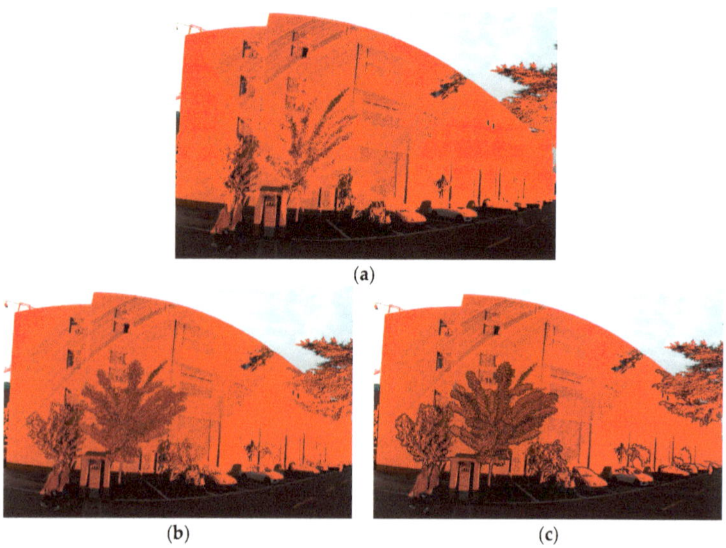

(a) (b) (c)

Figure 6. *Cont.*

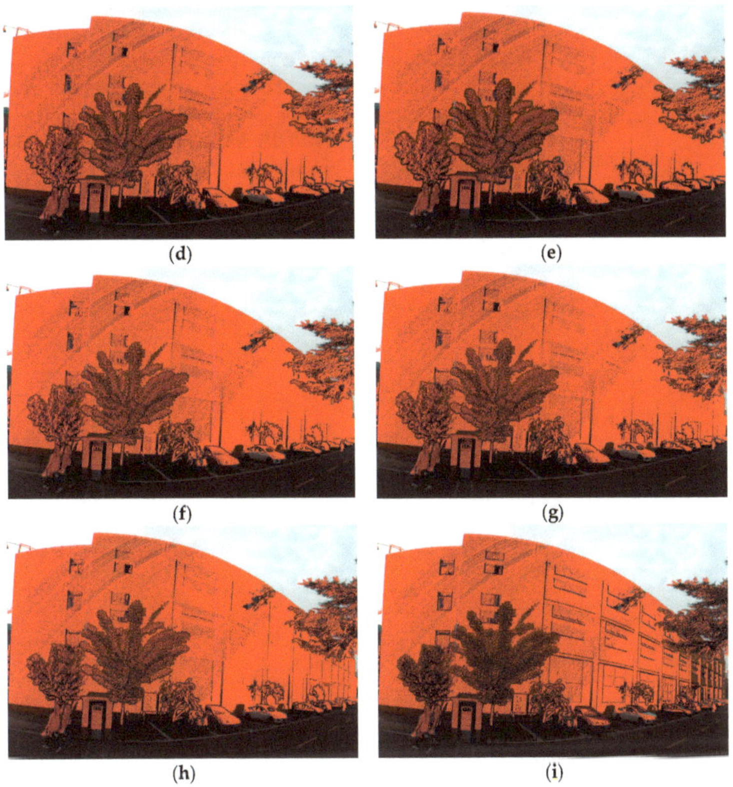

Figure 6. The visualization of occlusion removal under different parameter configurations: (a) The original effect of a point cloud projection. (b–i) Occlusion removal effect with different th_{angle} and r_{filter} at the following local positions: (b) 0.1–2, (c) 0.1–5, (d) 0.1–7, (e) 0.1–10, (f) 0.07–5, (g) 0.05–5, (h) 0.03–5, and (i) 0.01–5.

Each panoramic image is taken on the ground, and there are some point clouds that are significantly higher than or far away from the camera optical center. According to the algorithm principle, the vector angle does not change with the increase or decrease in the distance between the object and the camera optical center, so the selection of the threshold th_{angle} is not obviously affected by the point cloud distance. However, the width of the edge area resulting from the projection of adjacent objects is not affected by point cloud distance. For point clouds that are far away from the camera optical center, the edge area occupies a large proportion of the projection area, which may lose the effective spatial information. Therefore, when determining the value of r_{filter}, the utilization rate of the projection information for tall or far away objects should be considered. For the situation where the corresponding spatial information of these two types of objects needs to be retained, r_{filter} should not be set too large. On the contrary, the value of r_{filter} can be appropriately increased to make the projection discrimination of objects close to the camera optical center more obvious.

3.2.2. Visualization of the Processing Effect

In the visualization of sequence point cloud scene generation, a series of point cloud scene data extracted frame by frame is obtained through the designed workflow. Figure 7a,b shows the generation effect of the point cloud map and the extraction effect of the point cloud scene corresponding to a certain trajectory point. As shown in the figure, the device generates a 3D point cloud scene of the planned road section, with a large amount of noise and many insignificant points. After processing via the workflow, the valuable point clouds (such as buildings) in the scene are well preserved, the invalid points are mostly filtered, and the quality of the point cloud is generally good. The point cloud corresponds to the real scene around the panoramic camera at the time of shooting and can be used to match the panoramic image with the corresponding point cloud. Additionally, Figure 7c shows the generation effect of the point cloud coloring algorithm. As can be seen from the figure, this algorithm ensures the accurate assignment of real color information to the point cloud, without affecting the point cloud map data. The data benchmark of this method is point cloud information, and it lacks the capability to extract spatial information from images.

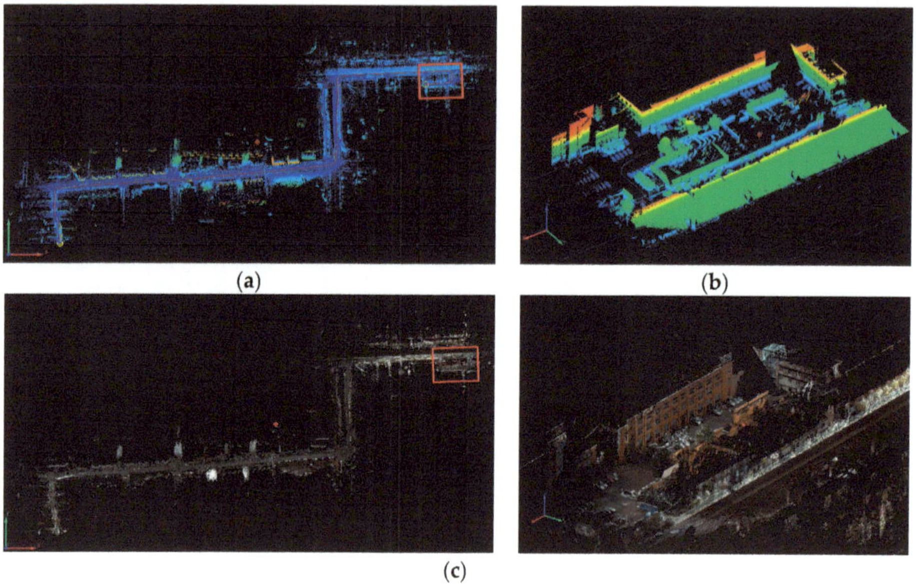

Figure 7. Visualization of point cloud processing. (**a**) Point cloud map generation. (**b**) Point cloud scene extraction. (**c**) Point cloud map coloring.

In the visualization of data matching, using point cloud scenes, trajectory data, and corresponding panoramic images as inputs for the matching module, th_{angle} is set to 0.1, r_{filter} is set to 5, and corresponding point cloud projections are generated on the image, acquiring the point cloud direct projection result using the point cloud map as input. Figure 8 shows the visual effects of point cloud direct projection and the matching method we proposed, before and after occlusion removal. It is observable that, upon transforming the point cloud onto a two-dimensional plane, the projection of the point cloud exhibits a good degree of coincidence with the corresponding objects in the panoramic imagery; however, some point clouds that should not appear in the image also generate projections. The point cloud direct projection result contains significant numbers of occluded point clouds, noise points, and ground point clouds. If one were to directly query the spatial coordinates corresponding to image pixels, representing objects of surveying value (such as building facades), based on this matching result, erroneous outcomes are likely to be obtained. In the matching method we proposed before occlusion removal, noise points and ground points were effectively removed, while the projection points of some occluded objects still persist, affecting the correct matching relationship between pixels and point clouds. Point cloud matching with occlusion removal significantly improves this phenomenon. Visual objects can be correctly assigned with point cloud information, and the point clouds that should not be displayed on the image are effectively filtered. The edge discrimination between the projections of different objects is obvious, and the spatial information of tall and far away objects is also well preserved. The matching results between 2D image pixels and 3D point cloud coordinates are more consistent with physical reality.

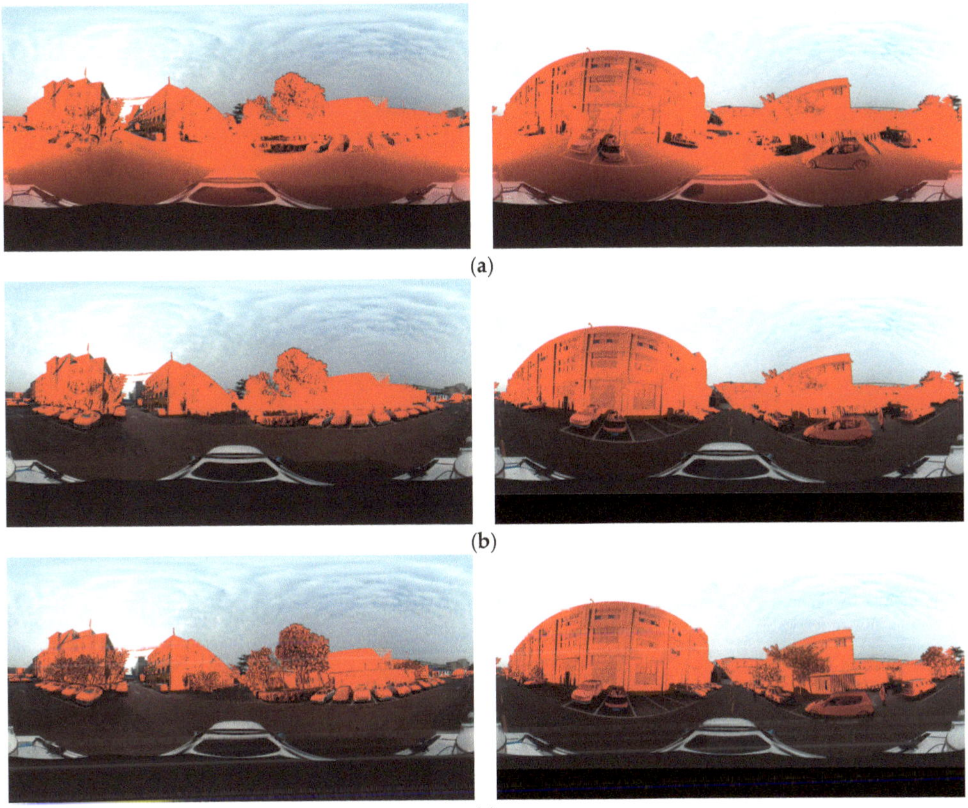

(a)

(b)

(c)

Figure 8. Visualization of different matching methods. (**a**) Point cloud direct projection. (**b**) Matching method before occlusion removal. (**c**) Matching method after occlusion removal.

3.2.3. Processing Efficiency of the Method

In terms of processing efficiency, we compared the designed method and two comparative algorithms in terms of data quantity and time metrics in our experiments, as shown in Table 3, revealing that the sequence point cloud generation workflow processed approximately 460 million data points in 4337.68 s, batch-generating files corresponding to 430 trajectory points and filtering out approximately three-quarters of insignificant points, thereby conserving computational resources for subsequent processing. The point cloud coloring algorithm consumed 3775 s to generate a colored point cloud map, which was saved as a map file; however, this commonly used matching method does not demonstrate significant efficiency advantages. In the matching result generation workflow, taking the trajectory points processed in the previous section as an example, 20,326,339 data points were read, and 1,866,860 data points were processed for matching, with a total time consumption of 60.62 s. The point cloud direct projection algorithm consumed 607.31 s due to using raw point cloud map as input, indicating relatively inefficient processing. In summary, the method proposed in this paper can perform serialization processing on large data volumes with relatively short time consumption and good processing efficiency.

Table 3. The processing efficiency of different methods.

Algorithm	Time Metrics	Values	Quantitative Metrics	Values
Sequence point cloud scene generation	/	/	Number of points read	459,358,534
	Downsampling	4.95 s	Number of points after downsampling	229,679,267
	Statistical filtering	1311.98 s	Number of points after statistical filtering	227,209,999
	CSF filtering	523.98 s	Number of points after CSF filtering	125,906,128
	Region segmentation and saving	2496.77 s	Number of files generated	430
	Total	4337.68 s	/	/
Point cloud map coloring	Total	3775 s	Number of colored points	459,358,534
Point cloud direct projection (Take the processing of a trajectory point)	/	/	Number of points read	459,358,534
	Total	607.31 s	Number of point clouds projected onto the image	9,744,335
Data matching (Take the processing of a trajectory point)	/	/	Number of points read	20,326,339
	Coordinate transformation	27.91 s	Number of point clouds projected onto the image	3,178,693
	Occlusion removal	32.71 s	Number of point clouds matching with pixels	1,866,860
	Total	60.62 s	/	/

3.3. *Analysis of Matching Accuracy*

To quantitatively analyze the matching error, 18 typical feature points in the industrial park were selected as control points. These control points were located 5–30 m from the center of the shooting area. The authentic world coordinates of the control points were obtained using a reflector-less total station, employing the WGS84 coordinate system, wherein the x, y, and z axes, respectively, represent the north coordinate, east coordinate, and geodetic height. On the bases of data acquisition and the processing results of the algorithm described in the previous section, the coordinates were used as a benchmark for comparative evaluation.

3.3.1. Generation Accuracy of Point Clouds

In order to evaluate the generation accuracy of point clouds, we manually selected the point cloud locations corresponding to the control points on the point cloud map and calculated the difference between the generated coordinates of the selected points and the true coordinates of the control points. The results are shown in Table 4 and Figure 9. As shown in the table, the average point cloud generation error is about 3 cm, the mean error in the plane is about 2.5 cm, and the maximum generation error is not above 5 cm. The source of error mainly includes the mechanical error of the sensor in data acquisition, the calibration error of the sensor, etc.; the error distribution of each point is relatively uniform, the overall generation accuracy is good, and the point cloud map can better restore the spatial information characteristics of the real scene.

Table 4. Accuracy error in point cloud generation.

Num	Control Points			Scanning Points			Errors			
	x/m	y/m	z/m	x/m	y/m	z/m	dx/m	dy/m	dz/m	d/m
1	513,033.881	406,5421.019	29.501	513,033.905	406,5421.024	29.519	0.024	0.005	0.018	0.030
2	513,034.079	406,5417.709	30.886	513,034.094	406,5417.698	30.905	0.015	−0.011	0.019	0.026
3	513,052.259	406,5431.14	34.345	513,052.272	406,5431.123	34.361	0.013	−0.017	0.016	0.027
4	513,052.256	406,5433.86	29.44	513,052.274	406,5433.854	29.462	0.018	−0.006	0.022	0.029
5	513,052.381	406,5429.156	28.316	513,052.396	406,5429.137	28.340	0.015	−0.019	0.024	0.034
6	513,079.245	406,5422.23	30.011	513,079.263	406,5422.253	30.021	0.018	0.023	0.010	0.031
7	513,056.006	406,5421.309	27.651	513,056.027	406,5421.324	27.667	0.021	0.015	0.016	0.030
8	513,051.238	406,5416.133	27.585	513,051.262	406,5416.131	27.623	0.024	−0.003	0.038	0.044
9	513,060.643	406,5385.074	44.224	513,060.657	406,5385.077	44.259	0.014	0.003	0.035	0.038
10	513,031.439	406,5388.888	37.06	513,031.462	406,5388.868	37.081	0.023	−0.020	0.021	0.037
11	513,013.924	406,5389.131	34.979	513,013.946	406,5389.114	35.005	0.022	−0.018	0.026	0.038
12	513,002.87	406,5388.761	37.571	513,002.877	406,5388.746	37.602	0.007	−0.015	0.031	0.035
13	513,050.191	406,5388.903	32.037	513,050.182	406,5388.917	32.024	−0.009	0.014	−0.013	0.021
14	513,043.428	406,5388.673	31.944	513,043.425	406,5388.689	31.969	−0.003	0.016	0.025	0.030
15	513,033.963	406,5418.855	29.508	513,033.984	406,5418.852	29.528	0.021	−0.003	0.020	0.029
16	513,089.411	406,5386.011	44.216	513,089.432	406,5386.045	44.214	0.021	0.034	−0.002	0.040
17	513,040.848	406,5426.86	27.589	513,040.861	406,5426.887	27.598	0.013	0.027	0.009	0.031
18	513,033.798	406,5423.09	28.772	513,033.832	406,5423.099	28.777	0.034	0.009	0.005	0.035
Mean square error							0.019	0.017	0.021	0.033
Mean square error in plane							0.025		Average value d	0.032

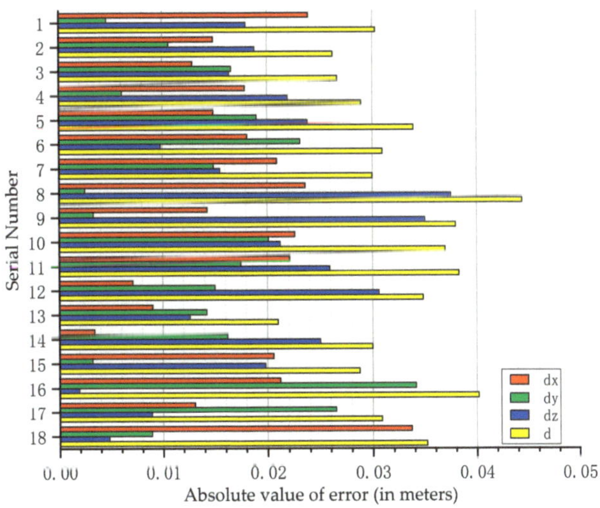

Figure 9. The histogram represents the accuracy error in point cloud generation.

3.3.2. Comparison of Pixel Matching Error

To evaluate the accuracy of the projection, 18 control points were manually labeled with pixel coordinates on the image. Using the real 3D coordinates of the control point as input and calculating the target pixel coordinates of the control point on the corresponding panoramic image based on the point cloud projection formula, the difference between the two coordinates was calculated as an evaluation of projection accuracy. As shown in Table 5 and Figure 10, the average error in pixel coordinates is 2.82, with a median error of 3.2 and a maximum error of around 6. The error source of pixel matching is mainly the accuracy error generated from the camera optical center trajectory, which is directly related to the calibration effect of the camera. Moreover, as the distance between the control point and the shooting center increases, the error tends to increase. The overall matching error is at a low level, and it can be considered that the matching effect between the point clouds and the images is relatively accurate.

Table 5. Pixel error in point cloud image matching.

Num	Control Points		Target Points		Pixel Errors		
	w	h	w′	h′	dw	dh	d
1	5717	2139	5717	2142	0	−3	3.00
2	5983	1953	5982	1955	1	−2	2.24
3	1922	1524	1922	1523	0	1	1.00
4	1540	2106	1537	2103	3	3	4.24
5	1479	2274	1479	2270	0	4	4.00
6	1970	2059	1970	2056	0	3	3.00
7	2437	2322	2437	2321	0	1	1.00
8	3197	2381	3197	2379	0	2	2.00
9	3549	1646	3548	1645	1	−1	1.41
10	4637	1827	4638	1832	−1	−5	5.10
11	4075	1617	4075	1620	0	−3	3.00
12	4600	1670	4599	1672	1	−2	2.24
13	3870	2029	3869	2029	1	0	1.00
14	3887	2033	3885	2037	2	−4	4.47
15	5477	2141	5478	2147	−1	−6	6.08
16	2916	1774	2916	1775	0	1	1.00
17	7044	2836	7044	2834	0	2	2.00
18	5977	2227	5977	2231	0	−4	4.00
Mean square error					1.03	3.03	3.20
Average error of d							2.82

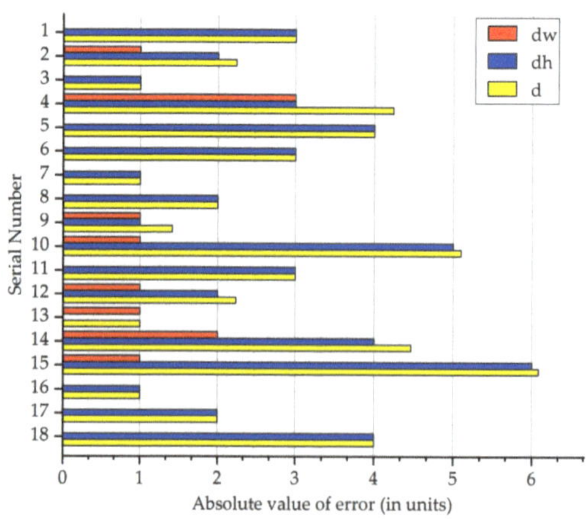

Figure 10. The histogram represents the pixel error in point cloud image matching.

3.3.3. Comparison of Position Matching Error

The matching results generated in this study correspond to the correspondence between the pixel points representing the primary targets in the panoramic images and their actual spatial coordinates. In order to evaluate the spatial matching errors of this method, the corresponding spatial coordinates were retrieved from the matching results for each control point, according to its pixel coordinates in the image. These coordinates were then compared with the actual coordinates of the control points to calculate the spatial matching errors, the results of which are presented in Table 6 and Figure 11. From the table, it can be seen that the mean square error in plane is about 3.2 cm, the distance error is about 4 cm, and the maximum distance error is no more than 6 cm. The accuracy of the position matching result is directly related to the accuracy of point cloud generation, the accuracy of the pixel matching result, and the filtering effect of the occluded data points. The overall error of this result is at a low level, and there is no obvious error fluctuation, indicating that there is no situation in which the wrong match point is queried. It can be considered that the matching relationship between the point cloud and the image generated using the method can better characterize the distribution characteristics of spatial information based on the images, and the corresponding position information can be obtained accurately for the pixels with spatial information.

Table 6. Spatial coordinate error of point cloud image matching.

Num	Control Points			Target Points			Errors of Spatial Coordinate			
	x/m	y/m	z/m	x/m	y/m	z/m	dx/m	dy/m	dz/m	d/m
1	513,033.881	406,5421.019	29.501	513,033.894	406,5421.048	29.541	−0.013	−0.029	−0.040	0.051
2	513,034.079	406,5417.709	30.886	513,034.088	406,5417.722	30.916	−0.009	−0.013	−0.030	0.034
3	513,052.259	406,5431.140	34.345	513,052.279	406,5431.118	34.299	−0.020	0.022	0.046	0.055
4	513,052.256	406,5433.860	29.440	513,052.293	406,5433.826	29.434	−0.037	0.034	0.006	0.051
5	513,052.381	406,5429.156	28.316	513,052.422	406,5429.128	28.302	−0.041	0.028	0.014	0.052
6	513,079.245	406,5422.230	30.011	513,079.257	406,5422.214	29.977	−0.012	0.016	0.034	0.040
7	513,056.006	406,5421.309	27.651	513,055.961	406,5421.318	27.650	0.045	−0.009	0.001	0.045
8	513,051.238	406,5416.133	27.585	513,051.208	406,5416.113	27.583	0.030	0.020	0.002	0.036
9	513,060.643	406,5385.074	44.224	513,060.606	406,5385.081	44.249	0.037	0.007	−0.025	0.045
10	513,031.439	406,5388.888	37.060	513,031.439	406,5388.918	37.085	0.000	−0.030	−0.025	0.039
11	513,013.924	406,5389.131	34.979	513,013.900	406,5389.116	35.027	0.024	0.015	0.048	0.056
12	513,002.870	406,5388.761	37.571	513,002.866	406,5388.745	37.612	0.004	0.016	−0.041	0.044
13	513,050.191	406,5388.903	32.037	513,050.155	406,5388.932	32.068	0.036	−0.029	−0.031	0.055
14	513,043.428	406,5388.673	31.944	513,043.406	406,5388.669	31.928	0.022	0.004	0.016	0.028
15	513,033.963	406,5418.855	29.508	513,033.967	406,5418.861	29.531	0.004	−0.006	−0.023	0.024
16	513,089.411	406,5386.011	44.216	513,089.412	406,5386.039	44.216	−0.001	−0.028	0.000	0.028
17	513,040.848	406,5426.860	27.589	513,040.835	406,5426.853	27.592	0.013	0.007	−0.003	0.015
18	513,033.798	406,5423.090	28.772	513,033.811	406,5423.106	28.788	−0.013	−0.016	−0.016	0.026
Mean square error							0.024	0.021	0.027	0.042
Mean square error in plane							0.032		Average value of d	0.040

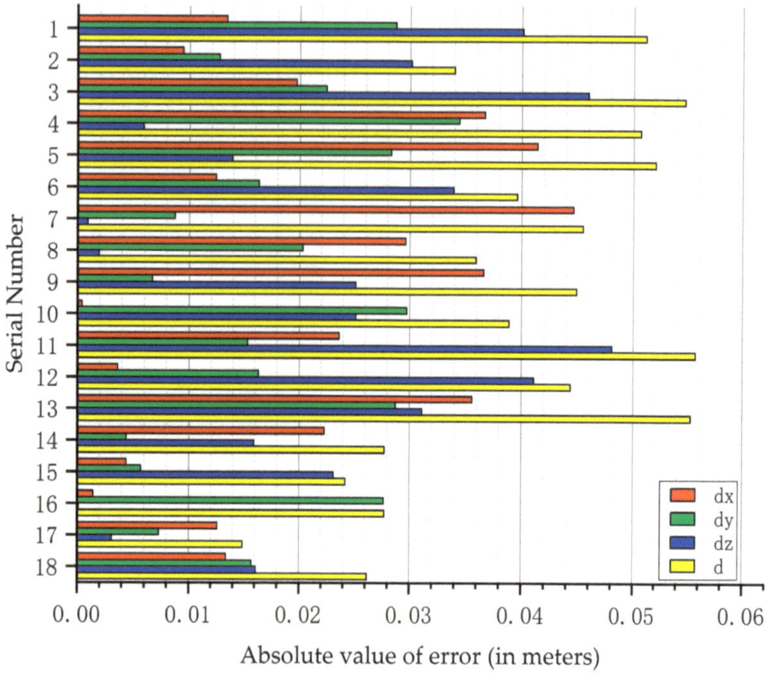

Figure 11. The histogram represents the spatial coordinate error in point cloud image matching.

4. Discussion

In this study, the proposed method automates the generation of a series of matching result files, based on point cloud and panoramic image data acquired using a vehicle-borne MMS. In experiments, the method demonstrated favorable outcomes in terms of visualization effects, processing efficiency, and matching accuracy. Compared to two other matching methods, the point cloud coloring algorithm has a relatively shorter processing time, but the resulting output is a colored point cloud map, which does not facilitate spatial information applications based on vision. The method of point cloud direct projection is simple, but the matching results contain a significant amount of noise points and projected points from occluded objects, which interfere with accuracy and require longer processing times. Overall, the method presented in this paper demonstrates certain advantages.

Additionally, the method is subject to certain limitations of use, which are summarized as follows:

1. The data for this method are derived from a vehicle-borne MMS; hence, the input necessitates point cloud maps and panoramic images captured from a ground perspective on flat roads. Should the acquisition method alter (such as handheld or unmanned aerial vehicle acquisition), the data processing logic may require corresponding adjustments.
2. In order to achieve the desired processing outcomes, the threshold of the algorithm should be adjusted according to the characteristics of the input data. The text discusses the method of adjusting the threshold based on point cloud density, which is the most significant data characteristic; however, this necessitates a certain level of prior knowledge and preliminary debugging. Consideration should be given to the intelligent enhancement of the process, enabling the threshold setting to adapt and adjust according to the actual circumstances.
3. With this method, the point cloud map exhibits a sparse distribution of points in the peripheral regions and the partial absence of point clouds for certain key tar-

gets, thereby affecting the comprehensiveness of spatial data. In order to acquire a comprehensive point cloud map, one may refer to the methods described in the literature [30,31], which involve acquiring point clouds multiple times for the same target to generate redundant data. Subsequently, an error model is employed to assess the quality of the point cloud, with only higher-quality data points retained, thus yielding a high-density, precise point cloud of the target object.

5. Conclusions

For this research, we studied a method for generating a matching relationship between point clouds and panoramic image data obtained from a vehicle-borne mobile mapping system based on occlusion removal. Through the designed point-cloud-processing workflow, the effective point clouds corresponding to the camera scenes were generated; based on the spherical projection model and the design logic of filtering occluded points using vector angles, the problem of filtering the projection of occluded points in the perspective of panoramic images was solved, and the method logic was streamlined with the ability to automate and batch process.

An experimental comparison of the characteristics between this method and two commonly used matching techniques—point cloud coloring and point cloud direct projection—was conducted to demonstrate our method's superiority in terms of processing efficiency and accuracy. Furthermore, an analysis was performed to examine the impacts of data characteristics, primarily point cloud density and distance of the points from the camera optical center, on the selection of algorithm thresholds. Under appropriate threshold values, accuracy experiments were also conducted, and the results show that, under the proposed process, the average generation error of point clouds is around 3 cm, with a maximum error of no more than 5 cm; the average pixel matching error is around 2.8 pixels, with a maximum error of about 6; additionally, the average position matching error is about 4 cm, with a maximum error of no more than 6 cm.

In light of the preceding discussion, the prospective research directions for this method ought to be concentrated on the following components:

1. Enhancing the applicability of algorithms under different acquisition methods and varying data characteristics. Devising multiple algorithmic logics in order to accomplish the processing of input data in diverse forms should be considered. Threshold control functions that utilize more fundamental variables, such as vehicle speed and the distance to the primary target, should be designed as inputs, enabling adaptive adjustment of thresholds to achieve the desired processing outcomes.
2. Acquisition of high-quality point cloud maps. Incorporating an optimization module for point cloud maps should be considered. Based on redundant point cloud data acquired multiple times, a filtering algorithm should be designed to retain high-quality data points, thereby generating an accurate point cloud of the primary target and supplementing the missing portions of the point cloud maps from conventional acquisition methods.
3. Enhancing registration accuracy and processing efficiency of data. Accurate matching results are derived from precise calibration parameters of the camera and laser scanner. Consideration should be given to employing registration algorithms that integrate point clouds with images to acquire more accurate coordinate transformation parameters between the two data types. Additionally, the computational speed of the method can be further enhanced through programming techniques such as parallelization and CUDA acceleration, thereby reducing the time consumption for generating matching results.

Overall, the proposed method yields satisfactory matching results, offering a reference-value-rich scheme for the fusion of multi-source data based on point clouds and panoramic images. The generated matching results can provide spatial information corresponding to pixel coordinates based on the image, offering robust support for algorithms that acquire object positional information through visual features. This method can be broadly applied in the fields of surveying, mapping, positioning, and navigation.

Author Contributions: Conceptualization, J.J. and W.W.; methodology, J.J. and H.B.; software and investigation, J.J.; validation, J.J., H.B. and W.W.; formal analysis, J.J. and Y.N.; resources and supervision, Y.R. and W.W.; data curation and project administration, W.W.; writing—original draft preparation and visualization, J.J.; writing—review and editing, Y.N. and J.J.; funding acquisition, W.W. and Y.R. All authors have read and agreed to the published version of the manuscript.

Funding: This work was financially supported by the Key Technology Research and Development Program of Shandong Province (2021SFGC0401), the Natural Science Foundation of China (42204011), and the Shandong Provincial Natural Science Foundation (ZR2021QD058).

Data Availability Statement: The data presented in this study are available on request from the corresponding author, due to legal reasons.

Acknowledgments: We thank the editors and reviewers for their hard work and valuable advice.

Conflicts of Interest: The authors declare no conflicts of interest.

References

1. Xu, S. Research on Calibration of Mobile Mapping System for Land Vehicle and Its Accuracy Assessment. Ph.D. Thesis, Wuhan University, Wuhan, China, 2016.
2. de Paula Pires, R.; Olofsson, K.; Persson, H.J.; Lindberg, E.; Holmgren, J. Individual tree detection and estimation of stem attributes with mobile laser scanning along boreal forest roads. *ISPRS J. Photogramm. Remote Sens.* **2022**, *187*, 211–224. [CrossRef]
3. Li, Q.; Yuan, P.; Lin, Y.; Tong, Y.; Liu, X. Pointwise classification of mobile laser scanning point clouds of urban scenes using raw data. *J. Appl. Remote Sens.* **2021**, *15*, 024523. [CrossRef]
4. Li, Y.; Cai, Y.; Malekian, R.; Wang, H.; Sotelo, M.A.; Li, Z. Creating navigation map in semi-open scenarios for intelligent vehicle localization using multi-sensor fusion. *Expert Syst. Appl.* **2021**, *184*, 115543. [CrossRef]
5. Lin, Y.-C.; Manish, R.; Bullock, D.; Habib, A. Comparative analysis of different mobile LiDAR mapping systems for ditch line characterization. *Remote Sens.* **2021**, *13*, 2485. [CrossRef]
6. Xu, M.; Zhong, X.; Huang, J.; Ma, H.; Zhong, R. A method for accurately extracting power lines and identifying potential intrusion risks from urban laser scanning data. *Opt. Lasers Eng.* **2024**, *174*, 107987. [CrossRef]
7. Paijitprapaporn, C.; Thongtan, T.; Satirapod, C. Accuracy assessment of integrated GNSS measurements with LIDAR mobile mapping data in urban environments. *Meas. Sens.* **2021**, *18*, 100078. [CrossRef]
8. Javed, Z.; Kim, G.-W. PanoVILD: A challenging panoramic vision, inertial and LiDAR dataset for simultaneous localization and mapping. *J. Supercomput.* **2022**, *78*, 8247–8267. [CrossRef]
9. Zhang, J.; Lin, X. Advances in fusion of optical imagery and LiDAR point cloud applied to photogrammetry and remote sensing. *Int. J. Image Data Fusion* **2017**, *8*, 1–31. [CrossRef]
10. Zhang, J.; Pan, L.; Wang, S. *Geo-Spatial Information Science*; Wuhan University Press: Wuhan, China, 2009.
11. Ravi, R.; Habib, A. Fully Automated profile-based calibration strategy for airborne and terrestrial mobile LiDAR systems with spinning multi-beam laser units. *Remote Sens.* **2020**, *12*, 401. [CrossRef]
12. Yao, L.; Wu, H.; Li, Y.; Meng, B.; Qian, J.; Liu, C.; Fan, H. Registration of vehicle-borne point clouds and panoramic images based on sensor constellations. *Sensors* **2017**, *17*, 837. [CrossRef]
13. Zhang, Y.; Cui, Z. Registration of terrestrial LiDAR and panoramic imagery using the spherical epipolar line and spherical absolute orientation model. *IEEE Sens. J.* **2022**, *22*, 13088–13098. [CrossRef]
14. Wang, Y.; Li, Y.; Chen, Y.; Peng, M.; Li, H.; Yang, B.; Chen, C.; Dong, Z. Automatic registration of point cloud and panoramic images in urban scenes based on pole matching. *Int. J. Appl. Earth Obs. Geoinf.* **2022**, *115*, 103083. [CrossRef]
15. Zhu, N.; Yang, B.; Dong, Z.; Chen, C.; Huang, X.; Xiao, W. Automatic registration of mobile mapping system lidar points and panoramic-image sequences by relative orientation model. *Photogramm. Eng. Remote Sens.* **2021**, *87*, 913–922.
16. Li, J.; Yang, B.; Chen, C.; Huang, R.; Dong, Z.; Xiao, W. Automatic registration of panoramic image sequence and mobile laser scanning data using semantic features. *ISPRS J. Photogramm. Remote Sens.* **2018**, *136*, 41–57. [CrossRef]
17. Wang, B.; Li, H.; Zhao, S.; He, L.; Qin, Y.; Yang, X. Automatic Registration of Panoramic Image and Point Cloud Based on the Shape of the Overall Ground Object. *IEEE Access* **2023**, *11*, 30146–30158. [CrossRef]
18. Liu, R.; Chai, Y.; Zhu, J. Accuracy analysis and optimization of panoramic depth image. *Sci. Surv. Mapp.* **2021**, *10*, 170–176.
19. Julin, A.; Kurkela, M.; Rantanen, T.; Virtanen, J.-P.; Maksimainen, M.; Kukko, A.; Kaartinen, H.; Vaaja, M.T.; Hyyppä, J.; Hyyppä, H. Evaluating the quality of TLS point cloud colorization. *Remote Sens.* **2020**, *12*, 2748. [CrossRef]
20. Yuan, C.; Pan, J.; Zhang, Z.; Qi, M.; Xu, Y. 3D-PCGR: Colored Point Cloud Generation and Reconstruction with Surface and Scale Constraints. *Remote Sens.* **2024**, *16*, 1004. [CrossRef]

21. Shinohara, T.; Xiu, H.; Matsuoka, M. Point2color: 3d point cloud colorization using a conditional generative network and differentiable rendering for airborne lidar. In Proceedings of the IEEE/CVF Conference on Computer Vision and Pattern Recognition, Nashville, TN, USA, 19–25 June 2021; pp. 1062–1071.
22. Liu, R.; Yang, J.; Ren, H.; Cong, B.; Chang, C. Research on a pavement pothole extraction method based on vehicle-borne continuous laser scanning point cloud. *Meas. Sci. Technol.* **2022**, *33*, 115204.
23. Xu, Z.; Xiang, Z.; Liang, F. A fusion method of LiDAR point cloud and ladybug panoramic image. *Bull. Surv. Mapp.* **2019**, 78–81. [CrossRef]
24. Ku, J.; Harakeh, A.; Waslander, S.L. In defense of classical image processing: Fast depth completion on the cpu. In Proceedings of the 2018 15th Conference on Computer and Robot Vision (CRV), Toronto, ON, Canada, 9–11 May 2018; pp. 16–22.
25. Xiang, R.; Zheng, F.; Su, H.; Zhang, Z. 3ddepthnet: Point cloud guided depth completion network for sparse depth and single color image. *arXiv* **2020**, arXiv:.09175.
26. Bai, L.; Zhao, Y.; Elhousni, M.; Huang, X. DepthNet: Real-time LiDAR point cloud depth completion for autonomous vehicles. *IEEE Access* **2020**, *8*, 227825–227833. [CrossRef]
27. Zhang, W.; Qi, J.; Wan, P.; Wang, H.; Xie, D.; Wang, X.; Yan, G. An easy-to-use airborne LiDAR data filtering method based on cloth simulation. *Remote Sens.* **2016**, *8*, 501. [CrossRef]
28. Liu, J.; Xu, W.; Jiang, T.; Han, X. Development of an Attitude Transformation Method From the Navigation Coordinate System to the Projection Coordinate System. *IEEE Geosci. Remote Sens. Lett.* **2019**, *17*, 1318–1322. [CrossRef]
29. Guo, M.; Zhu, L.; Huang, M.; Ji, J.; Ren, X.; Wei, Y.; Gao, C. Intelligent extraction of road cracks based on vehicle laser point cloud and panoramic sequence images. *J. Road Eng.* **2024**, *4*, 69–79. [CrossRef]
30. Vallet, B.; Soheilian, B.; Paparoditis, N. Uncertainty propagation for terrestrial mobile laser scanner. *Int. Arch. Photogramm. Remote Sens. Spat. Inf. Sci.* **2016**, *41*, 331–335.
31. Ozendi, M.; Akca, D.; Topan, H. A point cloud filtering method based on anisotropic error model. *Photogramm. Rec.* **2023**, *38*, 460–497. [CrossRef]

remote sensing

MDPI

Article

Image-Aided LiDAR Extraction, Classification, and Characterization of Lane Markings from Mobile Mapping Data

Yi-Ting Cheng [1], Young-Ha Shin [2], Sang-Yeop Shin [1], Yerassyl Koshan [1], Mona Hodaei [1], Darcy Bullock [1] and Ayman Habib [1,*]

[1] Lyles School of Civil Engineering, Purdue University, West Lafayette, IN 47907, USA; cheng331@purdue.edu (Y.-T.C.); shin337@purdue.edu (S.-Y.S.); ykoshan@purdue.edu (Y.K.); mhodaei@purdue.edu (M.H.); darcy@purdue.edu (D.B.)

[2] Department of Geoinformation Engineering, Sejong University, Seoul 05006, Republic of Korea; subzero@sju.ac.kr

* Correspondence: ahabib@purdue.edu

Abstract: The documentation of roadway factors (such as roadway geometry, lane marking retroreflectivity/classification, and lane width) through the inventory of lane markings can reduce accidents and facilitate road safety analyses. Typically, lane marking inventory is established using either imagery or Light Detection and Ranging (LiDAR) data collected by mobile mapping systems (MMS). However, it is important to consider the strengths and weaknesses of both camera and LiDAR units when establishing lane marking inventory. Images may be susceptible to weather and lighting conditions, and lane marking might be obstructed by neighboring traffic. They also lack 3D and intensity information, although color information is available. On the other hand, LiDAR data are not affected by adverse weather and lighting conditions, and they have minimal occlusions. Moreover, LiDAR data provide 3D and intensity information. Considering the complementary characteristics of camera and LiDAR units, an image-aided LiDAR framework would be highly advantageous for lane marking inventory. In this context, an image-aided LiDAR framework means that the lane markings generated from one modality (i.e., either an image or LiDAR) are enhanced by those derived from the other one (i.e., either imagery or LiDAR). In addition, a reporting mechanism that can handle multi-modal datasets from different MMS sensors is necessary for the visualization of inventory results. This study proposes an image-aided LiDAR lane marking inventory framework that can handle up to five lanes per driving direction, as well as multiple imaging and LiDAR sensors onboard an MMS. The framework utilizes lane markings extracted from images to improve LiDAR-based extraction. Thereafter, intensity profiles and lane width estimates can be derived using the image-aided LiDAR lane markings. Finally, imagery/LiDAR data, intensity profiles, and lane width estimates can be visualized through a web portal that has been developed in this study. For the performance evaluation of the proposed framework, lane markings obtained through LiDAR-based, image-based, and image-aided LiDAR approaches are compared against manually established ones. The evaluation demonstrates that the proposed framework effectively compensates for the omission errors in the LiDAR-based extraction, as evidenced by an increase in the recall from 87.6% to 91.6%.

Keywords: lane marking inventory; lane marking extraction; LiDAR; image; visualization/reporting; mobile mapping systems; lane marking characterization

Citation: Cheng, Y.-T.; Shin, Y.-H.; Shin, S.-Y.; Koshan, Y.; Hodaei, M.; Bullock, D.; Habib, A. Image-Aided LiDAR Extraction, Classification, and Characterization of Lane Markings from Mobile Mapping Data. *Remote Sens.* **2024**, *16*, 1668. https://doi.org/10.3390/rs16101668

Academic Editors: Kevin Tansey, Wanshou Jiang, San Jiang, Duojie Weng and Jianchen Liu

Received: 22 February 2024
Revised: 24 April 2024
Accepted: 4 May 2024
Published: 8 May 2024

1. Introduction

Several transportation engineers conduct road safety analyses to explore ways to decrease accidents by correlating crash data with human, roadway, or vehicle factors. As shown in Figure 1, out of the 95% of accidents related to human factors, 41% are the result of roadway features (highlighted in lavender in Figure 1) [1], which include (i) road surface conditions (i.e., potholes or cracks), (ii) lane marking classification (i.e., color/pattern/location),

(iii) lane marking retroreflectivity, (iv) roadway geometry, and (v) lane width [2–4]. Therefore, lane marking inventory documenting roadway factors could reduce accidents and facilitate road safety analyses. Furthermore, a practical reporting mechanism that can visualize such information is critical for comprehensive, easy-to-grasp lane marking inventory. Several researchers established inventory using MMS data through road surface identification, color/intensity enhancement, lane marking extraction, classification, and characterization, as well as reporting mechanisms, which are briefly explained below.

- **Road surface identification:** define road surface regions in imagery/LiDAR data;
- **Color/intensity enhancement:** enhance the utility of color/intensity information for road surface in imagery/LiDAR data;
- **Lane marking extraction:** detect lane markings in the enhanced road surface imagery/LiDAR data;
- **Lane marking classification:** assign varying labels based on color/pattern/location to detected lane markings according to the Federal Highway Administration (FHWA) standard [5];
- **Lane marking characterization:** derive lane marking attributes (e.g., visibility conditions of lane markings, intensity profiles, and lane width) using classified lane markings;
- **Reporting mechanisms:** visualize the derived lane marking results (e.g., extracted lane markings and their characteristics) based on imagery and/or LiDAR data.

It is important to acknowledge that while human, vehicle, and roadway factors play a significant role in accidents, other factors such as weather, lighting, and systemic transportation issues can also contribute to accidents. This study focused on the importance of human, vehicle, and roadway factors due to their notable influence on road safety analysis, as outlined by the Federal Highway Administration (FHWA) [6].

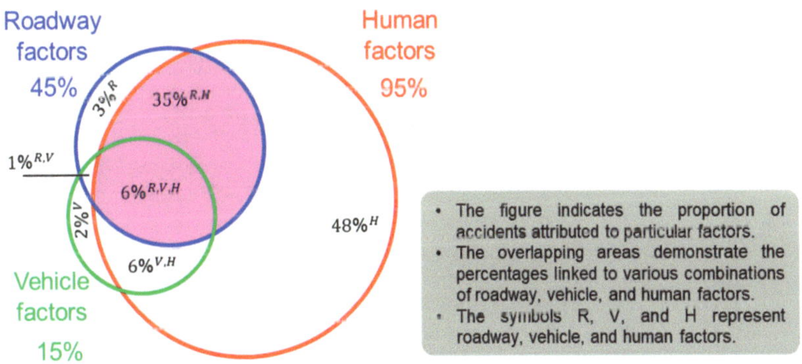

Figure 1. Distribution of how human, vehicle, and roadway factors result in accidents (adapted from Plankermann, K., 2014 [1]).

The majority of existing research has focused on steps a-d using either (1) imagery, (2) LiDAR, or (3) a combination of both. Image-based approaches [7–11] encompass road surface identification, color enhancement, and lane marking extraction. These approaches are reliable under reasonable weather and lighting conditions. On the other hand, LiDAR-based strategies [12–16] involve road surface identification, intensity enhancement, and lane marking extraction. LiDAR-based strategies demonstrate greater effectiveness in areas where there is a significant contrast between lane marking and pavement regions. Recently, several studies have investigated the potential of combining MMS imagery and LiDAR data for lane marking extraction. Huang et al. [17] proposed a lane marking extraction algorithm based on video and LiDAR datasets in urban areas. They used a 1D kernel filter to extract lane markings from each video frame and remove false positives using LiDAR-based road surface identification. Li et al. [18] utilized an MMS to detect the road surface by combining LiDAR and image data and subsequently extracted lane markings from the images. First, the road surface is determined using elevation changes in point clouds and color variability in RGB images. After that, road surface regions in all images are transferred into bird-eye-view imagery. Finally, Otsu thresholding and Hough transform strategies are applied to the bird-eye-view imagery for lane marking extraction. Shin et al. [19] presented a lane marking extraction approach based on imagery and LiDAR data, where road surface points are first identified by applying RANSAC-based plane fitting to point clouds. Next, lane marking points are extracted from road surface point clouds through intensity thresholding. Lane markings are also identified from images using a median filter and local thresholding. Finally, the LiDAR-based lane markings are projected onto images to eliminate false positives, considering only the lane markings detected by both LiDAR and images as correct detections. Gu et al. [20] proposed an algorithm to fuse imagery and LiDAR data for lane marking extraction. They identified road surface points based on the installation height of the used LiDAR scanner and projected them onto the corresponding images. In the road surface regions of each image, the RGB model is converted into HSV, and the saturation channel is replaced with LiDAR intensity. These road surface regions are used as input for training a convolutional neural network (CNN) model [21] for lane marking extraction. They labeled 30,000 images to train the model. Bai et al. [22] developed a CNN model for lane marking extraction using imagery and LiDAR data. They converted LiDAR point clouds into three channel images, with the first channel representing intensity and the second and third channels representing the elevations of the highest and lowest points within each discretization bin. After that, the images were fed to a ResNet50 model [23] to estimate road surface elevation. Based on the road surface elevation, all camera images are converted into bird-eye-view ones. Finally, LiDAR and camera images are used to train a CNN model for lane marking extraction.

Although the aforementioned approaches are capable of extracting lane markings, they have primarily been conducted or evaluated using an MMS equipped with a single camera and/or LiDAR unit within a limited study area (typically ranging from 1 to 25 km long). Furthermore, these approaches have mainly focused on extracting lane markings along the driving lane, with few studies addressing lane marking classification and characterization. In most image LiDAR-based approaches, imagery data have been predominantly used as the primary source, while LiDAR data are utilized to enhance lane marking extraction. Even when using imagery to enhance LiDAR-based extraction, the focus is mainly on addressing false positives in LiDAR-based lane markings. However, it is important to consider the strengths and weaknesses of both camera and LiDAR units when establishing lane marking inventory, as summarized in Table 1. Images are susceptible to weather and lighting conditions, and lane markings can be obstructed by neighboring traffic. They also lack 3D and intensity information, although color information is available. On the other hand, LiDAR data collection is not affected by adverse weather and lighting conditions, and it has minimal occlusions. One should note that occlusion minimization in LiDAR data depends on the type of used scanner. A single 2D LiDAR system would suffer from occlusions as it scans a given object location only once. Implementing two 2D LiDAR units will have fewer occlusions as the two units will scan an area from two locations, which could minimize occlusions. When using several multi-beam spinning LiDAR units, occlusions are greatly reduced by the fact that a given object region is scanned from several locations by the same and different units onboard the mobile mapping system. LiDAR data also provide 3D and intensity information. Moreover, relying solely on imagery or LiDAR data presents challenges in assessing the insufficient grinding of lane markings and temporary pavement markers. Figure 2 displays a situation where inadequate grinding of previous lane markings can still be visible in a LiDAR point cloud but not in an image. Conversely, temporary lane markings can be observed in imagery but not in LiDAR data if glass beads are not applied, as shown in Figure 3. Considering the complementary characteristics of camera and LiDAR units, an image-aided LiDAR framework would be highly advantageous for lane marking inventory purposes.

Table 1. Advantages and shortcomings of using either camera or LiDAR units onboard an MMS for lane marking extraction.

Sensor	Pros	Cons
Camera	• Color information is available • Massive existing image-processing strategies • Temporary lane markings might be detected	• Images are affected by weather and lighting conditions • Lots of occlusions due to neighboring traffic • No 3D and intensity information • Insufficient grinding of lane markings might not be detected
LiDAR scanner	• Data collection is not affected by adverse weather and lighting conditions • Minimal occlusions • 3D and intensity information is available • Insufficient grinding of lane markings might be detected	• No color information • Temporary lane markings might not be detected

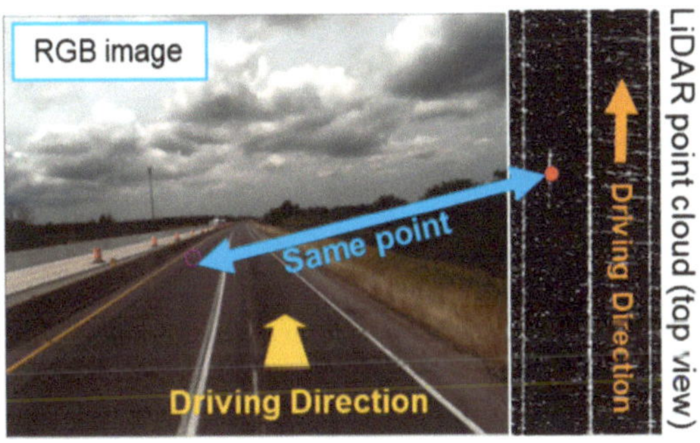

Figure 2. Insufficient grinding of lane markings invisible in RGB imagery (empty magenta circle) but visible in a LiDAR point cloud (red dot).

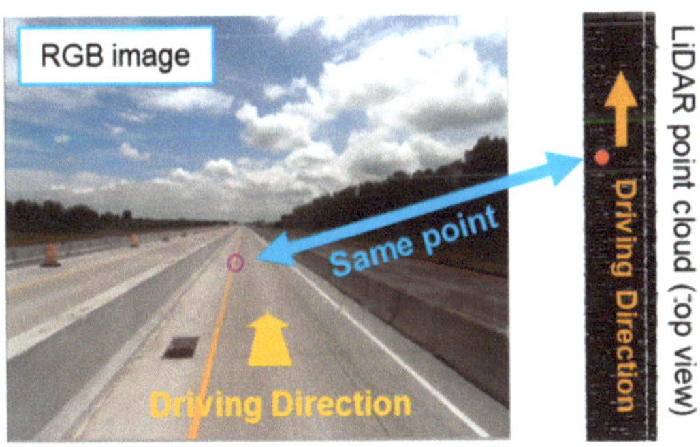

Figure 3. Temporary lane markings visible in RGB imagery (empty magenta circle) but invisible in a LiDAR point cloud (red dot).

For a reporting mechanism of lane marking inventory, several transportation agencies have utilized a geographic information system (GIS) [24,25]. These GIS-based reporting mechanisms involve overlaying various 2D bird-eye-view features on top of each other for visualization or spatial analysis [26]. However, with the advent of MMS platforms, existing GIS-based reporting mechanisms may not be adequate for displaying lane marking results derived from imagery/LiDAR data. Given that an MMS captures perspective-view images, the image-based lane marking results may require orthorectification before using them in GIS-based reporting mechanisms. However, the visualization of perspective-view images in a reporting mechanism raises another issue of integrating 2D imagery and 3D LiDAR data. This issue could be addressed through forward and backward projection. Forward projection allows the transformation from a 2D point on an image to the corresponding 3D point in LiDAR data. In contrast, backward projection enables the transformation from a 3D point in LiDAR data to the corresponding 2D point on an image. Additionally, forward/backward projection in a reporting mechanism is essential for effective quality control of the results obtained from lane marking extraction, as it allows for the assessment of the alignment of derived lane markings from multiple sensors or an MMS. Nevertheless, to the best of the authors' knowledge, no study has been found that utilizes the projection capability to develop reporting mechanisms for lane marking inventory.

This paper addresses the aforementioned challenges by developing an image-aided LiDAR framework for establishing lane marking inventory. The main contributions of this study can be summarized as follows:

- Propose an image-aided LiDAR framework for the following:
- Lane marking extraction/classification/characterization;
- Identifying all lane markings visible in imagery/LiDAR data (not only along the driving lane);
- Handling multiple imaging and LiDAR sensors onboard an MMS;
- Evaluate the performance of the proposed strategies using an MMS equipped with multiple camera and LiDAR units along extended road segments;
- Develop a reporting mechanism for visualizing imagery and LiDAR data together with extracted lane markings, as well as their characteristics (e.g., visibility conditions of lane markings, intensity profiles, and lane width).

The remainder of this paper is organized as follows. Section 2 introduces the used MMS and imagery/LiDAR data. Section 3 presents the models used for forward/backward projection, the proposed image-aided LiDAR lane marking inventory framework, and metrics for performance evaluation. The experiment results are reported in Section 4. Finally, the conclusions and scope for future work are summarized in Section 5.

2. Data Acquisition Systems and Dataset Description

To propose a framework that can handle multiple imaging and LiDAR sensors with different positions and orientations, an MMS was deployed for data acquisition along extended road segments in this study. The following subsections outline the MMS specifications and provide details about the study site and acquired datasets.

2.1. Mobile Mapping System

To address the research objectives, this study adopted an MMS equipped with multiple imaging and LiDAR sensors; namely, the Purdue wheel-based mobile mapping system–high accuracy (PWMMS-HA), as displayed in Figure 4. The PWMMS-HA is equipped with four multi-beam LiDAR scanners: three Velodyne HDL-32E and one Velodyne VLP-16 Hi-Res. The HDL-32E comprises 32 vertically aligned laser rangefinders with a total vertical FOV ranging from -30.67° to $+10.67^\circ$. The VLP-16 Hi-Res consists of 16 laser rangefinders with a vertical FOV ranging from -10° to $+10^\circ$. The HDL-32E has a point capture rate of 700,000 points per second [27], while the VLP-16 Hi-Res has a point capture rate of 300,000 points per second [28]. In addition, three FLIR Grasshopper 3 9.1MP GigE cameras are installed on the PWMMS-HA: two forward facing and one rear facing. These cameras have a maximum image resolution of 9.1 MP (3376 column pixels $\times$ 2704 row pixels) and are synchronized to capture images at a rate of 1 frame per second per camera. The two front cameras are installed at 2.3 m height, with a slight downward angle of approximately 1° from the horizontal plane at their locations, allowing the bottom row of the image to capture the ground roughly 1.5 m ahead of the camera position. Similarly, the rear camera is positioned at the same height but at a slightly steeper angle of approximately 7°, allowing the bottom row of the image to capture the ground roughly 1 m behind the camera position. The above sensors are directly georeferenced by an Applanix POS LV 220 GNSS/INS unit. The GNSS collection rate is 20 Hz, and the measurement rate of the IMU is 200 Hz [29].

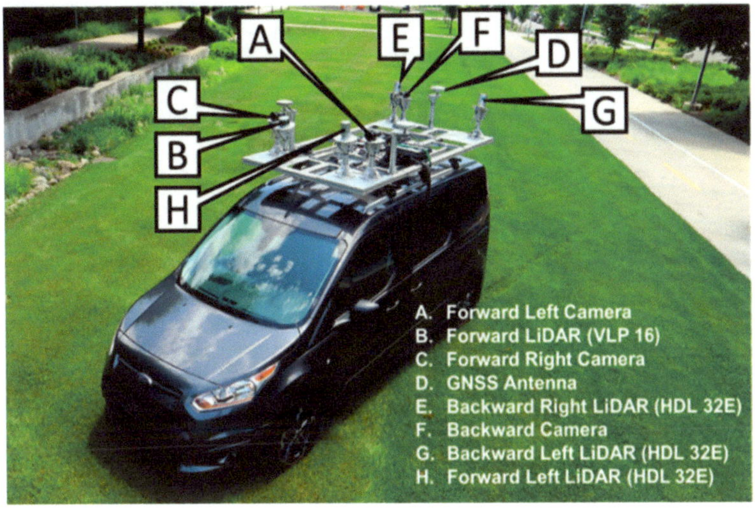

Figure 4. Illustrations of the Purdue wheel-based MMS–high accuracy system (PWMMS-HA).

After GNSS/INS post-processing, the positional accuracy is around ±2 cm, and the attitude accuracy is approximately ±0.020° [29]. The expected accuracy of the resulting point cloud, based on the LiDAR and navigation system specifications, is roughly 2–4 cm at a 30 m scanning range [30]. To achieve this expected accuracy, system calibration is conducted to estimate the mounting parameters relating the GNSS/INS unit to LiDAR and imaging sensors [31–33].

2.2. Study Site and Dataset Description

To demonstrate the capability of the proposed image-aided LiDAR framework, the PWMMS-HA is employed to collect a large set of imagery and LiDAR data. Specifically, a 110-mile section of I-465 in the United States, which includes both inner and outer loops, is surveyed. The trajectory and testing locations, where the performance of the proposed approaches (discussed later in Section 4.3) was evaluated, along I-465 are depicted in Figure 5. For the testing locations in Figure 5, locations 1–300 are situated on the inner loop, while locations 301–600 are located within the outer loop. The PWMMS-HA was able to capture up to five lanes per driving direction, denoted by location *i* in Figure 5. The specifications of the acquired imagery and LiDAR datasets along I-465 are provided in Table 2. The average local point spacing (LPS) of the road surface point clouds captured by the PWMMS-HA is 2.5 cm.

Table 2. Description of the imagery/LiDAR data collected by the PWMMS-HA along I-465 in the United States.

Highway	Date	Average Driving Speed (mph)	Imagery Data (# of Images)	LiDAR Data (# of Points)
I-465	11 July 2023	inner loop: 50.6 outer loop: 50.4	22,428 [1]	42,000 million

[1] Each camera captured a total of 7476 images.

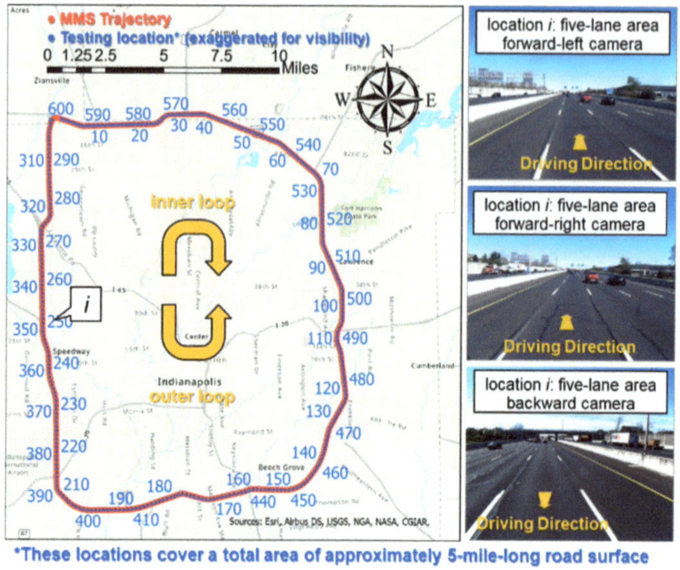

Figure 5. Study site, MMS trajectory (red dot), and 600 testing locations (blue dot, exaggerated for visibility) for performance evaluation along I-465, as well as sample RGB images capturing up to five lanes per driving direction in location *i*.

3. Methodology

To ensure consistent performance across datasets captured by multiple imaging and LiDAR sensors, this study proposes a geometric image-aided LiDAR framework. The geometric framework aims to overcome the potential inconsistency in performance that may arise from learning-based strategies when applied to datasets captured by different sensors [34]. The proposed image-aided LiDAR lane marking inventory framework is illustrated in Figure 6. This framework builds upon the forward/backward projection technique and previous research on LiDAR-based strategies for road surface identification, intensity normalization, and lane marking extraction/classification/characterization (interested readers can refer to previous studies [35] for more details regarding the above procedure). The subsequent subsections provide detailed explanations of the models used for the forward/backward projection between the imaging and LiDAR units onboard an MMS, the proposed image-aided LiDAR framework, and the metrics for performance evaluation.

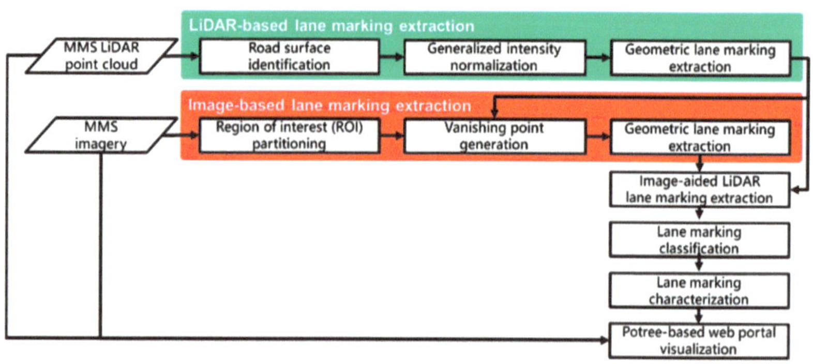

Figure 6. Flowchart of the proposed image-aided LiDAR lane marking inventory framework.

3.1. Point Positioning Models for Forward/Backward Projection

For the forward/backward projection in this study, the point-positioning equations for camera and LiDAR units onboard an MMS are established, as graphically explained in Figure 7. First, an imaging or laser beam ray relative to its sensor coordinate system needs to be defined. This can be achieved based on the imagery/LiDAR measurements (i.e., the image coordinates for a camera as well as the scanning range and direction for a LiDAR unit) and the Interior Orientation Parameters (IOPs) of the used sensors (i.e., parameters describing principal point coordinates, principal distance, and distortion parameters for a camera, as well as encoder mechanism for a LiDAR unit). After that, the position and orientation of the imaging or laser beam ray relative to the mapping frame can be established based on the Exterior Orientation Parameters (EOPs), which describe the position and orientation of the sensor relative to the mapping frame.

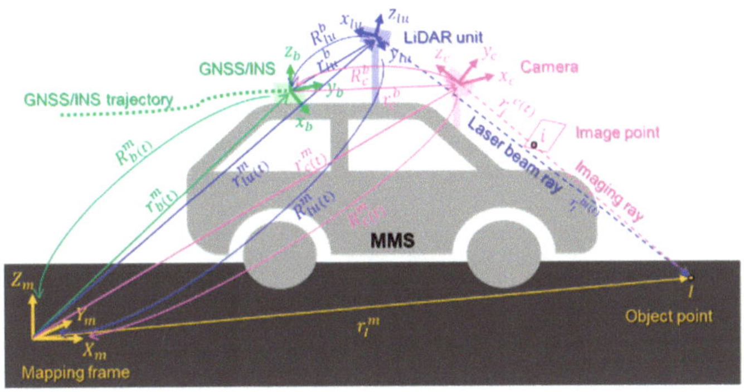

Figure 7. Schematic diagram of the point positioning models for GNSS/INS-assisted camera and LiDAR units onboard an MMS.

According to Figure 7, the point-positioning model of a camera onboard an MMS can be described by Equation (1), where $\lambda_{(i,c,t)}$ is a scaling factor associated with point i in the image captured by camera c at time t and $r_i^{c(t)}$ is the imaging ray for an image point i relative to the camera coordinate system at time t. Here, $r_i^{c(t)}$ is defined by the image coordinates of a point i (x_i and y_i) and the camera IOP, including the principal point coordinates (x_p and y_p) and principal distance (f), as well as distortions for the x and y coordinates ($dist_{x_i}$ and $dist_{y_i}$). As presented in Equation (2) and Figure 7, the position and orientation information of the camera frame relative to the mapping frame ($r_{c(t)}^m$ and $R_{c(t)}^m$) can be derived using the GNSS/INS trajectory information ($r_{b(t)}^m$ and $R_{b(t)}^m$) and mounting parameters between the camera frame and GNSS/INS body frame (r_c^b and R_c^b).

$$r_I^m = r_{c(t)}^m + \lambda_{(i,c,t)} R_{c(t)}^m r_i^{c(t)}, \quad r_i^{c(t)} = \begin{bmatrix} x_i - x_p - dist_{x_i} \\ y_i - y_p - dist_{y_i} \\ -f \end{bmatrix} \tag{1}$$

$$r_{c(t)}^m = r_{b(t)}^m + R_{b(t)}^m r_c^b \;\; \& \;\; R_{c(t)}^m = R_{b(t)}^m R_c^b \tag{2}$$

Similarly, the point-positioning model for a LiDAR unit onboard an MMS is described in Equation (3) and Figure 7. In Equation (3), $r_I^{lu(t)}$ represents the position of the footprint of a laser beam relative to the LiDAR unit frame at time t, while $r_{lu(t)}^m$ and $R_{lu(t)}^m$ are the position and orientation of the LiDAR unit frame relative to the mapping frame at time t. $r_I^{lu(t)}$ is determined based on the measurements of the LiDAR unit's range and pointing direction, as well as its IOP. In the case of a spinning multi-beam laser scanner, each laser beam is fired at a fixed vertical angle β, and the horizontal angle α is determined based on the rotation of the LiDAR unit. The range ρ is calculated based on the time it takes for the laser beam to travel from the firing point to the footprint. Accordingly, Equation (4) defines the coordinates of a 3D point relative to the LiDAR unit coordinate system. As shown in Figure 7, $r^m_{lu(t)}$ and $R^m_{lu(t)}$ can be estimated according to Equation (5), where $r^m_{b(t)}$ and $R^m_{b(t)}$ are the position and orientation of the GNSS/INS body frame relative to the mapping frame at time t and r^b_{lu}/R^b_{lu} represent the lever arm and boresight rotation matrix between the LiDAR unit and GNSS/INS body frame coordinate systems. Accordingly, for a LiDAR unit, the coordinates of an object point I in the mapping frame (r^m_I) can be computed in Equations (3)–(5).

$$r^m_I = r^m_{lu(t)} + R^m_{lu(t)} r^{lu(t)}_I \tag{3}$$

$$r^{lu(t)}_I = \begin{pmatrix} x \\ y \\ z \end{pmatrix} = \begin{pmatrix} \rho(t)cos\beta(t)cos\alpha(t) \\ \rho(t)cos\beta(t)sin\alpha(t) \\ \rho(t)sin\beta(t) \end{pmatrix} \tag{4}$$

$$r^m_{lu(t)} = r^m_{b(t)} + R^m_{b(t)} r^b_{lu} \quad \& \quad R^m_{lu(t)} = R^m_{b(t)} R^b_{lu} \tag{5}$$

Equations (1)–(5) indicate that accurate sensor measurements, trajectory, and system calibration parameters are crucial for obtaining properly georeferenced data from imaging and LiDAR units. In this study, a calibration procedure [32] is conducted to estimate the mounting parameters between the different LiDAR scanners and the GNSS/INS unit to reconstruct georeferenced, well-aligned point clouds. In addition, a simultaneous calibration [32,33] between the camera and LiDAR units is implemented to compute the mounting parameters of the onboard cameras, relative to the reference frame of the GNSS/INS unit. Based on the estimated trajectory and mounting parameters, forward and backward projection between point clouds and images can be established. Specifically, forward projection can identify the corresponding location in a LiDAR point cloud for a selected feature point in an image. On the other hand, backward projection allows for the determination of the corresponding point in an image where a 3D object point identified in a point cloud is selected. This forward/backward projection serves as the foundation for proposing an image-aided LiDAR lane marking inventory framework.

For forward projection, the corresponding object point coordinates to an image point (x_i, y_i) are estimated by finding the intersection between the imaging ray and a 3D LiDAR point cloud. To solve for the unknown scale factor $\lambda_{(i,c,t)}$ in Equation (1), an octree-based ray tracing algorithm [36] is adopted. This algorithm involves building an octree of the LiDAR points and generating a set of points at equal intervals along the imaging ray, denoted by $I_{ray,1} \sim I_{ray,8}$ in Figure 8. The closest LiDAR point to each point along the ray is identified, and the distance between the imaging ray and LiDAR points is computed. Starting from $I_{ray,1}$ (the closest point to the perspective center) in Figure 8, the first point (considered as the forward projected point) that satisfies the following two criteria can be determined: (i) the distance is smaller than a prespecified threshold (e.g., 0.2 m), and (ii) the distance is smaller than the distance for the next point. These criteria ensure the identification of the intersection between the imaging ray and the closest LiDAR surface (i.e., the visible surface). As shown in Figure 8, $I_{ray,3}$ is the first point that satisfies the criteria, with L_3 being the closest LiDAR point, while $I_{ray,4}$ and $I_{ray,7}$, although meeting the criteria, are not visible. By projecting L_3 onto the imaging ray (the yellow point in Figure 8), the closest LiDAR surface (i.e., the visible surface) can be determined. Finally, the forward projection solution can be obtained by the intersection between the projected point and the object's surface.

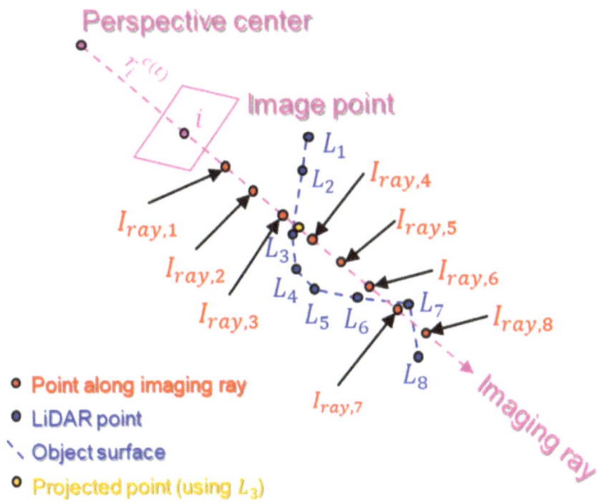

Figure 8. Schematic illustration of the forward projection algorithm: starting from $I_{ray,1}$ and $I_{ray,3}$ is the first point that satisfies the criteria, while $I_{ray,4}$ and $I_{ray,7}$ (which are not visible) also meet the criteria.

For backward projection, the corresponding image point for an object point I in a point cloud can be directly evaluated. The image point positioning equations—Equations (1) and (2)—can be reformulated as per Equation (6), expressing the image point coordinates as a function of known parameters, such as GNSS/INS trajectory information, camera IOP, camera mounting parameters, and ground coordinates of the object point, as well as the unknown scale factor $\lambda_{(i,c,t)}$. To eliminate this unknown scale factor, the first and second rows of Equation (6) are divided by the third row, resulting in Equations (7) and (8), which present the image point coordinates (x_i, y_i) corresponding to an object point I.

$$r_i^{c(t)} = \frac{1}{\lambda_{(i,c,t)}}\left[R_b^c R_m^{b(t)}\left(r_I^m - r_{b(t)}^m - R_{b(t)}^m r_c^b\right)\right] = \frac{1}{\lambda_{(i,c,t)}}\begin{bmatrix} N_x \\ N_y \\ D \end{bmatrix} \quad (6)$$

$$x_i = -c\frac{N_x}{D} + x_p + dist_{x_i} \quad (7)$$

$$y_i = -c\frac{N_y}{D} + y_p + dist_{y_i} \quad (8)$$

3.2. Image-Aided LiDAR Lane Marking Inventory Framework

The image-aided LiDAR lane marking inventory framework section starts by describing the LiDAR-based, image-based, and image-aided LiDAR strategies. Subsequently, the developed Potree-based web portal is introduced.

3.2.1. LiDAR-Based Lane Marking Extraction

According to Figure 6, the adopted LiDAR-based lane marking extraction includes road surface identification, generalized intensity normalization, and geometric lane marking extraction. For road surface identification, the cloth simulation filtering algorithm [37] is adopted to separate bare earth from above-ground objects. After that, the bare earth point cloud is divided into road surface blocks using the block length (L) along the driving direction and width (W) across the driving direction. In this study, L—e.g., 12.8 m—is chosen based on the minimum radius of curvature required for designing highways [38], which ensures that the lane markings are straight along the driving direction. The block width (W)—e.g., 18 m (with a 9 m extent on the right and left sides of the vehicle)—is determined by the average LPS of a point cloud, which ensures that the points representing lane markings are sufficiently dense for their extraction. To assess the LPS of the lane markings, a hypothesized lane marking is derived by applying the percentile intensity thresholding (Th_{int})—e.g., top 95th—to a small bare earth point cloud along the driving direction. The LPS of the hypothesized lane marking point cloud is then evaluated using the approach proposed by Lari and Habib [39]. This analysis indicates that if the width of road surface blocks exceeds 9 m on either side of the vehicle, the LPS of hypothesized lane markings is larger than the standard width of lane markings (i.e., lane markings cannot be extracted from LiDAR data at such distance).

The original intensity values of these road surface blocks are then normalized using a generalized normalization approach [35], which reduces intensity variation within and across different LiDAR units (i.e., normalizing intensity across all laser beams of a LiDAR scanner as well as all LiDAR units onboard an MMS). Finally, lane markings are extracted through five steps: (i) top 95th percentile intensity thresholding, (ii) scan line-based outlier removal, (iii) density-based spatial clustering [40], (iv) geometry-based outlier removal, and (v) local and global refinement. For a given road surface block, as shown in Figure 9a, hypothesized lane markings after step (i) are illustrated in Figure 9b. Next, for step (ii), the scan lines (which represent a sequence of points emitted by a LiDAR unit, denoted by *scan line* in Figure 9b) within these hypothesized lane markings are removed if they exceed a certain length, as shown in Figure 9c. This removal approach is based on the assumption that scan lines within a lane marking should not exceed a pre-defined threshold (Th_{line}), as lane markings have a specific width. The remaining hypothesized lane marking points are grouped into isolated segments using a density-based spatial clustering algorithm, as displayed in Figure 9d. After that, a geometry-based strategy is applied to remove non-linear segments and outlier points within linear segments, as shown in Figure 9e. Removing non-linear segments relies on each segment's principal component analysis (PCA), while removing outlier points within linear segments is based on a straight line that best fits each segment and random sample consensus (RANSAC). To connect isolated linear segments, two refinement strategies are employed: local refinement aims to connect small segments within each block and identify undetected lane marking points between small segments, while global refinement focuses on combining the same lane marking segments in successive blocks, as shown in Figure 9f,g.

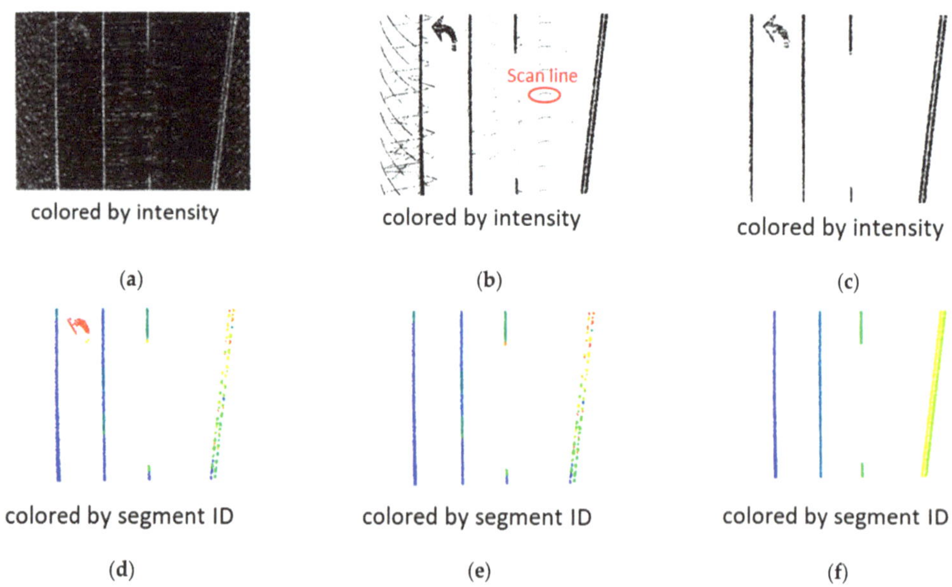

Figure 9. *Cont.*

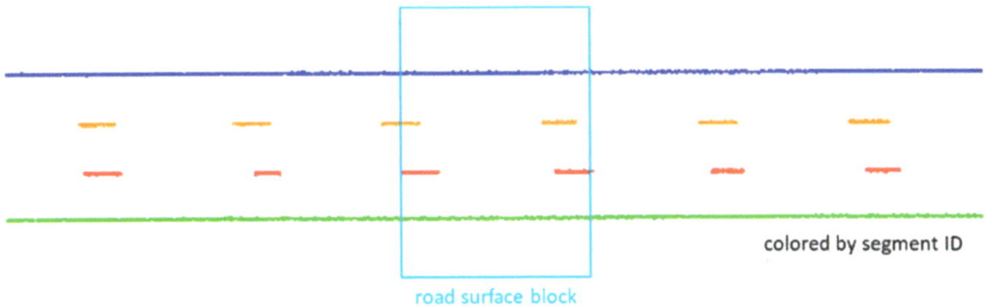

(g)

Figure 9. LiDAR-based lane marking extraction workflow: (**a**) road surface block, (**b**) hypothesized lane markings, (**c**) lane marking points after the scan line-based outlier removal, (**d**) lane marking segments after density-based spatial clustering, (**e**) lane marking segments after geometry-based outlier removal, lane marking segments after (**f**) local and (**g**) global refinements.

3.2.2. Image-Based Lane Marking Extraction

For images captured by an MMS, lane markings are typically found in the bottom half of an image, while the upper half contains elements, such as the sky and landmarks [41,42], in Figure 10. Thus, it is reasonable to focus on the bottom half of an image for image-based lane marking extraction. Nonetheless, merely applying a thresholding-based segmentation or straight line-based detection algorithm to the bottom half may not suffice due to several factors influencing the accuracy of image-based lane marking extraction. These factors include lighting conditions, lane marking geometry, and image resolution. For instance, lighting conditions may fluctuate within the bottom half of an image, leading to incorrect segmentation when using global thresholding. Although most lane markings close to the camera appear as straight lines in an image, their geometry may change to curved lines as they get further away. Additionally, as the distance from a camera to a lane marking increases, the image resolution at those lane markings decreases. To address these challenges, a region of interest (ROI) partitioning inspired by a prior study [43] is implemented. In this study, two ROIs—near and distant—within the bottom half of an image are established. The near ROI encompasses the area from the bottom row of an image up to a row above, denoted by a 10 m boundary, where a camera captures the road surface 10 m in front of the vehicle, as indicated by the red polygons in Figure 11. Conversely, the distant ROI covers the area beyond the 10 m boundary and extends to another row above, denoted by a 25 m boundary, where a camera captures the road surface 25 m in front of the vehicle, as represented by the blue polygons in Figure 11. One should note that the same ROI definition procedure applies to the rear camera, with the exception that the 10 m and 25 m distances are behind the vehicles. Applying different segmentation or detection algorithms to the two ROIs separately can reduce the impact of the aforementioned issues.

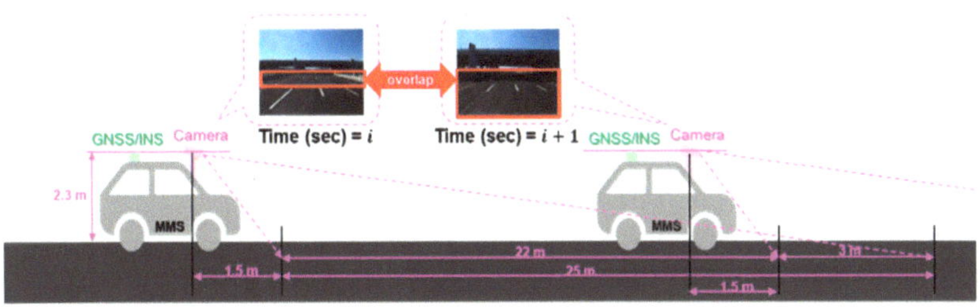

Figure 10. Schematic diagram of a front camera (capable of capturing one image per second; the bottom row of an image is capturing the ground roughly 1.5 m ahead of the camera position) onboard an MMS operating at 50 miles per hour (≈22 m per second) with two sample RGB images captured at epochs i seconds and $i + 1$ s.

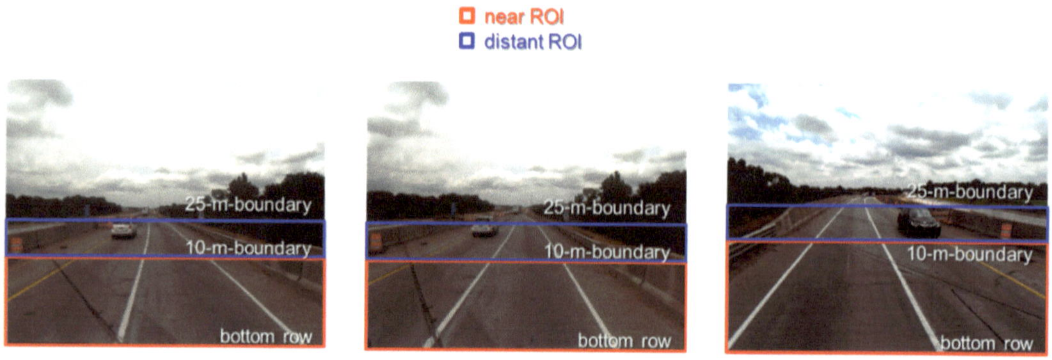

front-left camera front-right camera rear camera

Figure 11. Near and distant ROIs for sample images captured by front and rear cameras onboard the PWMMS-HA.

To define the 10 m boundary and 25 m boundary on an image, the LiDAR point cloud, as well as the position and orientation information of the used cameras (as described in Section 2.1), are utilized. For the 10 m boundary, the point along the road surface 10 m ahead of a camera is selected from the LiDAR data. This point is then backward projected (as described in Section 3.1) to find its corresponding point on an image, allowing for the determination of the 10 m boundary. The same procedure is implemented to determine the 25 m boundary. The boundaries, as described by their image rows, of the near/distant ROIs for each camera onboard the PWMMS-HA are listed in Table 3. One should note that the two front cameras are installed with similar downward angles, resulting in the same 10 and 25 m boundaries. However, the rear camera has a slightly steeper downward angle, leading to different 10 and 25 m boundaries. The choice of a 10 m boundary is based on a minimum design speed of 30 mph (≈13 m/s) and a chord length of 10 m, corresponding to an arc length of approximately 10.01 m for the minimum radius of curvature of 231 ft (≈70.4 m). Thus, the 10 m boundary ensures that lane markings are straight in a near ROI. In addition, the 25 m boundary guarantees sufficient overlap between successive images while the vehicle travels at speeds of approximately 50 mph (≈22 m/s). Given that the PWMMS-HA can capture one image per second, the gap between two successive image positions is about 22 m, which is less than the specified 25 m (resulting in around 3 m overlap between successive images), as shown in Figure 10. This study also assumes that lane markings are straight in the distant ROI, as the chord length and arc length are around 14.97 m and 15.0 m, respectively, for a minimum design speed of 30 mph.

Table 3. Boundary utilized for defining near and distant ROIs within the bottom half of images captured by the three cameras used in this study for image-based lane marking extraction.

Camera	Image Size (Pixel) #of Columns × # of Rows	Near ROI (Pixel) 10 m Boundary—Bottom Row	Distant ROI (Pixel) 25 m Boundary—10 m Boundary
front-left camera	3376 × 2704	1800–2704	1450–1800
front-right camera			
rear camera		1600–2704	1250–1600

Next, according to the proposed framework (Figure 6), vanishing point generation is conducted in this study. Vanishing points are crucial in extracting lane markings from images, as they provide geometric information for delineating lane markings along road surfaces. The concept of vanishing points relies on the fundamental principle of perspective in images—parallel lines in the real world converge to a single point in an image. For images captured by an MMS, lane markings are typically represented by parallel lines. By identifying the vanishing points corresponding to lane markings, one can infer the geometry information of the lane markings, which aids extraction.

Because lane markings are assumed to be straight in near/distant ROIs, two distinct vanishing points are generated using LiDAR-based lane markings within the respective ROIs in this study. Vanishing point generation is achieved by backward projecting the LiDAR-based lane markings (as explained in Section 3.1) onto each ROI in the images. Subsequently, the intersection point, determined by the projected lane markings in each ROI, can be considered a vanishing point, as shown in Figure 12. According to Yang et al. [44], the two vanishing points can be utilized to remove outliers within their respective ROIs. One should note that in cases where lane markings are not detected in LiDAR data, the vanishing points from the previous frame will be used for the current frame.

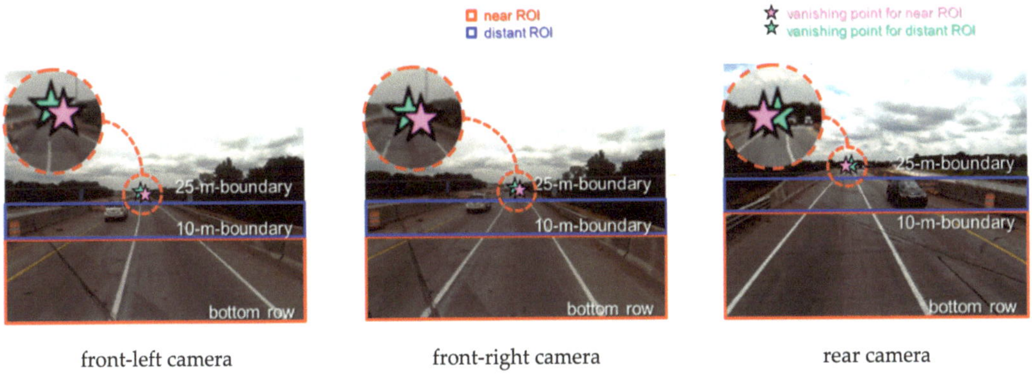

Figure 12. Vanishing points in near and distant ROIs for sample images captured by each camera onboard the PWMMS-HA.

Finally, for image-based lane marking extraction, color information of lane markings (e.g., white or yellow) is utilized, as they exhibit distinct hues compared to the surrounding pavement. According to Son et al. [10], lane markings exhibit consistent characteristics in the YCbCr model under different lighting conditions. Specifically, the Y layer can be utilized to identify most lane markings, as they consistently have higher values compared to other colors, regardless of lighting conditions. Conversely, the Cb layer can be used to distinguish non-white (e.g., yellow or red) lane markings, as they consistently have lower Cb values compared to other colors under various lighting conditions. Figure 13 illustrates an RGB image captured under poor lighting conditions and the corresponding Y and Cb layers. By taking advantage of the fact that lane markings are generally brighter than the surrounding pavement, most of them can be detected in the Y layer of an image, as shown in Figure 13a. Furthermore, the Cb layer can differentiate non-white lane markings, as shown in Figure 13b. One should note that the selection of the dark RGB image in Figure 13 was deliberate to illustrate the characteristics of lane markings in the Y and Cb layers under poor lighting conditions, and these images are not a consequence of printing issues.

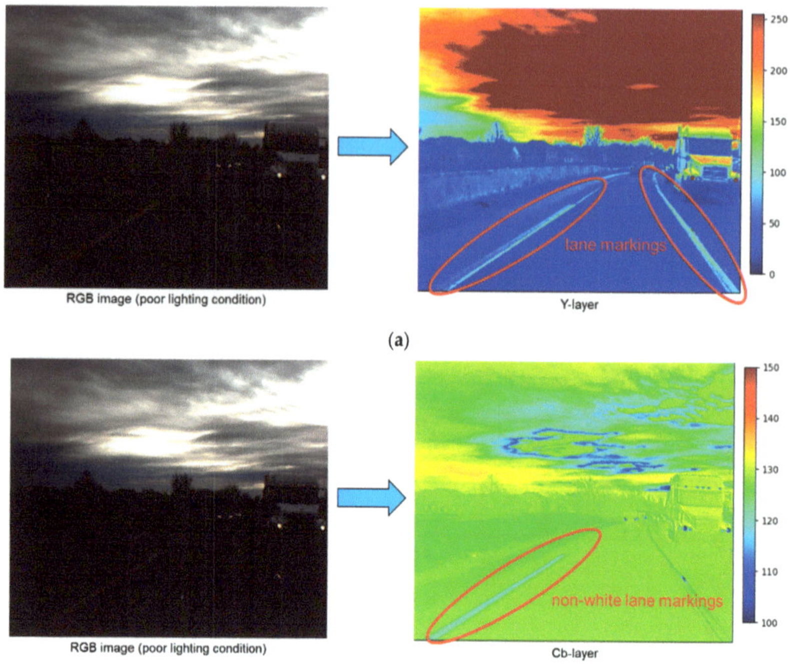

(**a**)

(**b**)

Figure 13. RGB-to-YCbCr conversion under poor lighting conditions: (**a**) Y layer with clear lane markings and (**b**) Cb layer with clear non-white lane markings.

Based on the ROIs, vanishing points, and characteristics of lane markings in Y/Cb layers, image-based lane marking extraction will be implemented through the following steps: (a) RGB-to-YCbCr conversion, (b) Y/Cb thresholding, (c) connected component labeling (CCL) [45], (d) contrast-based outlier removal, (e) vanishing point-based outlier removal, and (f) 3D width-based outlier removal. Figure 14a displays the Y and Cb layers overlaid with near/distant ROIs after step (a). Step (b) applies a Y percentile thresholding (Th_y)—e.g., top 97th—to the Y layers within each ROI since lane markings have higher Y values. In addition, a Cb percentile thresholding (Th_{cb})—e.g., lowest third—is applied to the Cb layers within each ROI since lane markings have lower Cb values. The selection of the top ninety-seventh and lowest third thresholding is based on the experiments conducted by Son et al. [10]. The results after Y/Cb thresholding are shown as binary images in Figure 14b. Subsequently, the detected pixels in each binary image are grouped into several segments based on four connectivity-based CCL in step (c), as presented in Figure 14c. After that, step (d) identifies segments with low contrast compared to their neighboring area and eliminates them as false positives, as displayed in Figure 14d. In step (d), each lane marking segment is divided into smaller segments using a length parameter (S)—e.g., 100 pixels—along the column direction of an image, as shown in Figure 15. Then, a buffer is created around each small segment by a buffer parameter (B)—e.g., 20 pixels—along the row direction of an image, and the average Y/Cb value within the buffer is calculated. Additionally, the average Y/Cb value within each small segment is computed. If the difference between the Y/Cb averages of a small segment and its surrounding buffer is less than a pre-defined threshold ($Th_{contrast}$), the small segment is deemed a false positive and subsequently removed. Step (e), as shown in Figure 14e, eliminates segments whose directional vector is not parallel to the vector formed by connecting its centroid to the corresponding vanishing point (the correspondence is determined based on near and distant ROIs). Step (f) involves checking the average 3D distance between the two edges across the main direction of each segment to remove outliers. The two edges are forward projected onto LiDAR data (as explained in Section 3.1), and the average distance between the projected edges is computed. Segments whose width does not meet the standard size of lane markings (Th_{3D}) are removed. The lane marking segments after step (f) are illustrated in Figure 14f. Finally, the results from the Cb layers can be used to differentiate between white and non-white lane markings extracted from the Y layers.

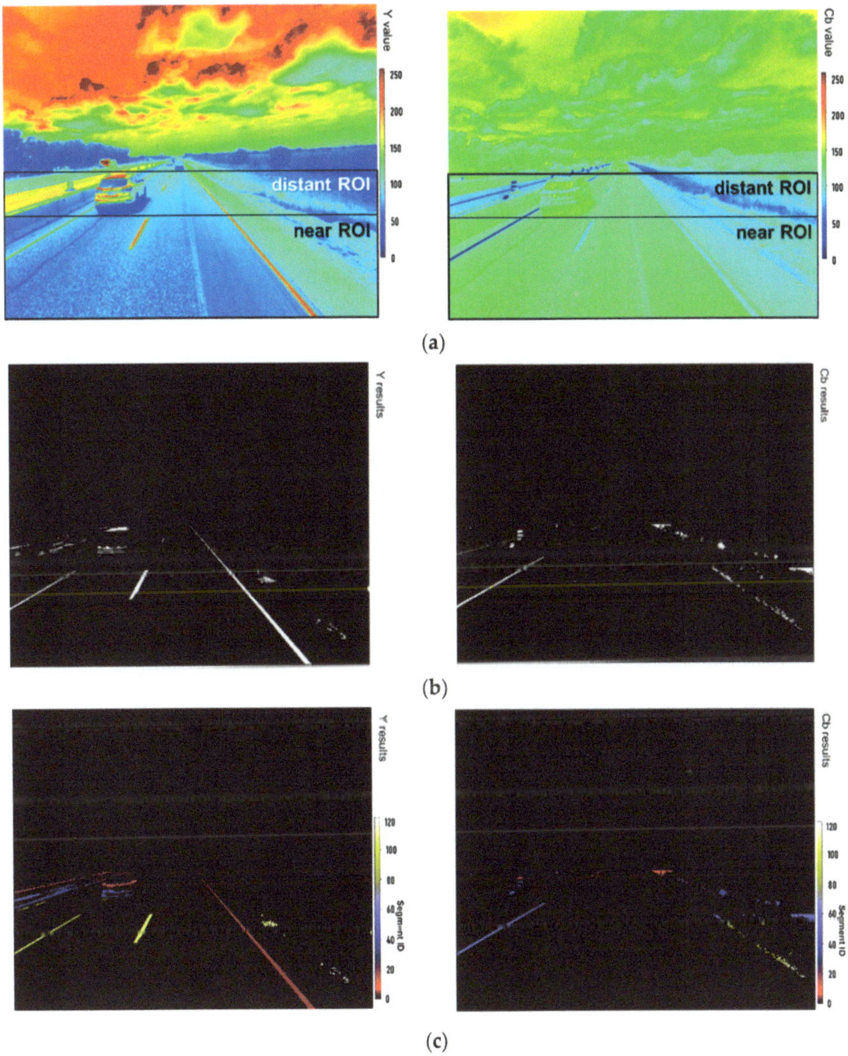

Figure 14. *Cont.*

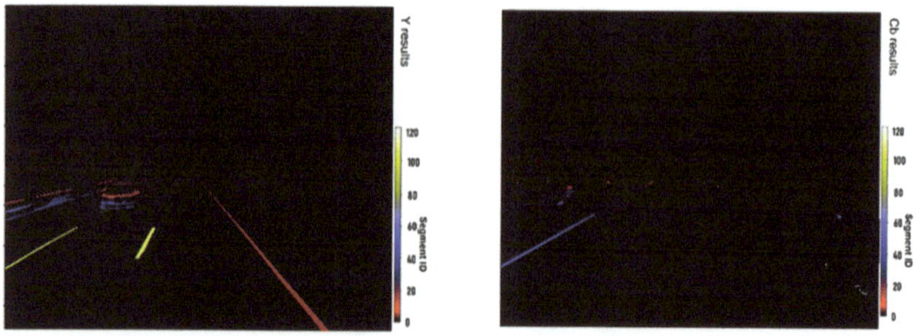

(**d**)

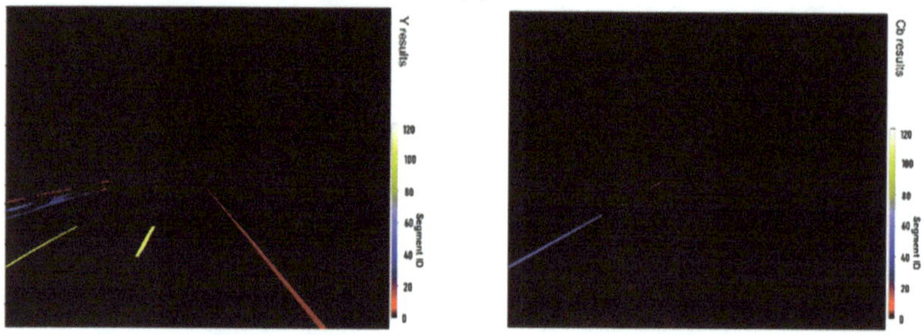

(**e**)

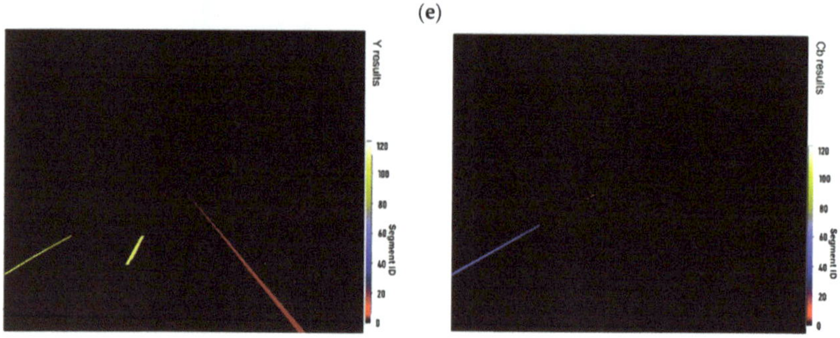

(**f**)

Figure 14. Image-based lane marking extraction algorithm: (**a**) Y and Cb layers overlaid with near/distant ROIs after (**b**) applying top 97th percentile thresholding to the Y layers within each ROI and lowest 3rd percentile thresholding to the Cb layers within each ROI, (**c**) four connectivity-based connected component labelings, (**d**) contrast-based outlier removal, (**e**) vanishing point-based outlier removal, and (**f**) 3D width-based outlier removal.

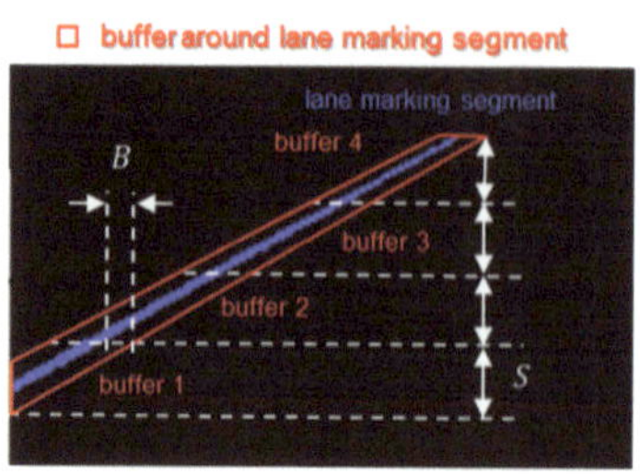

Figure 15. Schematic diagram of a lane marking segment in an image and the corresponding buffer determined by a length parameter (*S*) and a buffer parameter (*B*).

3.2.3. Image-Aided LiDAR Lane Marking Extraction/Classification/Characterization

Based on the lane markings derived from the LiDAR-based and image-based strategies described in Sections 3.2.1 and 3.2.2, respectively, image-aided LiDAR lane marking extraction can be conducted. First, all lane marking segments extracted from images are forward projected (as explained in Section 3.1) onto LiDAR data. In this study, the results obtained from all the images captured by the three cameras are projected onto LiDAR data. Figure 16 illustrates examples of the centerline points (at intervals of 10 pixels) of 2D image-based lane markings and the corresponding forward projected points onto the LiDAR data. Thereafter, considering the complementary characteristics of camera and LiDAR units (as mentioned in Section 1), LiDAR-based lane markings can be refined based on the correspondence between image-based and LiDAR-based lane markings. In this study, the correspondence between image-based and LiDAR-based lane markings is determined by a distance threshold (Th_{aid})—e.g., 20 cm—as follows:

- If projected image-based lane markings are within 20 cm of LiDAR-based lane markings, the LiDAR-based extraction will be colored according to the image-based results and FHWA standard colors [5]. For instance, if image-based lane markings are white, LiDAR-based ones will be colored using RGB values of 247, 241, and 227.
- If no projected image-based lane markings are within a 20 cm neighborhood of LiDAR-based lane markings (i.e., no corresponding LiDAR-based extraction in point clouds), the image-based extraction will be utilized to extract lane markings in point clouds. First, the top 95th percentile intensity thresholding is applied to a road surface point cloud to derive hypothesized lane markings. In the hypothesized lane marking point cloud, points within 20 cm (Th_{aid}) of projected image-based lane markings are extracted. The resultant lane markings will also be colored according to the abovementioned procedure. One should note that this study aims to utilize image information to refine LiDAR-based lane markings for establishing inventory, including intensity profiles for evaluating retroreflectivity. To prevent misrepresentation of the intensity profiles, areas with no intensity contrast in the LiDAR data will not be utilized to extract lane markings in point clouds.

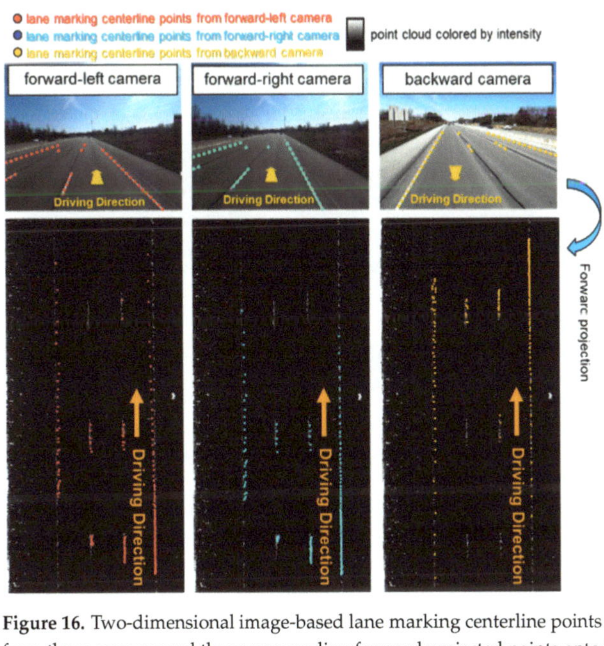

Figure 16. Two-dimensional image-based lane marking centerline points (at intervals of ten pixels) from three cameras and the corresponding forward projected points onto LiDAR data.

Once the image-aided LiDAR extraction is completed, lane markings can be further classified according to their pattern/location information through three steps: (I) length-based classification, (II) spline-based clustering, and (III) reference-based clustering. For step (I), as depicted in Figure 17a, each lane marking is categorized as either part of a solid, dash, or dotted line based on the FHWA standard length [5]. Thereafter, lane markings of the same type are grouped based on their relative position information for step (II), as shown in Figure 17b. Spline fitting [46] is applied to each lane marking within a specific type, and if the distance between any two splines meets a pre-defined criterion ($Th_{cluster}$), these two lane markings are grouped together. Here, the distance between two splines is estimated using an approach proposed by D'Errico [47]. In step (III), the lane marking clusters derived from step (II) are further grouped, as displayed in Figure 17c. Starting from the longest lane marking cluster, spline fitting is applied, and the spline is used as a reference. Spline fitting is then applied to the other lane marking clusters, and the distances between the reference spline and the splines from the other lane markings are estimated. If any splines from the other lane markings meet a pre-defined criterion ($Th_{cluster}$), these lane markings are grouped together. Throughout the subsequent discussion, we will refer to the groups of lane markings derived from the classification as "lane marking clusters".

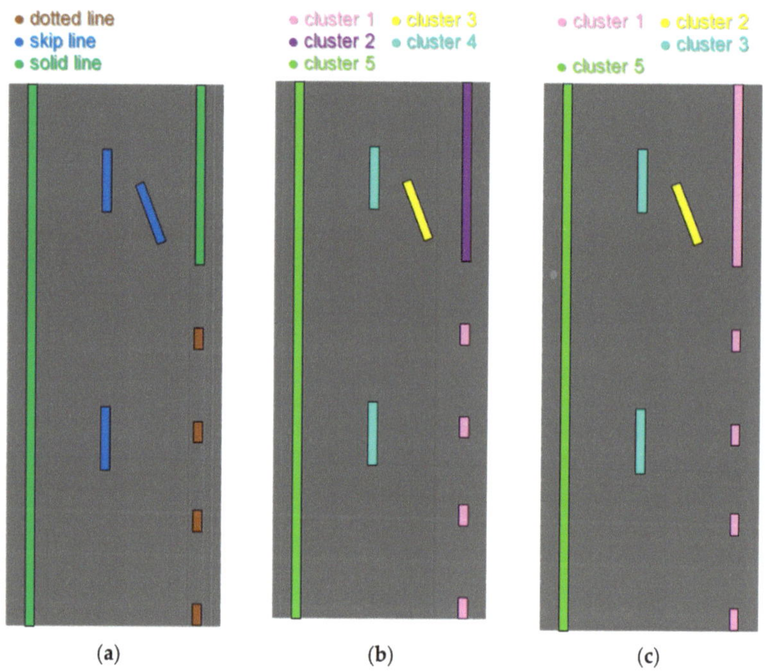

Figure 17. Lane marking classification algorithm: (**a**) length-based classification, (**b**) spline-based clustering, and (**c**) reference-based clustering.

After the classification, the resultant lane marking clusters can be utilized to derive intensity and lane width information. For intensity profile generation, each lane marking cluster (as shown in Figure 18a) is divided into 20 cm portions along the driving direction. The average intensity values are computed for each portion, and a graph can be created to display the average intensity against the traveled distance, as shown in Figure 18b. Lane width estimation starts with lane marking clusters. First, centerline points are generated at intervals of 20 cm along each cluster, as displayed in Figure 19a. In cases with missing lane markings for the gaps between dashed or dotted lines, a linear interpolation is performed to fill the gap between two consecutive centerline clusters, as shown in Figure 19b. One should note that this study adopts linear interpolation as per the FHWA standard [5], which stipulates that a straight line should connect two dashed/dotted lane markings. However, to avoid linear interpolation on curved roads, interpolation is not applied if the gap exceeds a pre-defined threshold (Th_{gap}). Subsequently, lane width can be computed using opposite centerline points from different lane marking clusters, as presented in Figure 19c. Finally, a plot can be generated to display the lane width against the traveled distance, as shown in Figure 19d.

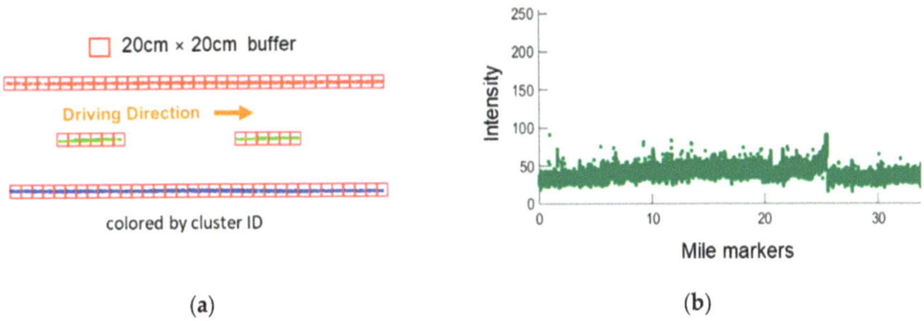

Figure 18. Illustrations of (**a**) 20 cm portions of each lane marking cluster and (**b**) an intensity profile plot.

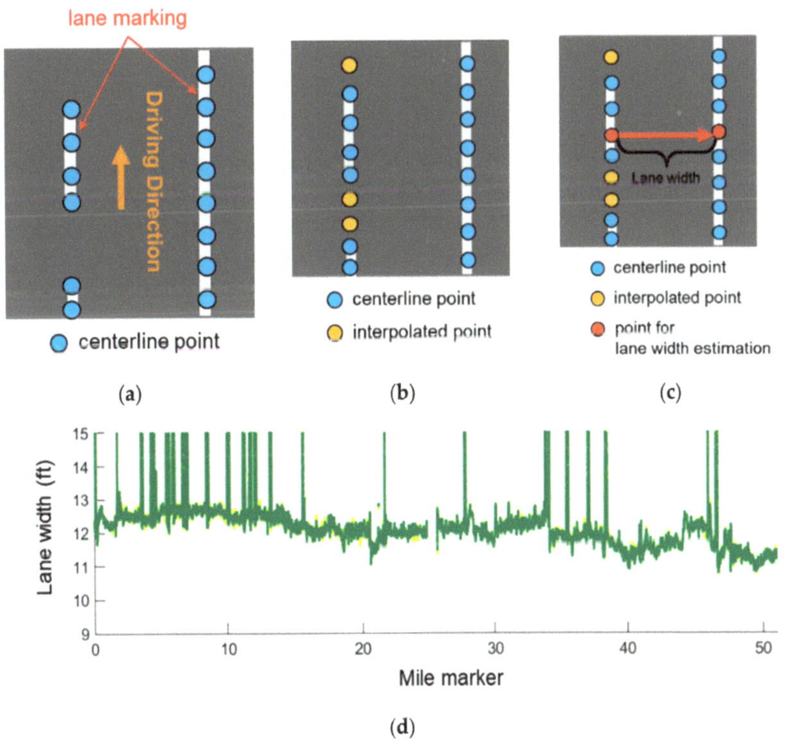

Figure 19. Illustrations of (**a**) centerline points generation, (**b**) linear interpolation, (**c**) lane width computation, and (**d**) a lane width plot.

3.2.4. Potree-Based Web Portal Visualization

In this study, Potree (http://www.potree.org, accessed on 1 January 2023) [48] is adopted to create a prototype visualization web portal. The architecture of the developed web portal, as well as the forward/backward projection functions and intensity profile/lane width displaying tools, are discussed in the following paragraphs.

Potree-based web portal

The architecture of the established Potree-based web portal is depicted in Figure 20 The front end, which is the graphical user interface, is used to display georeferenced imagery/LiDAR data (such as satellite imagery and LAS/LAZ files). The back end consists of various functions that allow users to manipulate the georeferenced data. The imagery and LiDAR data are stored in a database. Figure 21 displays the image placeholders (which indicate the position/orientation of the images) and LiDAR point clouds captured by the PWMMS-HA on top of a Cesium base map (https://cesium.com/, accessed on 25 September 2023) at Exit 25 on I-465. Thanks to the georeferencing parameters obtained from the GNSS/INS trajectory and system calibration procedures, the images are properly positioned and oriented relative to the point clouds, denoted by the yellow ovals in Figure 22. The back end receives client requests from the front end and processes them by interacting with the database using visualization and/or computational functions. For instance, in this study, the back end facilitates forward/backward projection functions in coordination with the front end and database.

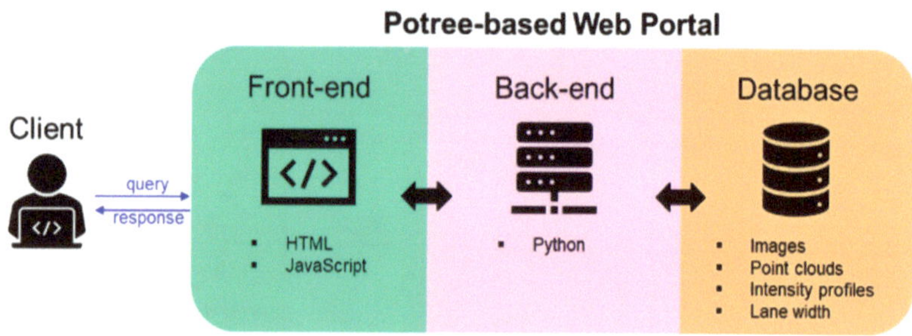

Figure 20. Architecture of the Potree-based web portal established in this study.

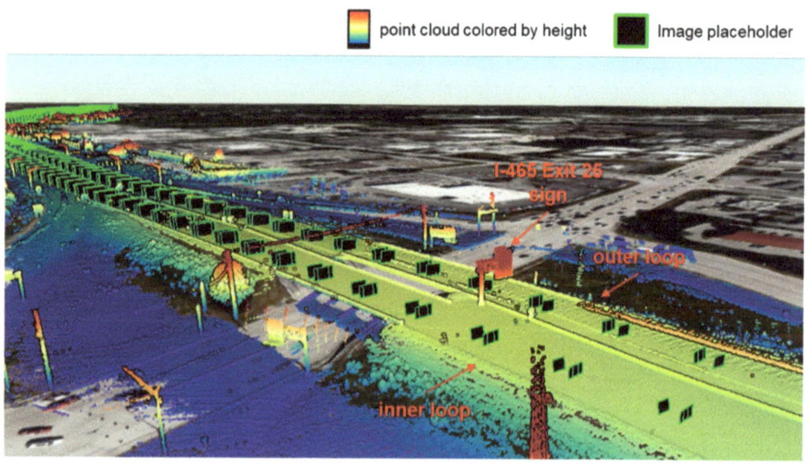

Figure 21. Illustration of the base map overlaid with a LiDAR point cloud and image placeholders (black polygons with a green boundary) at Exit 25 on I-465 in the Potree-based web portal.

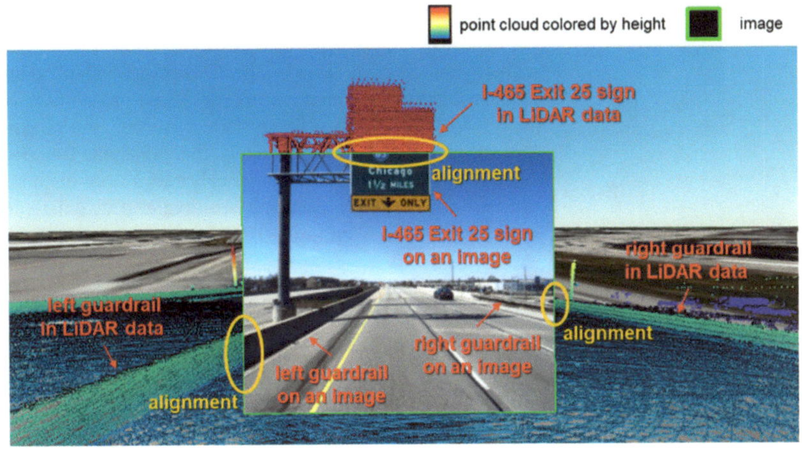

Figure 22. Illustration of an image properly aligned with a LiDAR point cloud, as highlighted by the yellow ovals, on top of a base map at Exit 25 on I-465 in the Potree-based web portal.

In addition, the Potree-based web portal provides users with built-in functionality for measuring distance, angles, and area in LiDAR data. Each measurement is visually represented by a sequence of user-selected vertices and labels, adhering to the International System of Units (SI). For example, a distance measurement can be represented by a vector connecting two vertices and a label showing the distance, as displayed in Figure 23a. On the other hand, an angle or area measurement necessitates a minimum of three vertices selected by a user, after which the angles or lengths of each edge and the area will be displayed in the web portal, as depicted in Figure 23b,c.

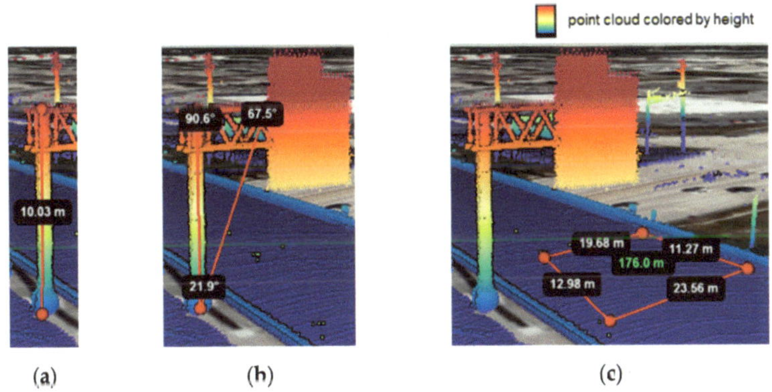

Figure 23. Illustrations of built-in measurement functionality in the Potree-based web portal: (**a**) measuring distance using two user-selected vertices (red dot) together with the corresponding label (white on black), (**b**) measuring angles using three user-selected vertices (red dot) together with the corresponding labels (white on black), and (**c**) measuring area using four user-selected vertices (red dot) together with the corresponding labels for the length of each edge (white on black) and area (green on black).

Forward/backward projection functions

To facilitate the visualization of corresponding features in imagery and LiDAR data, forward/backward projection functions (as explained in Section 3.1) are developed. As illustrated in Figure 24, the forward projection function projects a selected point from an image onto the corresponding LiDAR point cloud, with the blue placemark in the former appearing as a red dot (with a white-on-black label for the 3D coordinates) in the latter. Figure 25 shows that the backward projection function projects an object point (red dot with a white-on-black label for the 3D coordinates) in a point cloud onto the corresponding images, denoted by a blue placemark. These projection functions enable users to visualize georeferenced imagery/LiDAR data captured simultaneously or at different times by the same or various MMS. Additionally, this projection function can be employed to assess the accuracy of trajectory and system calibration. For instance, LiDAR-based lane markings can be backward projected onto an image to visualize any discrepancies between lane markings derived from multi-modal data.

Figure 24. Illustration of a selected point on an image (blue placemark) is forward projected onto the corresponding LiDAR point cloud (a red dot with a white-on-black label for the 3D coordinates) in the Potree-based web portal.

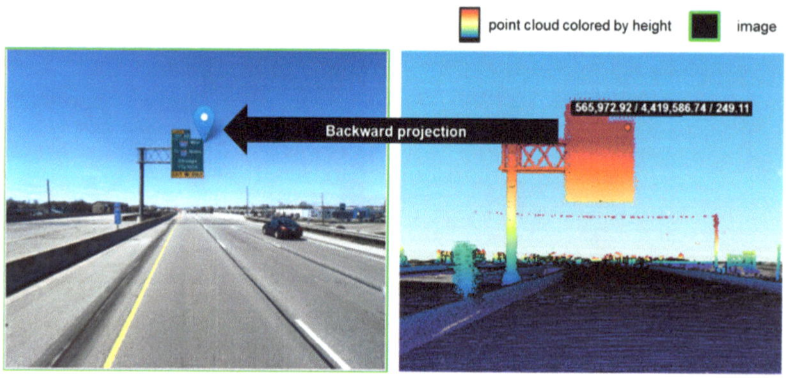

Figure 25. Illustration of a selected point in a LiDAR point cloud (a red dot with a white-on-black label for the 3D coordinates) is backward projected onto the corresponding image (blue placemark) in the Potree-based web portal.

Displaying tools for intensity profiles and lane width

To facilitate the visualization of the intensity profiles and lane width plots derived from the proposed lane marking characterization (as discussed in Section 3.2.3), tools for displaying these products are developed in this study. Figure 26 illustrates an intensity profile viewer within the Potree-based web portal. This viewer allows users to select a point of interest (crosshair cursor) in a profile, which will then display the corresponding point (red dot with a white-on-black label for the 3D coordinates) on the point cloud. Similarly, the lane width plot can also be visualized by a viewer, as shown in Figure 27. Users can select a point of interest (crosshair cursor) in lane width estimates, and then the corresponding point pair (connected by two red dots with a white-on-black label for the lane width) on the point cloud will be shown on the web portal, as shown in Figure 27. Once the points of interest are projected onto point clouds, users can further utilize the backward projection function to locate the corresponding points in an image.

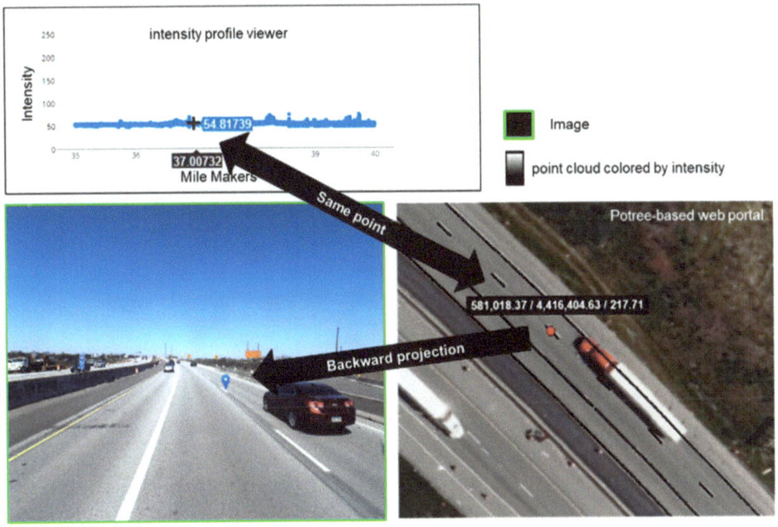

Figure 26. Illustration of a selected point in an intensity profile (crosshair cursor) and the corresponding point in a LiDAR point cloud (a red dot with a white-on-black label for the 3D coordinates), as well as the backward projected point (blue placemark), on an image.

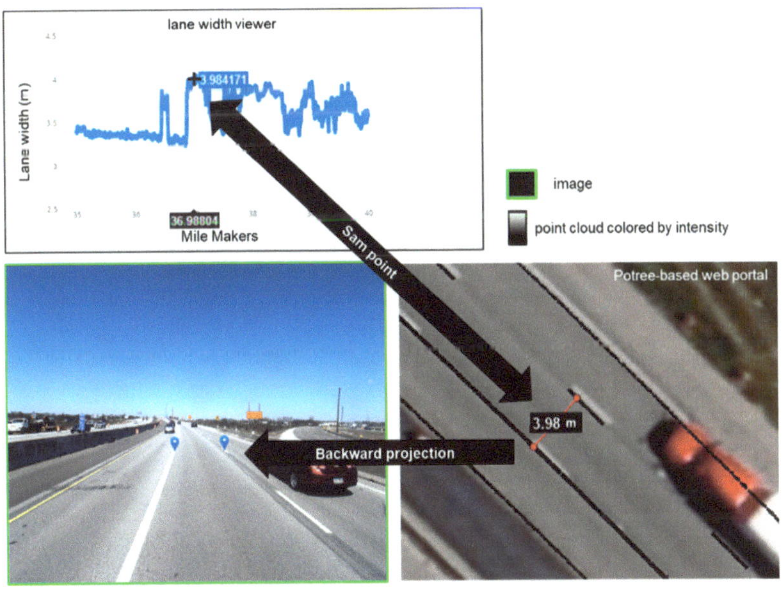

Figure 27. Illustration of a selected point in lane width plot (crosshair cursor) and the corresponding point pair in a LiDAR point cloud (connected by two red dots with a white-on-black label for the lane width), as well as the backward projected points (blue placemark), on an image.

3.3. Performance Evaluation

Given the minimal occlusions in LiDAR data compared to images, this study assumes that all lane markings are visible in point clouds. Consequently, this study manually annotated the point clouds captured by the PWMMS-HA to generate reference data for performance evaluation. Figure 5 shows the locations where LiDAR data were annotated. The annotated LiDAR data cover 600 road surface blocks (each block spans 12.8 m along the driving direction, covering a total area of approximately a 5-mile-long road surface), with 300 blocks located along the inner loop and another 300 blocks along the outer loop. As mentioned previously in Section 3.2.1, each LiDAR road surface block has a width of 18 m, which can capture up to five lanes (assuming a lane width of 3.6 m). However, not all 600 road surface blocks capture five lanes due to variations in road geometry, lane width, and the number of lanes on I-465. Additionally, even in a five-lane area, not all lanes can be captured if the vehicle is not driving in the middle lane. Thus, a total of 600 road surface blocks along I-465 are selected to ensure the evaluation of performance under different scenarios described above.

To compare the extracted lane markings with the reference LiDAR data, centerline points were created along each detected/annotated lane marking through the following steps. For the LiDAR-based extraction, each lane marking is divided into 3-meter-long segments to represent curved solid lines using straight segments. Centerline points are then computed along each segment at 20 cm intervals. The above procedure is also implemented to generate centerline points for the image-aided LiDAR and reference lane marking point clouds. On the other hand, for the image-based extraction, the lane marking segments in all images are forward projected onto LiDAR data (as explained in Section 3.1). Again, each projected lane marking is divided into 3-meter-long segments, and then 20-centimeter-interval centerline points are created along each segment.

Precision, recall, and F1-score are used as metrics to evaluate the performance of the LiDAR-based, image-based, and image-aided LiDAR lane marking extraction strategies. These metrics are calculated using Equations (9)–(11), where true positives, false positives, and false negatives are denoted by TP, FP, and FN, respectively. True positives refer to the lane markings correctly identified by a particular approach, while false positives (also known as commission errors) happen when a lane marking is mistakenly identified, even though it does not exist in the actual scene. False negatives (also known as omission errors) are the lane markings that an approach fails to identify. Accordingly, if an extracted centerline point is within 20 cm of a reference centerline point, it is considered a true positive. Conversely, if there is no reference point within the 20 cm neighborhood of an extracted point, it is considered a false positive. Similarly, if there is no extracted point within the 20 cm neighborhood of a reference point, it is considered a false negative. Precision measures the proportion of correctly detected lane markings out of detected ones, while recall presents the proportion of correctly detected lane markings out of reference ones. Lastly, F1-score, which provides an overall assessment of performance, is a combination of precision and recall using a harmonic mean.

$$Precision = \frac{TP}{TP + FP} \quad (9)$$

$$Recall = \frac{TP}{TP + FN} \quad (10)$$

$$F1 - score = \frac{2 \times Precision \times Recall}{Precision + Recall} \quad (11)$$

4. Experimental Results and Discussion

For the proposed image-aided LiDAR lane marking inventory framework, the thresholds/parameters used in this study are summarized in Table 4. These thresholds/parameters are applied consistently across imagery/LiDAR data obtained from different MMS sensors.

The experimental results and discussion section begins by presenting the products (i.e., extracted lane markings, intensity profiles, and lane width estimates) generated by the image-aided LiDAR lane marking inventory framework. Subsequently, the performance of LiDAR-based/image-based/image-aided LiDAR lane marking extraction strategies is evaluated through qualitative and quantitative analyses. Finally, the products and performance evaluation are discussed.

Table 4. Thresholds/parameters used for LiDAR-based/image-based/image-aided LiDAR strategies in this study.

Threshold/ Parameter	Description	Strategy (Section)	Value
L	Length of road surface blocks	LiDAR-based lane marking extraction (3.2.1)	12.8 m
W	Width of road surface blocks	LiDAR-based lane marking extraction (3.2.1)	18 m
Th_{int}	Percentile intensity threshold for lane marking extraction from point clouds	LiDAR-based lane marking extraction (3.2.1)	top 95th %
Th_{line}	Length threshold for scan line-based outlier removal	LiDAR-based lane marking extraction (3.2.1)	25 cm
Th_y	Percentile Y value threshold for lane marking extraction from images	Image-based lane marking extraction (3.2.2)	top 97th %
Th_{cb}	Percentile Cb value threshold for lane marking extraction from images	Image-based lane marking extraction (3.2.2)	lowest 3rd %
S	Length for dividing a segment in an image	Image-based lane marking extraction (3.2.2)	100 pixels
B	Number of pixels for creating buffers around a segment in an image	Image-based lane marking extraction (3.2.2)	20 pixels
$Th_{contrast}$	Y/Cb value threshold for contrast-based outlier removal in an image	Image-based lane marking extraction (3.2.2)	5 Y/Cb values
Th_{3D}	Lane marking width threshold for 3D width-based outlier removal in an image	Image-based lane marking extraction (3.2.2)	15 cm
Th_{aid}	Distance threshold for determining the correspondence between image-based and LiDAR-based lane markings, as well as extracting lane markings in point clouds using image-based results	Image-aided LiDAR lane marking extraction (3.2.3)	20 cm
$Th_{cluster}$	Distance threshold for grouping splines	Image-aided LiDAR lane marking classification (3.2.3)	75 cm
Th_{gap}	Gap threshold for avoiding linear interpolation on curved roads	Image-aided LiDAR lane marking characterization (3.2.3)	40 m

4.1. Products from Image-Aided LiDAR Lane Marking Inventory Framework

The proposed image-aided LiDAR lane marking inventory framework was applied to mobile data spanning 110 miles. Table 5 lists the execution time for the different lane marking extraction approaches. Figure 28 demonstrates the Cesium base maps overlaid with the point clouds and image placeholders along I-465. The portal is able to render the LiDAR and imagery datasets for both inner and outer loops in around ten seconds. Furthermore, extracted lane markings can also be visualized through the portal, as shown in Figure 29. Users can interact with the rendered data using the built-in functions, such as rotation, zooming in/out, and panning, without experiencing any delays, as demonstrated by the red zoom-in boxes in Figures 28 and 29. Additionally, the web portal allows users to visualize intensity profile/lane width plots along I-465. By selecting a specific portion based on mile markers, users can view the corresponding intensity profile/lane width plot in a viewer window, as shown in Figure 30.

Table 5. Processing time for LiDAR-based, image-based, and image-aided LiDAR strategies based on one-mile-long lane marking extraction.

Approach	Time Taken (Seconds) for One-Mile-Long Lane Marking Extraction	Platform
LiDAR based	~450 [1]	32 GB RAM computer
Image based	~5070 [2]	12.7 GB RAM (GPU) Google Collaboratory
Image-aided LiDAR	~5970 [3]	32 GB RAM computer and 12.7 GB RAM (GPU) Google Collaboratory

[1] Time is estimated using four LiDAR units. [2] Time is estimated using three cameras, and each individual image requires approximately 45 s. [3] Time includes the duration required for LiDAR-based and image-based strategies.

(a)

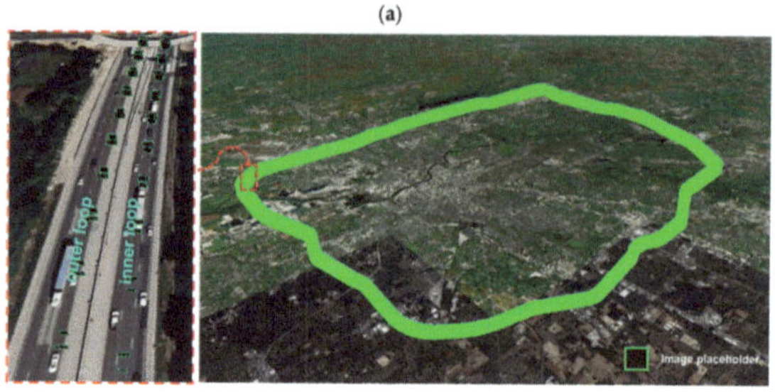

(b)

Figure 28. Illustrations of the base map overlaid with (**a**) LiDAR point clouds and (**b**) image placeholders and their zoom-in windows (red dotted polygon) in the Potree-based web portal.

Figure 29. Illustration of the base map overlaid with lane marking point clouds and a zoom-in window (red dotted polygon) in the Potree-based web portal.

Figure 30. Illustration of the Potree-based web portal viewers for intensity profiles and lane width estimates, allowing users to select a specific portion of the highway (red polygon) based on mile markers for visualization.

4.2. Qualitative Evaluation Using Potree-Based Web Portal Visualization

Through the built-in functionality and tools developed in this study for the web portal, the extracted lane markings and intensity profile/lane width results, as well as the alignment of lane markings derived from imaging/LiDAR units, can be qualitatively evaluated. For evaluating LiDAR-based/image-based/image-aided LiDAR extraction strategies, LiDAR-based lane marking point clouds are imported into the web portal. The centerline points generated using the forward projected image-based lane markings (as explained in Section 3.3) are also imported. Finally, the image-aided LiDAR and reference lane markings are imported for evaluation. Figure 31 presents a region where lane markings were not detected using the image-based approach but were successfully identified using the LiDAR-based and image-aided LiDAR strategies. In contrast, Figure 32 demonstrates a region where the LiDAR-based approach failed but the image-based and image-aided LiDAR strategies were effective. Figures 33 and 34 display the LiDAR-based and image-based lane marking centerline points as well as the corresponding images with the 2D extraction at the same locations as Figures 31 and 32.

As depicted in Figure 33, the inability of the image-based strategy to extract lane markings is attributed to the excessive change in lighting conditions. To investigate the failure of the LiDAR-based lane marking extraction, the intensity profiles in the same locations as Figure 32 were examined using the intensity profile display tool in the web portal. The intensity values for the lane markings were not detected by the LiDAR-based approach, and their surrounding lane markings are depicted in Figure 34. The decrease in intensity values from 52 to 47, as shown in Figure 34, could potentially explain the failure of the LiDAR-based extraction. The lane markings with intensity values lower than the surrounding ones might not be extracted using the LiDAR-based approach. This finding also suggests that the incorporation of image information can enhance the extraction of lane markings that are not detected by the LiDAR-based approach. Furthermore, Figures 32 and 34 serve as evidence that the proposed framework is capable of extracting lane markings within a five-lane region.

For the qualitative evaluation of intensity profiles, Figure 35 displays a worn-out lane marking region with an intensity value of 45, along with the corresponding extracted lane marking point cloud and image. Figure 36 shows another region with an intensity value of 55 for well-preserved lane markings and its corresponding point cloud and image. These intensity values are consistent with the lane marking conditions in the corresponding images. For the qualitative evaluation of lane width estimates, Figure 37 presents a region with a lane width estimate of 3.60 m (yellow oval) and the corresponding point cloud and image. The red dots within the yellow oval in Figure 37 are correctly positioned on the opposite lane markings, and the estimate is similar to the manual measurements of 3.61 m (cyan oval), which is close to the value of the estimate, obtained through the built-in functionality of the web portal. All the placemarks in all the camera images in Figures 35–37 are derived by backward projecting the intensity profile/lane width points in LiDAR data.

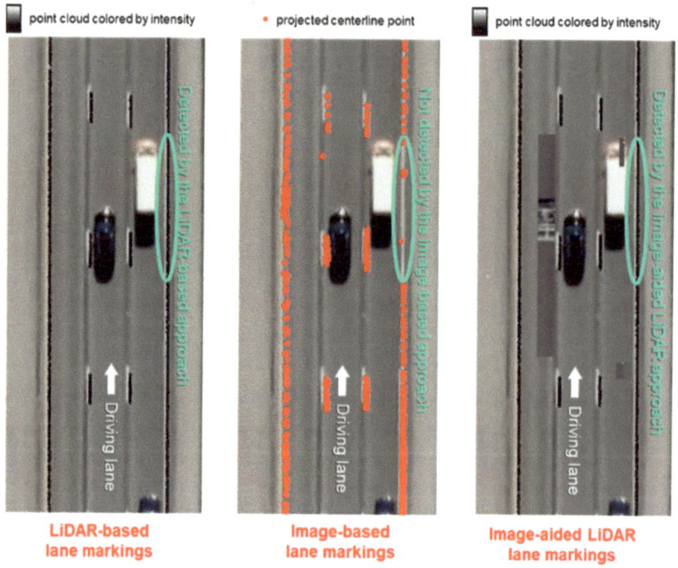

Figure 31. Illustration of lane markings derived through LiDAR-based, image-based, and image-aided LiDAR approaches (cyan ovals show a region where lane markings were not detected using the image-based approach but were successfully identified using the LiDAR-based and image-aided LiDAR strategies).

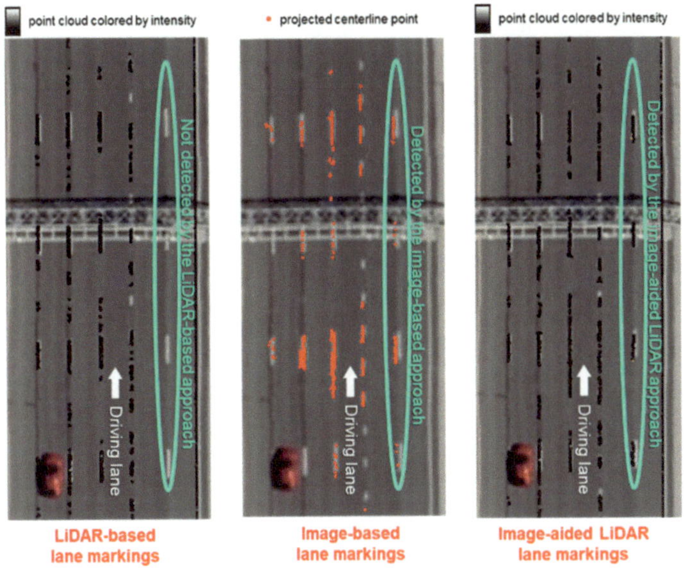

Figure 32. Illustration of lane markings derived through LiDAR-based, image-based, and image-aided LiDAR approaches (cyan ovals show a region where lane markings were not detected using the LiDAR-based approach but were successfully identified using the image-based and image-aided LiDAR strategies).

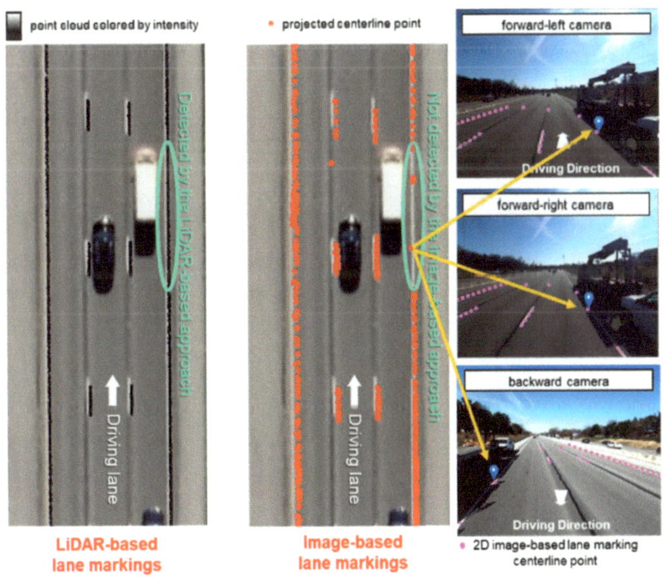

Figure 33. Illustration of the LiDAR-based and image-based lane markings and the corresponding images with 2D extraction (lavender point) for lane markings that were not detected in the imagery data (cyan oval shows a region where lane markings were not detected using the image-based approach but were successfully identified using the LiDAR-based and image-aided LiDAR strategies).

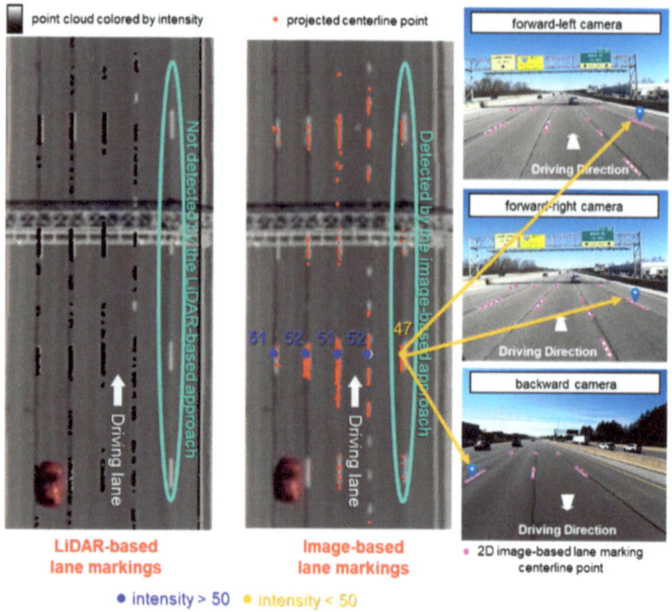

Figure 34. Illustration of the LiDAR-based and image-based lane markings with LiDAR intensity values (blue text) highlighting an area (yellow text) showing the relative lane marking extraction performance (cyan ovals show a region where lane markings were not detected using the LiDAR-based approach but were successfully identified using the image-based and image-aided LiDAR strategies) and the corresponding images with 2D extraction (lavender point).

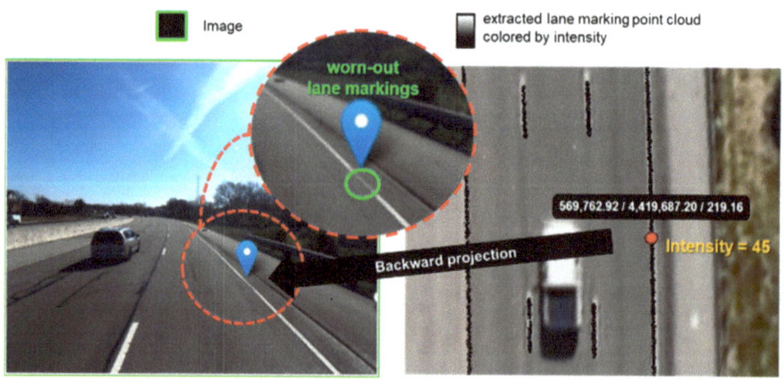

Figure 35. Illustration of an intensity value of 45 (red dot) and the corresponding point cloud and image (overlaid by a blue placemark, showing the backward projected intensity point) with a zoom-in window (red dotted circle) for worn-out lane markings (green oval).

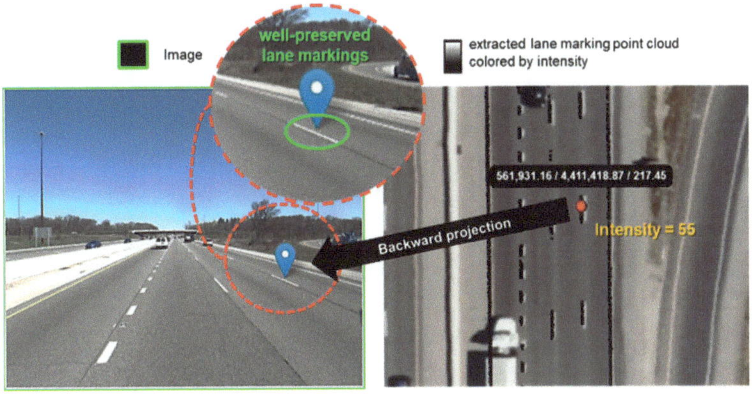

Figure 36. Illustration of an intensity value of 55 (red dot) and the corresponding point cloud and image (overlaid by a blue placemark, showing the backward projected intensity point) with a zoom-in window (red dotted circle) for well-preserved lane markings (green oval).

To evaluate the alignment of lane markings derived from imagery and LiDAR data, Figure 38 displays the forward projection of two points along 2D image-based lane marking centerlines onto the corresponding LiDAR data. Figure 39 shows the backward projection function of two points along LiDAR-based lane markings onto the corresponding images. These figures can be used to assess the quality of the current trajectory and system calibration. It is noted that when an image-based object/LiDAR point is close to the camera, no significant discrepancies are observed between the image-based and LiDAR-based lane markings. However, as the distance between an image-based object/LiDAR point and the camera increases, slight discrepancies between the image-based and LiDAR-based lane markings become apparent.

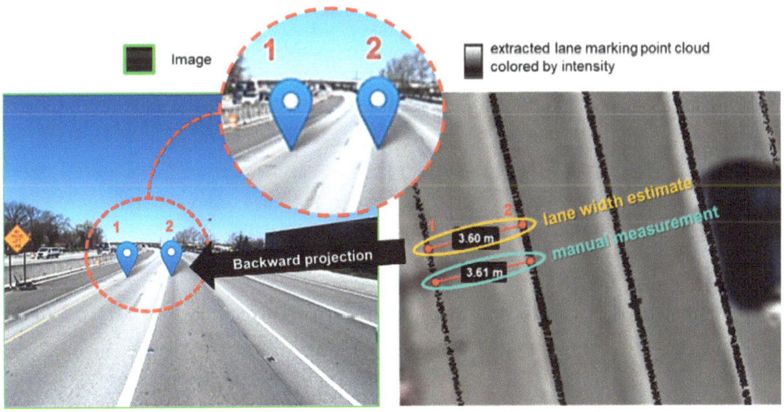

Figure 37. Illustration of a lane width estimate of 3.60 m (yellow oval) and the corresponding point cloud and image (overlaid by two blue placemarks, showing the backward projected lane width points) with a zoom-in window (red dotted circle) as well as a manual measurement of 3.61 m (cyan oval) in LiDAR data.

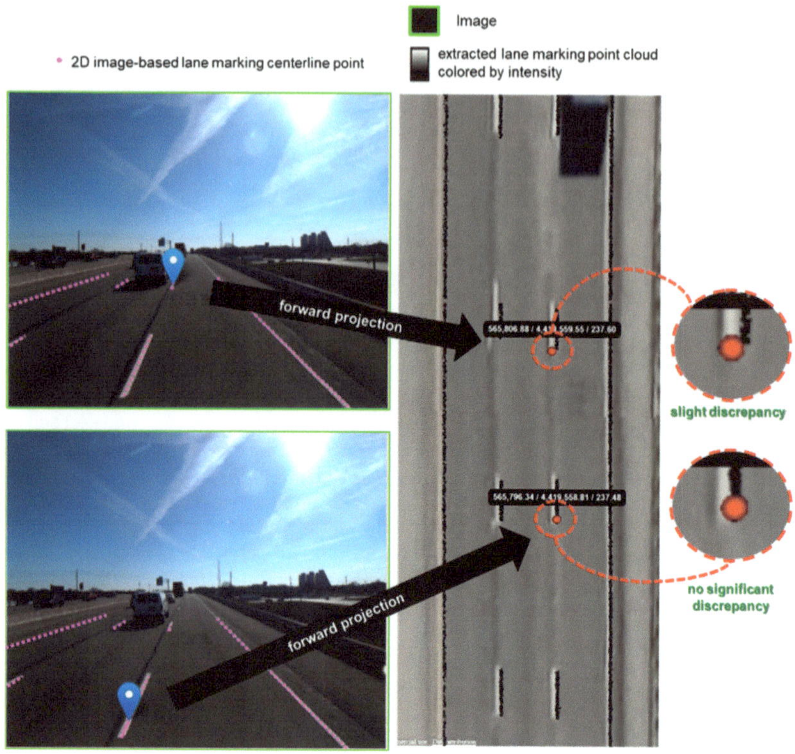

Figure 38. Illustration of two points (blue placemark) along 2D image-based lane marking centerlines (lavender point) that are forward projected onto the corresponding LiDAR point cloud (red dot) with their zoom-in windows (red dotted circles) in the Potree-based web portal.

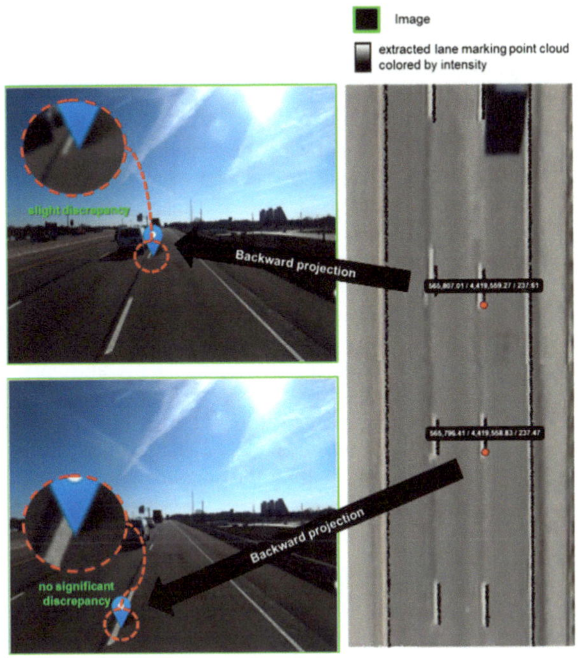

Figure 39. Illustration of two points (red dot) along LiDAR-based lane markings that are backward projected onto the corresponding image (blue placemark) with their zoom-in windows (red dotted circle) in the Potree-based web portal.

4.3. *Quantitative Evaluation*

Based on the centerline points derived (as explained in Section 3.3) from the three lane marking extraction strategies—LiDAR based, image based, and image-aided LiDAR—the performance is evaluated by comparing them to the reference data. The evaluation metrics for the 600 locations, as shown in Figure 5, are summarized in Table 6. Figures 40–44 display samples of centerline points and corresponding images for each approach. The main findings are categorized based on (1) LiDAR based, (2) image based, and (3) image-aided LiDAR.

LiDAR-based lane marking extraction

The LiDAR-based approach demonstrates slightly lower precision (93.2%), recall (87.6%), and F1-score (90.3%) compared to the image-aided LiDAR strategy. This is expected, as LiDAR sensors are not affected by adverse weather and lighting conditions and have minimal occlusions, allowing them to outperform the image-based extraction. However, there is room for improvement in refining some lane markings (denoted by the red ovals and blue ovals/placemarks in Figures 40 and 41), where image-aided LiDAR performs better. Overall, the LiDAR-based extraction has few commission (as represented by the precision) and omission (as represented by the recall) errors, and most extracted lane markings are true positives.

Image-based lane marking extraction

The image-based approach exhibits the lowest precision (88.5%), recall (69.4%), and F1-score (77.8%) compared to the other two strategies. The significantly lower recall is caused by the inevitable influence of excessive change in lighting conditions (denoted by the blue ovals/placemarks in Figure 42) and/or traffic occlusions (denoted by the blue ovals/placemarks in Figure 43). Additionally, the low resolution of the images significantly limits the image-based extraction approach, making it challenging to identify dash/dotted lines located beyond the driving lane on either side (denoted by the blue ovals/placemarks in Figure 44).

Image-aided LiDAR lane marking extraction

The image-aided LiDAR approach achieves the highest precision (93.4%), recall (91.6%), and F1-score (92.5%) compared to the other two strategies. The recall increases from 87.6% (LiDAR based) to 91.6% (image-aided LiDAR), surpassing the improvement from 93.2% (LiDAR based) to 93.4% (image-aided LiDAR) in precision. This suggests that the enhancement in lane marking extraction is more pronounced when addressing the omission errors in the LiDAR-based approach (as shown in Figure 32, Figure 40, and Figure 41) rather than compensating for commission errors. The F1-score also shows an increase from 90.3% (LiDAR based) to 92.5% (image-aided LiDAR), indicating that the image-aided LiDAR approach is capable of extracting most lane markings, and the image information indeed enhances the LiDAR-based extraction, as per the discussion in Section 4.2.

In summary, the image-aided LiDAR lane marking extraction achieves the best performance, followed by the LiDAR-based approach. The image information is particularly effective in compensating for the omission errors in the LiDAR-based approach. These findings align with the complementary nature of camera and LiDAR units emphasized in this study.

Table 6. Performance metrics for different lane marking extraction strategies.

Lane Marking Extraction	Precision (%)	Recall (%)	F1-Score (%)
LiDAR based	93.2	87.6	90.3
Image based	88.5	69.4	77.8
Image-aided LiDAR	93.4	91.6	92.5

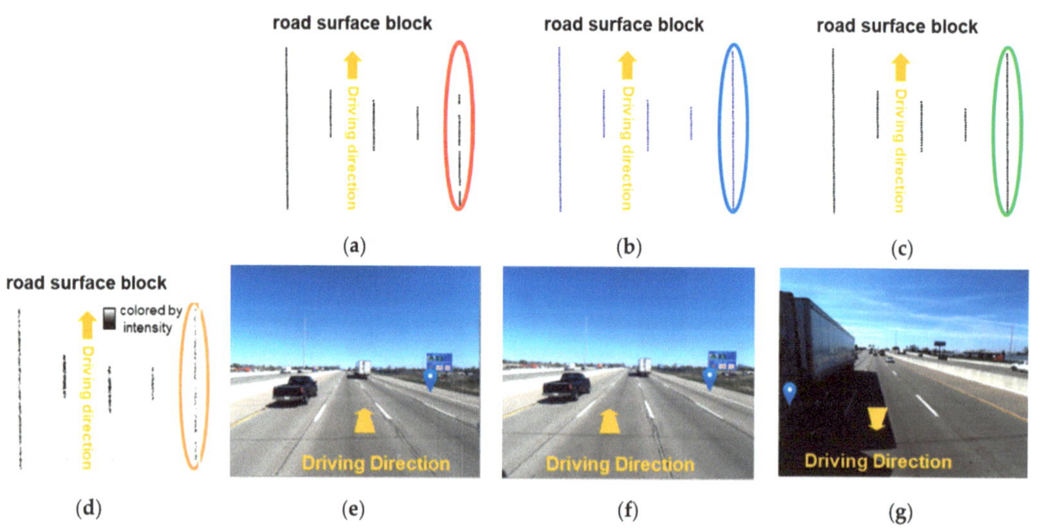

Figure 40. Illustrations of centerline points derived through (**a**) LiDAR-based, (**b**) image-based, and (**c**) image-aided LiDAR lane marking extraction in areas where lane markings were not detected using the LiDAR-based approach (red oval)—due to low point density in (**d**) hypothesized lane markings (yellow oval)—but were successfully identified using the image-based (blue oval/placemark) and image-aided LiDAR strategies (green oval), as well as the corresponding images captured by (**e**) front-left, (**f**) front-right, and (**g**) rear cameras onboard the PWMMS-HA.

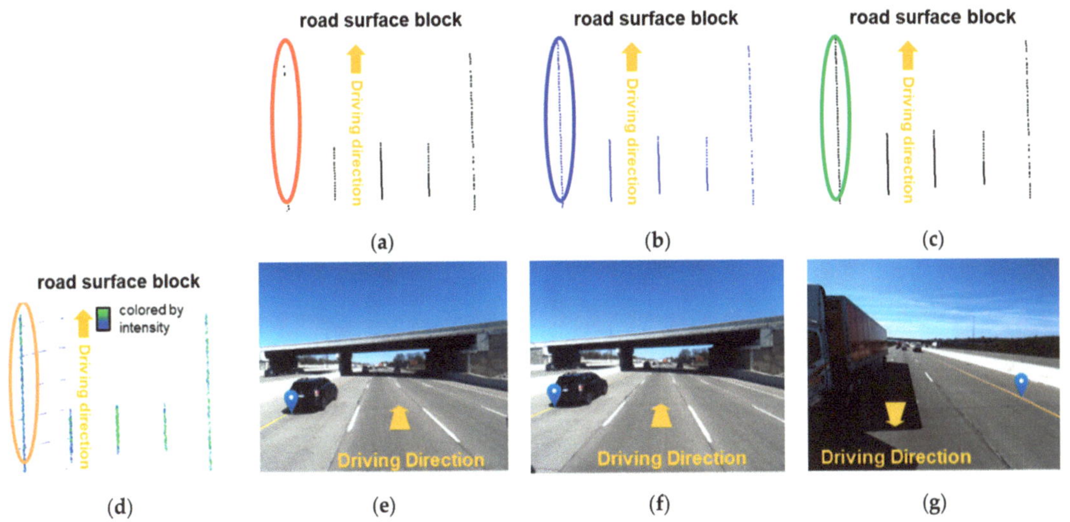

Figure 41. Illustrations of centerline points derived through (**a**) LiDAR-based, (**b**) image-based, and (**c**) image-aided LiDAR lane marking extraction in areas where lane markings were not detected using the LiDAR-based approach (red oval)—due to low intensity in (**d**) hypothesized lane markings (yellow oval)—but were successfully identified using the image-based (blue oval/placemark) and image-aided LiDAR strategies (green oval), as well as the corresponding images captured by (**e**) front-left, (**f**) front-right, and (**g**) rear cameras onboard the PWMMS-HA.

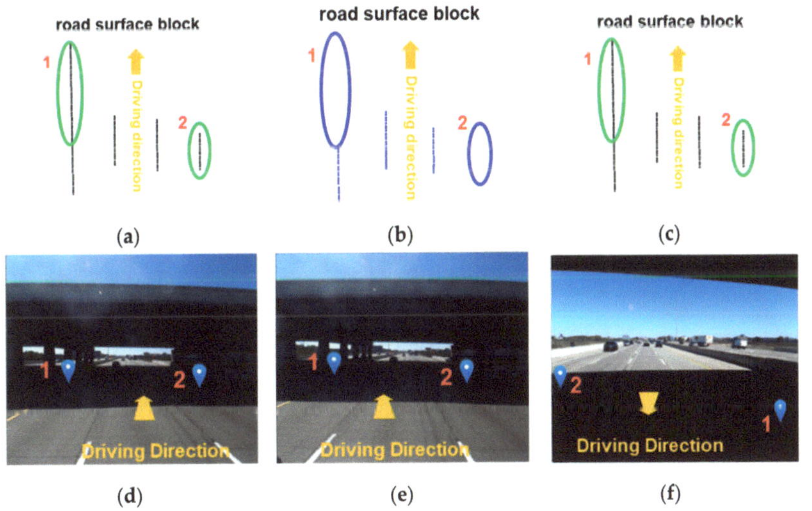

Figure 42. Illustrations of centerline points derived through (**a**) LiDAR-based, (**b**) image-based, and (**c**) image-aided LiDAR lane marking extraction in areas where lane markings were not detected using the image-based approach (blue oval)—due to excessive change in lighting conditions in imagery (blue placemark)—but were successfully identified using the LiDAR-based and image-aided LiDAR strategies (green oval), as well as the corresponding images captured by (**d**) front-left, (**e**) front-right, and (**f**) rear cameras onboard the PWMMS-HA.

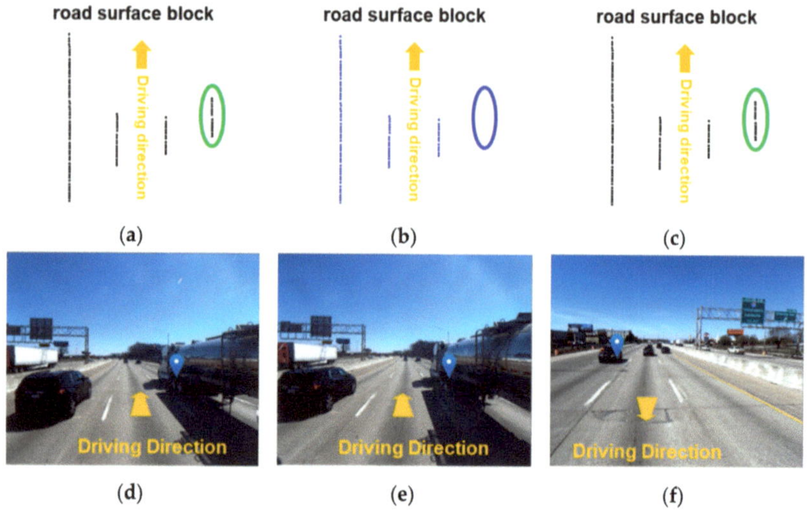

Figure 43. Illustrations of centerline points derived through (**a**) LiDAR-based, (**b**) image-based, and (**c**) image-aided LiDAR lane marking extraction in areas where lane markings were not detected using the image-based approach (blue oval)—due to traffic occlusion in imagery (blue placemark)—but were successfully identified using the LiDAR-based and image-aided LiDAR strategies (green oval), as well as the corresponding images captured by (**d**) front-left, (**e**) front-right, and (**f**) rear cameras onboard the PWMMS-HA.

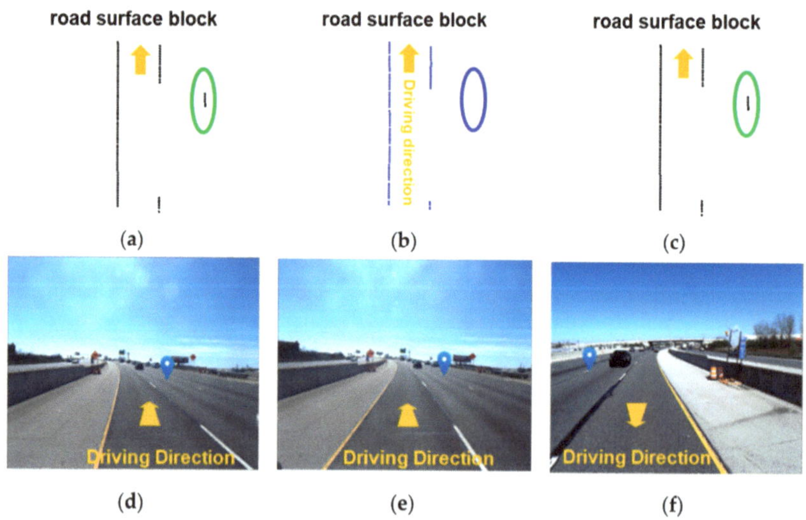

Figure 44. Illustrations of centerline points derived through (**a**) LiDAR-based, (**b**) image-based, and (**c**) image-aided LiDAR lane marking extraction in areas where lane markings were not detected using the image-based approach (blue oval)—due to insufficient resolution for representing a dotted lane marking in imagery (blue placemark)—but were successfully identified using the LiDAR-based and image-aided LiDAR strategies (green oval), as well as the corresponding images captured by (**d**) front-left, (**e**) front-right, and (**f**) rear cameras onboard the PWMMS-HA.

4.4. Discussion

In this study, the processing time for image-aided LiDAR lane marking extraction approaches was approximately 100 min per mile (including the duration required for

LiDAR-based—8 min per mile—and image-based—85 min per mile—strategies) of the point clouds from four LiDAR units and the images from three cameras. For instance, on a 100-mile-long highway, the extraction of lane markings using the image-aided LiDAR framework would take approximately 168 h (7 days), which is less than the estimated service times for lane markings (white: 10.4~22.6 months; yellow: 15.6~39.7 months) on highways [49]. Thus, the proposed framework has the potential to provide routine updates to lane marking inventory throughout the lifespan of lane markings. One should note that the processing time in this study was not executed in parallel, and the use of parallel processing techniques could significantly reduce processing time.

Overall, the image-aided LiDAR lane marking extraction (F1-score: 92.5%) outperforms the LiDAR-based (F1-score: 90.3%) and image-based (F1-score: 77.8%) approaches. Nevertheless, LiDAR-based and image-based approaches have strengths and weaknesses for various potential applications. Image-based approaches may have inferior performance compared to LiDAR-based strategies for establishing lane marking inventory. However, image-based approaches are more suitable for autonomous vehicle applications when only lane markings along the driving lane need to be identified, owing to recent advancements in machine learning technology and camera affordability.

5. Conclusions and Recommendations for Future Research

This paper presents an image-aided LiDAR framework for establishing lane marking inventory. The framework utilizes lane markings extracted from images to enhance the accuracy of LiDAR-based extraction. Thereafter, intensity profiles and lane width estimates can be derived using image-aided LiDAR lane markings. The proposed image-aided LiDAR framework can handle lane markings within a 9-meter-extent on either side of the vehicle, as well as multiple imaging and LiDAR sensors mounted on an MMS. Additionally, this study developed a Potree-based web portal for visualizing imagery/LiDAR data. A Potree-based web portal was developed to include a function that facilitates the projection between 2D images and 3D point clouds, as well as tools for displaying intensity profiles and lane width estimates.

The performance of the proposed framework was evaluated using a dataset of 22,428 images and approximately 42,000 million LiDAR points collected along I-465 in the United States. Lane markings spanning around 110 miles (55-mile-long inner and outer loops) were extracted using the image-aided LiDAR approach, requiring an average of 100 min per mile for processing. The proposed framework improves the performance of lane marking extraction, as evidenced by the highest F1-score (92.5%) of the image-aided LiDAR approach, outperforming the LiDAR-based (90.3%) and image-based (77.8%) ones. Specifically, the recall increase of 0.4%—from 87.6% (LiDAR based) to 91.6% (image-aided LiDAR)—surpasses the slight improvement in the precision of 0.2%—from 93.2% (LiDAR based) to 93.4% (image-aided LiDAR). These findings indicate that the enhancement in LiDAR-based extraction is more pronounced when addressing omission errors rather than compensating for commission errors. On the other hand, the web portal can render the LiDAR datasets along I-465 in around ten seconds and visualize intensity profiles and lane width estimates. Additionally, users can select points of interest in an intensity profile/lane width plot, which will then be highlighted as points in the corresponding LiDAR data. Furthermore, the highlighted points can be projected onto the corresponding images where they are visible.

Future work will focus on leveraging the lane markings derived across different imaging and LiDAR sensors to enhance the alignment of imagery and LiDAR data. The lane markings derived from LiDAR point clouds can be backward projected onto corresponding images to identify any discrepancy between the LiDAR-based and image-based lane markings. By minimizing the discrepancy between conjugate lane markings extracted from multiple modalities, it will be possible to enhance the trajectory information of an MMS, as well as the calibration parameters of imaging and LiDAR sensors. Furthermore, future efforts will seek to reduce the execution time by utilizing parallel processing techniques. Reducing the execution time is particularly crucial to ensure that the proposed framework can offer regular updates for lane marking inventory over the lifespan of pavement markers.

Author Contributions: Conceptualization, D.B. and A.H.; formal analysis, methodology, and validation, Y.-T.C., D.B. and A.H.; investigation, Y.-T.C., Y.-H.S., S.-Y.S., Y.K., M.H., D.B. and A.H.; software, Y.-T.C., Y.-H.S. and S.-Y.S.; writing—original draft preparation, Y.-T.C.; writing—review and editing, Y.-T.C. and A.H.; supervision, D.B. and A.H. All authors have read and agreed to the published version of the manuscript.

Funding: This work was supported by the Joint Transportation Research Program administered by the Indiana Department of Transportation and Purdue University (grant Nos. SPR-4741 and SPR-4742). The contents of this paper reflect the views of the authors, who are responsible for the facts and accuracy of the data presented herein and do not necessarily reflect the official views or policies of the sponsoring organizations or data vendors. These contents do not constitute a standard, specification, or regulation.

Data Availability Statement: Data sharing is not applicable to this paper.

Conflicts of Interest: The authors declare no conflicts of interest.

References

1. Plankermann, K. Human Factors as Causes for Road Traffic Accidents in the Sultanate of Oman under Consideration of Road Construction Designs. Ph.D. Dissertation, Universität Regensburg, Regensburg, Germany, 2014.
2. Chen, S.; Saeed, T.U.; Alinizzi, M.; Lavrenz, S.; Labi, S. Safety sensitivity to roadway characteristics: A comparison across highway classes. *Accid. Anal. Prev.* **2019**, *123*, 39–50. [CrossRef]
3. Zegeer, C.V.; Deacon, J.A. Effect of lane width, shoulder width, and shoulder type on highway safety. *State Art Rep.* **1987**, *6*, 1–21.
4. Stein, W.J.; Neuman, T.R. *Mitigation Strategies for Design Exceptions*; Federal Highway Administration, Office of Safety: Washington, DC, USA, 2007.
5. FHWA. *Manual on Uniform Traffic Control Devices 2009*; US Department of Transportation, Federal Highway Administration: Washington, DC, USA, 2009.
6. Highway Safety Improvement Program Manual. 2011. Available online: https://safety.fhwa.dot.gov/hsip/resources/fhwasa09029/sec3.cfm (accessed on 6 April 2024).
7. Sebsadji, Y.; Tarel, J.-P.; Foucher, P.; Charbonnier, P. Robust road marking extraction in urban environments using stereo images. In Proceedings of the 2010 IEEE Intelligent Vehicles Symposium (IV), La Jolla, CA, USA, 21–24 June 2010; pp. 394–400.
8. Foucher, P.; Sebsadji, Y.; Tarel, J.P.; Charbonnier, P.; Nicolle, P. Detection and recognition of urban road markings using images. In Proceedings of the 2011 14th International IEEE Conference on Intelligent Transportation Systems (ITSC), Washington, DC, USA, 5–7 October 2011; IEEE: Piscateville, NJ, USA; pp. 1747–1752.
9. Jung, S.; Youn, J.; Sull, S. Efficient lane detection based on spatiotemporal images. *IEEE Trans. Intell. Transp. Syst.* **2015**, *17*, 289–295. [CrossRef]
10. Son, J.; Yoo, H.; Kim, S.; Sohn, K. Real-time illumination invariant lane detection for lane departure warning system. *Expert Syst. Appl.* **2015**, *42*, 1816–1824. [CrossRef]
11. Xu, S.; Wang, J.; Wu, P.; Shou, W.; Wang, X.; Chen, M. Vision-based pavement marking detection and condition assessment—A case study. *Appl. Sci.* **2021**, *11*, 3152. [CrossRef]
12. Chen, X.; Kohlmeyer, B.; Stroila, M.; Alwar, N.; Wang, R.; Bach, J. Next generation map making: Geo-referenced ground-level LIDAR point clouds for automatic retro-reflective road feature extraction. In Proceedings of the 17th ACM SIGSPATIAL International Conference on Advances in Geographic Information Systems, Seattle, WA, USA, 4–6 November 2009; pp. 488–491.
13. Guan, H.; Li, J.; Yu, Y.; Wang, C.; Chapman, M.; Yang, B. Using mobile laser scanning data for automated extraction of road markings. *ISPRS J. Photogramm. Remote Sens.* **2014**, *87*, 93–107. [CrossRef]
14. Cheng, M.; Zhang, H.; Wang, C.; Li, J. Extraction and classification of road markings using mobile laser scanning point clouds. *IEEE J. Sel. Top. Appl. Earth Obs. Remote Sens.* **2016**, *10*, 1182–1196. [CrossRef]
15. Yu, Y.; Li, J.; Guan, H.; Jia, F.; Wang, C. Learning hierarchical features for automated extraction of road markings from 3-D mobile LiDAR point clouds. *IEEE J. Sel. Top. Appl. Earth Obs. Remote Sens.* **2014**, *8*, 709–726. [CrossRef]
16. Yan, L.; Liu, H.; Tan, J.; Li, Z.; Xie, H.; Chen, C. Scan line based road marking extraction from mobile LiDAR point clouds. *Sensors* **2016**, *16*, 903. [CrossRef]
17. Huang, A.S.; Moore, D.; Antone, M.; Olson, E.; Teller, S. Finding multiple lanes in urban road networks with vision and lidar. *Auton. Robot.* **2009**, *26*, 103–122. [CrossRef]
18. Li, Q.; Chen, L.; Li, M.; Shaw, S.L.; Nüchter, A. A sensor-fusion drivable-region and lane-detection system for autonomous vehicle navigation in challenging road scenarios. *IEEE Trans. Veh. Technol.* **2013**, *63*, 540–555. [CrossRef]

19. Shin, S.; Shim, I.; Kweon, I.S. Combinatorial approach for lane detection using image and LIDAR reflectance. In Proceedings of the 2015 12th International Conference on Ubiquitous Robots and Ambient Intelligence (URAI), Goyangi, Republic of Korea, 28–30 October 2015; IEEE: Piscateville, NJ, USA.
20. Gu, X.; Zang, A.; Huang, X.; Tokuta, A.; Chen, X. Fusion of color images and LiDAR data for lane classification. In Proceedings of the 23rd SIGSPATIAL International Conference on Advances in Geographic Information Systems, Seattle, WA, USA, 3–6 November 2015; pp. 1–4.
21. LeCun, Y.; Bengio, Y.; Hinton, G. Deep learning. *Nature* **2015**, *521*, 436–444. [CrossRef] [PubMed]
22. Bai, M.; Mattyus, G.; Homayounfar, N.; Wang, S.; Lakshmikanth, S.K.; Urtasun, R. Deep multi-sensor lane detection. In Proceedings of the 2018 IEEE/RSJ International Conference on Intelligent Robots and Systems (IROS), Madrid, Spain, 1–5 October 2018; IEEE: Piscateville, NJ, USA; pp. 3102–3109.
23. He, K.; Zhang, X.; Ren, S.; Sun, J. Deep residual learning for image recognition. In Proceedings of the IEEE Conference on Computer Vision and Pattern Recognition, Las Vegas, NV, USA, 26 June–1 July 2016; pp. 770–778.
24. Pavement Marking Management System—FHWA Office of Safety. 1999. Available online: https://safety.fhwa.dot.gov/roadway_dept/night_visib/pavement_visib/pmms/docs/ref_manual.pdf (accessed on 22 July 2022).
25. Pavement Marking Inventory. 2022. Available online: https://solutions.arcgis.com/local-government/help/pavement-marking-inventory/ (accessed on 28 July 2022).
26. Budzyński, M.; Kustra, W.; Okraszewska, R.; Jamroz, K.; Pyrchla, J. The use of GIS tools for road infrastructure safety management. *E3S Web Conf.* **2018**, *26*, 00009. [CrossRef]
27. Velodyne LiDAR. HDL-32E User Manual and Programming Guide. 2012. Available online: https://s3.us-east-2.amazonaws.com/nclt.perl.engin.umich.edu/manuals/HDL-32E_manual.pdf (accessed on 1 January 2021).
28. Velodyne LiDAR. VLP-16 User Manual and Programming Guide. 2015. Available online: https://usermanual.wiki/Pdf/VLP162 0User20Manual20and20Programming20Guide2063924320Rev20A.1947942715/view (accessed on 1 January 2021).
29. Applanix. POSLV Specifications. 2015. Available online: https://www.applanix.com/downloads/products/specs/POS-LV-Datasheet.pdf (accessed on 1 January 2021).
30. Habib, A.; Lay, J.; Wong, C. *Specifications for the Quality Assurance and Quality Control of Lidar Systems*; Base Mapping and Geomatic Services of British Columbia: Victoria, BC, Canada, 2006.
31. Kuçak, R.A.; Erol, S.; Erol, B. The strip adjustment of mobile LiDAR point clouds using iterative closest point (ICP) algorithm. *Arab. J. Geosci.* **2022**, *15*, 1017. [CrossRef]
32. Ravi, R.; Lin, Y.-J.; Elbahnasawy, M.; Shamseldin, T.; Habib, A. Bias impact analysis and calibration of terrestrial mobile LiDAR system with several spinning multibeam laser scanners. *IEEE Trans. Geosci. Remote Sens.* **2018**, *56*, 5261–5275. [CrossRef]
33. Ravi, R.; Lin, Y.J.; Elbahnasawy, M.; Shamseldin, T.; Habib, A. Simultaneous system calibration of a multi-lidar multicamera mobile mapping platform. *IEEE J. Sel. Top. Appl. Earth Obs. Remote Sens.* **2018**, *11*, 1694–1714. [CrossRef]
34. Romano, J.D.; Le, T.T.; Fu, W.; Moore, J.H. Is deep learning necessary for simple classification tasks? *arXiv* **2020**, arXiv:2006.06730.
35. Cheng, Y.-T.; Lin, Y.-C.; Habib, A. Generalized LiDAR Intensity Normalization and Its Positive Impact on Geometric and Learning-Based Lane Marking Detection. *Remote Sens.* **2022**, *14*, 4393. [CrossRef]
36. Revelles, J.; Urena, C.; Lastra, M. An Efficient Parametric Algorithm for Octree Traversal. 2000. Available online: http://wscg.zcu.cz/wscg2000/Papers_2000/X31.pdf (accessed on 21 February 2024).
37. Lin, Y.-C.; Habib, A. Quality control and crop characterization framework for multi-temporal UAV LiDAR data over mechanized agricultural fields. *Remote Sens. Environ.* **2021**, *256*, 112299. [CrossRef]
38. AASHTO. *A Policy on Geometric Design of Highways and Streets*; American Association of State Highway and Transportation Officials: Washington, DC, USA, 2018.
39. Lari, Z.; Habib, A. New approaches for estimating the local point density and its impact on lidar data segmentation. *Photogramm. Eng. Remote Sens.* **2013**, *79*, 195–207. [CrossRef]
40. Ester, M.; Kriegel, H.P.; Sander, J.; Xu, X. A density-based algorithm for discovering clusters in large spatial databases with noise. *KDD* **1996**, *96*, 226–231.
41. Foedisch, M.; Takeuchi, A. Adaptive real-time road detection using neural networks. In Proceedings of the 7th International IEEE Conference on Intelligent Transportation Systems, Washington, DC, USA, 3–6 October 2004; IEEE: Piscateville, NJ, USA, 2004.
42. Glaser, S.; Mammar, S.; Sentouh, C. Integrated driver–vehicle–infrastructure road departure warning unit. *IEEE Trans. Veh. Technol.* **2010**, *59*, 2757–2771. [CrossRef]
43. Wang, Q.; Wei, Z.; Wang, J.; Chen, W.; Wang, N. Curve recognition algorithm based on edge point curvature voting. *Proc. Inst. Mech. Eng. Part D J. Automob. Eng.* **2020**, *234*, 1006–1019. [CrossRef]
44. Yang, Q.; Ma, Y.; Li, L.; Gao, Y.; Tao, J.; Huang, Z.; Jiang, R. A fast vanishing point detection method based on row space features suitable for real driving scenarios. *Sci. Rep.* **2023**, *13*, 3088. [CrossRef]
45. Resonfeld, A.; Pfaltz, J. Sequential operations in digital image processing. *JACM* **1966**, *13*, 471–494. [CrossRef]
46. De Boor, C. *A Practical Guide to Splines*; Springer: New York, NY, USA, 1978; Volume 27.
47. D'Errico, J. Distance2curve. 2023. Available online: https://www.mathworks.com/matlabcentral/fileexchange/34869-distance2curve (accessed on 27 October 2022).

48. Schütz, M. *Potree: Rendering Large Point Clouds in Web Browsers*; Vienna University of Technology: Vienna, Austria, 2016.
49. Jiang, Y. *Durability and Retro-Reflectivity of Pavement Markings (Synthesis Study)*; Indiana Department of Transportation and Purdue University: West Lafayette, IN, USA, 2008.

remote sensing

Article

High-Precision Map Construction in Degraded Long Tunnel Environments of Urban Subways

Cheng Li [1,2], Wenbo Pan [2,*], Xiwen Yuan [2], Wenyu Huang [2], Chao Yuan [2], Quandong Wang [2] and Fuyuan Wang [2]

1. Institute of Rail Transit, Tongji University, Shanghai 201804, China; licheng5@csrzic.com
2. CRRC Zhuzhou Institute Co., Ltd., 169 Shidai Road, Zhuzhou 412001, China; yuanxw@csrzic.com (X.Y.); huangwy@csrzic.com (W.H.); yuanchao5@csrzic.com (C.Y.); wangqd@csrzic.com (Q.W.); wangfy4@csrzic.com (F.W.)

* Correspondence: panwb1@csrzic.com

Abstract: In response to the demand for high-precision point cloud mapping of subway trains in long tunnel degradation scenarios in major urban cities, we propose a map construction method based on LiDAR and inertial measurement sensors. This method comprises a tightly coupled frontend odometry system based on error Kalman filters and backend optimization using factor graphs. In the frontend odometry, inertial calculation results serve as predictions for the filter, and residuals between LiDAR points and local map plane point clouds are used for filter updates. The global pose graph is constructed based on inter-frame odometry and other constraint factors, followed by a smoothing optimization for map building. Multiple experiments in subway tunnel scenarios demonstrate that the proposed method achieves robust trajectory estimation in long tunnel scenes, where classical multi-sensor fusion methods fail due to sensor degradation. The proposed method achieves a trajectory consistency of 0.1 m in tunnel scenes, meeting the accuracy requirements for train arrival, parking, and interval operations. Additionally, in an industrial park scenario, the method is compared with ground truth provided by inertial navigation, showing an accumulated error of less than 0.2%, indicating high precision.

Keywords: urban subway; multi-sensor integration; simultaneous localization and mapping; degraded environments

Citation: Li, C.; Pan, W.; Yuan, X.; Huang, W.; Yuan, C.; Wang, Q.; Wang, F. High-Precision Map Construction in Degraded Long Tunnel Environments of Urban Subways. *Remote Sens.* **2024**, *16*, 809. https://doi.org/10.3390/rs16050809

Academic Editors: San Jiang, Duojie Weng, Jianchen Liu and Wanshou Jiang

Received: 15 October 2023
Revised: 22 January 2024
Accepted: 24 January 2024
Published: 26 February 2024

1. Introduction

Urban rail transit, as a critical infrastructure and major livelihood project, plays a pivotal role as the arterial system of urban transportation. After more than a century of development, major metropolises around the world have evolved into cities on rails. Traditional urban rail transit trains primarily rely on automatic block signaling technology provided by the communication signal system to avoid collisions, enabling trains to be isolated from each other on different sections. This system uses beacons to obtain discontinuous positions of trains, lacking efficiency and accuracy in real-time applications. Moreover, it requires massive civil construction investment for building and continuous maintenance, hindering the technological upgrade and widespread development of urban rail transit. Therefore, it is crucial to use autonomous perception technology to achieve environmental information in tunnel scenes and the pose information of trains.

With the development of intelligent and unmanned technologies, the field of rail transit is gradually introducing intelligent driving systems to enhance operational efficiency and safety. However, commonly used Global Navigation Satellite Systems (GNSSs) in autonomous driving provide flexibility and accurate positioning in open areas but are not suitable for tunnel scenes in large urban subways. To obtain accurate pose data of trains and surrounding environmental information, it is necessary to construct a high-precision point cloud map of the train operating area, providing rich a priori information for positioning and environmental perception. Many previous works based on Mobile Mapping

Systems (MMSs) have adopted this approach [1,2]. MMSs can provide direct georeferencing but require a series of post-processing and expensive measurement instruments. Therefore, they are not suitable for real-time positioning of urban subway vehicles and large-scale deployment.

With the continuous maturation of Simultaneous Localization and Mapping (SLAM) technology, new opportunities have arisen for the construction of high-precision maps for subway environments and applications based on high-precision maps. However, there is currently a lack of methods for high-precision map construction for long subway tunnel features in degraded scenes. The main technical challenges can be summarized as follows:

1. Cumulative Errors in Long Tunnel Environments: Subway tunnels in large cities are often long and lack reference information like GNSSs for ground truth vehicle pose estimation. This leads to increased positioning errors with distance, making it challenging to meet the accuracy requirements for train pose estimation during station stops.
2. Degraded Scenarios with Repetitive Features: Inside tunnels, the most observable features are repetitive tunnel walls, tracks, and power supply systems. This presents challenges for existing SLAM methods designed for urban scenes.
3. Lack of Loop Closure Opportunities: SLAM typically corrects accumulated drift over detected loop closures. However, trains lack revisit locations, making loop closure detection difficult.
4. Narrow-Field, Non-Repetitive Scan LiDARs: Solid-state LiDARs with limited fields of view can easily fail in scenarios with insufficient geometric features.

To address these issues, we propose a system for the precise positioning and mapping of rail vehicles in tunnel environments. This system tightly integrates multimodal information from LiDARs and IMUs in a coupled manner. The main contributions of our work can be summarized as follows:

1. We develop a compact positioning and mapping system that tightly integrates LiDARs and IMUs.
2. In response to tunnel degradation scenarios, a high-dimensional, multi-constraint framework is proposed, integrating a frontend odometry based on an error state Kalman filter and a backend optimization based on a factor graph.
3. Leveraging geometric information from sensor measurements, we mitigate accumulated pose errors in degraded tunnel environments by introducing absolute pose, iterative closest point (ICP), and Landmark constraints.
4. The algorithm's performance is validated in urban subway tunnel scenarios and industrial park environments.

The rest of this paper is organized as follows. Section 2 reviews related work. Section 3 introduces the specific algorithms used in our system. Section 4 presents experimental results. Finally, Section 5 concludes the paper and outlines future research directions.

2. Related Work

Train positioning based on query/response systems, commonly referred to as a "Balise transmission system", is a prevalent method in rail transportation. Typically, it comprises onboard interrogators, ground beacons, and trackside electronic units. Ground beacons are strategically placed along the railway line at specific intervals. As a train passes each ground beacon, the onboard interrogator retrieves stored data, enabling point-based train positioning [3,4]. However, this method provides only point-based positioning, leading to conflicts between beacon spacing and investment requirements. Consequently, hybrid positioning methods have been widely adopted, involving distance accumulation through wheel encoders and error correction using query/response systems. Nevertheless, this approach can introduce significant cumulative errors in scenarios involving changes in wheel diameter, slipping, or free-wheeling. Furthermore, onboard interrogators rely on ground beacons and trackside electronic units, making trains incapable of self-locating in cases of ground system failures. Given the substantial capital investment required and the issues related to low positioning efficiency and the inability of onboard equipment to self-locate, researchers have explored solutions using onboard sensors [5–7] or feature matching-based methods [8].

In the field of intelligent transportation, both domestic and international scholars have proposed numerous methods for constructing point cloud maps [9–16]. Among these methods, LIO mapping stands out as a real-time technique for 3D pose estimation and mapping. This method successfully achieves tight coupling between IMUs and LiDAR technology. However, it comes with a high computational cost and lacks backend global pose optimization, resulting in substantial cumulative errors over long distances [14]. LiDAR-inertial odometry and mapping (LIOM), on the other hand, presents a method for correcting distortion in LiDAR point clouds using IMUs and employs nearest-neighbor techniques for semantic segmentation of point clouds in urban road conditions to mitigate the influence of moving objects. Nevertheless, its frontend adopts a loosely coupled design, leading to reduced performance in feature-sparse degraded scenarios [15]. VINS-MONO introduces a tightly coupled method that combines vision and IMUs, offering advantages such as high real-time performance and insensitivity to external parameters. However, it exhibits insensitivity to measurement scales, rendering it unsuitable for tunnel areas with suboptimal lighting conditions [16]. With the rapid advancement of LiDAR hardware, solid-state LiDARs have gained renown for their cost-effectiveness and compliance with automotive regulations, making them widely adopted in autonomous driving and robotics technologies [17–19]. However, their limited field of view makes them susceptible to failures in degraded environments lacking distinctive features [20]. To address this limitation, integrating LiDAR with other sensors proves effective in enhancing the system's robustness and accuracy [21–24].

In the realm of rail transportation, O Heirich and others from Germany have proposed a synchronous mapping and localization method based on track geometry information [25]. However, it exhibits low accuracy and is unsuitable for relocalization. In China, Y Wang et al., for instance, have introduced a mapping and localization method for outdoor rail transportation scenes based on a tightly coupled LiDAR-vision-GNSS-IMU system [26]. This method offers advantages such as high accuracy and robustness. Nevertheless, it necessitates GNSS integration and has not been optimized to address the specific challenges posed by long subway tunnel degradation scenarios.

Therefore, this paper addresses the need for high-precision offline point cloud map construction in subway tunnel environments with degraded features. It presents a mapping method designed for long-distance feature-degraded scenarios, relying on a tightly coupled LiDAR IMU frontend inter-frame odometry and backend global graph optimization. First, it introduces a framework that incorporates an error state Kalman filter (ESKF)-based frontend odometry and a factor graph-based backend optimization. This framework facilitates the establishment of frontend point-plane residual constraints using local maps updated after backend pose refinement. Second, to tackle the challenges posed by degraded tunnel features, this paper introduces absolute pose constraints, iterative closest point (ICP) constraints, and Landmark constraints to the backend factor graph constraints, effectively reducing pose accumulation errors. Finally, the algorithm's performance is validated in rail transit tunnel scenarios.

3. Materials and Methods

Common options for positioning, mapping, and target perception sensors include GNSS, IMU, LiDAR, and cameras. However, the tunnel's suboptimal lighting conditions significantly affect cameras, and their contribution to improving mapping accuracy in tunnel scenes is limited [27,28]. Additionally, GNSS signals cannot be received underground. Therefore, this study primarily employs LiDAR and IMU units as the main sensors. LiDAR can be further categorized into mechanical LiDAR and solid-state LiDAR. Mechanical LiDAR, with its large size and high cost, contrasts with solid-state LiDAR, which is lightweight, cost-effective, and more suitable for mass applications. However, solid-state LiDAR also introduces new challenges to algorithms, including a small field of view (FOV) that leads to degradation in scenes with fewer features. Due to differences in the LiDAR's scanning method, traditional point cloud feature extraction algorithms need adaptation based on the scanning method. Moreover, compared to the rotational scanning of mechanical LiDAR, the laser point sampling time of solid-state LiDAR varies and is challenging to compensate for using kinematic equations. All these factors pose challenges to the mapping and positioning applications of solid-state LiDAR [20]. This paper proposes a universal frontend odometry that eliminates the commonly used point cloud feature extraction module. The algorithm is agnostic to the scanning method and principles of the LiDAR. The workflow of the algorithm is illustrated in Figure 1 and can be broadly divided into five modules: hardware drivers, data preprocessing, frontend odometry, backend graph optimization, and map maintenance.

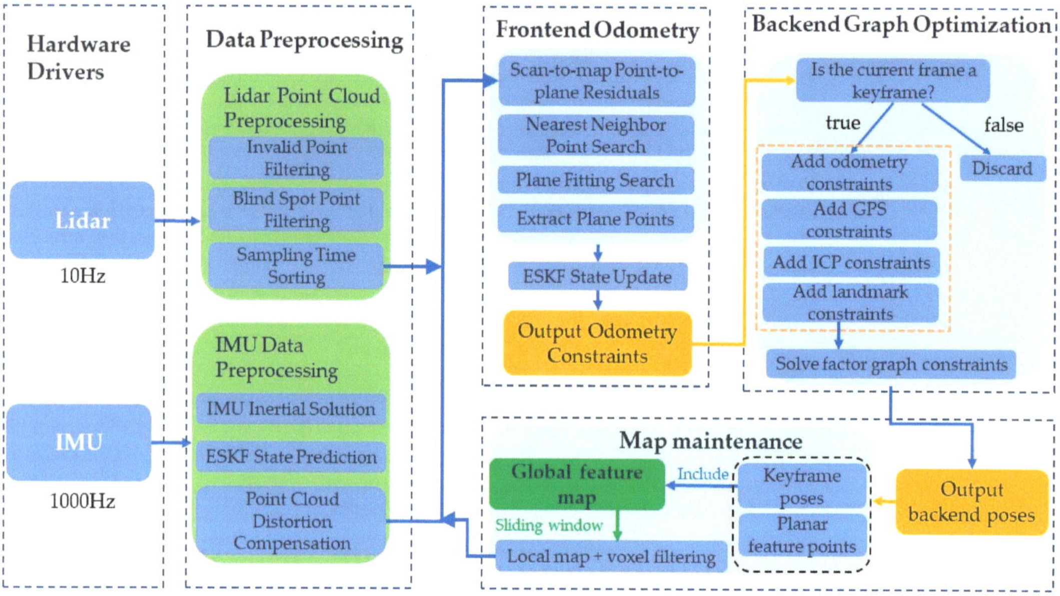

Figure 1. Algorithm overall flowchart.

3.1. Frontend Odometry

The frontend odometry module is responsible for calculating the relative pose relationship between consecutive LiDAR frames, providing pose constraints between adjacent LiDAR frames. As LOAM-series frontend odometry relies on the computation of point-to-line features [29], and the extraction of line-plane features in solid-state LiDAR is related to the LiDAR's scanning method. This paper adopts the idea from FastLio, proposing a tightly coupled frontend odometry that does not depend on traditional point cloud curvature calculation for extracting line features [18,30]. The algorithm is modified to suit the rail transportation environment and the requirements of offline map construction.

The algorithm is based on an ESKF filter for the tightly coupled LiDAR-IMU method [30]. During initialization, the system is required to remain stationary for a period, utilizing collected data to initialize the gravity vector, IMU biases, and noise, among other parameters. When the algorithm is running, raw data from the LiDAR are input into the LiDAR point cloud preprocessing module. Invalid points and points in close proximity are filtered out, and the remaining points are sorted based on sampling time in ascending order. This sorting facilitates distortion compensation based on IMU preintegration results. The preintegration method is then used to perform inertial navigation on the raw IMU data. Based on the inertial navigation results, motion distortion in the point cloud is compensated, and the prediction phase of the ESKF filter is executed. The temporal flow of LiDAR and IMU data is illustrated in Figure 2 [20,31].

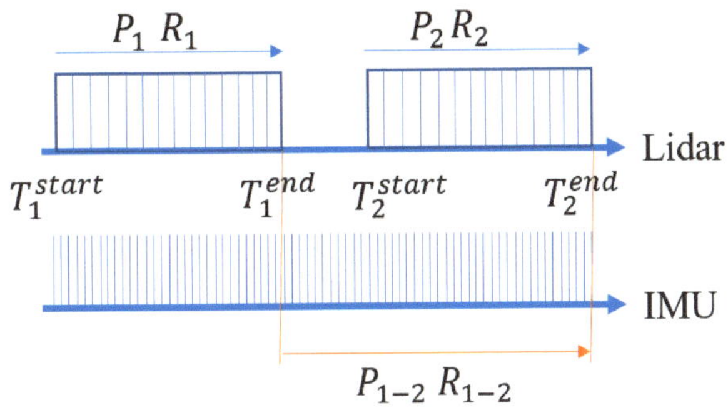

Figure 2. Schematic diagram of time flow for LiDAR and IMU.

Figure 2 depicts two scans of the LiDAR, denoted as T_1 and T_2, with the starting time as the start and the ending time as the end. During a scan, the pose transformation of the LiDAR within the time interval T_1^{start} to T_1^{end} is represented by P_1 and R_1. Therefore, all point clouds within the time interval T_1^{start} to T_1^{end} are transformed to the T_1^{end} moment to compensate for the motion distortion in the original point cloud. Simultaneously, the frontend odometry needs to output the inter-frame pose transformation between two scans. In the prediction phase of the ESKF filter, the inertial navigation results P_{1-2} and R_{1-2} within the time interval T_1^{end} to T_2^{end} are directly used as the filter's prediction input.

The state variables and kinematic equations used in the ESKF filter are represented by Equations (1) and (2), where all variables are denoted with superscript "I" for the IMU coordinate system and "G" for the Earth coordinate system.

$$x = \begin{bmatrix} R_I^G & p_I^G & v_I^G & b_\omega & b_a & G_g \end{bmatrix} \tag{1}$$

$$\begin{cases} \dot{p}_I^G = v_I^G \\ \dot{v}_I^G = R_I^G(a_m - b_a - n_a) + G_g \\ G_{\dot{g}} = 0 \\ \dot{R}_I^G = R_I^G \lfloor \omega_m - b_\omega - n_\omega \rfloor_\wedge \\ \dot{b}_\omega = n_{b_\omega} \\ \dot{b}_a = n_{b_a} \end{cases} \tag{2}$$

p_I^G—position in the Earth coordinate system; v_I^G—velocity in the Earth coordinate system; R_I^G—rotation matrix of the attitude in the Earth coordinate system; a_m—accelerometer measurement; b_a—accelerometer bias; n_a—accelerometer noise; G_g—gravity vector; ω_m—gyroscope measurement; b_ω—gyroscope bias; n_ω—gyroscope noise; n_{b_ω}—gyroscope bias random walk noise; n_{b_a}—accelerometer bias random walk noise.

In the map maintenance module, a sliding window is maintained based on the current position of the LiDAR, and a local map is output for scan-to-map matching. The raw LiDAR point cloud undergoes motion compensation and voxel filtering down-sampling. The ESKF filter establishes the point-to-plane constraint relationship. Using kd-tree nearest-neighbor search, the five nearest points ($P_1 \quad P_2 \quad P_3 \quad P_4 \quad P_5$) to the current point (P) are selected from the local map. This decision is primarily based on the structural characteristics of the point cloud within the tunnel environment. Opting for five points in the plane-fitting process ensures accurate fitting of the ground plane and other features present in the tunnel, such as installed signs and road edge planes. Choosing more than five points might result in a scarcity of plane points, leading to significant solving errors, while selecting fewer than five points could result in larger residuals in the fitted plane, causing fitting inaccuracies. The plane equation is then fitted using Principal Component Analysis (PCA) as follows [32]:

Each dimension of the data is subtracted by its mean value. After transformation, the mean value of each dimension becomes zero. Compute the covariance matrix for the three coordinates. The covariance matrix C is defined as follows:

$$C = \begin{bmatrix} \text{cov}(x,x)\text{cov}(x,y)\text{cov}(x,z) \\ \text{cov}(y,x)\text{cov}(y,y)\text{cov}(y,z) \\ \text{cov}(z,x)\text{cov}(z,y)\text{cov}(z,z) \end{bmatrix} \tag{3}$$

where $\text{cov}(x,x)$ represents the covariance between the x and y coordinates, and $\text{cov}(x,x)$ is the variance of the x coordinate. The covariance calculation is defined by Equation (4), where x_i, y_i are the coordinates of the centered points:

$$\text{cov}(x,y) = \frac{\sum_{i=1}^{n} x_i y_i}{n-1} \tag{4}$$

The eigenvalues and eigenvectors of the covariance matrix C are computed. The calculated eigenvalues, sorted in descending order, are denoted as λ_1, λ_2, λ_3, with corresponding eigenvectors ξ_1, ξ_2, ξ_3. Clearly, the eigenvectors ξ_1, ξ_2 corresponding to the two largest eigenvalues form a set for the plane to be fitted, while ξ_3 represents the normal vector of the fitting plane, with components a, b, c. If the fitting plane passes through the point $P(x_0, y_0, z_0)$, the equation of the fitted plane is given by Equation (5):

$$a(x - x_0) + b(y - y_0) + c(z - z_0) = 0 \tag{5}$$

The curvature-based feature extraction method has the advantage of rapidly extracting line and surface features, but it is challenging to achieve comprehensive and accurate feature extraction in long tunnel scenarios lacking distinct features. This method is prone to degradation in the driving direction. To prevent the ESKF filter from diverging in scenes with fewer features, a method for constructing plane point constraints is proposed. This method uses the following two conditions to determine whether a point can be used to construct a constraint relationship as a planar point:

1. The distance from each of the five points (P_1 P_2 P_3 P_4 P_5) to the fitted plane is less than 0.1 m.

2. The threshold is set to $s = 1 - 0.9 \times \frac{pd}{pl}$, where pd is the distance from point P to the fitted plane, and pl is the distance from point P to the center of the LiDAR. As pd is much smaller than pl between any two frames, to filter measurement errors from exceptional plane points, the constructed plane constraint is considered valid only when $s \geq 0.9$.

Finally, the ESKF filter is updated based on the point-to-plane residual constraints, and the optimal estimate of the state variables is output as the output of the inter-frame frontend odometry. The covariance matrix is updated, and the ESKF filter is iterated.

3.2. Backend Graph Optimization

In the context of the backend optimization problem based on the pose graph, each node in the factor graph represents a pose to be optimized. The edges between any two nodes represent spatial constraints between the corresponding poses, including relative pose relationships and their associated covariances. The relative pose relationships between nodes can be computed using frontend odometry, IMU pre-integration, frame-to-frame matching, and other methods. Given the utilization of the tightly coupled LiDAR-IMU approach in the frontend odometry, frame-to-frame IMU pre-integration constraints are not employed in the backend optimization.

Addressing the challenges of solving high-dimensional constraints, this paper proposes a framework with high dimensionality and multiple constraints, as illustrated in Figure 1. The framework leverages ESKF in the frontend odometry to provide high-frequency position updates. In the backend, a graph optimization constraint-solving approach is employed, integrating various constraints. The ESKF frontend odometry provides high-frequency position updates, and the backend uses graph optimization constraints that fuse various constraints. The key constraints integrated into the factor graph include frame-to-frame odometry factors, absolute pose factors, ICP factors, and Landmark factors, forming the factor graph depicted in Figure 3. In the optimization process after adding each new keyframe, the initial values for solving are provided by the frontend odometry.

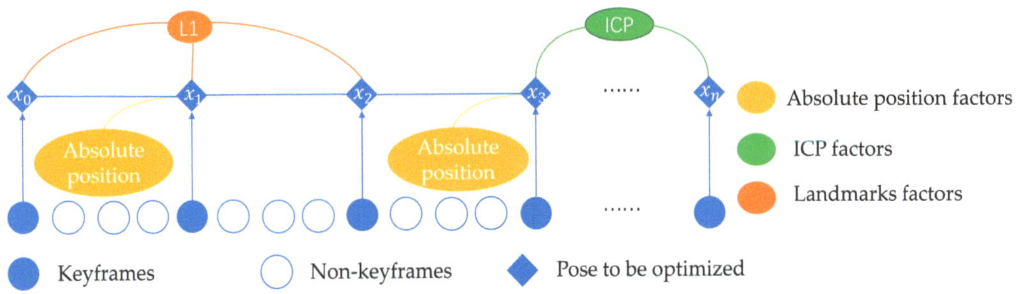

Figure 3. Algorithm flowchart for backend graph optimization.

To batch optimize historical keyframe poses $x = \{x_0\ x_1\ x_2\ \cdots x_i\}$, this paper employs a factor graph optimization method, where each keyframe pose x_i is a vertex in the graph. Through the computation of frontend odometry and point cloud matching results, edges are constructed between adjacent keyframe poses or any two keyframe poses. Additionally, for extra observations such as absolute pose constraints or Landmark constraints, edges connecting vertices are added to the factor graph.

3.2.1. Frame-to-Frame Odometry

The inter-frame constraints for adjacent keyframes in the backend graph optimization are provided by the frontend odometry module. To select keyframes for optimization, the current frame x_{i+1} is compared to the state of the previous keyframe x_i. When the pose change exceeds a threshold, the current frame is chosen as a keyframe. In the factor graph, the newly selected keyframe x_{i+1} is associated with the previous state node x_i. LiDAR scans between two keyframes are discarded to maintain a relatively sparse factor graph while balancing map density and memory consumption, suitable for map construction. Ultimately, the relative pose transformation $\Delta T_{i,i+1}$ between x_i and x_{i+1} is obtained. In practical testing, considering the field of view of the LiDAR and ensuring offline mapping accuracy in degraded scenarios, the thresholds for positional and rotational changes to identify keyframes are set to 0.5 m and 1 degree, respectively.

3.2.2. Absolute Pose Factors

In degraded scenarios, relying solely on long-term pose estimates from IMU and LiDAR will accumulate errors. To address this issue, the backend optimization system needs to incorporate sensors providing absolute pose measurements to eliminate cumulative errors. Absolute pose correction factors mainly include two types:

1. GPS-Based Factors

Absolute poses are obtained from GPS sensors in the current state and transformed into the local Cartesian coordinate system. As shown in Figure 3, an absolute pose factor has already been introduced at keyframe x_1. After adding new keyframes and other constraints to the factor graph, due to the slow growth of accumulated errors from the frontend odometry, introducing absolute pose constraints too frequently for backend optimization can lead to difficulty in constraint solving and poor real-time algorithm performance. Therefore, a new GPS factor is added to keyframe x_3 only when the position change between keyframe x_3 and keyframe x_1 exceeds a threshold. The covariance matrix of the absolute pose depends on the precision of the sensor used and the quality of satellite signal reception.

2. Control Point-Based Factors in GPS-Limited Environments

In environments lacking satellite signals, such as tunnels, GPS sensors cannot be directly used for pose correction. In such cases, control points' absolute coordinates are obtained in advance using surveying equipment like total stations. When the LiDAR moves near the relevant control points, the target perception algorithm outputs the control points' relative coordinates in the LiDAR coordinate system. Using Equation (6), the LiDAR's absolute pose coordinates are then determined. The covariance matrix of the absolute pose depends on the covariance of the target positions output by the Kalman tracking algorithm in the perception algorithm.

$$P_{point} = R \times P_{rel} + P_{lidar} \tag{6}$$

where P_{point}—absolute coordinates of the control point; R—rotation matrix representing the LiDAR's pose; P_{rel}—relative coordinates of the control point in the LiDAR coordinate system; P_{lidar}—absolute coordinates of the LiDAR.

In the actual process, absolute pose factors are only introduced into the system for global optimization when the pose covariance output by the frontend odometry is significantly larger than the received absolute pose covariance.

3.2.3. ICP Factors

The ICP factor involves solving the relative pose transformation between point clouds corresponding to any two keyframes using the ICP algorithm. In the factor graph shown in Figure 3, when keyframe x_n is added to the factor graph, a set of ICP constraints is constructed between keyframes x_3 and x_n. The backend optimization factor graph adds ICP factors in the following two situations:

1. Loop Closure Detection

When a new keyframe x_{i+1} is added to the factor graph, it first searches for the keyframe x_k in the Euclidean space that is closest to x_{i+1}. An ICP factor is added to the factor graph only if x_k and x_{i+1} are within a spatial distance threshold Δd and a temporal separation greater than a threshold Δt. In practical experiments, due to the difficulty of forming loop constraints in the unidirectional movement of subways, loop constraints are constructed in platform areas of both up and down directions on the same route.

2. Low-Speed or Stationary Conditions

In degraded scenarios, the IMU zero offset estimates in the frontend odometry can accumulate significant errors during prolonged low-speed or stationary vehicle conditions, leading to drift in the frontend odometry. Therefore, additional constraints need to be added in such scenarios to avoid pose drift during prolonged stops. Subway trains typically stop only at platforms in tunnel scenes, where point cloud features are abundant, providing sufficient geometric information for ICP constraint solving. When the system detects low-speed or stationary states, it re-caches every keyframe in this state. Whenever a new keyframe x_{i+1} is added to the factor graph, a constraint relationship is established between x_{i+1} and the keyframe x_k, which is the furthest in time from the current keyframe.

When the conditions for adding ICP factors are met, the system searches for the n closest keyframes in the historical keyframes to establish a local point cloud map. This local point cloud map is then used for ICP constraint solving with x_{i+1}, ultimately obtaining a set of relative pose transformation relationships between x_{i+1} and x_k. The covariance matrix of the ICP factor is calculated based on the goodness of fit output during the ICP solving process.

3.2.4. Landmark Factors

The establishment and solving of Landmark factors adopt the Bundle Adjustment (BA) optimization concept commonly used in visual SLAM. As shown in Figure 3, when keyframes x_0, x_1, and x_2 all observe the same landmark point L_1, and since the absolute coordinates of L_1 remain constant, constraint relationships between x_0—x_1, x_1—x_2, and x_0—x_2 can be established based on Equation (7).

$$P_{L1} = R_0 \times P_{r0} + P_{l0} \tag{7}$$

where P_{L1}—absolute coordinates of landmark point $L1$, not directly solved during the process; R_i—attitude rotation matrix of the LiDAR in keyframe x_i; P_{ri}—relative coordinates of the landmark point in keyframe x_i in the LiDAR coordinate system; P_{li}—absolute coordinates of the LiDAR in keyframe x_i.

Therefore, the key to adding Landmark factors lies in how to obtain real-time observations of the position and attitude of the same landmark point. The selection of landmark points is crucial, ensuring continuous observations over a short period and maintaining relatively constant shape and size during the observation to avoid abrupt changes in the object's center of mass. In urban scenes, road signs are chosen as landmark points, while in tunnel scenes of rail transportation, mileposts alongside the track are selected as landmark points.

3.3. *Map Update*

After completing the global optimization for each keyframe, it is necessary to update the stored global map based on the optimized keyframe poses. Furthermore, considering the LiDAR's position in the map, a local feature map is extracted from the global map. This local feature map serves as input to the frontend odometry for scan-to-map matching. In the process of updating the local feature map, this paper implements a position-based sliding window approach. It involves extracting information from the nearest n sub-keyframes to the current LiDAR position, focusing on plane point clouds. Subsequently, the concatenated map undergoes voxel filtering and downsampling to reduce computational load during the matching process.

4. Experimental Results and Discussion

Considering the difficulty in obtaining real-time ground truth poses in the tunnel environment of rail transportation, the proposed offline mapping method with a tightly coupled frontend and graph optimization backend was experimentally validated in both urban road outdoor scenes and rail transportation scenes.

Taking into account the challenge of obtaining real-time ground truth poses in the tunnel environment of rail transit, the mapping method proposed in this paper, featuring a tightly coupled frontend and a graph optimization backend, has not only been experimentally validated in subway scenarios but has also been compared with RTK+IMU integrated navigation in industrial park building obstruction environments. This additional comparison aims to further assess the cumulative error of the proposed method.

4.1. *Experimental Equipment*

The mapping data acquisition system uses the RS-LiDAR-M1, an automotive-grade solid-state LiDAR. It operates with a 905 nm wavelength laser, providing a maximum range of 200 m and an accuracy ranging within ±5 cm. The LiDAR has a horizontal field of view of 120° with a resolution of 0.2°, a vertical field of view of 25° with a resolution of 0.2°, and the ability to output up to 750,000 points per second in single-echo mode. The selected IMU model is the STIM300, with an accelerometer resolution of 1.9 μg, bias instability of 0.05 mg, gyroscope resolution of 0.22°/h, and gyroscope bias instability of 0.3°/h. The sensor installation and layout diagrams for the subway environment and the industrial park environment are illustrated in Figure 4a,b, respectively.

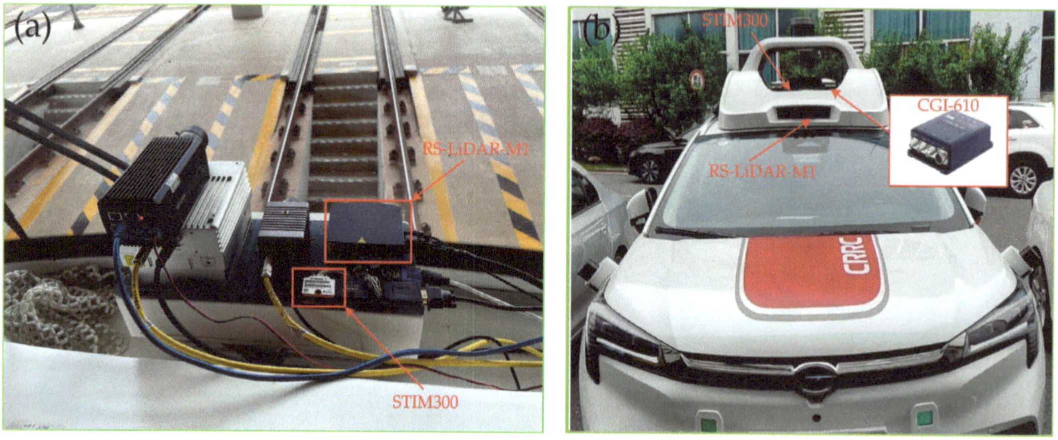

Figure 4. Physical installation and arrangement of sensors. (**a**) Sensors on the train. (**b**) Sensors on the autonomous driving platform vehicle.

4.2. *Subway Tunnel Scene*

Given the degraded nature of tunnel scenes, where the number of feature points in LiDAR point clouds for matching is limited and prone to misalignment, the covariance of point-to-plane residuals needs to be increased when detecting degradation in the frontend odometry. Simultaneously, in the ESKF filter, the covariance of IMU inertial solutions is reduced. During the backend optimization process, considerations include addressing drift in low-speed stationary train scenarios and selecting appropriate landmarks.

4.2.1. Low-Speed Stationary Scenario

During the map data collection process, the train normally stops in the platform area, which is rich in features. There are enough planar feature points for inter-frame matching and the addition of ICP constraints, as shown in Figure 5. Features such as the tunnel wall and the train stop sign can be used for matching.

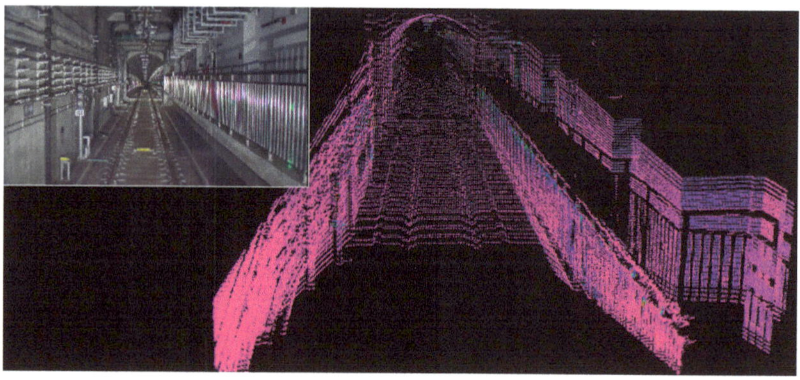

Figure 5. Point cloud effect in normal station platform parking (underground subway parking).

However, in situations where the train is stationary in the curved section of the tunnel or when the ICP factor is turned off in the algorithm, significant drift can occur when there is a large change in train speed during stationary periods, as illustrated in Figure 6. The output trajectory exhibits a backward movement when the train is stationary, emphasizing the need to avoid abrupt acceleration, deceleration, and stops in severely degraded scenes during map data collection.

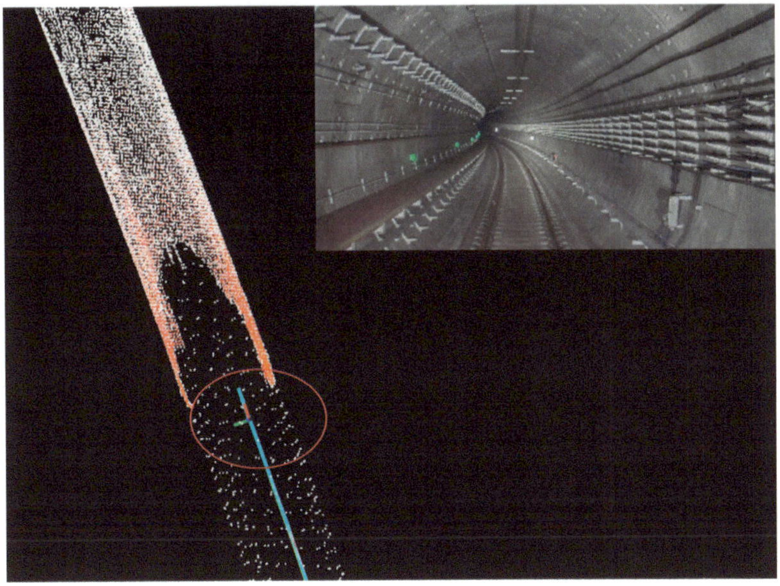

Figure 6. Drift in train parking trajectory in tunnel. The trajectory exhibits a phenomenon of moving backward within the red circle.

4.2.2. Landmark Selection

Due to the limited number of extractable landmarks in subway tunnel scenes, the intensity information of point clouds and the arrangement of signs inside the tunnel are considered. The recognition of hundred-meter markers is chosen as a landmark for constraints, as shown in Figure 7. Since the hundred-meter markers are made of metal, the intensity information is substantial, allowing for direct extraction of relevant point clouds based on intensity filtering. The final step involves extracting the centroid coordinates of the relevant point clouds and incorporating them into the factor graph for optimization and solving.

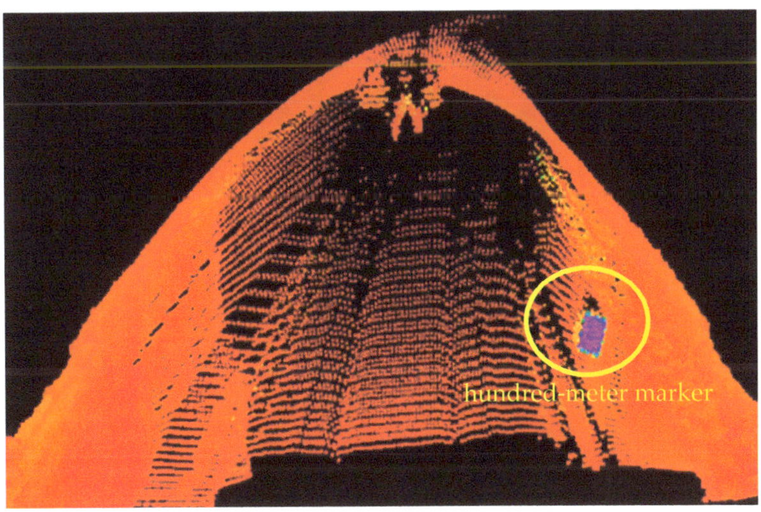

Figure 7. Detection of hundred-meter markers in the tunnel.

4.2.3. Landmark Selection

In the subway tunnel scenario, quantitative analysis is challenging due to the lack of ground truth. Therefore, the evaluation is based on multiple data collections in the same subway tunnel scene, comparing the consistency of trajectories and focusing on the assessment in platform areas and tunnel sections. The three-dimensional point cloud results of the mapping are shown in Figures 8 and 9. In the original point cloud, the tracks and tunnel walls are clearly visible, indicating that our algorithm achieves high accuracy in local areas.

Figure 8. Mapping results of subway tunnel platform.

Figure 9. Mapping results of subway tunnel curve.

To evaluate the consistency of data collected, we used the trajectory from the initial mapping session as ground truth and analyzed the error in the overlap between the trajectories of subsequent mapping sessions. To address the challenge of ensuring a consistent starting point for each data collection, we utilized the evo tool to align the trajectories of multiple sessions, as illustrated in Figure 10. Developed by Michael Grupp, the evo tool is a Python package designed for assessing odometry and SLAM results. It provides functionalities, including aligning and comparing trajectories, computing errors, and generating visualizations, facilitating a comprehensive evaluation of localization and mapping performance. The maximum Absolute Pose Error (APE) recorded was 0.1 m, with an average of 0.04 m and a Root Mean Square Error (RMSE) of 0.05 m.

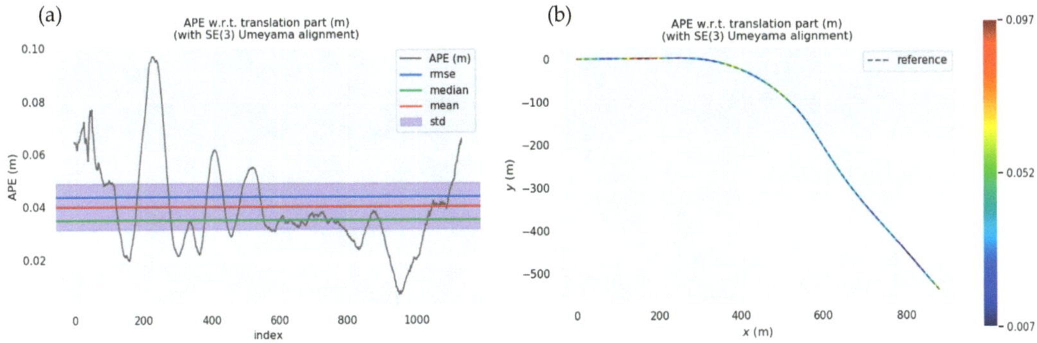

Figure 10. Consistency results for data collected at different times in the same subway tunnel scene. (**a**) APE, RMSE, median, mean, std. (**b**) APE in the xy-plane.

Additionally, due to the significant distance between subway stations in tunnels, mapping within the tunnels involves a higher number of keyframes and the use of unsampled point clouds for stitching. This results in elevated computational and memory requirements for offline map construction. To address these challenges, we propose a multi-map stitching approach, creating a map for each platform interval and then concatenating maps from multiple intervals. Since subway platforms exhibit rich features, we choose to stop and concatenate maps at these locations. We record the last frame pose of the previous map as the initial pose for the current map and manually adjust constraints at the platform using the interactive SLAM method to reduce cumulative errors [33].

4.3. Industrial Park Building Obstructed Environment

To simulate tunnel environments as much as possible and provide a comparison with RTK + IMU combined navigation positioning as ground truth, the experiments were conducted in an industrial park scene. In this scene, the LiDAR's horizontal field of view was obstructed by buildings, but it still received satellite positioning signals. The sensors were mounted on the roof of the vehicle, as shown in Figure 4a. The addition of GPS factors in the backend optimization was constrained only at the starting and ending positions of the trajectory. The ground truth trajectory during mapping was provided by a high-precision RTK + IMU combined navigation device, and the established point cloud map is shown in Figure 11. In the 3D point cloud map, vehicles and signs are clearly visible, indicating that our algorithm has high precision in local areas.

We compared the keyframe trajectories output by our algorithm after backend optimization with the ground truth provided by the combined inertial navigation (RTK + IMU) to quantitatively evaluate the accuracy of the mapping algorithm. The trajectory curves in the x, y, and z directions are plotted in Figure 12. The blue curve represents the ground truth trajectory provided by the RTK + IMU combined navigation device, and the gray dashed line represents the keyframe trajectory output by the mapping algorithm. It can be observed that the trajectory error is small in the horizontal direction, while in the vertical direction, the altitude error from the RTK + IMU combined navigation is relatively larger compared to the errors in the horizontal direction. The altitude trajectory curve output by our mapping algorithm is smoother and generally consistent with the ground truth trend.

In the quantitative assessment of algorithm accuracy, we selected the APE of the trajectory as the evaluation metric, focusing only on position error and neglecting orientation error. Therefore, the calculated APE results are in units of meters. The computed APE results are shown in Figures 13 and 14.

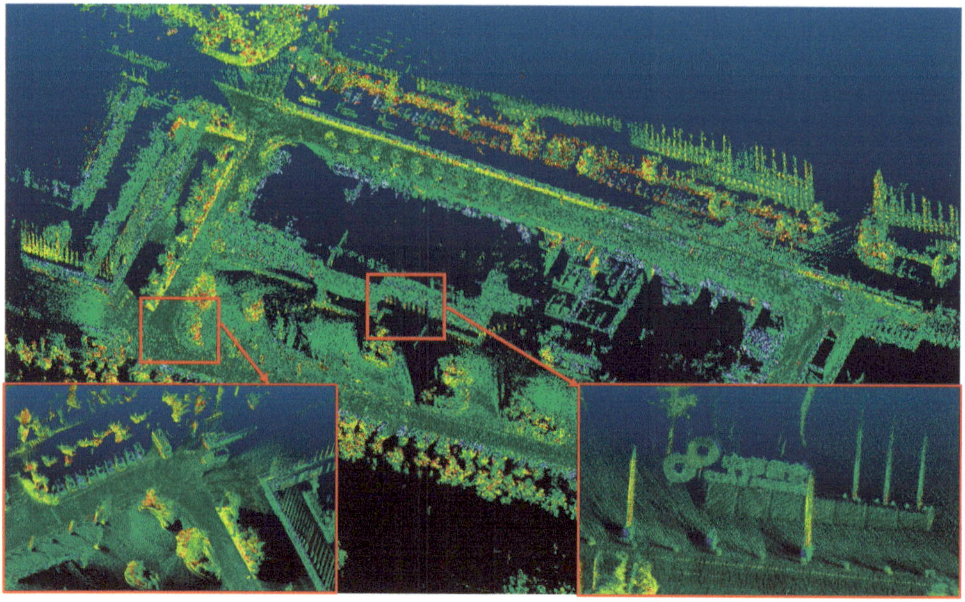

Figure 11. Point cloud map in an industrial park scene. The red boxes are displays that have been partially enlarged.

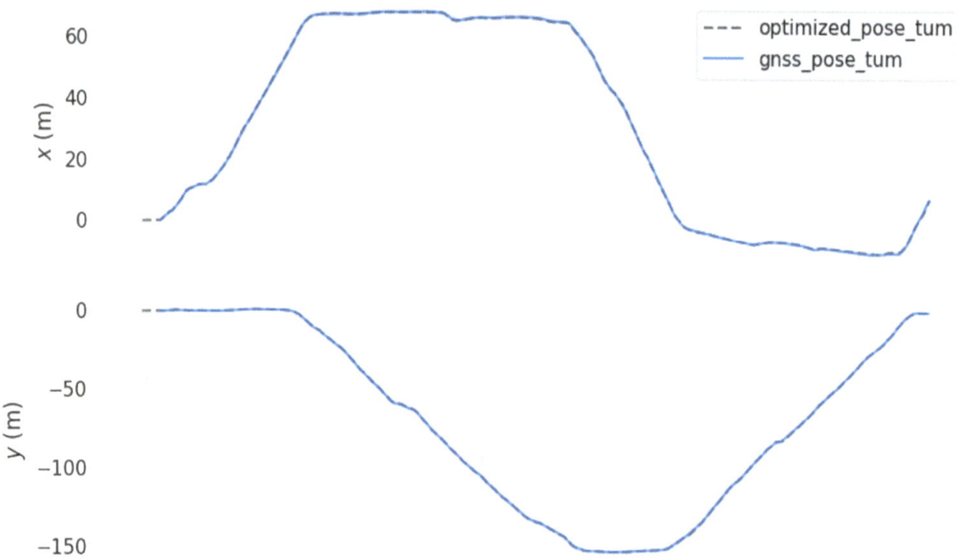

Figure 12. *Cont.*

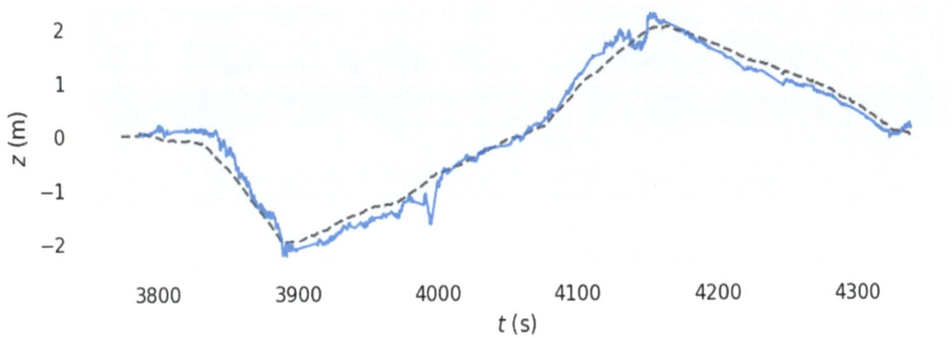

Figure 12. Position comparison in x, y, and z directions.

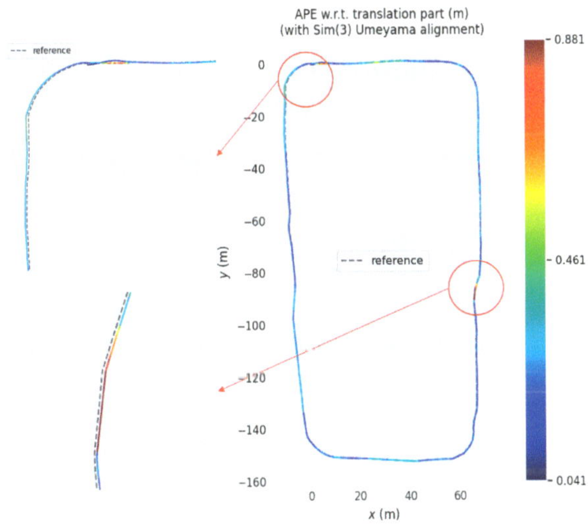

Figure 13. APE trajectory curve. Red arrows and circles are employed for locally enlarged displays.

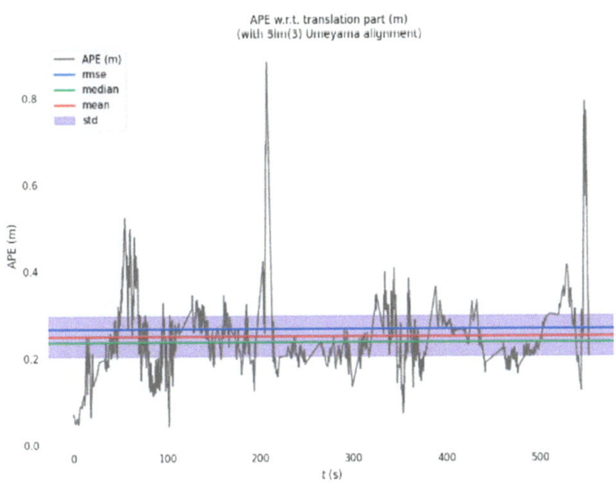

Figure 14. APE statistical results.

Due to the inclusion of GPS factors only at the starting and ending points during the backend optimization process and the use of trajectory alignment methods, the APE is larger at the starting point. The increased error at turning points is attributed to calibration errors between the LiDAR and IMU, with the calibration error causing more noticeable APE as the turning speed increases. Additionally, partial occlusion by buildings results in a decrease in the accuracy of the RTK + IMU combined navigation used as ground truth, further contributing to an increase in APE values.

The statistical results of APE in Figure 14 are as follows: maximum value (max) = 0.88 m, minimum value (min) = 0.04 m, mean = 0.23 m, and root mean square error = 0.26 m. The quantitative analysis results demonstrate the algorithm's high trajectory accuracy.

4.4. Discussion

In comparison to mainstream SLAM algorithms, such as Fast-LIO [18,30], our proposed frontend odometer and backend optimization framework focuses on mapping. This approach addresses the challenges of pose estimation in degraded tunnel environments. The design of our framework is specifically tailored to the structural characteristics of tunnel scenes and the operational requirements of trains in tunnel intervals, resulting in high-precision point cloud construction.

In the subway mapping and localization process, the absence of a GNSS as ground truth may result in cumulative errors. Additionally, during the initial wake-up phase of the train, without GNSS signals for providing the initial position, the system faces challenges in initialization.

To address these limitations, we propose utilizing visual recognition of mileposts and their unique identifiers alongside the tracks. The unique identifiers of mileposts can be leveraged for calibrating subway positions. Moreover, incorporating visual methods can enhance the success rate of initialization, thus increasing the robustness of the subsequent localization system.

5. Conclusions

In this paper, we proposed a high-precision point cloud map construction method based on LiDAR and IMU. The approach utilizes a tightly coupled frontend odometry with an ESKF for inter-frame pose estimation, and a backend global pose optimization employing graph optimization theory. Absolute pose factors, ICP factors, and Landmark factors are incorporated into the optimization process based on real-world scenarios. In the context of long urban subway tunnels, the algorithm introduces detection for degraded scenes in the frontend odometry and emphasizes the inclusion of ICP factors in low-speed stationary situations and the selection of Landmark points in the backend graph optimization.

The algorithm's performance is evaluated by assessing the consistency of trajectories using different data collected on the same route, with a particular focus on platform and tunnel areas. The trajectory alignment error is consistently below 0.11 m. No degradation anomalies were observed throughout the entire tunnel section. Additionally, in the experimental setup in an industrial park scenario, the optimized trajectory is compared with the ground truth provided by the integrated navigation system, yielding an RMSE of 0.26 m for the APE and an accumulated error of less than 0.2%. It is evident that the proposed algorithm achieves high-precision map construction in tunnels and obstructed environments. The next step will be to address the pose initialization problem in degraded environments, particularly in long tunnel scenarios.

Author Contributions: Conceptualization, C.L. and W.P.; methodology, C.L. and W.P.; software, W.H. and F.W.; validation, W.P. and W.H.; formal analysis, W.H.; investigation, C.L. and X.Y.; data curation, C.Y. and Q.W.; writing—original draft preparation, W.P. and W.H.; writing—review and editing, W.H. and X.Y.; visualization, X.Y. and F.W. All authors have read and agreed to the published version of the manuscript.

Funding: This work was funded by the National Key Research and Development Project of China (2022YFB4300400).

Data Availability Statement: Data are contained within the article.

Acknowledgments: All authors would like to thank the editors and anonymous reviewers for their valuable comments and suggestions, which improved the quality of the manuscript. Additionally, we would like to express our gratitude to Shanghai Huace Navigation Technology Co., Ltd. for their technical assistance and support during testing.

Conflicts of Interest: All authors are employed by the company CRRC Zhuzhou Institute Co., Ltd. All authors declare that the research was conducted in the absence of any commercial or financial relationships that could be construed as a potential conflict of interest.

References

1. Sun, H.; Xu, Z.; Yao, L.; Zhong, R.; Du, L.; Wu, H. Tunnel monitoring and measuring system using mobile laser scanning: Design and deployment. *Remote Sens.* **2020**, *12*, 730. [CrossRef]
2. Foria, F.; Avancini, G.; Ferraro, R.; Miceli, G.; Peticchia, E. ARCHITA: An innovative multidimensional mobile mapping system for tunnels and infrastructures. *MATEC Web Conf.* **2019**, *295*, 01005. [CrossRef]
3. Cheng, R.; Song, Y.; Chen, D.; Chen, L. Intelligent localization of a high-speed train using LSSVM and the online sparse optimization approach. *IEEE Trans. Intell. Transp. Syst.* **2017**, *18*, 2071–2084. [CrossRef]
4. Wu, Y.; Weng, J.; Tang, Z.; Li, X.; Deng, R.H. Vulnerabilities, attacks, and countermeasures in balise-based train control systems. *IEEE Trans. Intell. Transp. Syst.* **2016**, *18*, 814–823. [CrossRef]
5. Wang, Z.; Yu, G.; Zhou, B.; Wang, P.; Wu, X. A train positioning method based-on vision and millimeter-wave radar data fusion. *IEEE Trans. Intell. Transp. Syst.* **2021**, *23*, 4603–4613. [CrossRef]
6. Otegui, J.; Bahillo, A.; Lopetegi, I.; Diez, L.E. Evaluation of experimental GNSS and 10-DOF MEMS IMU measurements for train positioning. *IEEE Trans. Instrum. Meas.* **2018**, *68*, 269–279. [CrossRef]
7. Buffi, A.; Nepa, P. An RFID-based technique for train localization with passive tags. In Proceedings of the IEEE International Conference on RFID (RFID), Phoenix, AZ, USA, 9–11 May 2017; pp. 155–160.
8. Daoust, T.; Pomerleau, F.; Barfoot, T.D. Light at the end of the tunnel: High-speed lidar-based train localization in challenging underground environments. In Proceedings of the 2016 13th Conference on Computer and Robot Vision (CRV), Victoria, BC, Canada, 1–3 June 2016.
9. Liu, H.; Pan, W.; Hu, Y.; Li, C.; Yuan, X.; Long, T. A Detection and Tracking Method Based on Heterogeneous Multi-Sensor Fusion for Unmanned Mining Trucks. *Sensors* **2022**, *22*, 5989. [CrossRef]
10. Pan, W.; Fan, X.; Li, H.; He, K. Long-Range Perception System for Road Boundaries and Objects Detection in Trains. *Remote Sens.* **2023**, *15*, 3473. [CrossRef]
11. Wang, J.; Chen, W.; Weng, D.; Ding, W.; Li, Y. Barometer assisted smartphone localization for vehicle navigation in multilayer road networks. *Measurement* **2023**, *211*, 112661. [CrossRef]
12. Gao, F.; Wu, W.; Gao, W.; Shen, S. Flying on point clouds: Online trajectory generation and autonomous navigation for quadrotors in cluttered environments. *J. Field Robot.* **2019**, *36*, 710–733. [CrossRef]
13. Wang, J.; Weng, D.; Qu, X.; Ding, W.; Chen, W. A Novel Deep Odometry Network for Vehicle Positioning Based on Smartphone. *IEEE Trans. Instrum. Meas.* **2023**, *72*, 2505512. [CrossRef]
14. Ye, H.; Chen, Y.; Liu, M. Tightly coupled 3d lidar inertial odometry and mapping. In Proceedings of the IEEE International Conference on Robotics and Automation (ICRA), Montreal, QC, Canada, 20–24 May 2019.
15. Zhao, S.; Fang, Z.; Li, H.; Scherer, S. A Robust Laser-Inertial Odometry and Mapping Method for Large-Scale Highway Environments. In Proceedings of the 2019 IEEE/RSJ International Conference on Intelligent Robots and Systems (IROS), Macau, China, 4–8 November 2019; pp. 1285–1292. [CrossRef]
16. Qin, T.; Li, P.; Shen, S. VINS-Mono: A Robust and Versatile Monocular Visual-Inertial State Estimator. *IEEE Trans. Robot.* **2018**, *34*, 1004–1020. [CrossRef]
17. Liu, X.; Zhang, F.Z. Extrinsic calibration of multiple lidars of small fov in targetless environments. *IEEE Robot. Autom. Lett.* **2021**, *6*, 2036–2043. [CrossRef]
18. Xu, W.; Zhang, F. Fast-lio: A fast, robust lidar-inertial odometry package by tightly-coupled iterated kalman filter. *IEEE Robot. Autom. Lett.* **2021**, *6*, 3317–3324. [CrossRef]
19. Liu, Z.; Zhang, F. BALM: Bundle adjustment for lidar mapping. *IEEE Robot. Autom. Lett.* **2021**, *6*, 3184–3191. [CrossRef]
20. Lin, J.; Zhang, F. Loam livox: A fast, robust, high-precision lidar odometry and mapping package for lidars of small fov. In Proceedings of the 2020 IEEE International Conference on Robotics and Automation (ICRA), Paris, France, 31 May–31 August 2020; pp. 3126–3131.
21. Lin, J.; Zhang, F. R 3 LIVE: A Robust, Real-time, RGB-colored, LiDAR-Inertial-Visual tightly-coupled state Estimation and mapping package. In Proceedings of the IEEE International Conference on Robotics and Automation (ICRA), Philadelphia, PA, USA, 23–27 May 2022.

22. Bao, Z.; Hossain, S.; Lang, H.; Lin, X. A review of high-definition map creation methods for autonomous driving. *Eng. Appl. Artif. Intell.* **2023**, *122*, 106125. [CrossRef]
23. Zhou, H.; Yao, Z.; Lu, M. Uwb/lidar coordinate matching method with anti-degeneration capability. *IEEE Sens. J.* **2020**, *21*, 3344–3352. [CrossRef]
24. Zhuang, Y.; Sun, X.; Li, Y.; Huai, J.; Hua, L.; Yang, X.; Cao, X.; Zhang, P.; Cao, Y.; Qi, L.; et al. Multi-sensor integrated navigation/positioning systems using data fusion: From analytics-based to learning-based approaches. *Inf. Fusion* **2023**, *95*, 62–90. [CrossRef]
25. Heirich, O.; Robertson, P.; Strang, T. RailSLAM—Localization of Rail Vehicles and Mapping of Geometric Railway Tracks. In Proceedings of the IEEE International Conference on Robotics and Automation (ICRA), Karlsruhe, Germany, 6–10 May 2013.
26. Wang, Y.; Song, W.; Lou, Y.; Zhang, Y.; Huang, F.; Tu, Z.; Liang, Q. Rail Vehicle Localization and Mapping with LiDAR-Vision-Inertial-GNSS Fusion. *IEEE Robot. Autom. Lett.* **2022**, *7*, 9818–9825. [CrossRef]
27. Tschopp, F.; Schneider, T.; Palmer, A.W.; Nourani-Vatani, N.; Cadena, C.; Siegwart, R.; Nieto, J. Experimental comparison of visual-aided odometry methods for rail vehicles. *IEEE Robot. Autom. Lett.* **2019**, *4*, 1815–1822. [CrossRef]
28. Wang, Y.; Song, W.; Lou, Y.; Huang, F.; Tu, Z.; Zhang, S. Simultaneous Location of Rail Vehicles and Mapping of Environment with Multiple LiDARs. *arXiv* **2021**, arXiv:2112.13224.
29. Shan, T.; Englot, B. LeGO-LOAM: Lightweight and Ground-Optimized Lidar Odometry and Mapping on Variable Terrain. In Proceedings of the IEEE/RSJ International Conference on Intelligent Robots and Systems (IROS), Madrid, Spain, 1–5 October 2018; IEEE: Piscataway, NJ, USA, 2018.
30. Xu, W.; Cai, Y.; He, D.; Lin, J.; Zhang, F. Fast-lio2: Fast direct lidar-inertial odometry. *IEEE Trans. Robot.* **2022**, *38*, 2053–2073. [CrossRef]
31. Fan, X.; Chen, Z.; Liu, P.; Pan, W. Simultaneous Vehicle Localization and Roadside Tree Inventory Using Integrated LiDAR-Inertial-GNSS System. *Remote Sens.* **2023**, *15*, 5057. [CrossRef]
32. Feng, C.; Taguchi, Y.; Kamat, V.R. Fast plane extraction in organized point clouds using agglomerative hierarchical clustering. In Proceedings of the IEEE International Conference on Robotics and Automation (ICRA), Hong Kong, China, 31 May–7 June 2014.
33. Koide, K.; Miura, J.; Yokozuka, M.; Oishi, S.; Banno, A. Interactive 3D graph SLAM for map correction. *IEEE Robot. Autom. Lett.* **2020**, *6*, 40–47. [CrossRef]

remote sensing

MDPI

Article

A Novel Automatic Registration Method for Array InSAR Point Clouds in Urban Scenes

Chenghao Cui [1,2], Yuling Liu [1], Fubo Zhang [1,*], Minan Shi [1,2], Longyong Chen [1], Wenjie Li [1,2] and Zhenhua Li [1,2]

1 National Key Laboratory of Microwave Imaging Technology, Aerospace Information Research Institute, Chinese Academy of Sciences, Beijing 100190, China; cuichenghao21@mails.ucas.ac.cn (C.C.); liyuling22@mails.ucas.ac.cn (Y.L.); shiminan21@mails.ucas.ac.cn (M.S.); chenly@aircas.ac.cn (L.C.); liwenjie21@mails.ucas.ac.cn (W.L.); lizhenhua19@mails.ucas.ac.cn (Z.L.)

2 School of Electronic, Electrical and Communication Engineering, University of Chinese Academy of Sciences, Beijing 100049, China

* Correspondence: zhangfb@aircas.ac.cn

Abstract: The array interferometric synthetic aperture radar (Array InSAR) system resolves shadow issues by employing two scans in opposite directions, facilitating the acquisition of a comprehensive three-dimensional representation of the observed scene. The point clouds obtained from the two scans need to be transformed into the same coordinate system using registration techniques to create a more comprehensive visual representation. However, the two-point clouds lack corresponding points and exhibit distinct geometric distortions, thereby preventing direct registration. This paper analyzes the error characteristics of array InSAR point clouds and proposes a robust registration method for array InSAR point clouds in urban scenes. It represents the 3D information of the point clouds using images, with pixel positions corresponding to the azimuth and ground range directions. Pixel intensity denotes the average height of points within the pixel. The KAZE algorithm and enhanced matching approach are used to obtain the homonymous points of two images, subsequently determining the transformation relationship between them. Experimental results with actual data demonstrate that, for architectural elements within urban scenes, the relative angular differences of registered facades are below 0.5°. As for ground elements, the Root Mean Square Error (RMSE) after registration is less than 1.5 m, thus validating the superiority of the proposed method.

Keywords: array interferometric synthetic aperture radar (Array InSAR); KAZE; point clouds registration; flattened phase error; RANSAC

Citation: Cui, C.; Liu, Y.; Zhang, F.; Shi, M.; Chen, L.; Li, W.; Li, Z. A Novel Automatic Registration Method for Array InSAR Point Clouds in Urban Scenes. *Remote Sens.* **2024**, *16*, 601. https://doi.org/10.3390/rs16030601

Academic Editor: Fabio Rocca

Received: 10 January 2024
Revised: 3 February 2024
Accepted: 4 February 2024
Published: 5 February 2024

1. Introduction

Recently, 3D imaging techniques have witnessed rapid development. Compared with 2D images, point clouds have the ability to capture the precise spatial structures and geometric features of objects. By analyzing and processing point clouds, valuable information such as distances, angles, and occlusion relationships between objects can be extracted, making it of great significance in applications such as robot navigation [1], map creation [2], environmental reconstruction [3], and virtual reality [4].

Laser scanning technology [5,6], photogrammetric stereo matching [7,8], and array InSAR [9] are the primary methods for acquiring point clouds. In contrast to optical sensors, SAR exhibits excellent imaging capabilities even under adverse weather conditions. By deploying multiple antennas in the across-track direction, array InSAR enables multi-angle observations of the target scene. It effectively addresses the problem of overlap between targets and terrain in 2D images, significantly enhancing the capabilities of target detection, identification, and detailed interpretation [10].

In urban scenes, microwaves emitted by radar are often obstructed by artificial facilities, leading to incomplete point clouds generated from a single scan. In practical applications, point cloud registration techniques are typically required to match and fuse point clouds acquired from different scans. Figure 1 presents a schematic diagram illustrating the acquisition of complete 3D information of urban scenes through two scans.

Figure 1. Airborne array InSAR acquires complete 3D information of an urban area through two flight tests.

Two flight tests were conducted from opposing directions to image and generate point clouds. Firstly, disparities in shadow positions and the anisotropy of scatterers result in a lack of corresponding points between the two scans. Additionally, the SAR point cloud contains a substantial number of outliers, attributed partly to multiple scattering effects and partly originating from the super-resolution imaging algorithm. Subsequently, the SAR point cloud requires a transformation from the azimuth-range-elevation coordinate system to the azimuth-ground range-height coordinate system. Discrepancies in the selection of reference heights lead to conspicuous vertical and ground range offsets in the point clouds, as well as a stretching effect along the ground range direction [11], resulting, as illustrated in Figure 2. Lastly, airborne array InSAR exhibits significant changes in local incidence angles within the spatial domain, introducing supplementary geometric approximation errors [12]. In a word, the registration of array InSAR point clouds faces substantial challenges.

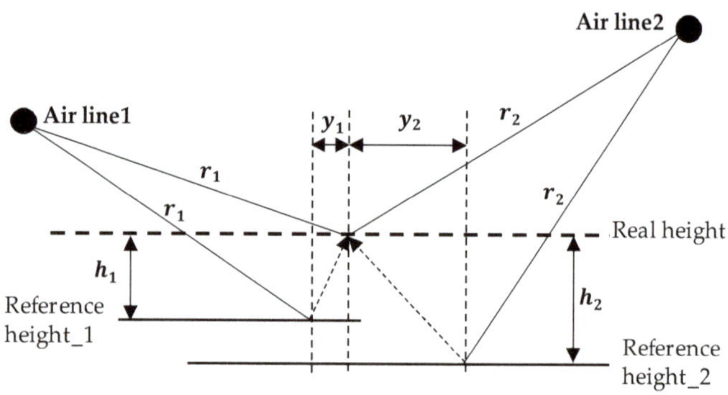

Figure 2. Capture geometry of the two tracks.

Conventional point cloud registration typically follows a strategy of coarse registration followed by fine registration [13]. The purpose of coarse registration is to find a suitable initial transformation that serves as a foundation for subsequent fine registration. Fine registration involves refining the initial transformation matrix through multiple iterative optimization steps to achieve a global optimum.

Coarse registration of point clouds typically involves extracting geometric features from the point cloud. These features can be categorized as point-based, line-based, and surface-based. Barnea applied the Scale-Invariant Feature Transform (SIFT) to laser point cloud registration [14]. Aiger introduced a method called Four-Point Congruent Sets (4PCS), which utilized the invariant property of the ratio of lines formed by four coplanar points, achieving global point cloud registration [15]. Compared to points, lines possess stronger geometric topological characteristics and are easier to extract. Jaw proposed a line-based registration method, where the matching of 3D line features is constrained by angle and distance [16]. Cheng Liang presented a hierarchical registration method based on 3D road networks and building outlines [17]. Lee extracted line features by utilizing the intersection points of adjacent planes and adjusted the differences between overlapping data using these line features [18]. Surface features contain more information compared to point or line features and are less affected by noise. Researchers generally use methods such as least squares, random sample consensus (RANSAC), and principal component analysis (PCA) for surface fitting. The minimum sum of squared Euclidean distances between surfaces is taken for the objective function [19].

One of the most classical methods for fine registration is the Iterative Closest Point (ICP) algorithm [20]. Through iterative optimization, the ICP algorithm aims to align the positions of two sets of point clouds as closely as possible. K. AL-Durgham combined the RANSAC method with the SIFT operator, effectively addressing the registration problem without local features [21]. Eijiro proposed the Normal Distribution Transform (NDT) method [22], which converts point clouds in a 3D grid into probability distribution functions. The probability distribution of each position measurement sample in the grid follows a normal distribution. By optimizing the normal distribution probabilities of two point clouds using the Hessian matrix method, fine registration is achieved. These methods assume that one point set is a subset of the other. When this assumption is invalid, it leads to false matches [23].

In recent years, the success of deep learning in advanced visual tasks has extended to the domain of point cloud processing. PointNet [24] and PointNet++ [25] represent two significant milestones. PointNet generates a descriptor for each point, while PointNet++ is a key technology for extracting local information from point clouds. The crucial stage involves the set abstraction module, composed of sampling, grouping, and PointNet components. Subsequently, numerous researchers have adopted learning-based techniques [26–29] for point cloud registration. The objective of these techniques is to extract features from 3D points and find accurate corresponding points, followed by the estimation of transformations using these corresponding points.

The aforementioned methods are widely applied in the registration of laser point clouds. However, for array InSAR point clouds, it is a challenge to extract matching features from the 3D information of point clouds. Dr. Zhu proposed an approach to extract the L-shaped structures of buildings in tomographic SAR point clouds and achieve automatic registration of point clouds from different scans [30]. Dr. Tong from Tongji University proposed a method that utilizes the constraint of parallel building facades to match specific pairs of building facades [31]. However, the bottom scenes of buildings have holes due to occlusion, and there is a large amount of noise below the building facades due to third-order scattering [32]. The fitted building facades exhibit large errors. Additionally, the stretching phenomenon within the ground range of the point clouds has not been taken into account. To address these challenges, this paper proposes a novel method for the registration of array InSAR point clouds.

In this study, we first correct the flattened phase error caused by the differences in local incidence angles. For point clouds of large urban scenes, the ground range can be several hundred meters or more, and the flattened phase error caused by the differences in local incidence angles cannot be ignored. The height variation of the ground points is relatively flat, which allows us to easily calculate the relationship between point cloud height and ground range and correct the flattened phase error. Next, we project the corrected point cloud onto the x–y plane and divide the plane into grids, which serve as pixels for generating grayscale images. The pixel intensity is represented by the average height of the points falling within each grid. The quality of the generated images is subpar, and utilizing traditional image-matching methods makes it challenging to attain the transformation relationship between the two images. We utilize the KAZE [33] algorithm to extract feature points from both the original and blurred images. The stable feature point refers to a feature point in the original image for which there exists a feature point in the blurred image that is sufficiently close to it. Next, we filter matching point pairs from the stable feature points in the images. The transformation relationship between point clouds in the azimuth and ground range directions is calculated based on the positional relationship of the matching points. The height offset between point clouds is represented by the average intensity difference of the matching points. In summary, this method makes two main contributions:

1. An analysis was conducted on the height errors in airborne array InSAR point clouds caused by local incidence angle variations, followed by their subsequent correction.
2. The KAZE algorithm was introduced into the point cloud registration problem, and a method for selecting robust feature points was proposed to address the registration of array InSAR point clouds.

2. Methods

The main challenge in effectively fusing array InSAR point clouds lies in the inability to extract stable feature points and determine true corresponding matching points in the 3D information. The proposed workflow for point cloud registration is shown in Figure 3. Firstly, the flattened phase error caused by local incidence angle differences is corrected. Then, the point cloud is projected onto the ground to generate a grayscale image, where the pixel intensity represents the average height of the points within the pixel. To obtain stable feature points, the KAZE algorithm is employed to extract feature points from both the grayscale image and the image with applied defocus blur, and a distance threshold is set to select stable feature points. Subsequently, the nearest neighbor distance ratio (NNDR) strategy and vector consistency are employed to determine the matching points between the two images. The position of the matching points is used to determine the transformation relationship in the azimuth and ground range directions of the two flight test point clouds. The pixel intensity of the matching points is utilized to determine the height offset between the two point clouds.

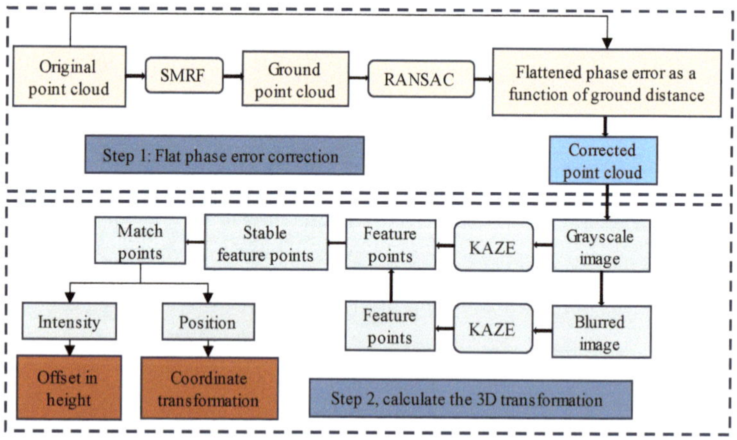

Figure 3. Algorithm flow.

2.1. Flattened Phase Error Correction

The multi-channel images of the airborne array InSAR are obtained simultaneously, so there is no temporal decoherence factor, and it is only sensitive to the target elevation. The phase component of airborne array InSAR is composed of flat earth effect, height, and system noise [12]. Figure 4a illustrates the geometric configuration of radar interferometry in relation to the flattened phase. In the process of interferometry, a reference object is essential to mitigate the impact of the flat earth effect. Due to the nature of radar imaging, it becomes challenging to distinguish scatterers that are equidistant from the radar. Thus, considering a point p with a relative height of h, an equivalent point r is specified to calculate the flattened phase, local incident angle, and perpendicular baseline. Then h can be defined as follows:

$$h = R\cos\theta_r - R\cos\theta_p \tag{1}$$

where R is the slant range and θ_p is the local incident angle. θ_r is the equivalent incident angle for calculating flat earth effect.

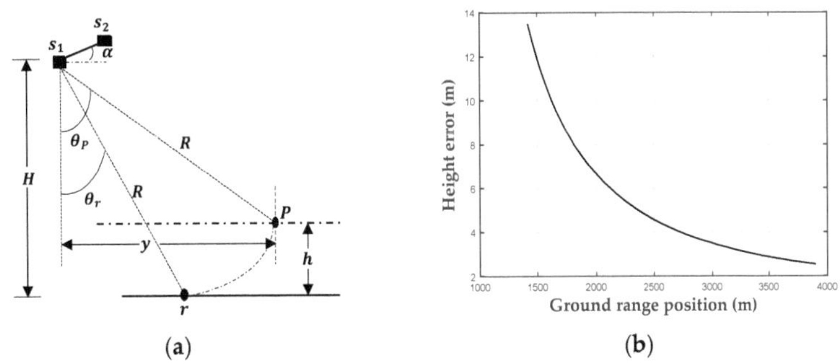

Figure 4. Calculate target height using local incidence angle of reference point. (**a**) InSAR system geometric model. (**b**) Simulation of the relationship between the height error and ground range position.

In fact, the actual local incident angle θ_p cannot be obtained. In conventional processing, the local incident angle is substituted for the equivalent incident angle θ_r on the reference body. At this time, h can be expressed as

$$h = \frac{\lambda R \sin\theta_r}{4\pi B\cos(\theta_r - \alpha)}(\phi_p - \phi_r) \tag{2}$$

where ϕ_p and ϕ_r are the interference phases of point p and point r, respectively, and B denotes the length of baseline. The following can be obtained through the combined calculation of Equations (1) and (2):

$$\Delta h = R\cos\theta_r - R\cos\theta_p - \frac{\lambda R \sin\theta_r}{4\pi B\cos(\theta_r - \alpha)}(\phi_p - \phi_r) \tag{3}$$

And according to the geometric relationship shown in Figure 4a, ϕ_p and ϕ_r can be represented as

$$\phi_p = \frac{4\pi(R - \sqrt{B^2 + R^2 - 2BR\sin(\theta_p - \alpha)})}{\lambda} \tag{4}$$

$$\phi_r = \frac{4\pi(R - \sqrt{B^2 + R^2 - 2BR\sin(\theta_r - \alpha)})}{\lambda} \tag{5}$$

The ground range position of the point p is y, according to geometric relationships, y can be represented by R and ϕ_p.

$$y = R \cdot \sin\theta_p \tag{6}$$

By substituting Equations (4)–(6) into Equation (3), setting the baseline inclination angle α to 0, the relationship between Δh and y can be obtained as follows:

$$\Delta h = h - \frac{\sqrt{(H-h)^2 + y^2} \cdot \left(\sqrt{B^2 - 2B\sqrt{h^2 - 2Hh + y^2} + (H-h)^2 + y^2} - \sqrt{B^2 - 2By + (H-h)^2 + y^2}\right) \cdot \sqrt{h^2 - 2Hh + y^2}}{H} \tag{7}$$

The Δh is related to the radar platform height H, target height h, and local incidence angle θ_p (corresponding to y). For a point cloud generated from a single flight, the height of the radar platform remains constant, thereby exerting an equal influence on the measurement errors. The error impact caused by the height factor of the same target is equivalent between two flights. Hence, we solely consider the influence of the local incidence angle on height errors and aim to establish the relationship between height errors and ground range positions. Based on Equation (3) and the simulation parameters from Table 1, the relationship between height error and ground range position is simulated, as shown in Figure 4b.

Table 1. Simulation parameters.

$H/(m)$	$\lambda/(cm)$	$h/(m)$	$B/(m)$	$\theta_P/(^\circ)$
4000	2	100	2	20–45

Expanding Equation (7) in a Taylor series, where the series is finite, and the highest power term is a 5th-order term. The magnitudes of the third, fourth, and fifth-order terms are 10^{-8}, 10^{-11} and 10^{-14}, respectively. In this paper, we can neglect terms of the third order and higher. In response to the Δh, we assume that the urban terrain is a flat plane. The plan is to extract the ground portion and fit a quadratic function to model the relationship between height and ground distance. According to the analysis in [34], among the various filtering algorithms, morphology-based filters have demonstrated the best performance in extracting the ground in urban scenes. Morphology-based filters primarily rely on two fundamental operations: dilation and erosion. These operations, in combination, give rise to opening and closing operations, which are employed for point cloud filtering. The method rasterizes the original point cloud based on the lowest points within a given window size and subsequently processes it using an opening operation. Points for which the height difference before and after the operation is less than a specified tolerance are labeled as ground points. It is evident that the performance of this filtering technique is greatly influenced by the choice of window size, making it challenging to strike a balance between removing large-sized objects and retaining detailed ground features. As shown in Figure 5, the progressive morphological filters proposed in [35,36] address this issue by gradually increasing the window size and height threshold.

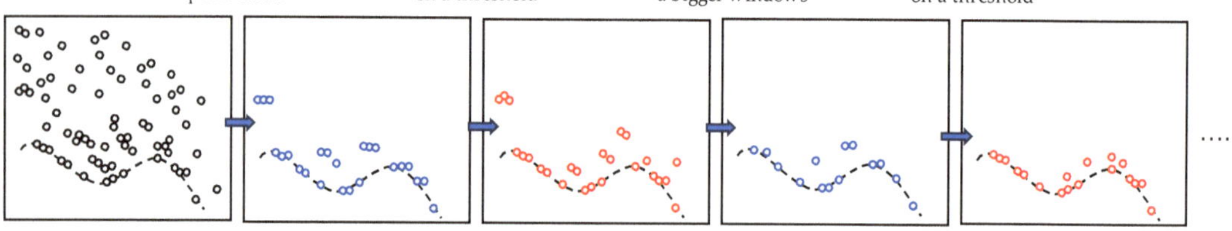

Figure 5. SMRF flow.

In this paper, the ground extraction is performed using the simple morphological filter (SMRF) proposed in [35]. Subsequently, we divide the ground range from the original point cloud into sub-intervals, project the ground points onto each sub-interval, and calculate the average height of the points within each sub-interval. The RANSAC method is then employed to fit a quadratic function that models the relationship between the average height and the position of the ground range. For a single flight-acquired point cloud, using the center position along the ground distance axis as a reference, we calculate the required upward or downward adjustment in height for each point based on its distance from the center along the ground distance axis and its relationship with the fitted quadratic curve. This allows us to correct the overall height of the point cloud, ensuring that each point is adjusted appropriately to align with the desired height. The flowchart of point cloud height correction is shown in Figure 6.

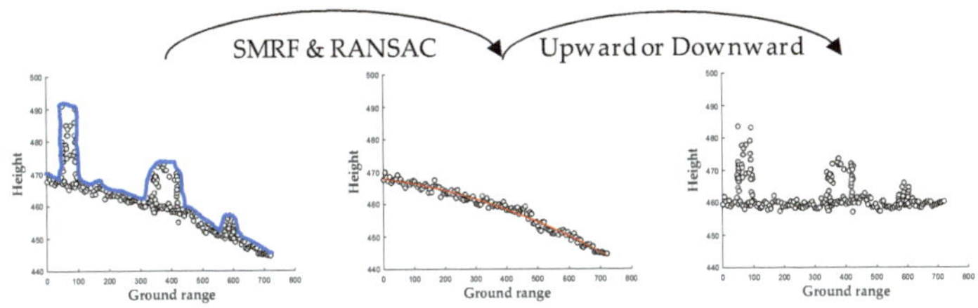

Figure 6. The flowchart of point cloud height correction.

2.2. Obtain Matching Points with KAZE

2.2.1. Generate Grayscale Image

In this study, the point cloud is projected onto the x–y plane, where the x-axis represents the azimuth direction and the y-axis represents the ground range direction. A 2D matrix is created by dividing the x–y plane into grids with a specified step size along the x and y axes, with a step size of 0.8 m. The average height of the points that fall within each grid cell is computed and assigned as the corresponding element of the matrix.

2.2.2. Feature Point Extraction

Traditional feature detection methods employ Gaussian linear scale-space downsampling to detect feature points. Visually, the matching points between two images are typically found along the edges and certain details within the scene. However, Gaussian filtering can cause edge blurring and loss of fine details. As a result, using linear scale-space feature detection algorithms for image registration in this study yielded unsatisfactory results. The KAZE algorithm uses nonlinear diffusion filtering to construct a scale space, which effectively reduces image edge blur and detail loss [33]. It retains higher local accuracy and distinguishability while maintaining scale invariance. The KAZE algorithm mainly includes the following steps:

1. Constructing Nonlinear Scale Space:

The KAZE algorithm constructs a nonlinear scale space through the utilization of nonlinear diffusion filtering and the Additive Operator Splitting (AOS) algorithm. The nonlinear diffusion filtering method interprets the variations in image brightness at different scales as the divergence of a certain form of flow function, which can be described by nonlinear partial differential equations:

$$\frac{\partial L}{\partial t} = div(c(x, y, t) \cdot \nabla L) \tag{8}$$

where L represents the image brightness, c denotes the conductivity function, and t represents the scale parameter. The conductivity function determines the extent to which the diffusion process in an image adapts to its local structure. The expression for c is as follows:

$$c(x, y, t) = g(|\nabla L_\sigma(x, y, t)|) \tag{9}$$

where L_σ is the gradient of the image after Gaussian smoothing. In this study, we adopt the g_2 function as proposed in [33].

$$g_2 = \frac{1}{1 + \frac{|\nabla L_\sigma|^2}{k^2}} \tag{10}$$

Due to the lack of specific analytical solutions for the partial differential equation of nonlinear diffusion filtering, numerical methods are required to estimate the solution of the differential equation. The linear implicit scheme is a feasible discretization method, and the equation is as follows:

$$\frac{L^{i+1} - L^{i}}{\tau} = \sum_{l=1}^{m} A_l(L^i)L^{i+1} \tag{11}$$

A_l represents the matrix representation of the image in different dimensions. The solution L^{i+1} of the equation is represented as follows:

$$L^{i+1} = \left(I - \tau\sum_{l=1}^{m} A_l(L^i)\right)^{-1} L^i \tag{12}$$

The aforementioned steps constitute the fundamental construction scheme for nonlinear scale space.

2. Feature point detection:

Since nonlinear diffusion filtering is based on the theory of heat conduction, its model is formulated in terms of time units. Therefore, it is necessary to perform a conversion between image pixel units and time units. This conversion can be represented by Equation (13), where t_i is referred to as the evolution time.

$$t_i = \frac{1}{2}\sigma_i^2 \tag{13}$$

Using the AOS scheme, the nonlinear scale space can be represented as follows:

$$L^{i+1} = \left(I - (t_{i+1} - t_i)\sum_{l=1}^{m} A_l(L^i)\right)^{-1} L^i \tag{14}$$

The feature point detection in KAZE is achieved by searching for local maxima using the Hessian matrix:

$$L_{Hessian} = \sigma^2(L_{xx}L_{yy} - L_{xy}^2) \tag{15}$$

Each pixel is compared with the pixels in a 3×3 neighborhood window at its current scale as well as the scales above and below. If the pixel value is greater than all the pixels in the neighborhood window, it is considered a feature point. Subpixel-level localization of feature points is achieved by employing a Taylor expansion in the scale space.

3. Feature descriptor:

For feature points with a scale parameter of σ_i, a window of size $24\sigma_i \times 24\sigma_i$ is taken on the gradient image, centered at the feature point. The window is divided into a grid of 4×4 sub-scenes, each with a size of $9\sigma_i \times 9\sigma_i$. Adjacent sub-scenes have an overlap strip of width $2\sigma_i$. Each sub-scene is weighted using a Gaussian kernel with a standard deviation of $\sigma_1 = 2.5\sigma_i$. A sub-scene descriptor vector of length 4 is computed for each sub-scene. These sub-scene descriptors are then weighted using another Gaussian window of size 4×4 with a standard deviation of $\sigma_2 = 1.5\sigma_i$. Finally, the descriptors are normalized to obtain a 64-dimensional descriptor vector.

2.2.3. Feature Matching Method

Traditional feature point matching algorithms typically compute the Euclidean distance between feature vectors and utilize the NNDR strategy to determine whether two feature points are a match. After applying the NNDR, RANSAC methods are often employed to determine the final set of matched point pairs.

The generated images from the point cloud exhibit a significant number of unstructured holes with an unordered distribution. The application of the KAZE algorithm leads to the detection of numerous unstable feature points, and many of these feature points have very similar descriptors. Increasing the threshold in the NNDR algorithm does not yield better matching results; instead, it may even result in the elimination of correctly matched point pairs.

This study proposes the construction of a circular scene mean filter to perform filtering on the original image. The filtering process aims to eliminate small holes present in the original image while also resulting in increased blurring along the image boundaries. Subsequently, the KAZE is employed to detect feature points separately in both the original and filtered images. For a particular feature point $p = (x, y)$ in the original image, if there exists a feature point $q = (x', y')$ in the filtered image and it satisfies condition $|p - q| \leq \varepsilon$, the point p is considered a stable feature point, where the size of ε is one pixel length. Subsequently, we utilize the NNDR to find matching point pairs. In this study, there is no rotation transformation between the two images. Only displacements exist in the azimuthal and ground range directions, with a certain level of scaling in the ground range direction. To further eliminate false matching points, the angle between the spatial vector of the matched point pairs and the horizontal vector is computed. After applying the NNDR, let us denote the set of feature points in the target image as $A = \{a_1, a_2, \ldots, a_n\}$, with individual points represented as $a_n = (x_n, y_n)$, and the set of feature points in the registration image as $B = \{b_1, b_2, \ldots, b_n\}$, with individual points represented as $b_n = (w_n, k_n)$. We can calculate the angle between the distance vector and the horizontal vector (1,0).

$$\theta_n = \arccos\left(\frac{w_n - x_n}{\sqrt{(w_n - x_n)^2 + (k_n - y_n)^2}}\right) \tag{16}$$

The probability distribution of θ_n is depicted in Figure 7. It can be observed that after NNDR, θ_n is concentrated around a prominent peak, which exhibits a triangular shape. To eliminate matching point pairs that deviate from the main peak, a threshold is set. The purpose of this threshold is to select matching point pairs that satisfy the condition of θ being within the triangular peak. The threshold is determined as follows:

$$\delta_\theta = \frac{1}{\max(PDF_\theta)} \tag{17}$$

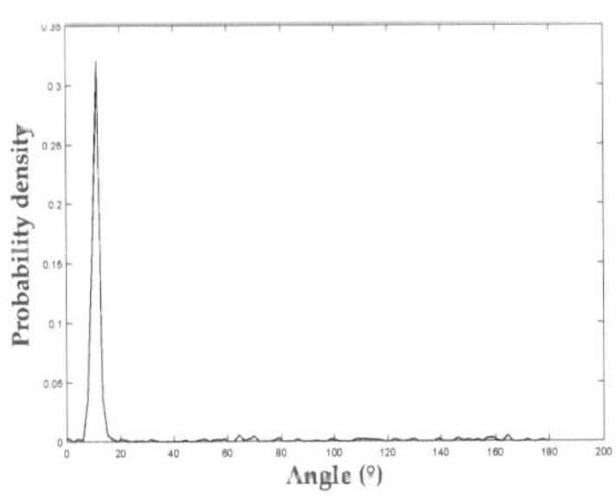

Figure 7. The probability distribution of θ_n.

2.3. Calculate 3D Transformations

The airborne array interferometric SAR system incorporates a high-precision position and orientation system (POS), consequently yielding minimal errors in the azimuthal direction between the point clouds obtained from two consecutive flights. In the ground range direction, apart from a certain displacement, there was also scaling. In the vertical direction, after the flattened phase error correction, only displacement was evident. The positions of the feature points in the image correspond to the coordinates in the azimuth and ground range directions of the point cloud, while the intensity of the feature point pixels corresponds to the average height of the point cloud. We computed the angular deviation between matching points and performed a statistical analysis to examine the probability distribution of these deviations, as illustrated in Figure 8a. We employed a quantile-quantile (Q-Q) plot to assess the adherence of this dataset to a Gaussian distribution, aiming to determine its normality.

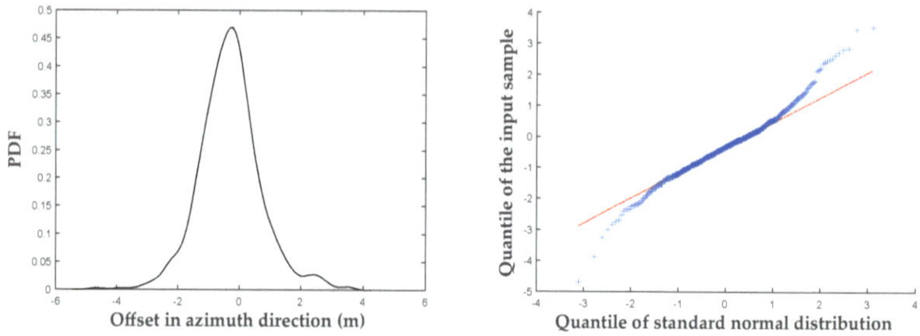

(**a**) (**b**)

Figure 8. Statistics of azimuth offset between matching points. (**a**) Probability distribution of offset; (**b**) Q-Q plot (sample data-standard normal).

Within the Q-Q plot, a significant number of points align along a straight line while demonstrating some curvature on the tails. This curvature phenomenon can be attributed to the existence of upper and lower limits in the actual data. Therefore, the deviation of azimuthal orientations between point clouds can be confirmed as the offset corresponding to the maximum probability density.

The directional offsets in ground distance and height corresponding to the matching points of the two images are depicted in Figure 9a,b, respectively. In Figure 9a, the abscissa represents the ground distance coordinates corresponding to the matching points in the source image. By fitting these coordinates into a straight line using the least squares method, the slope of the red line reflects the stretching effect in the ground distance direction between the two acquired point clouds. For the source point cloud, the offset value is determined based on the relationship between the ground distance coordinate of each point and the fitted line. The intensity differences data between matched points of the two images is divided into four segments. The value of 0 is observed when no point cloud falls into the matched pixel in either of the two images. The outliers in the upper and lower sections are caused by pixel intensities of 0 in only one of the two matched images. In this study, we only consider the real values from the middle section, where the height offset fluctuations within ± 2 m, as shown in Figure 9b. The average value of these points is taken as the offset in the height direction between point clouds.

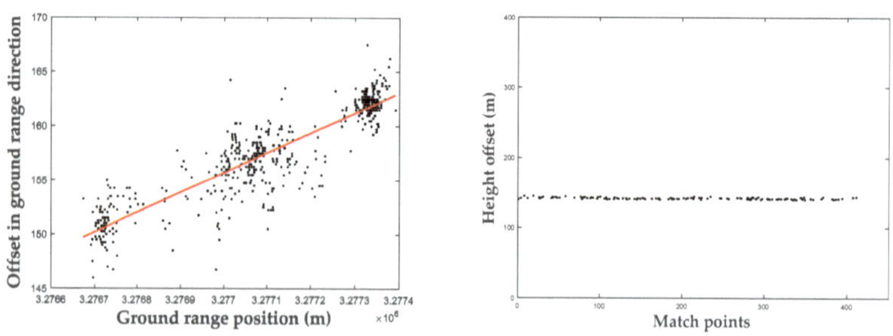

(**a**) (**b**)

Figure 9. The offsets in the ground distance direction and height direction between the matching points. (**a**) Offset of ground distance; (**b**) Height offset.

3. Results

3.1. Experimental Data

To conduct experimental validation in this study, we utilized point cloud data obtained from actual flight tests. These flight tests were conducted in Sichuan Province in 2022. The radar images are presented in Figures 10a and 10b, respectively. The flight-related parameters are listed in Table 2, where S_a represents azimuth resolution, S_r represents range resolution, and S_h represents elevation resolution.

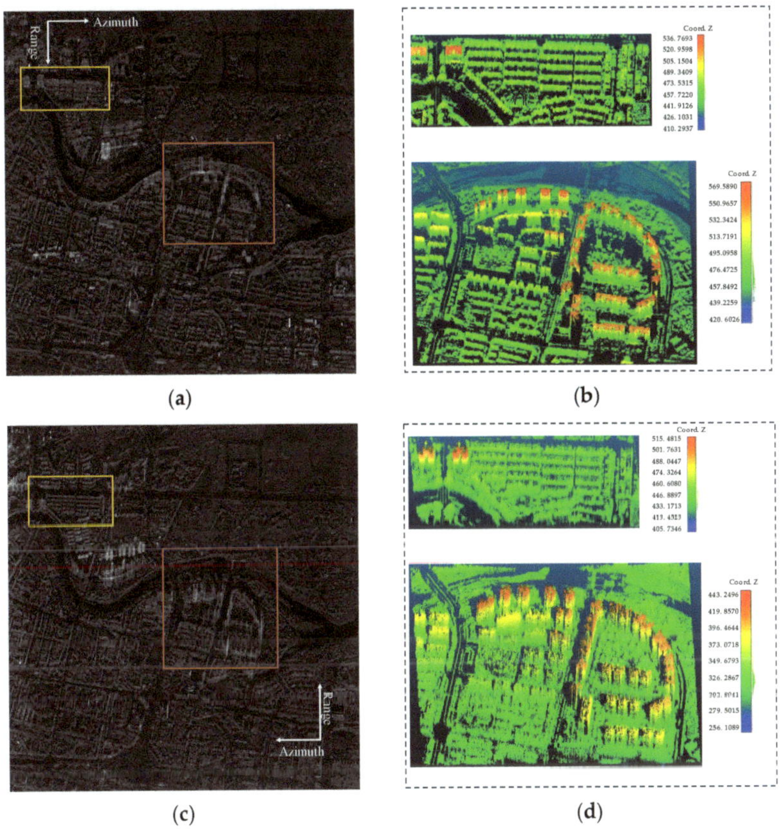

Figure 10. Intensity SAR images: (**a**) The SAR image obtained from the first flight (platform moving from west to east); (**b**) Point clouds of the two scenes generated from the first flight; (**c**) The SAR image obtained from the second flight (platform moving from east to west); (**d**) Point clouds of the two scenes generated from the second flight.

Table 2. Flight parameters.

H/(m)	Band	α/(°)	B/(m)	S_a/(m)	S_r/(m)	S_h/(m)
4500	Ku	0	1.986	0.237	0.1875	1.357

The area of experimental scene 1 is 0.22 square kilometers, with a ground range length of 0.31 km. The area of experimental scene 2 is 0.72 square kilometers, with a ground range length of 0.83 km. Scene 2 has a larger area with more diverse elements, including clear roads, bridges, and riverbanks. This contributes to the registration task for the image. Scene 1, on the other hand, has a smaller area, with only a prominent road on the left side. The generated images in this scene exhibit a simpler composition, primarily aimed at verifying the applicability of the proposed method. Point clouds of both scenes depict urban scenes adhering to the assumption of a level ground surface, as posited in this study.

3.2. Evaluation Criterion

Traditional evaluation metrics for point cloud registration methods include Root Mean Square Error (RMSE), mutual information, entropy, and point cloud overlap. RMSE measures the distance difference between point pairs in point clouds. It is calculated by computing the distances between corresponding points in the point clouds, taking the square of each distance, averaging them, and then taking the square root to obtain the RMSE value. Mutual information is calculated to assess the similarity between two point clouds. Entropy is used to measure the uncertainty of point distribution within a point cloud and can evaluate the consistency of its structure. Point cloud overlap evaluates the registration quality by calculating the proportion of the overlap scene between two point clouds.

Due to the low overlap between the SAR point clouds obtained from two flight experiments, these metrics cannot directly evaluate the effectiveness of SAR point cloud registration. Therefore, we adopt the metrics proposed by [31]. Ref. [31] utilizes the constraint of parallel relative facades of the same building to extract the building facades from the fused point cloud. For effective registration methods, the directions of the two relative facades should be parallel. Hence, as illustrated in Figure 11, ref. [31] calculate the angular difference θ between the two normal vectors of each facade pair.

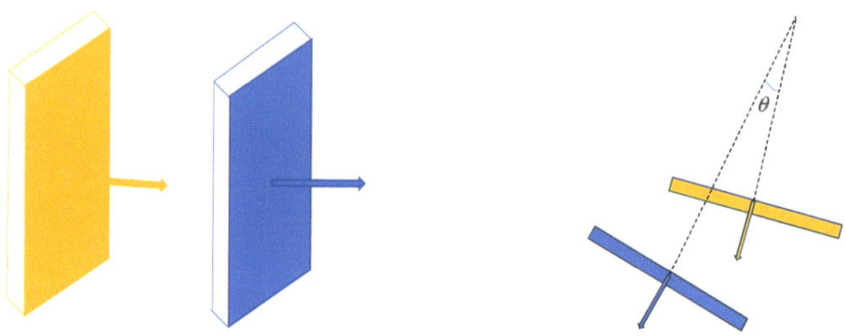

(a) (b)

Figure 11. Evaluation index for registration results. (**a**) Two parallel facades; (**b**) The angle difference of normal vector from the source façade center to the normal extension of the target façade.

The evaluation metrics proposed in [31] only capture the effectiveness of building point registration. In this study, we manually selected certain road point clouds and considered them to be overlapping between two point clouds. The RMSE between these road point clouds was computed as an evaluation metric. The definition of RMSE is as follows:

$$\mathrm{RMSE} = \sqrt{\sum_{i=1}^{n} \frac{(x_i - y_i)^2}{n}} \tag{18}$$

where X and Y represent two point clouds, N represents the number of corresponding points, x_i is the $i-th$ point in X, and y_i is the corresponding point in Y for x_i,

The locations of the road are illustrated in the red box in Figure 12. On the other hand, we adopted the correntropy proposed in [37] as an additional evaluation metric. Correntropy effectively alleviates the impact of outliers and noise, and its definition is as follows:

$$V(X,Y) = \frac{1}{N}\sum_{i=1}^{N} \exp(\frac{-|x_i - y_i|^2}{2\sigma^2}) \tag{19}$$

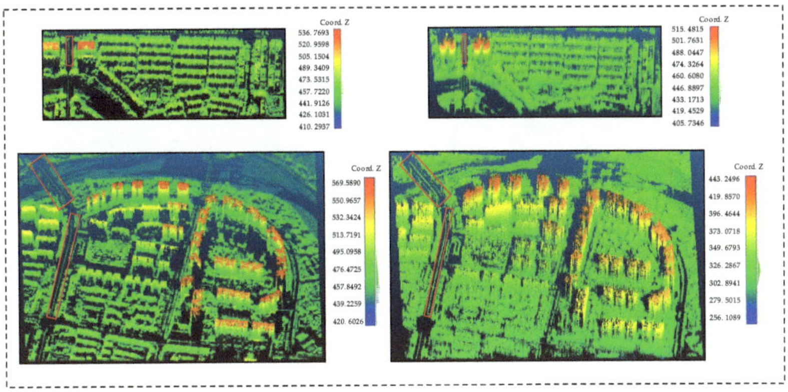

Figure 12. The positions of control points.

The definition of parameters is the same as Formula (18) and σ takes the value of 1 in this paper. A larger correntropy indicates a better registration performance.

3.3. Experimental Results

To validate the claimed superiority, in this subsection, we apply our proposed method alongside the approach outlined in [31] and the classical ICP algorithm to the point cloud fusion task of two distinct scenes. In scene one, the two-point clouds consist of 1,672,216 and 1,682,162 points, respectively. After applying simple morphological filtering and outlier removal using the RANSAC method, the ground points for scene 1 were obtained. The relationship between the average height of ground points and ground distance for the two corresponding point clouds in scene 1 is shown in Figure 13.

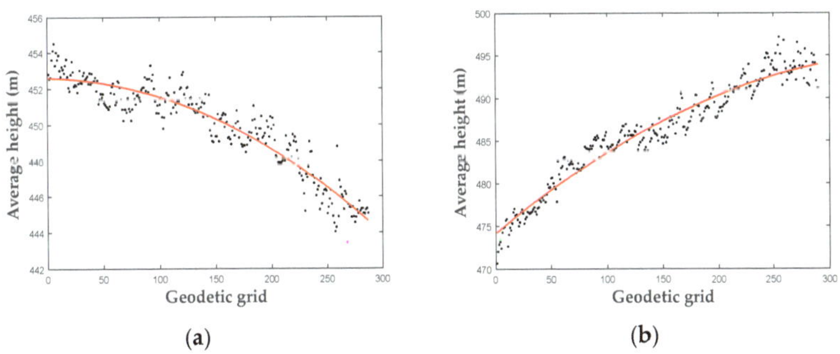

Figure 13. The average height of the ground as a function of the ground distance position. (a) Source point cloud; (b) Target point cloud.

After height calibration of the point clouds, a two-dimensional image was generated using the method mentioned in Section 2.2.1. The KAZE algorithm was employed to extract key points from the image, and matching points were obtained using our proposed method, as illustrated in Figure 14.

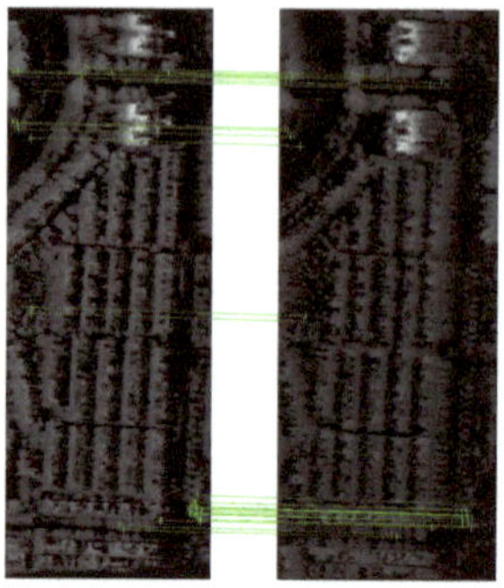

Figure 14. Results of image matching algorithms.

In scene 1, the area is relatively small, with the majority of the scene being comprised of buildings. The left side of Figure 14 corresponds to a small area in Figure 10b. Due to occlusion caused by buildings, there are significant shadows present near the riverbank adjacent to the building area. As a result, the majority of the matching points are concentrated in the road area above the image and in the vicinity of the bridges spanning the river.

The results of point cloud registration are depicted in Figure 15. The three fused results demonstrate the extraction of building facades using the density threshold filtering method. Excluding no corresponding facades, there are 22 pairs of building facades corresponding to each other.

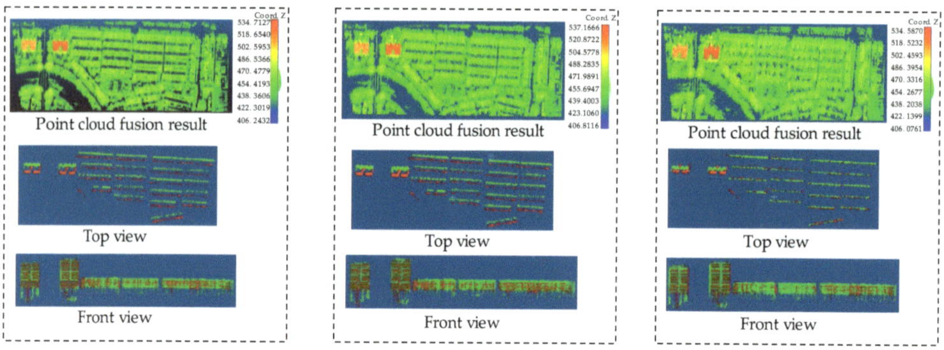

(**a**) (**b**) (**c**)

Figure 15. The effect of registration in scene 1 after using (**a**) our proposed method, (**b**) the method of [31], and (**c**) ICP.

The ICP algorithm tends to maximize the alignment of two-point clouds. From the extracted building facades, it can be observed that the two opposing building facades almost completely overlap. The algorithm is essentially ineffective in the task of SAR point cloud fusion. The method in [31] first extracts the building facade and calculates the transformation relationship from the source point cloud to the target point cloud using the constraint of two opposite facades of the same building being parallel to each other. Visually, there is no significant difference between the two methods for extracting building point clouds after registration. Table 3 presents the quantified results.

Table 3. Evaluation of registration accuracy for scene 1.

Method	RMSE (m)	Correntropy	Mean θ (deg)	Time (s)
ICP	5.3275	0.1074	0.5372	74.56183
[25]	4.0469	0.2282	0.4263	250.6351
Proposed	1.4773	0.2640	0.4292	3.8633

From the third quantitative indicator in Table 3, it seems that our method does not have superiority over the method proposed in [31]. However, the effect of SAR point cloud registration should not be solely considered from the constraints of extracting parallel building facades. The algorithm in [31] minimizes the angle between the planes fitted by the building facade point cloud and the correspondence between the center points, naturally resulting in better indicators. In the Euclidean distance metric of point-to-point, our method is significantly superior to the method proposed in [31]. As shown in Figure 16, our proposed approach exhibits superior accuracy. In our approach, 81.91% of the nearest neighbor distances fall within the range of 0 to 5 m, whereas the corresponding value for the comparative method is 76.05%.

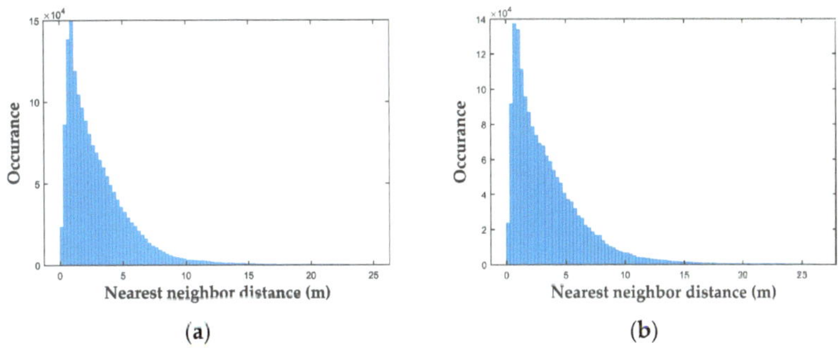

Figure 16. Histogram of nearest neighbor distance of (**a**) our proposed approach and (**b**) the method of [31].

For scene 2, the number of points obtained during the two flights is 3,536,789 and 4,554,655, respectively. Some results of using our method to process the point cloud in scene 2 have been shown in the second part. Figure 17 shows the matching point pairs.

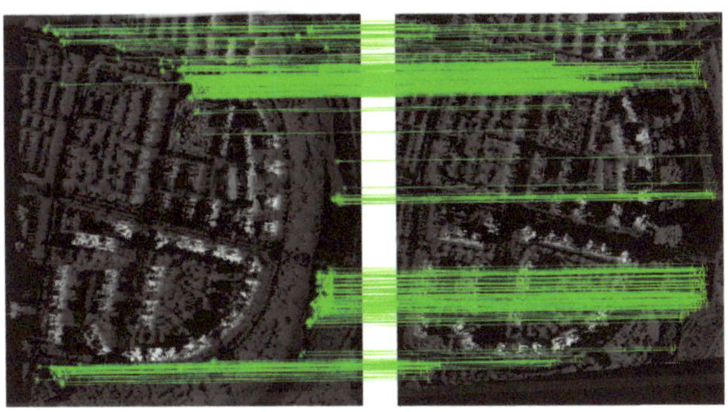

Figure 17. Results of image matching algorithms.

Scene 2 has a large area and rich contents, and more matching point pairs were obtained using the KAZE algorithm compared to scene 1. When integrating the point clouds of scene 2 using the algorithm proposed in [31], there was a significant difference in the extracted sets of building facade point clouds from the two point clouds. In some cases, only one of the point clouds captured the facade corresponding to the same building, and there were substantial disparities in the relative facades of most buildings. The coarse registration method employed in [31] faced challenges in determining which facades corresponded to each other. Consequently, we manually selected several facades with better extraction results and used the algorithm in [31] to fuse them in order to compare the registration performance of our proposed algorithm against that of [31].

The registration results are depicted in Figure 18. Observably, the results of the fusion using the ICP algorithm display misaligned architectural structures, with considerable fusion errors evident in ground-level roads. The method proposed in [31] exhibits poor fusion results for the ground above the scene, where the two point clouds fail to align adequately. Conversely, our proposed method demonstrates superior fusion outcomes for both ground and architectural points in the SAR point cloud.

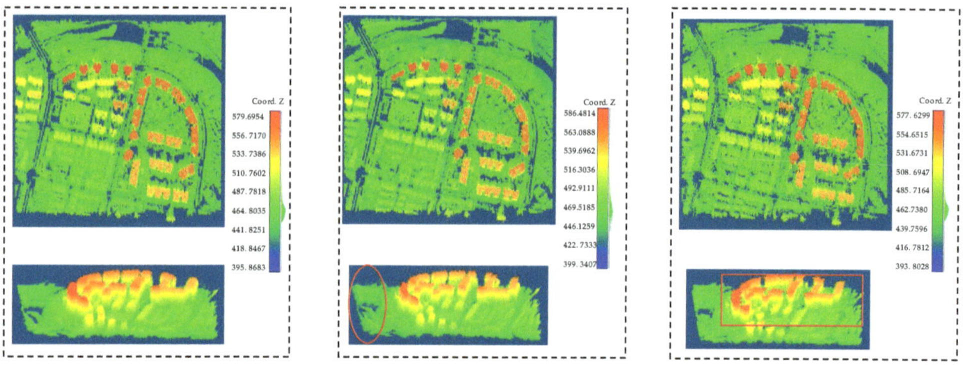

(**a**) (**b**) (**c**)

Figure 18. The effect of registration in scene 2 after using (**a**) our proposed method, (**b**) the method of [31], and (**c**) ICP.

For a more detailed analysis, we employed a density threshold method to extract the buildings within the scene, as depicted in Figure 19. The results indicate that concerning the reconstruction of architectural structures, there is no significant difference between our proposed method and the approach outlined in [31]. However, the ICP algorithm merely aligns the two point clouds without adequately reconstructing the architectural elements. Table 4 presents the quantified results.

Table 4. Evaluation of registration accuracy for scene 2.

Method	RMSE (m)	Correntropy	Mean θ (deg)	Time (s)
ICP	8.357	0.0725	0.9382	453.3789
[31]	5.863	0.0910	0.3873	120.3572
Proposed	1.035	0.2239	0.3892	16.8694

As shown in Figure 20, compared to scene 1, the application of our approach in scene 2 demonstrates a more pronounced advantage. In our approach, 69.14% of the nearest neighbor distances fall within the range of 0 to 5 m, whereas the corresponding value for the comparative method is 44.71%.

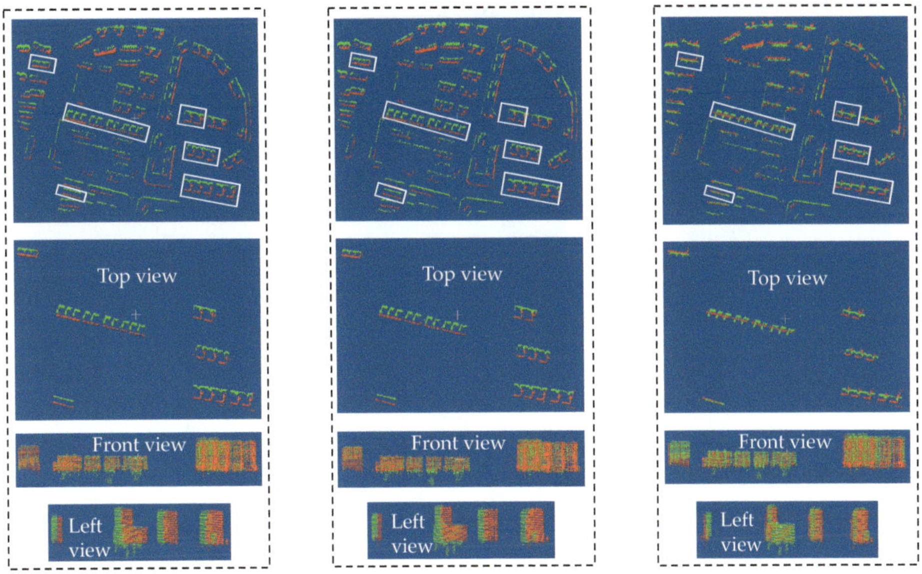

(**a**) (**b**) (**c**)

Figure 19. The registration effect on the buildings within scene 2 after using (**a**) our proposed method, (**b**) the method of [31], and (**c**) ICP.

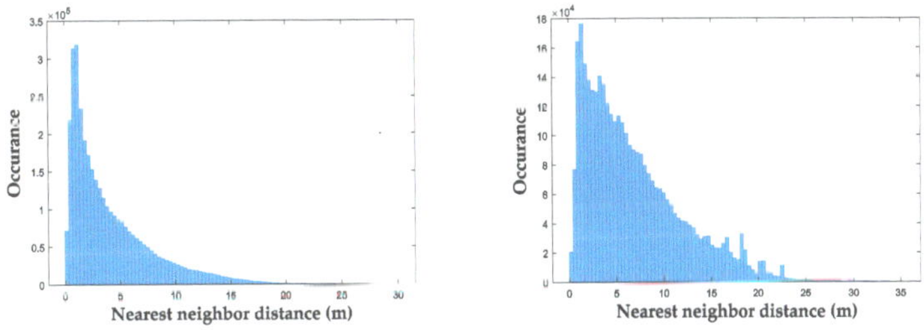

(**a**) (**b**)

Figure 20. Histogram of nearest neighbor distance of (**a**) our proposed approach and (**b**) the method of [31].

3.4. Time Performance

To evaluate the efficiency of our proposed method, we computed the time cost for each registration process. Our method was implemented using MATLAB 2021, and all experiments were conducted on a computer with an AMD R7 5800H processor and 16 GB of memory. The ICP algorithm requires multiple iterations to select the closest points between two point clouds and calculate the transformation relationship. As the number of points in the point cloud increases, the computation time also increases. The method proposed in [31] involves extracting building facades from the point cloud and fitting these facades to generate corresponding parameters. The processing time is related to the number of buildings in the scene. However, in large-scale scenarios, significant differences might exist between the extracted building facades from two-point clouds. This dissimilarity sometimes prevents the automatic determination of which facades belong to the same building, thereby limiting its application. The time required by our proposed method is primarily dependent on the size of the scene. Our proposed method demonstrates clear advantages in terms of efficiency.

4. Discussion

In this study, a proposed approach is presented to address the point cloud registration problem of array InSAR, which contains a large number of noisy points and exhibits significant errors. The approach involves utilizing image registration methods to achieve point cloud registration. The analysis focuses on the height errors along the ground range direction in a single flight experiment and the scale variations in the ground range direction between two consecutive flight experiments. Unlike traditional point cloud registration tasks that compute a rotation matrix and translation vector as transformation parameters, the registration of array InSAR point clouds primarily involves error correction and computation of the displacement between the two point clouds.

The urban scenes under consideration predominantly consist of building areas, but they also contain several features that are beneficial for point cloud registration tasks, such as road lines, bridges, and structured artificial facilities. In a specific scenario, referred to as scene 1, with a relatively small area, there are only noticeable common features on the left side of the two point clouds. Although the number of computed matching point pairs is limited, it does not affect the accuracy of point cloud registration, as these matching points can be considered true correspondences.

To achieve high-precision point cloud registration, this study relies on subpixel-level accurate image registration algorithms to calculate the offsets between the azimuth and ground range directions of the point clouds. Additionally, the study reveals that the majority of image-matching points are concentrated in the unobstructed ground scenes. The building facade points directly beneath contain a significant amount of clutter caused by triple scattering. Additionally, due to interference from high-angle sidelobes, the unstructured ground scene also presents some artifacts in the vertical dimension. By utilizing the average height of the point cloud to represent the pixel intensity of the image and using the pixel intensity difference of the matching points as the offset in the height direction, the registration accuracy in the height direction can be ensured to be lower than the height resolution of the array InSAR point cloud.

In contrast to previous work, which innovatively utilized the angles between the extracted normal vectors of building facades and the distances from the facade centers to the extended normal vectors of opposing facades as evaluation metrics, this study found that accurately extracting building facades from array InSAR point clouds is challenging. The simple application of density threshold filtering methods tends to filter out low-rise buildings, and some extracted facades are incomplete, resulting in significant differences between opposing facades of the same building and making it difficult to fit the facades correctly. Moreover, the presence of clutter generated by triple scattering at the bottom of the buildings hinders the accurate correspondence of the fitted facade center heights. As for the classic ICP algorithm, it is entirely unsuitable for SAR point cloud registration tasks because the two point clouds lack matching points. The approach of manually annotating control points and using the Euclidean distances between them as evaluation metrics also has limitations, as the true correspondences of the manually annotated points cannot be determined. Therefore, for the registration task of array InSAR point clouds, it is necessary to define more comprehensive metrics to evaluate the accuracy of building facade extraction and point cloud registration.

5. Conclusions

This paper proposes an automatic image-based registration method for array InSAR point cloud registration. It analyzes the height errors present in array InSAR point clouds and describes the entire process of point cloud registration.

According to the InSAR system model, an analysis of the relationship between the height errors in point clouds and their ground range positions is conducted. Initially, the SMRF algorithm is employed to extract the ground portion of the point cloud, which is utilized for fitting the relationship between height errors and ground range. Subsequently, the height-corrected point clouds are projected onto the azimuth-ground range plane to generate images, where the pixel intensity is represented by the average height of all points falling within the pixel. Finally, the KAZE algorithm, along with an angular threshold, is employed to extract matching points between two images. The transformation relationship between the two point clouds is then calculated based on the positions and intensity differences of the matching points.

Previous research on array InSAR point cloud registration is limited, and this paper primarily compares the proposed method with the approach presented in [31]. Experimental results using real data demonstrate the high robustness of the proposed method in two different scenarios. For the architectural elements within the scene, the average angular difference between their respective facades is less than 0.5°. As for the ground portions within the scene, the RMSE after registration is less than 1.5 m. These results are considered acceptable for SAR point clouds. Compared to previous methods that extract and fuse building facades, our approach addresses point cloud registration from the perspective of image registration. It involves fewer steps, is more efficient, and consumes only 14% of the time required by the method proposed in [31].

In future work, for array InSAR point cloud registration, we consider utilizing deep learning methods after obtaining a large dataset to achieve the task of point cloud registration.

Author Contributions: Conceptualization, C.C. and F.Z.; methodology, C.C.; software, Y.L.; validation, C.C. and M.S.; formal analysis, C.C.; investigation, C.C.; resources, F.Z.; data curation, Y.L.; writing—original draft preparation, C.C.; writing—review and editing, F.Z.; visualization, W.L.; supervision, Z.L.; project administration, L.C.; funding acquisition, L.C. All authors have read and agreed to the published version of the manuscript.

Funding: This research received no external funding.

Data Availability Statement: The data presented in this study are available on request from the corresponding author.

Conflicts of Interest: The authors declare no conflict of interest.

References

1. Wang, X.; Mizukami, Y.; Tada, M.; Matsuno, F. Navigation of a mobile robot in a dynamic environment using a point cloud map. *Artif. Life Robot.* **2021**, *26*, 10–20. [CrossRef]
2. Chen, S.; Liu, B.; Feng, C.; Vallespi-Gonzalez, C.; Wellington, C. 3d point cloud processing and learning for autonomous driving: Impacting map creation, localization, and perception. *IEEE Signal Process. Mag.* **2020**, *38*, 68–86. [CrossRef]
3. Fuhrmann, S.; Langguth, F.; Goesele, M. Mve-a multi-view reconstruction environment. *GCH* **2014**, *3*, 4.
4. Blanc, T.; El Beheiry, M.; Caporal, C.; Masson, J.-B.; Hajj, B. Genuage: Visualize and analyze multidimensional single-molecule point cloud data in virtual reality. *Nat. Methods* **2020**, *17*, 1100–1102. [CrossRef]
5. Dong, Z.; Liang, F.; Yang, B.; Xu, Y.; Zang, Y.; Li, J.; Wang, Y.; Dai, W.; Fan, H.; Hyyppä, J. Registration of large-scale terrestrial laser scanner point clouds: A review and benchmark. *ISPRS J. Photogramm. Remote Sens.* **2020**, *163*, 327–342. [CrossRef]
6. Pu, S.; Vosselman, G. Knowledge based reconstruction of building models from terrestrial laser scanning data. *ISPRS J. Photogramm. Remote Sens.* **2009**, *64*, 575–584. [CrossRef]
7. Barazzetti, L.; Scaioni, M.; Remondino, F. Orientation and 3D modelling from markerless terrestrial images: Combining accuracy with automation. *Photogramm. Rec.* **2010**, *25*, 356–381. [CrossRef]
8. Simon, L.; Teboul, O.; Koutsourakis, P.; Van Gool, L.; Paragios, N. Parameter-free/pareto-driven procedural 3d reconstruction of buildings from ground-level sequences. In Proceedings of the 2012 IEEE Conference on Computer Vision and Pattern Recognition, Providence, RI, USA, 16–21 June 2012; pp. 518–525.
9. Zhu, X.X.; Bamler, R. Tomographic SAR inversion by L_1-norm regularization—The compressive sensing approach. *IEEE Trans. Geosci. Remote Sens.* **2010**, *48*, 3839–3846. [CrossRef]
10. Zeng, Z.; Sun, J.; Han, Z.; Hong, W. SAR automatic target recognition method based on multi-stream complex-valued networks. *IEEE Trans. Geosci. Remote Sens.* **2022**, *60*, 1–18. [CrossRef]
11. Gernhardt, S.; Cong, X.; Eineder, M.; Hinz, S.; Bamler, R. Geometrical fusion of multitrack PS point clouds. *IEEE Geosci. Remote Sens. Lett.* **2011**, *9*, 38–42. [CrossRef]
12. Hu, F.; Wang, F.; Ren, Y.; Xu, F.; Qiu, X.; Ding, C.; Jin, Y. Error analysis and 3D reconstruction using airborne array InSAR images. *ISPRS J. Photogramm. Remote Sens.* **2022**, *190*, 113–128. [CrossRef]
13. Ge, X.; Hu, H.; Wu, B. Image-guided registration of unordered terrestrial laser scanning point clouds for urban scenes. *IEEE Trans. Geosci. Remote Sens.* **2019**, *57*, 9264–9276. [CrossRef]

14. Barnea, S.; Filin, S. Registration of terrestrial laser scans via image based features. *Int. Arch. Photogramm. Remote Sens. Spat. Inf. Sci.* **2007**, *36*, 32–37.
15. Aiger, D.; Mitra, N.J.; Cohen-Or, D. 4-points congruent sets for robust pairwise surface registration. *ACM Trans. Graph.* **2008**, *27*, 1–10. [CrossRef]
16. Jaw, J.J.; Chuang, T.Y. Registration of ground-based LiDAR point clouds by means of 3D line features. *J. Chin. Inst. Eng.* **2008**, *31*, 1031–1045. [CrossRef]
17. Cheng, L.; Wu, Y.; Tong, L.; Chen, Y.; Li, M. Hierarchical registration method for airborne and vehicle lidar point cloud. *Remote Sens.* **2015**, *7*, 13921–13944. [CrossRef]
18. Lee, J.; Yu, K.; Kim, Y.; Habib, A.F. Adjustment of discrepancies between LIDAR data strips using linear features. *IEEE Geosci. Remote Sens. Lett.* **2007**, *4*, 475–479. [CrossRef]
19. Gruen, A.; Akca, D. Least squares 3D surface and curve matching. *ISPRS J. Photogramm. Remote Sens.* **2005**, *59*, 151–174. [CrossRef]
20. Besl, P.J.; McKay, N.D. Method for registration of 3-D shapes. In Proceedings of the Sensor Fusion IV: Control Paradigms and Data Structures, Boston, MA, USA, 12–15 November 1992; pp. 586–606.
21. Al-Durgham, K.; Habib, A.; Kwak, E. RANSAC approach for automated registration of terrestrial laser scans using linear features. *ISPRS Ann. Photogramm. Remote Sens. Spat. Inf. Sci* **2013**, *2*, 13–18. [CrossRef]
22. Takeuchi, E.; Tsubouchi, T. A 3-D scan matching using improved 3-D normal distributions transform for mobile robotic mapping. In Proceedings of the 2006 IEEE/RSJ International Conference on Intelligent Robots and Systems, Beijing, China, 9–13 October 2006; pp. 3068–3073.
23. Fusiello, A.; Castellani, U.; Ronchetti, L.; Murino, V. Model acquisition by registration of multiple acoustic range views. In Proceedings of the Computer Vision—ECCV 2002: 7th European Conference on Computer Vision, Copenhagen, Denmark, 28–31 May 2002; pp. 805–819.
24. Qi, C.R.; Su, H.; Mo, K.; Guibas, L.J. Pointnet: Deep learning on point sets for 3d classification and segmentation. In Proceedings of the IEEE Conference on Computer Vision and Pattern Recognition, Honolulu, HI, USA, 21–26 July 2017; pp. 652–660.
25. Qi, C.R.; Yi, L.; Su, H.; Guibas, L.J. Pointnet++: Deep hierarchical feature learning on point sets in a metric space. *arXiv* **2017**, arXiv:1706.02413.
26. Deng, H.; Birdal, T.; Ilic, S. 3D local features for direct pairwise registration. In Proceedings of the IEEE/CVF Conference on Computer Vision and Pattern Recognition, Long Beach, CA, USA, 15–20 June 2019; pp. 3244–3253.
27. Yang, J.; Zhao, C.; Xian, K.; Zhu, A.; Cao, Z. Learning to fuse local geometric features for 3D rigid data matching. *Inf. Fusion* **2020**, *61*, 24–35. [CrossRef]
28. Valsesia, D.; Fracastoro, G.; Magli, E. Learning localized representations of point clouds with graph-convolutional generative adversarial networks. *IEEE Trans. Multimed.* **2020**, *23*, 402–414. [CrossRef]
29. Huang, X.; Mei, G.; Zhang, J. Feature-metric registration: A fast semi-supervised approach for robust point cloud registration without correspondences. In Proceedings of the IEEE/CVF Conference on Computer Vision and Pattern Recognition, Seattle, WA, USA, 13–19 June 2020; pp. 11366–11374.
30. Wang, Y.; Zhu, X.X. Automatic feature-based geometric fusion of multiview TomoSAR point clouds in urban area. *IEEE J. Sel. Top. Appl. Earth Obs. Remote Sens.* **2014**, *8*, 953–965. [CrossRef]
31. Tong, X.; Zhang, X.; Liu, S.; Ye, Z.; Feng, Y.; Xie, H.; Chen, L.; Zhang, F.; Han, J.; Jin, Y. Automatic Registration of Very Low Overlapping Array InSAR Point Clouds in Urban Scenes. *IEEE Trans. Geosci. Remote Sens.* **2022**, *60*, 1–25. [CrossRef]
32. Cheng, R.; Liang, X.; Zhang, F.; Guo, Q.; Chen, L. Multiple-bounce scattering of Tomo-SAR in single-pass mode for building reconstructions. *IEEE Access* **2019**, *7*, 124341–124350. [CrossRef]
33. Alcantarilla, P.F.; Bartoli, A.; Davison, A.J. KAZE features. In Proceedings of the Computer Vision–ECCV 2012: 12th European Conference on Computer Vision, Florence, Italy, 7–13 October 2012; pp. 214–227.
34. Chen, C.; Guo, J.; Wu, H.; Li, Y.; Shi, B. Performance comparison of filtering algorithms for high-density airborne LiDAR point clouds over complex LandScapes. *Remote Sens.* **2021**, *13*, 2663. [CrossRef]
35. Pingel, T.J.; Clarke, K.C.; McBride, W.A. An improved simple morphological filter for the terrain classification of airborne LIDAR data. *ISPRS J. Photogramm. Remote Sens.* **2013**, *77*, 21–30. [CrossRef]
36. Zhang, K.; Chen, S.-C.; Whitman, D.; Shyu, M.-L.; Yan, J.; Zhang, C. A progressive morphological filter for removing nonground measurements from airborne LIDAR data. *IEEE Trans. Geosci. Remote Sens.* **2003**, *41*, 872–882. [CrossRef]
37. Zhang, X.; Jian, L.; Xu, M. Robust 3D point cloud registration based on bidirectional Maximum Correntropy Criterion. *PLoS ONE* **2018**, *13*, e0197542. [CrossRef]

remote sensing

Article

A Multi-Level Robust Positioning Method for Three-Dimensional Ground Penetrating Radar (3D GPR) Road Underground Imaging in Dense Urban Areas

Ju Zhang [1,2], Qingwu Hu [3,*], Yemei Zhou [1], Pengcheng Zhao [3] and Xuzhe Duan [3]

1 School of Engineering and Architecture, Wuhan City Polytechnic, Wuhan 430072, China; zhangju@whu.edu.cn (J.Z.); 02009001@whcp.edu.cn (Y.Z.)
2 Key Laboratory of National Geographic Census and Monitoring, Ministry of Natural Resources, Wuhan 430072, China
3 School of Remote Sensing and Information Engineering, Wuhan University, Wuhan 430079, China; pengcheng.zhao@whu.edu.cn (P.Z.); duanxz@whu.edu.cn (X.D.)
* Correspondence: huqw@whu.edu.cn

Abstract: Three-Dimensional Ground Penetrating Radar (3D GPR) detects subsurface targets non-destructively, rapidly, and continuously. The complex environment around urban roads affects the positioning accuracy of 3D GPR. The positioning accuracy directly affects the data quality, as inaccurate positioning can lead to distortion and misalignment of 3D GPR data. This paper proposed a multi-level robust positioning method to improve the positioning accuracy of 3D GPR in dense urban areas in order to obtain more accurate underground data. In environments with good GNSS signals, fast and high-precision positioning can be achieved based on GNSS data using differential GNSS technology; in scenes with weak GNSS signals, high-precision positioning of subsurface data can be achieved by using GNSS and IMU as well as using GNSS/INS tightly coupled solution technology; in scenes with no GNSS signals, SLAM technology is used for positioning based on INS data and 3D point cloud data. In summary, this method ensures a positioning accuracy of 3D GPR better than 10 cm and high-quality 3D images of underground urban roads in any environment. This provides data support for urban road underground structure surveys and has broad application prospects in underground disease detection and prevention.

Keywords: multi-level robust positioning method; 3D ground penetrating radar; 3D mobile survey system; laser SLAM positioning

Citation: Zhang, J.; Hu, Q.; Zhou, Y.; Zhao, P.; Duan, X. A Multi-Level Robust Positioning Method for Three-Dimensional Ground Penetrating Radar (3D GPR) Road Underground Imaging in Dense Urban Areas. *Remote Sens.* **2024**, *16*, 1559. https://doi.org/10.3390/rs16091559

Academic Editor: David Gomez-Ortiz

Received: 28 February 2024
Revised: 18 April 2024
Accepted: 20 April 2024
Published: 27 April 2024

1. Introduction

Ground Penetrating Radar (GPR) is one of the non-destructive measurement techniques that uses electromagnetic waves to locate objects or interfaces buried within visually opaque material or underground. GPR transmits a regular sequence of low-power electromagnetic energy to the material or ground and receives and surveys weak reflected signals from buried objects. GPR uses electromagnetic waves to respond to changes in the electromagnetic properties of the shallow subsurface. The propagation velocity of electromagnetic waves is the main controlling factor in generating reflections; it is determined by the relative dielectric constant contrast between the background material and the object. The GPR method is a rapid, nondestructive, high-accuracy, continuous, and high-resolution method for subsurface target detection.

Three-dimensional (3D)-GPR is a new type of non-destructive detection equipment that can reconstruct underground 3D structure detection information. Compared with 2D-GPR, the 3D-GPR array antenna is able to acquire huge amounts of seamlessly stitched radar data without resulting in a lack of subsurface information. The 3D array antenna realizes true 3D acquisition, which makes the underground target imaging clear and accurate, and can display any depth horizontal slice of the underground target [1]. There are now many commercially available 3D-GPR devices, and the scope and capabilities of the technology are gradually evolving. GPR has also been successfully used to provide forensic information during criminal investigations [2,3], to detect buried mines [4–6], to survey roads [7–9], to detect utilities [10,11], to measure geophysical strata [12–14], and in other areas [15,16].

The 3D GPR data have high requirements for positioning accuracy because of the high sampling density. Large positioning errors may cause distortion of GPR data, as in Figure 1a. Only 3D GPR systems with centimeter-level positioning accuracy can collect high-quality 3D GPR data. In addition, with regard to the surface area required to perform covered underground detection, due to the limited single detection width of the 3D GPR system, it is usually necessary to operate in strips, and the positioning accuracy of 3D GPR directly affects the position alignment effect between channels and strips, as shown in Figure 1b, which in turn affects the data quality of underground remote sensing detection. Thus, 3D GPR positioning affects the quality and accuracy of underground remote sensing detection data. Therefore, the accurate positioning of the GPR system is crucial.

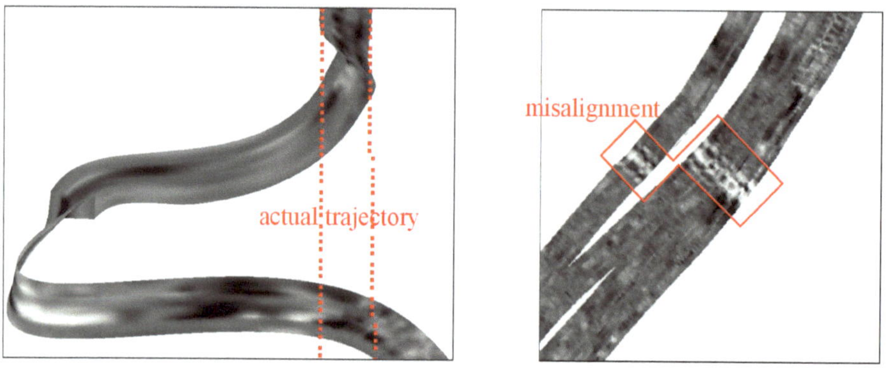

(a) Image distortion (b) Band misalignment

Figure 1. Results due to 3D GPR positioning errors.

The 3D-GPR positioning is generally achieved by laying out acquisition grids in the early stages [17,18], and the positioning accuracy is improved by modifying the encoder [19]. These methods have significant limitations, as the measurement needs to be kept straight. In addition, since no height information is available, the ground should not have too much undulation, and the measurement area should be smooth in topography. The positioning method without recording elevation information is obviously not suitable for 3D GPR, which requires x, y, and z coordinates.

Later, with the development of survey technology, various positioning techniques began to be used to improve 3D GPR positioning accuracy. At present, 3D GPR mainly uses GNSS for positioning. However, the positioning accuracy is affected by the quality of satellite signals. In practical applications, 3D GPR is used to survey a variety of environments, including roads, woods, and other areas surrounded by tall buildings or trees. These areas experience difficulty in receiving a sufficient number of satellite signals. The positioning accuracy of the 3D GPR system cannot be guaranteed simply by using a GNSS solution, and thus the quality of 3D GPR data cannot be guaranteed. In the environment with no satellite signals, such as tunnels or underground mines, these devices have no access to position information, which in turn does not allow for a reconstruction of the 3D underground space.

In addition, self-tracking systems, such as self-tracking robotic terrestrial positioning systems (TPSs) [20] and self-tracking total stations (TTSs) [21,22], were introduced in ground-penetrating radar positioning. These can achieve centimeter-level positioning accuracy. But, in dense urban areas, the signal tracked by the self-tracking system is lost when a clear line of sight is not available.

Other applications include the GPR positioning algorithm based on video recordings and special marker recognition [23], as well as a high-precision handheld GPR positioning system using an ultra-wideband (UWB) radio module [24]. In these positioning methods, with the help of RLPS and other mapping equipment and technology, positioning accuracy is significantly improved. However, these GPR data positioning methods need to set up one or more pieces of positioning equipment in the detection area and need to specify the detection route and delineate the detection range. When the ground is undulating and the shape of the detection area is irregular, one cannot lay positioning instruments in complex detection scenarios, limiting the possibility of flexible and convenient detection.

The Mobile Laser Scanner (MLS) has long been used in the field of land surveying. Studies [25,26] integrate MLSs into GPR. MLSs obtain ground point clouds of the ground that can be constructed in 3D space and correct the elevation of the GPR data. However, it is not possible to obtain accurate x, y coordinates. The portable rotary laser positioning system (RLPS) was applied to the GPR real-time positioning solution by Grasmueck and Viggiano [27]; this could obtain accurate centimeter-level x, y, and z coordinates. These positioning methods are aimed at small-scale underground detection, requiring GPR acquisition instruments in the range that other positioning equipment can capture, so they are not suitable for a large range of underground detection tasks, such as kilometer-level roadbed detection, large areas of underground pipeline detection, and other tasks.

There have also been some studies [28–30] that used SLAM algorithms to assist in GPR positioning. They integrated existing commercial mobile measurement systems with GPR and used SLAM algorithms to accomplish positioning in areas where GNSS signals were not available. However, the SLAM algorithm was not targeted to improve the GPR, which resulted in a GPR offset in the z-direction. In addition, some studies [29,30] do not add RTK GNSS, so if a survey area is wide and flat with no features, the laser scanner will not be able to obtain a valid point cloud to participate in the positioning. Moreover, in areas where GNSS signals are available, adding GNSS to participate in positioning can improve the positioning accuracy.

This paper proposes a high-precision positioning method with multi-level and multi-sensor fusion for 3D GPR integrated aboveground and underground remote sensing surveys. For the challenges of high-precision positioning of 3D GPR underground data and seamless splicing of multiple bands, an integrated aboveground and underground 3D mobile survey system is proposed and designed. It realizes the synchronous acquisition of an aboveground 3D laser point cloud, GNSS/IMU positioning and attitude, and underground 3D spatial data. The mobile survey module and the GPR control module were designed and developed with smaller hardware size, integrated acquisition and control, and more autonomy in solving the positioning data. Based on the multi-source data acquired by the system, a multi-level and multi-source data fusion positioning method is proposed for underground 3D GPR data. In areas without GNSS signals, this paper proposes a new and improved SLAM algorithm, which makes full use of the ground constraints through the double-threshold ground filtering algorithm and can effectively control the drift of the SLAM system in the z-direction. In areas with good GNSS signals, the tightly coupled GNSS/INS is used for positioning, and the positioning accuracy is higher compared with the SLAM algorithm. Through GNSS/INS tightly coupled positioning and laser SLAM positioning, this method realizes a positioning accuracy within 10 cm and a full spatial survey aboveground and underground, in an environment with or without good GNSS signals.

2. Materials and Methods

In order to enable 3D GPR to achieve multi-scene and multi-level high-precision positioning, an aboveground and underground integrated 3D mobile survey system is firstly designed. The GNSS receiver collects the GNSS signals for obtaining the position information of the system. The IMU acquires inertial data for obtaining the attitude and acceleration of the system. Attitude and acceleration from the inertial data are required for the GNSS/INS tightly coupled solution and SLAM algorithms. The 3D laser scanning system collects point cloud data for participating in SLAM positioning while constructing the 3D spatial structure on the ground. The 3D GPR acquires the underground 3D spatial data. In scenes with good GNSS signals, such as open squares, fast and high-precision positioning can be achieved based on GNSS data using differential GNSS technology. In scenes with weak GNSS signals, such as roads obscured by tall buildings or border trees, high-precision positioning of subsurface data can be achieved by using GNSS and IMU, combining position and attitude information, and using GNSS/INS tightly coupled solution technology. In scenes where GNSS signals are completely absent, such as tunnels and underground mines, SLAM technology is used for positioning based on INS data and 3D point cloud data. Ultimately, the underground data can be positioned with high precision in any scene, as shown in Figure 2.

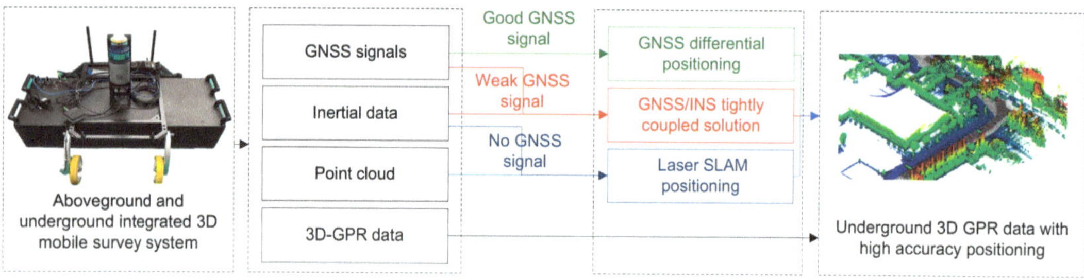

Figure 2. Multi-level and high-precision positioning of 3D GPR underground remote sensing detection.

In terms of hardware, the aboveground and underground integrated 3D mobile survey system is equipped with a GNSS receiver module, an inertial measurement unit module (IMU), a 3D laser scanner, and a 3D GPR. A GNSS receiver capable of receiving more than four satellites continuously can be characterized as being in an environment with good signals. With good GNSS signals, the combined positioning of GNSS and IMU can improve the positioning accuracy of 3D GPR. When the number of satellites is insufficient or even zero for a long time, the 3D LiDAR active positioning with the laser SLAM algorithm ensures 3D GPR positioning accuracy.

2.1. *Aboveground and Underground Integrated 3D Mobile Survey System*

The aboveground and underground integrated 3D survey system designed in this paper consists of a 3D mobile survey system and a GPR system. The GPR system explores the subsurface, and the 3D mobile survey system is used to locate the GPR and acquire the ground point cloud data. The configuration of the aboveground and underground integrated 3D survey system is shown in Figure 3. A cart is used as a carrier platform, where the 3D mobile survey system is mounted on the GPR system to obtain high-precision positioning information and ground point cloud data while the GPR obtains subsurface data. Figure 4 shows a simplified layout of the 3D survey system. The control units of the 3D mobile survey system and the GPR system are connected via an ethernet cable to the control PC, which controls both areas of data acquisition using the operating software.

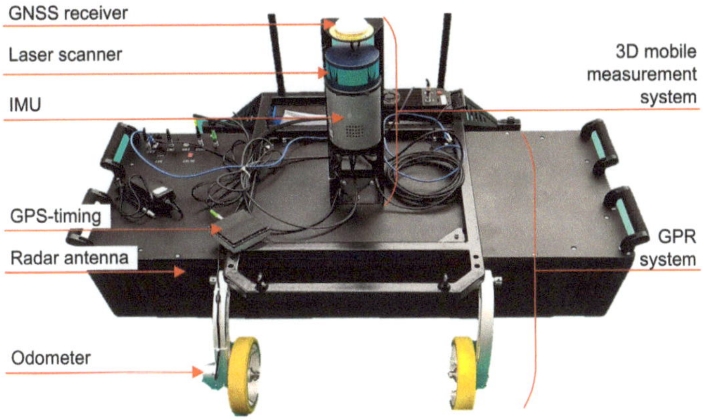

Figure 3. Aboveground and underground integrated 3D survey system configuration.

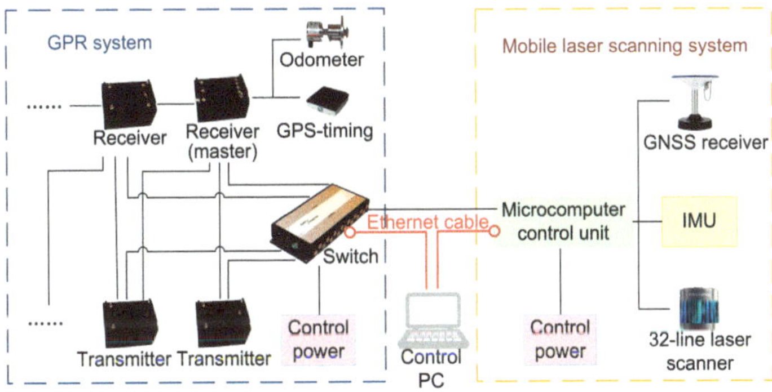

Figure 4. Simplified system layout with the aboveground and underground integrated 3D survey system.

2.2. Multilevel Positioning Framework

The aboveground and underground integrated 3D survey system gives the position attitude information of 3D laser scanning mobile survey system to 3D GPR system through GNSS time synchronization to realize the high-precision positioning of 3D GPR underground remote sensing detection. The positioning process of the multi-level multi-sensor fusion is shown in Figure 5.

The aboveground and underground integrated 3D survey system collects GNSS signal data, INS inertial data, and laser point cloud data, based on which a multi-level ground-penetrating radar remote sensing detection and positioning method applicable to different measurement environments can be realized. In scenarios with good GNSS signals but still having some time GNSS data missing (such as general urban roads), the positioning information is obtained by a tightly coupled GNSS + INS solution, which can not only overcome the transient GNSS signals being missing but also obtain smoother and more accurate attitude trajectory information. In the measurement environment where GNSS signals are weak or GNSS is completely unavailable (e.g., in tunnels, under buildings), the combination of laser scanner + INS inertial unit is used to realize the system self-positioning by using laser SLAM technology, which does not rely on GNSS for operation and greatly improves the system's applicability. In summary, the multi-level ground-penetrating radar remote sensing detection and positioning method can ensure that the aboveground and belowground integrated 3D survey system in this paper has the capability of system positioning and 3D mine detection mapping applicable in any environment.

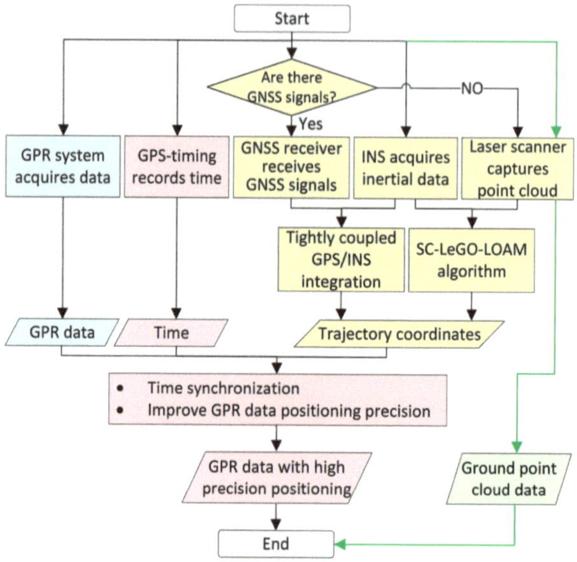

Figure 5. Flow chart of GPR data acquisition and positioning processing.

2.2.1. GNSS Differential Positioning with Good GNSS Signals

The GPR system is equipped with GNSS receivers; when the GNSS signal quality is good, fast positioning can occur based on the signal received from the GNSS receiver using the GNSS differential technique, as shown in Figure 6. GNSS differential technology is a method that can effectively reduce measurement errors to improve positioning accuracy. The measurements of satellite signals received by GNSS receivers all contain a certain amount of error, and some of these error terms are correlated in time and space, which is the fundamental reason for being able to use GNSS differential technology.

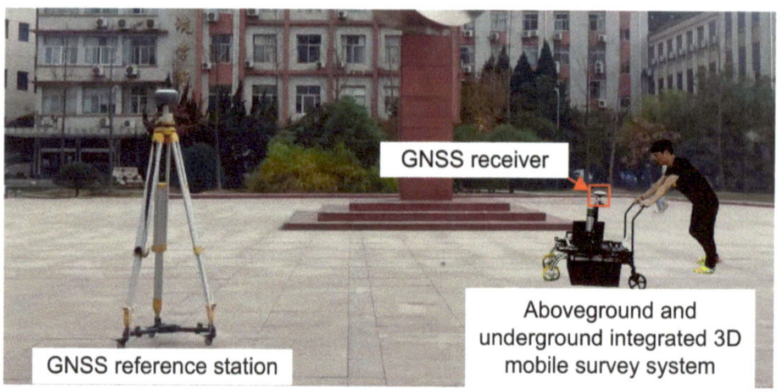

Figure 6. GNSS differential positioning in scene with good GNSS signals.

The deployment of additional reference stations is required to use differential technology, which is different from GNSS receivers. In addition to tracking visible satellite signals, the reference station has the function of transmitting signals. At a certain moment, the position coordinates of the reference station are precisely known, so its distance to the satellite is also precisely known. The reference station also measures the pseudo-range carrier phase measurements at this time. The difference value between the measured value and the actual value is the measurement error at the reference station at the current moment. The reference station continuously sends out the calculated measurement errors. The GNSS receivers within its signal coverage area can correct the measured pseudo-range carrier phase measurements. Positioning errors are reduced with the help of received differential corrections. The closer the distance to the reference station, the higher the correlation between the measurement errors and the better the effect of differential positioning.

2.2.2. GNSS/INS Tightly Coupled Positioning with Weak GNSS Signals

The pure GNSS differential algorithm positioning method has a simple workflow and is easy to implement, but it has a low positioning density and relatively low accuracy and is also completely dependent on the quality of the GNSS signals. Once the GNSS receiver is blocked or jammed, the GNSS signals will experience loss of lock, and thus the complete positioning cannot be achieved. However, GPR detection environments are complex and diverse; for example, detection on roads may be blocked by tall buildings and border trees on both sides of the road, detection in woods may be blocked by the dense tree canopy, and detection in tunnels may not even receive GNSS signals at all. In order to locate and increase the positioning accuracy even when the GNSS signals are weak or are out of lock in the short term, this paper combines GNSS and INS by using a tightly coupled GNSS/INS solution based on the pseudo range and the pseudo-range rate for positioning, as shown in Figure 7.

Figure 7. GNSS/INS tightly coupled positioning in scene with weak GNSS signal.

The basic principle of the GNSS/INS tightly coupled solution positioning algorithm is as follows. The pseudo range and pseudo-range rate output from the GNSS receiver are used as the reference information for the combined GNSS/INS solution. The calculated pseudo range and pseudo-range rate between the carrier and the satellite are used as the measurement information for the combined GNSS/INS solution. In addition, the difference between the two is used as the observation information of the system. The error information of INS (misalignment angle, velocity error, position error) and the clock error information of GNSS receiver are estimated by Kalman filtering; then, the system is corrected by open-loop output or closed-loop feedback [20]. The flow chart of GNSS/INS tightly coupled positioning is shown in Figure 8.

The tightly coupled results are smoothed after the solution is finalized. When there are breakpoints in the positioning results, processing with the smoothing algorithm not only reduces position, velocity, and attitude errors caused by GNSS signals' loss of lock, but also smooths the trajectory. GNSS/INS tight coupling can output high-update rate position information due to the INS output frequency of 500 Hz. The tight coupling provides continuous, high-accuracy, high-update rates and smooth positioning results, even when the GNSS receiver is tracking less than four satellites.

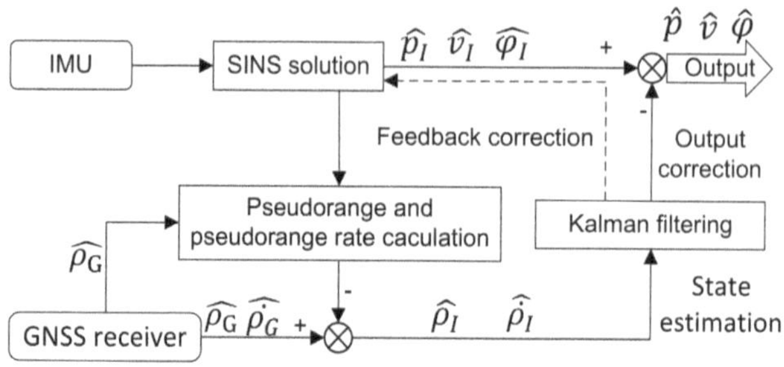

Figure 8. Flow chart of GNSS/INS tightly coupled positioning.

2.2.3. Laser SLAM Positioning with No GNSS Signal

Lidar is used for autonomous positioning in environments where GNSS signals are out of lock for a long time or even where there is no signal. In this paper, we use the tightly coupled iterative Kalman filter of FAST-LIO and FAST-LIO2 to implement the laser odometry module. Based on ScanContext [21] and LegoLOAM [22], we add Scan Context loopback detection to achieve the mapping optimization module.

Laser odometry high-frequency real-time operation is used to track the real-time motion. Forward propagation is performed for IMU pre-integration to obtain the prediction state and prediction error. The point cloud after ground segmentation is motion-compensated by backward propagation to obtain an in-frame distortion-free point cloud. The point-to-face distance is calculated as the residual, and the state is updated by iterative Kalman filtering until convergence, when the odometer is output.

The mapping optimization low frequency operates for closed-loop detection and optimization. The odometer estimated by the state estimation module will be added to a factor map in the form of factors; also introduced are the closed-loop factors obtained by scan matching. The odometer information is used to provide constraints for adjacent scans to ensure the accuracy of local maps, and the closed-loop information is used to provide constraints for global maps to ensure that large-scale map building can be performed properly. The flow chart of laser SLAM positioning is shown in Figure 9.

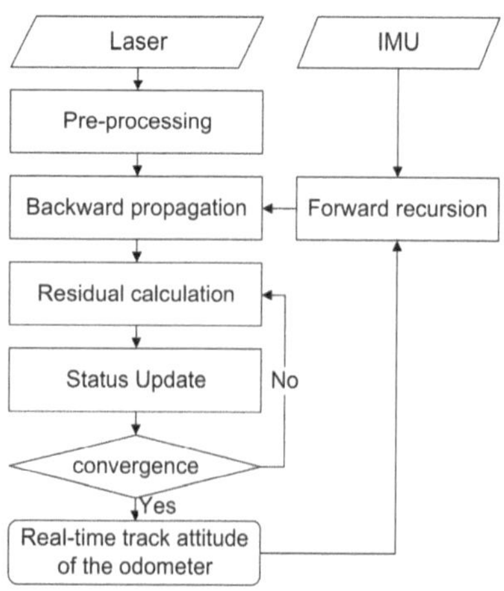

Figure 9. Flow chart of laser SLAM positioning processing.

Based on this framework for laser SLAM positioning, this paper improves the adaptability of adjacent frame matching and point cloud motion estimation in the framework.

1. Point cloud adjacency frame matching

In this paper, the ground point is filtered using a point cloud feature point matching method based on double-threshold ground filtering. The feature points are extracted by using a curvature-based point cloud feature extraction algorithm for non-ground points, and the feature points are aligned.

The specific algorithm steps are as follows.

Step 1: Double-threshold ground filtering process to filter out non-ground points

(1) Project p_{fk} onto the reference plane of the grid M. (2) Refine the roughly determined ground point G_{rough} as the determined ground point G using the RANSAC method. (3) Remove the NaN (Not a number) point cloud of non-ground point NG and points with too-close distance measurement results. The rocess is as in Algorithm 1.

Algorithm 1: Double-threshold ground filtering algorithm

Input: k moment point cloud $p_k{}^{(i)}$
Output: non-ground points NG
//Minimum distance $ring_{radius}$; Max distance $ghost_{radius}$; Height threshold δh_1, δh_2

While ($p_k{}^{(i)} \in p_k$) do
  if $ring_{radius} * ring_{radius} < Distance < ghost_{radius} * ghost_{radius}$
    $p_k{}^{(i)} \in p_{fk}$

While ($p_{fk} \in M_i$) do
  for ($p_{fk}{}^{(i)} \in p_{fk}$) do
    if $h_k - h^{(i)}{}_{min} < \delta h_1$ and $h^{(i)}{}_{min} - h^{(i)}{}_{minmin} < \delta h_2$
      $p_{fk}{}^{(i)} \in G_{rough}$

if $p_k{}^{(i)} \notin G$
  $p_k{}^{(i)} \in NG$

Step 2: Non-ground point cloud feature extraction based on curvature

Calculate the smoothness c of the Lidar points p_i in each frame of non-ground point NG for which curvature is to be found. Rank all the data according to the magnitude of the smoothness c. Classify the feature points into two categories, edge points ε_k and plane points H_k. Calculate using Equation (1).

$$c = \frac{1}{|S| \cdot \left\| X^L_{(k,i)} \right\|} \left\| \sum_{j \in S, j \neq i} \left(X^L_{(k,i)} - X^L_{(k,j)} \right) \right\| \tag{1}$$

where S in Equation (1) is the set of continuous points of I returned by the laser scanner in the same scan, and there is a point i in the coordinate system $\{L_k\}$ whose origin is located at the geometric center of the Lidar; $i \in p_k$ is the coordinate of a point in the point cloud sensed during scan k as $X^L_{(k,i)}$.

Edge points ε_k and face points H_k feature points are obtained within each scan line based on the edge point features with larger discrete curvature and face point features with smaller discrete curvature extracted from the discrete curvature of the single-frame Lidar point cloud.

Step 3: Feature matching based on edge points and face points

The point cloud p_k obtained during scan k is projected to the timestamp t_{k+1} to obtain $\overline{P}_k$. In the set ε_{k+1} of edge points in the feature points, the associated features of points are edges in $\overline{P}_k$. In the set H_{k+1} of plane points in the feature points, the associated features of the points are plane blocks in $\overline{P}_k$. For edge points, their association features are lines, and for planar points, their association features are faces. The distances from the two types of feature points to their associated features are calculated separately, which will be used in the spatial 3D building section to estimate the motion of the Lidar.

2. Point cloud motion estimation

The motion attitude of the optical radar is calculated using the Lidar odometry method, and finally, the Lidar building module is used to refine the trajectory for 3D building to obtain an accurate trajectory and point cloud map. The specific steps are as follows.

Step 1: Lidar motion estimation

The set of edge and plane points, ε_{k+1} and H_{k+1}, obtained from p_{k+1} are obtained by curvature-based non-ground point cloud feature extraction, and $\widetilde{\varepsilon}_{k+1}$ and $\widetilde{H}_{k+1}$ are the point sets projected to t_{k+1}. The transformation relationship between ε_{k+1} and $\widetilde{\varepsilon}_{k+1}$ or H_{k+1} and $\widetilde{H}_{k+1}$ needs to be found to estimate the motion of the Lidar. Using $T_{k+1}^{L} = [t_x, t_y, t_z, \theta_x, \theta_y, \theta_z]$ to represent the motion attitude of the Lidar, Equation (2) can be derived.

$$X_{(k+1,i)}^{L} = R\widetilde{X}_{(k+1,i)}^{L} + T_{(k+1,i)}^{L}(1:3) \tag{2}$$

where $X_{(k+1,i)}^{L}$ is the coordinate of point i in ε_{k+1} or H_{k+1}, $\widetilde{X}_{(k+1,i)}^{L}$ is the coordinate of the corresponding point in $\widetilde{\varepsilon}_{k+1}$ or $\widetilde{H}_{k+1}$, $T_{(k+1,i)}^{L}(1:3)$ is the first to third set of $T_{(k+1,i)}^{L}$, and R is the rotation matrix defined by the Rodriguez formula.

Step 2: The motion attitude of the Lidar calculated by the Lidar odometry method

The input value of the Lidar mileage calculation method is the undistorted point cloud $\overline{P}_k$. The point cloud p_{k+1} is obtained during $k + 1$, and the attitude T_{k+1}^{L} is obtained with respect to the Lidar odometer.

By the distances between the points in $\widetilde{\varepsilon}_{k+1}$ and $\widetilde{H}_{k+1}$ and their associated features, the geometric relationship between the edge points in ε_{k+1} and the corresponding edge lines can be derived, as shown in Equation (3).

$$f\varepsilon\left(X_{(k+1,i)}^{L}, T_{k+1}^{L}\right) = d\varepsilon, i \in \varepsilon_{k+1} \tag{3}$$

Similarly, the geometric relationship between the points in H_{k+1} and their associated planar blocks is

$$fH\left(X_{(k+1,i)}^{L}, T_{k+1}^{L}\right) = dH,\ i \in H_{k+1} \tag{4}$$

Next, the Levenberg–Marquardt method is used to estimate the motion of the Lidar. For each feature point in ε_{k+1} and H_{k+1} using the derived Equations (3) and (4), a nonlinear function, such as specified in Equation (5), can be obtained.

$$f\left(T_{k+1}^{L}\right) = d \tag{5}$$

Equation (5) is solved by minimizing the distance between each feature point and its associated feature to zero in a nonlinear iteration, as shown in Equation (6).

$$T_{k+1}^{L} \leftarrow T_{k+1}^{L} - \left(J^T J + \lambda \mathrm{diag}\left(J^T J\right)\right)^{-1} J^T d \tag{6}$$

λ is a factor determined using the Levenberg–Marquardt method. Double-squared weights are assigned to each feature point in this process, and iterations are performed to update the motion pose T_{k+1}^{L} of the Lidar for nonlinear optimization until the end of the iteration; the results are input to the Lidar map building module for processing.

3. Results

3.1. Experiment Area and Data

3.1.1. Experiment with Good GNSS Signals

The orthophoto of the measurement area of the experiment in the good GNSS environment is shown in Figure 10. The experiment was conducted at Wuhan University, where we surveyed the underground area of Zhuoer Gymnasium Ring Road, with a measured length of 872 m. The road was wide, with no trees or tall buildings blocking the road on both sides, and the GNSS signal quality was good. The number of satellites tracked by the GNSS receiver in the survey area is shown in Figure 11. Only in the middle few seconds of the time is the number of tracked satellites less than 4; the other moments have good satellite observation. We placed 13 metal plates of 35 cm × 35 cm on the road as positioning targets in order to calculate the system positioning accuracy.

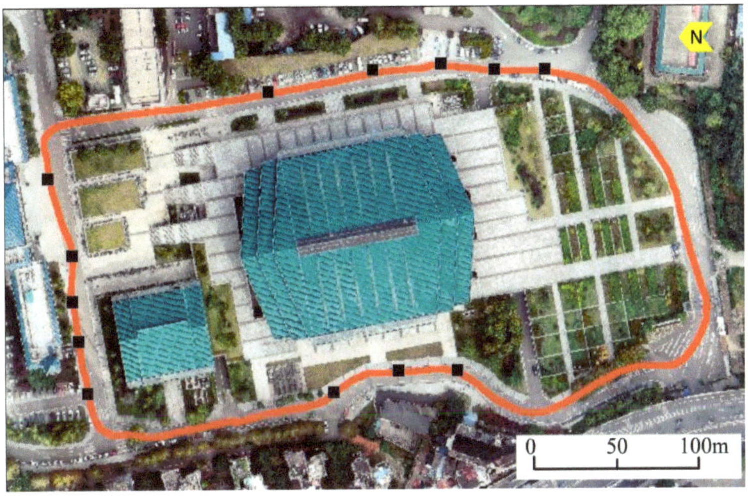

Figure 10. DOM, measurement trajectory and positioning target distribution of the good GNSS environment.

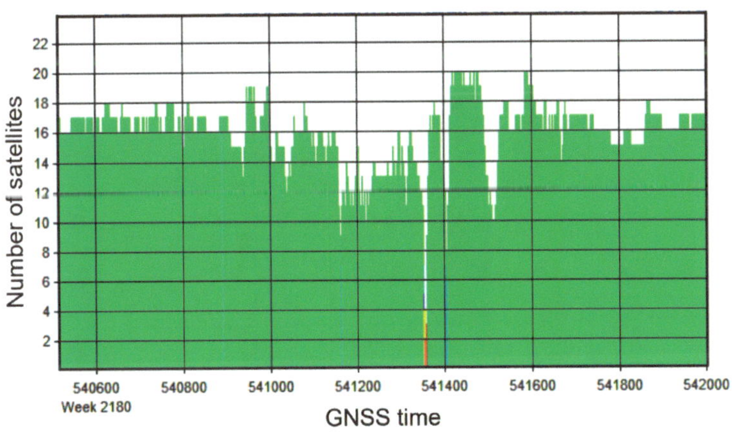

Figure 11. Number of satellites tracked by GNSS receivers in the good GNSS environment.

3.1.2. Experiment with Weak GNSS Signals

The orthophoto of the measurement area of the experiment in the partly loss-of-lock GNSS environment is shown in Figure 12. The experiment surveyed the underground area of the road around the playground of the Department of Informatics of Wuhan University; the length of the road is 630 m, the width of the road is about 5 m, the road is surrounded by dense trees and tall buildings and twice traverses the internal space of the building up to 20 m, and the environmental GNSS signals are seriously obscured. The number of satellites tracked by the GNSS receiver in the survey area is shown in Figure 13; excluding the good condition of GNSS satellites at the beginning and end of the static convergence phase of the measurement as well as the ability to track four satellites for part of the measurement process, there was an insufficient number of satellites or even zero satellites for a large part of the measurement time. As shown in Figure 12, 10 metal plates of 35 cm $\times$ 35 cm were evenly placed on the road as positioning targets to evaluate the positioning accuracy of the aboveground and underground integrated 3D survey system.

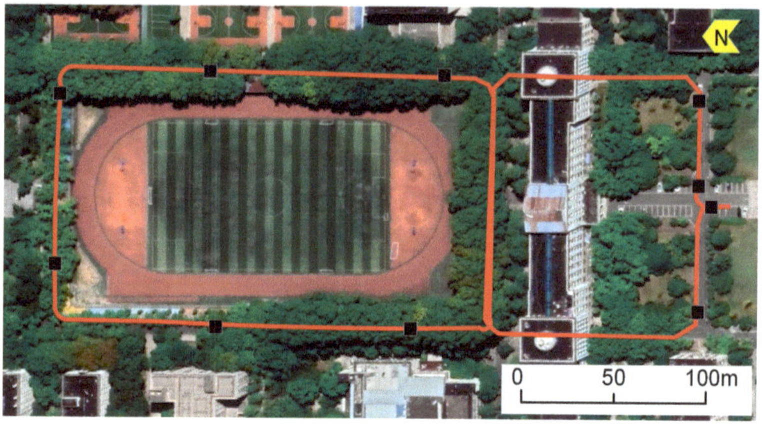

Figure 12. DOM, measurement trajectory, and positioning target distribution of the partly loss-of-lock GNSS environment.

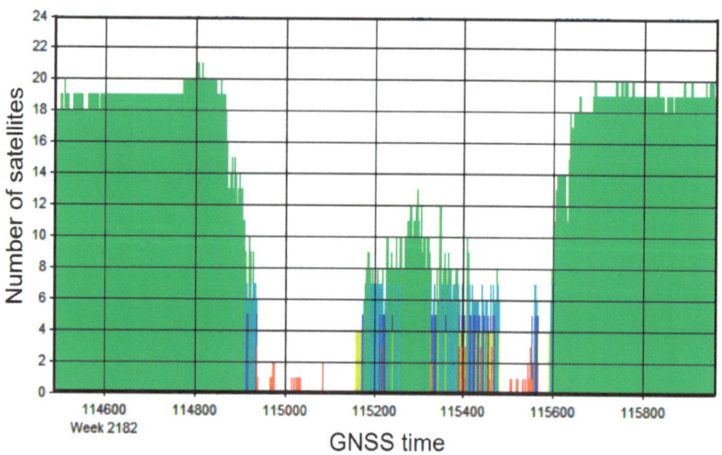

Figure 13. Number of satellites tracked by GNSS receivers in the partly loss-of-lock GNSS environment.

3.2. Positioning Results and Accuracy Analysis

3.2.1. Experiment with Good GNSS Signals

1. Trajectory results

As shown in Figure 14, the trajectory results of the GNSS differential solution, GNSS/INS tightly coupled solution, and laser SLAM autonomous positioning are obtained under the good GNSS signal environment. It can be seen that there are two interruptions in the trajectory of GNSS differential decomposition, while the trajectories of GNSS/INS tightly coupled decomposition and laser SLAM autonomous positioning are continuous, without interruption, and relatively smooth.

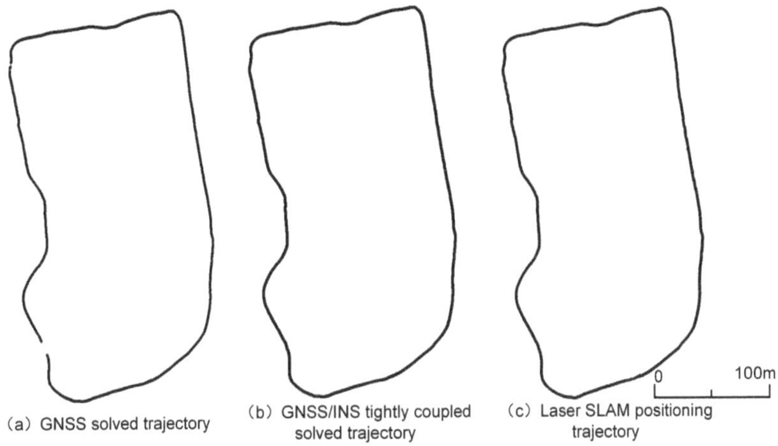

Figure 14. The results of solving the trajectory by three positioning methods in the GNSS good environment.

2. Positioning accuracy evaluation

The cart passed through 13 metal positioning targets during the measurement of experiment in the GNSS good environment. The coordinates of the center points of the targets can be calculated and extracted from each of the three positioning methods: the GNSS differential positioning corresponding to imaging, GNSS/INS tightly coupled positioning corresponding to imaging, and laser SLAM autonomous positioning. It is possible to count the errors between the coordinates and the true values at the 13 control points for the three positioning methods. Table 1 shows the statistics of positioning precision of the three positioning strategies for the experiment in the GNSS good environment at the 13 control points.

Table 1. Statistics of positioning precision of three positioning strategies for the GNSS good environment.

Methodology	Direction	MIN (m)	MAX (m)	AVE (m)	S.D.	RMSE (m)
GNSS	E	0.007	0.222	0.067	0.077	0.099
	N	0.003	0.107	0.036	0.037	0.050
GNSS/IMU	E	0.001	0.043	0.014	0.012	0.018
	N	0.000	0.066	0.014	0.018	0.022
SLAM	E	0.001	0.112	0.054	0.038	0.057
	N	0.000	0.087	0.032	0.028	0.037

From Table 1, it can be seen that the highest positioning accuracy is achieved by the GNSS/INS tightly coupled decomposition positioning method, with the average positioning error being less than 0.01 m in both the east and north directions and the RMSE around 0.02 m, and the minimum, maximum, mean, standard deviation and RMSE of the errors all at a minimum. Only the GNSS differential decomposition algorithm positioning method and the laser SLAM self-localization method performed comparably, with the mean error and RMSE higher than 0.05 m in the east direction. The experiment in the GNSS good environment shows that the system can obtain the best positioning effect by using the GNSS/INS tightly coupled decomposition positioning method in the case of good GNSS signals.

The GNSS positioning frequency is about 1–10 hz, and the IMU sampling frequency in the system is up to 500 hz. The IMU does not lose information during high-speed sampling, which can improve the sampling and positioning accuracy. It can also obtain high-precision attitude information for correcting the positioning attitude. With the combined GNSS/INS tightly coupled positioning, even if there is a short time quality degradation of the GNSS signals, the high-precision IMU can still provide continuous high-precision position reference. As a result, the positioning accuracy is not affected.

3.2.2. Experiment with Weak GNSS Signals

1. Trajectory results

Figure 15 shows the trajectory results of GNSS differential decomposition, GNSS/INS tightly coupled decomposition, and laser SLAM autonomous positioning under the weak GNSS signal environment. The trajectory calculated by GNSS differential decomposition has good trajectory quality, except for the southernmost and northernmost ends, and the reliable GNSS signals cannot be tracked on the east, west, and south sides of the playground because of thick trees and building obstruction. Because GNSS has a long, uninterrupted out-of-lock period (the first out-of-lock time is about 3 min, and the second out-of-lock time is about 2 min), the trajectory has a wide range of interruptions, and it is basically impossible to determine the results. In contrast, the GNSS/INS tightly coupled solution can provide a short-term continuous high-accuracy position reference without affecting the positioning accuracy, even if the quality of GNSS signals is degraded for a short period of time due to the participation of INS in the calculation; the solved trajectory is continuous without interruption and with high solution accuracy.

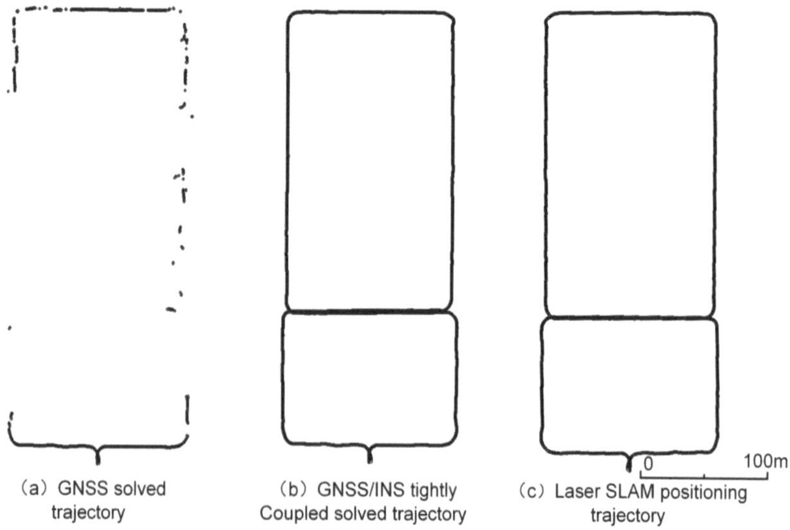

Figure 15. The results of solving the trajectory by three positioning methods in the GNSS partly loss-of-lock environment.

2. Positioning accuracy evaluation

The coordinates of the center point of the target can be calculated and extracted from the three positioning methods to obtain the coordinates of the 10 points. The positioning coordinates of the three positioning methods at the 10 control points were calculated, as was the error between them and the real values. Statistical analysis of the positioning errors was carried out; see Table 2.

Table 2. Statistics of positioning precision of different positioning strategies for the GNSS partly loss-of-lock environment.

Methodology	Direction	MIN (m)	MAX (m)	AVE (m)	S.D.	RMSE (m)
GNSS	E	0.033	20.531	2.325	6.400	6.502
	N	0.013	3.530	0.710	1.041	1.216
GNSS/IMU	E	0.002	0.282	0.059	0.086	0.101
	N	0.000	0.218	0.075	0.078	0.105
SLAM	E	0.004	0.104	0.071	0.036	0.079
	N	0.002	0.130	0.089	0.037	0.095

As can be seen from Table 2, the accuracy of using pure GNSS positioning is very poor in scenarios where GNSS signals are weak or even absent. The maximum value of the error in the north direction is greater than 3 m, the average error value is 0.7 m, the RMSE is greater than 1 m, and the standard deviation of the error is greater than 1. The error fluctuation is large. The positioning accuracy of the east direction is slightly higher, while the error average and RMSE are tens of centimeters, and the accuracy is lower than that of the GNSS/INS combined positioning and SLAM positioning algorithms.

The difference in positioning accuracy between the GNSS/INS post-solution method and the SLAM algorithm is not significant. The RMSEs of the GNSS/INS combined positioning method are slightly greater than 10 cm for the east and north directions, while the RMSEs of the SLAM algorithm are less than 10 cm for both the east and north directions. The error standard deviation of SLAM shows smaller error fluctuations, and the maximum value of the error is smaller than that of the combined GNSS/INS positioning. GNSS signals are not used in the SLAM algorithm, so an accuracy better than 10 cm can be obtained using the laser SLAM positioning algorithm, whether or not GNSS signals are available.

3.3. 3D GPR Imaging Results

3.3.1. Experiment with Good GNSS Signals

Figure 16I shows an example of 3D GPR data positioning imaging in a good GNSS signal environment. The 3D GPR data represent a horizontal section located at 0.2 m below ground level. Figure 16II shows the local enlargements of the imaging corresponding to the GNSS differential positioning, the imaging corresponding to the GNSS/INS tightly coupled positioning, and the imaging corresponding to the laser SLAM autonomous positioning in the overall image.

Because of the good quality of GNSS signals in the environment, the overall shapes of the subsurface GPR data imaged based on the three positioning methods were basically the same, and no obvious deformation occurred. Comparing with the local zoomed-in figure (Figure 11), the GPR data based on GNSS differential decomposition localization showed trajectory jitter or slight deformation in individual areas, as in part (a), (b), and (e) of the figure. In contrast, the GPR data based on GNSS/INS tightly coupled decomposition and laser SLAM self-localization have smooth trajectories throughout, and no deformation phenomenon occurs.

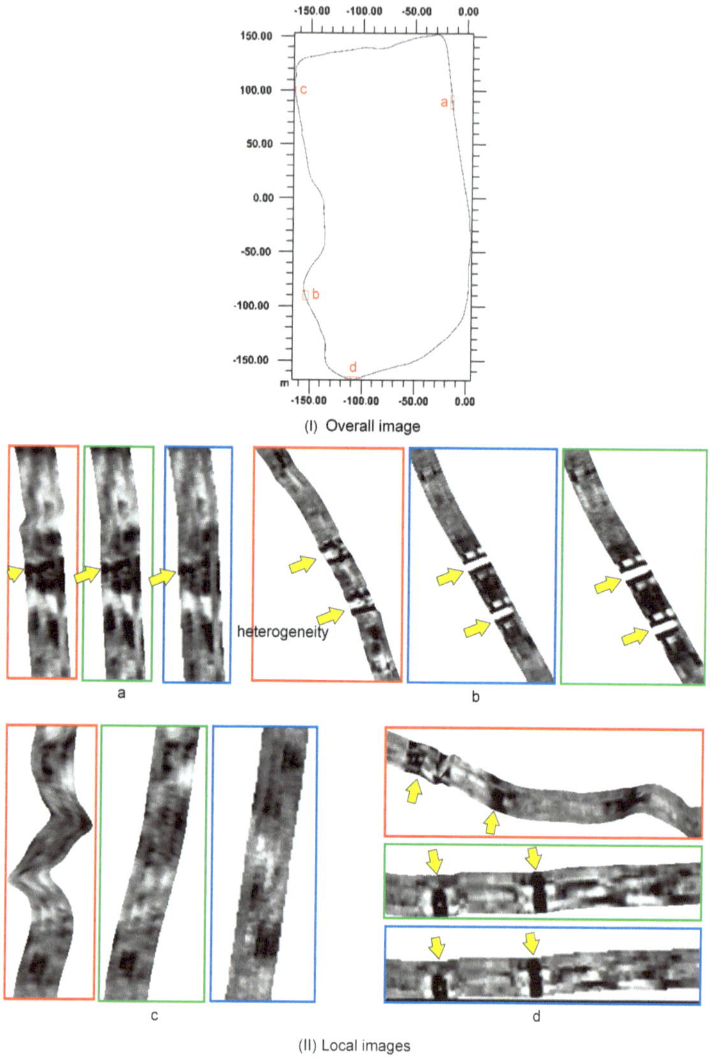

Figure 16. Overall and local images (horizontal section) of 3D GPR data imaged by different positioning methods in the GNSS good environment, with local enlarged images of GNSS differential positioning (red border), GNSS/INS tightly coupled positioning (green border), and laser SLAM autonomous positioning (blue border). The local images respectively belong to regions a, b, c, and d of the overall image. (The yellow arrows point to heterogeneity).

3.3.2. Experiment with Weak GNSS Signals

Figure 17 shows an example of 3D GPR data positioning imaging in the GNSS partly loss-of-lock environment, in which the 3D GPR data are located at 0.2 m below the ground level in the horizontal section. In the experiment, except for the north and south ends, which receive a small number of satellite signals, the region receives less than four satellites and even zero satellites. GPR data based on GNSS differential decomposition positioning as a whole have a dramatic deformation and serious track drift; thus, GPR data cannot be imaged properly. The overall shapes of subsurface GPR data imaging based on GNSS/INS tightly coupled decomposition and laser SLAM-based self-localization methods are basically the same.

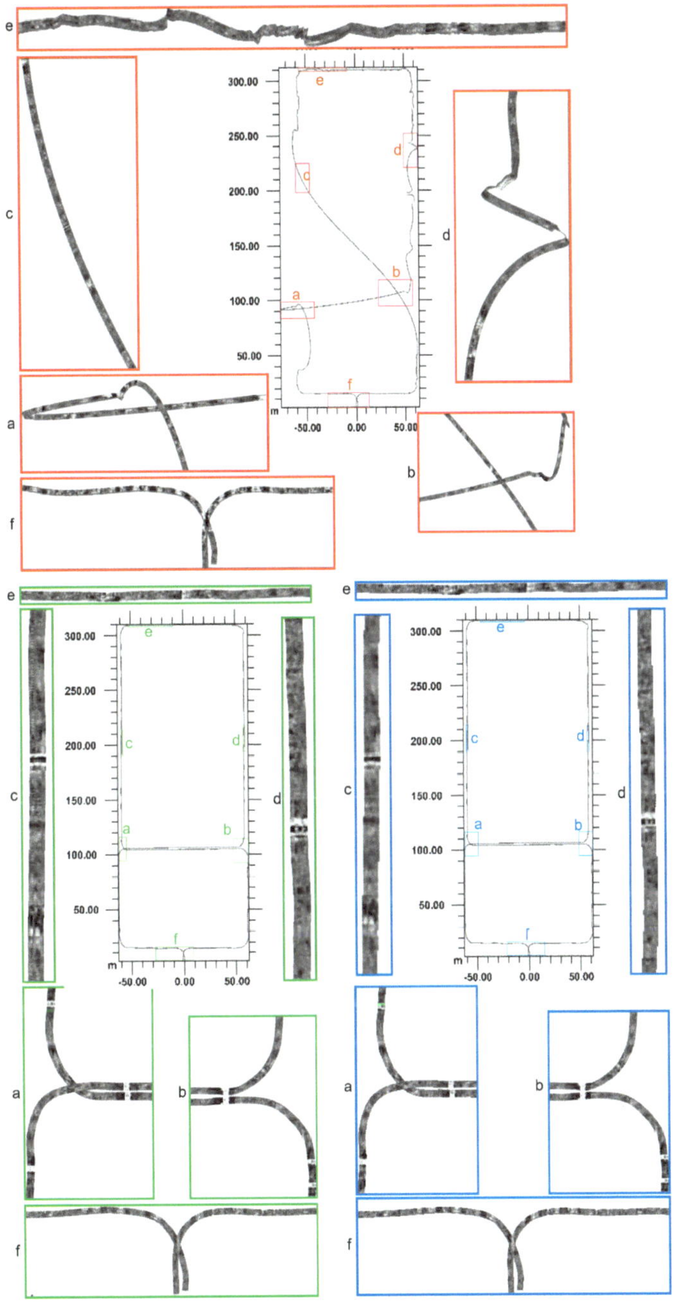

Figure 17. Overall and local images (horizontal section) of 3D GPR data imaged by different positioning methods in the GNSS partly loss-of-lock environment; GNSS differential positioning corresponding to 3D GPR image (red border), GNSS/INS tightly coupled positioning corresponding to 3D GPR image (green border), and laser SLAM autonomous positioning corresponding to 3D GPR image (blue border) in a locally enlarged view. The local images respectively belong to regions a, b, c, and d of the overall image.

Figure 17 shows the local enlargements of the imaging corresponding to GNSS differential positioning, GNSS/INS tightly coupled positioning, and laser SLAM autonomous positioning. From the local view, the subsurface GPR data based on GNSS/INS tight-coupling solution and laser SLAM-based autolocalization methods show no obvious deformation in each part, and normal data interpretation and decoding can be performed.

4. Discussion

This paper proposes a 3D GPR positioning method based on a GNSS differential solution, GNSS/INS tightly coupled solution and LiDAR SLAM method, and proves the accuracy and effectiveness of the algorithm through underground detection tests in different scenes. Compared with the traditional positioning by GNSS only, we add the GNSS/INS tightly coupling algorithm and SLAM algorithm to cope with different survey environments. We carry out good and weak GNSS signal experiments and compare the positioning accuracy using the GNSS algorithm, GNSS/INS tightly coupled algorithm, and SLAM algorithm. It is demonstrated that this multi-level positioning method has high accuracy and good robustness.

In scenes with good GNSS signals, the 3D GPR positioning can be achieved quickly by using the GNSS differential solution method, and positioning accuracy within 10 cm can be achieved. Compared with the GNSS differential solution, the GNSS/INS tightly coupled solution is more complicated, but the positioning accuracy is significantly improved. Therefore, in a scene with good GNSS signals, 3D GPR should choose the GNSS differential solution for fast positioning and the GNSS/INS tight coupling solution for high positioning accuracy. The SLAM method has no obvious advantage at this time.

In the condition of weak GNSS signals, the positioning will be offset by using the GNSS differential solution method, while using the GNSS/INS tightly coupled solution method can still provide reliable positioning. The GNSS positioning frequency is about 1–10 hz, while the sampling frequency of IMU in the aboveground and underground integrated 3D survey system is up to 500 hz, which can improve the sampling and positioning accuracy without losing information in the high-speed sampling process. The addition of IMU can also obtain high-precision attitude information for correcting the positioning attitude. With the combined GNSS/INS positioning method, even if the GNSS signals have a quality degradation problem in a short period of time, the high-precision IMU can still provide a continuous high-precision position reference in a short period of time without affecting the positioning accuracy.

In the case where GNSS signals demonstrate loss of lock for a long time or no GNSS signals, the GNSS receiver cannot receive GNSS signals, and neither the GNSS differential solution nor the GNSS/INS tightly coupled solution can achieve positioning. Aboveground and underground integrated 3D mobile survey systems use Lidar, IMU, and odometers for positioning by laser the SLAM algorithm. The IMU obtains the prediction state and prediction error, and the motion compensation of the point cloud acquired by the Lidar obtains a distortion-free point cloud. The odometer is calculated and output, and the mapping is optimized to achieve closed-loop detection. The odometer information is used to provide constraints for adjacent scans to ensure the accuracy of local positioning, and the closed-loop information is used to provide constraints for global maps to ensure that large-scale positioning can be completed. Adjacent frame matching and point cloud motion estimation are adaptively improved to achieve high-precision autonomous positioning by laser SLAM. The laser SLAM autonomous positioning algorithm can obtain positioning results with an accuracy better than 10 cm. The SLAM method combines IMU and laser point cloud features at the level of primary observations, realizes joint nonlinear optimization of multi-source data, and achieves accurate positioning optimization using laser point cloud precision matching [31–33]. This enables this aboveground and underground integrated 3D survey system to acquire 3D underground medium distribution data with high-precision positioning information in environments without GNSS signals (e.g., underground mines and tunnels).

5. Conclusions

This paper proposed a high-precision positioning method of multi-level and multi-sensor fusion for 3D GPR aboveground and underground integrated detection. Through the designed aboveground and underground integrated 3D survey system, the underground medium distribution is detected, and the aboveground 3D spatial structure is measured at the same time, to realize the rapid integrated measurement of aboveground and underground space. The survey system is able to achieve high-precision positioning of 3D GPR in environments with or without GNSS signals.

Compared with GNSS solved positioning, in the case of good GNSS signals, the aboveground and underground integrated 3D survey system collects INS data, has higher sampling frequency and accurate attitude information, has more accurate positioning, and can be applied to high-speed measurement scenarios. In scenarios where GNSS signals are weak or interrupted, the system is able to ensure continuous positioning output due to the use of a tightly coupled GNSS and INS solver positioning method. Such environments are the main working scenarios of GPR systems and include roads, bridges, and woods; these contribute to the stable and reliable use of 3D GPR. In scenarios without GNSS signals, the system uses Lidar sensors for active positioning, and the experiments prove that the positioning accuracy is better than 10 cm, which means the 3D GPR can be used for underground mine safety inspection, long tunnel construction detection, etc. In addition, the point cloud data obtained by the laser scanner in the system can generate the 3D spatial structure of the ground space. This multi-source data of the integrated spatial structure above and below ground is useful for spatial display, comprehensive analysis, and decision making.

In conclusion, the aboveground and underground integrated 3D detection multi-level, multi-sensor fusion high-precision positioning method proposed in this paper can achieve integrated aboveground and underground rapid measurement in any environment and ensure better than 10 cm positioning accuracy, ensuring that the 3D GPR can complete accurate detection and large scale survey and can provide data security for imaging and interpretation of underground data.

Author Contributions: Conceptualization, J.Z. and Q.H.; methodology, J.Z. and Q.H.; software, X.D.; validation, J.Z.; formal analysis, J.Z.; investigation, J.Z.; resources, J.Z.; data curation, J.Z.; writing—original draft preparation, J.Z.; writing—review and editing, J.Z. and Q.H.; visualization, J.Z.; supervision, P.Z.; project administration, Y.Z.; funding acquisition, Y.Z. and P.Z. All authors have read and agreed to the published version of the manuscript.

Funding: This research was funded by the Knowledge Innovation Program of Wuhan—Basic Research, grant number 2022010801010431; the Open Foundation of the Key Laboratory of National Geographic Census and Monitoring, Ministry of Natural Resources, grant number 2023NGCM08.

Data Availability Statement: The raw data supporting the conclusions of this article will be made available by the authors on request.

Conflicts of Interest: The authors declare no conflicts of interest. The funders had no role in the design of the study; in the collection, analyses, or interpretation of data; in the writing of the manuscript; or in the decision to publish the results.

References

1. Grasmueck, M.; Weger, R.; Horstmeyer, H. Full-resolution 3D GPR imaging. *Geophysics* **2005**, *70*, K12–K19. [CrossRef]
2. Novo, A.; Lorenzo, H.; Rial, F.I.; Solla, M. 3D GPR in forensics: Finding a clandestine grave in a mountainous environment. *Forensic Sci. Int.* **2011**, *204*, 134–138. [CrossRef]
3. Kelly, T.; Angel, M.; O'Connor, D.; Huff, C.; Morris, L.; Wach, G. A novel approach to 3D modelling ground-penetrating radar (GPR) data–A case study of a cemetery and applications for criminal investigation. *Forensic Sci. Int.* **2021**, *325*, 110882. [CrossRef]
4. Sato, M.; Yokota, Y.; Takahashi, K.; Grasmueck, M. Landmine detection by 3D GPR system. In Proceedings of the Detection and Sensing of Mines, Explosive Objects, and Obscured Targets XVII, Baltimore, MD, USA, 23–27 April 2012; Volume 8357, p. 835710.
5. Klesk, P.; Kapruziak, M.; Olech, B. Statistical moments calculated via integral images in application to landmine detection from Ground Penetrating Radar 3D scans. *Pattern Anal. Appl.* **2018**, *21*, 671–684. [CrossRef]

6. Ebrahim, S.M.; Medhat, N.; Mansour, K.K.; Gaber, A. Examination of soil effect upon GPR detectability of landmine with different orientations. *NRIAG J. Astron. Geophys.* **2018**, *7*, 90–98. [CrossRef]
7. Lee, S.-H.; Jang, I.-H. A Sudy on the Underground Condition of Road Using 3D-GPR Exploration. *J. Korean Geoenviron. Soc.* **2019**, *20*, 49–58.
8. Kang, M.-S.; Kim, N.; Im, S.B.; Lee, J.-J.; An, Y.-K. 3D GPR image-based UcNet for enhancing underground cavity detectability. *Remote Sens.* **2019**, *11*, 2545. [CrossRef]
9. Liu, Z.; Wu, W.; Gu, X.; Li, S.; Wang, L.; Zhang, T. Application of Combining YOLO Models and 3D GPR Images in Road Detection and Maintenance. *Remote Sens.* **2021**, *13*, 1081. [CrossRef]
10. Emilsson, J.; Viberg, A.; Gustafsson, J.; Langton, M.; Friborg, J. Efficient State-of-Art HDR 3D GPR Compared to 2D Traditional Utility Investigations. In Proceedings of the NSG2020 26th European Meeting of Environmental and Engineering Geophysics; European Association of Geoscientists & Engineers, Online, 7–8 December 2020; Volume 2020, pp. 1–4.
11. Feng, J.; Yang, L.; Wang, H.; Song, Y.; Xiao, J. Gpr-based subsurface object detection and reconstruction using random motion and depthnet. In Proceedings of the 2020 IEEE International Conference on Robotics and Automation (ICRA), Paris, France, 31 May–31 August 2020; pp. 7035–7041.
12. Luo, G.; Cao, Y.; Xu, H.; Yang, G.; Wang, S.; Huang, Y.; Bai, Z. Research on typical soil physical properties in a mining area: Feasibility of three-dimensional ground penetrating radar detection. *Environ. Earth Sci.* **2021**, *80*, 92. [CrossRef]
13. Zajc, M.; Urbanc, J.; Pečan, U.; Glavan, M.; Pintar, M. Using 3D GPR for determining soil conditions in precision agriculture. In Proceedings of the 18th International Conference on Ground Penetrating Radar; Society of Exploration Geophysicists, Golden, CO, USA, 14–19 June 2020; pp. 291–294.
14. Ercoli, M.; Di Matteo, L.; Pauselli, C.; Mancinelli, P.; Frapiccini, S.; Talegalli, L.; Cannata, A. Integrated GPR and laboratory water content measures of sandy soils: From laboratory to field scale. *Constr. Build. Mater.* **2018**, *159*, 734–744. [CrossRef]
15. Gaballah, M.; Grasmueck, M.; Sato, M. Characterizing subsurface archaeological structures with full resolution 3D GPR at the early dynastic foundations of Saqqara Necropolis, Egypt. *Sens. Imaging* **2018**, *19*, 23. [CrossRef]
16. Ozkan-Okay, M.; Kadioglu, S.; Samet, R. Enhancement of 2D/3D GPR Data Imaging by the Proposed TAEF Technique: Displaying Archeological Remains under the Colonnade Road in Anavarza Ancient City, Adana, Turkey. In Proceedings of the Geoinformatics, European Association of Geoscientists & Engineers, Online, 11–14 May 2021; Volume 2021, pp. 1–7.
17. Pipan, M.; Baradello, L.; Forte, E.; Prizzon, A.; Finetti, I. 2-D and 3-D processing and interpretation of multi-fold ground penetrating radar data: A case history from an archaeological site. *J. Appl. Geophys.* **1999**, *41*, 271–292. [CrossRef]
18. McMechan, G.A.; Gaynor, G.C.; Szerbiak, R.B. Use of ground-penetrating radar for 3-D sedimentological characterization of clastic reservoir analogs. *Geophysics* **1997**, *62*, 786–796. [CrossRef]
19. Grasmueck, M. 3-D ground-penetrating radar applied to fracture imaging in gneiss. *Geophysics* **1996**, *61*, 1050–1064. [CrossRef]
20. Šarlah, N.; Podobnikar, T.; Ambrožič, T.; Mušič, B. Application of Kinematic GPR-TPS model with high 3D georeference accuracy for underground utility infrastructure mapping: A case study from urban sites in Celje, Slovenia. *Remote Sens.* **2020**, *12*, 1228. [CrossRef]
21. Boniger, U.; Tronicke, J. On the potential of kinematic GPR surveying using a self-tracking total station: Evaluating system crosstalk and latency. *IEEE Trans. Geosci. Remote Sens.* **2010**, *48*, 3792–3798. [CrossRef]
22. Trinks, I.; Johansson, B.; Gustafsson, J.; Emilsson, J.; Friborg, J.; Gustafsson, C.; Nissen, J.; Hinterleitner, A. Efficient, large-scale archaeological prospection using a true three-dimensional ground-penetrating radar array system. *Archaeol. Prospect.* **2010**, *17*, 175–186. [CrossRef]
23. Sukhanov, D.Y.; Ponomarev, O.; Zavyalova, K.; Khmelev, V.; Roslyakov, S. Radar with a local positioning video-system. In Proceedings of the 2017 Progress in Electromagnetics Research Symposium-Spring (PIERS), St. Petersburg, Russia, 22–25 May 2017; pp. 3723–3728.
24. Kaniewski, P.; Kraszewski, T.; Pasek, P. UWB-Based Positioning System for Supporting Lightweight Handheld Ground-Penetrating Radar. In Proceedings of the 2019 IEEE International Conference on Microwaves, Antennas, Communications and Electronic Systems (COMCAS), Tel-Aviv, Israel, 4–6 November 2019; pp. 1–4.
25. Ciampoli, L.B.; Calvi, A.; Di Benedetto, A.; Fiani, M.; Gagliardi, V. Ground Penetrating Radar (GPR) and Mobile Laser Scanner (MLS) technologies for non-destructive analysis of transport infrastructures. In Proceedings of the Earth Resources and Environmental Remote Sensing/GIS Applications XII, Online, 13–18 September 2021; Volume 11863, pp. 166–174.
26. Merkle, D.; Frey, C.; Reiterer, A. Fusion of ground penetrating radar and laser scanning for infrastructure mapping. *J. Appl. Geod.* **2021**, *15*, 31–45. [CrossRef]
27. Grasmueck, M.; Viggiano, D.A. Integration of ground-penetrating radar and laser position sensors for real-time 3-D data fusion. *IEEE Trans. Geosci. Remote Sens.* **2006**, *45*, 130–137. [CrossRef]
28. Hui, C.K.; Luo, T.X.; Lai, W.W.; Chang, R.K. GPR mapping with mobile mapping sensing and tracking technologies. *Tunn. Undergr. Space Technol.* **2022**, *122*, 104362. [CrossRef]
29. Hjartarson, K. Dynamic Path Planning, Mapping, and Navigation for Autonomous GPR Survey Robots. Master's Thesis, Umeå University, Umeå, Sweden, 2023.
30. Ogunniyi, S.; Withey, D.; Marais, S.; Crafford, G. LiDAR-based 3D mapping and localisation system for ground penetrating radar. In Proceedings of the 2020 International SAUPEC/RobMech/PRASA Conference, Cape Town, South Africa, 29–31 January 2020; pp. 1–6.

31. Wang, X.; Li, X.; Liao, J. Tightly-coupled stereo visual-inertial-LiDAR SLAM based on graph optimization. *Acta Geod. Cartogr. Sin.* **2022**, *51*, 1744.
32. Shan, T.; Englot, B.; Meyers, D.; Wang, W.; Rus, D. LIO-SAM: Tightly-coupled Lidar Inertial Odometry via Smoothing and Mapping. In Proceedings of the 2020 IEEE/RSJ International Conference on Intelligent Robots and Systems (IROS), Las Vegas, NV, USA, 24 October–24 January 2021.
33. Zhao, S.; Zhang, H.; Wang, P.; Nogueira, L.; Scherer, S. Super Odometry: IMU-centric LiDAR-Visual-Inertial Estimator for Challenging Environments. In Proceedings of the 2021 IEEE/RSJ International Conference on Intelligent Robots and Systems (IROS), Prague, Czech Republic, 27 September–1 October 2021.

MDPI AG
Grosspeteranlage 5
4052 Basel
Switzerland
Tel.: +41 61 683 77 34

Remote Sensing Editorial Office
E-mail: remotesensing@mdpi.com
www.mdpi.com/journal/remotesensing

www.ingramcontent.com/pod-product-compliance
Lightning Source LLC
Chambersburg PA
CBHW041426210926
51850CB00013B/3363
* 9 7 8 3 7 2 5 8 2 2 1 3 3 *